Mathematical Tables and Formulas

Compiled by Robert D. Carmichael

and Edwin R. Smith

Dover Publications, Inc., New York

Published in Canada by General Publishing Company, Ltd., 30 Lesmill Road, Don Mills, Toronto, Ontario.
Published in the United Kingdom by Constable and Company, Ltd., 10 Orange Street, London WC2H 7EG.

This Dover edition, first published in 1962, is an unabridged, slightly corrected republication of the work originally published by Ginn and Company in 1931.

Standard Book Number: 486-60111-0
Library of Congress Catalog Card Number: 62-51392

Manufactured in the United States of America
Dover Publications, Inc.
31 East 2nd Street
Mineola, N.Y. 11501

PREFACE

These tables and formulas have been compiled for the use of students in mathematics courses and in other courses which require numerical computation or involve processes based on mathematics up to and including calculus. Part I contains the usual tables, to five places of decimals, which are necessary in the study of college algebra and trigonometry; Part II includes additional numerical tables which are frequently used in computation but are not generally accessible to students of college mathematics; Part III contains numerous carefully selected mathematical formulas for ready reference, together with tables of integrals and series.

The linear arrangement has been adopted in the tables of logarithms of numbers. We believe that this arrangement of a five-place table has distinct advantages for the experienced as well as for the inexperienced computer. The table of natural trigonometric functions corresponds in arrangement and in completeness of detail with the table of logarithmic trigonometric functions. The use of calculating machines has increased the importance of this table.

The numerical tables have been carefully computed or else compiled from tables of known accuracy. Every possible precaution has been taken to guard against errors.

R. D. C.

E. R. S.

CONTENTS

PART I. TABLES I–XIV

PART II. TABLES XV–XIX

PART III. FORMULAS AND GRAPHS FOR REFERENCE

MATHEMATICAL TABLES
AND FORMULAS

PART I. TABLES I–XIV

INTRODUCTION

1. Interpolation. The purpose of a mathematical table is to enable one to obtain readily the numerical value of a function which corresponds to a given value of the variable, or the value of the variable which corresponds to a given value of the function. It is obviously impossible to tabulate the values of the function for all values of the variable; only a limited number of values may be selected from a limited range of the variable.

The process of finding from a numerical table the value of a function which corresponds to a given value of the variable between two tabulated values is called *interpolation*.

In general, in interpolation one disregards the actual law which defines the relation of the values of the variable to the values of the function; the interpolated value is made to depend only on the tabulated values. Interpolation cannot therefore be expected always to give exact results, but usually gives approximations which are sufficiently close to the actual values for purposes of numerical computation.

The simplest method of interpolation is based on an assumption which may be stated as follows:

Let x_1 and x_2 be consecutive tabular values of a variable, and let y_1 and y_2 be the corresponding tabulated values of the function. Let x be any given value of the variable lying between x_1 and x_2, and let y be the corresponding value of the function. Then

$$\frac{x - x_1}{x_2 - x_1} = \frac{y - y_1}{y_2 - y_1}.$$

The foregoing assumption is equivalent to saying that the differences between the values of the function are proportional to the corresponding differences between the values of the variable, within the intervals between the consecutive tabular values.

1

When a given value is assigned to x, the corresponding value of y may be obtained; and when a given value is assigned to y, the corresponding value of x may be obtained. Thus, for a given x, we have the interpolation formula

$$y = y_1 + \frac{y_2 - y_1}{x_2 - x_1}(x - x_1),$$

and for a given value of y the formula

$$x = x_1 + \frac{x_2 - x_1}{y_2 - y_1}(y - y_1).$$

Example. Find the interpolated value of log 326.74, given log 326.7 = 2.51415 and log 326.8 = 2.51428.

Introducing the notation of the interpolation formulas, we have $x_1 = 326.7$, $x_2 = 326.8$, $x = 326.74$, $y_1 = 2.51415$, and $y_2 = 2.51428$. Substituting these values, we have

$$
\begin{aligned}
y &= \log 326.74 \\
&= 2.51415 + \frac{2.51428 - 2.51415}{326.8 - 326.7}(326.74 - 326.7) \\
&= 2.51420.
\end{aligned}
$$

2. Logarithms. Let a be a fixed positive number different from unity, and let M be a second positive number. Then there is a number x such that $a^x = M$. This exponent x is said to be the logarithm of M to the base a. The definition which is implied here may be formally stated in the following terms:

The logarithm of a given positive number M to the base a, where a is a positive number different from unity, is the exponent x of the power a^x of a such that $a^x = M$.

We denote this logarithm by $\log_a M$, and write

$$\log_a M = x.$$

The number M is said to be the *antilogarithm* of x, when M is the number whose logarithm is x. We write

$$M = \text{antilog } x.$$

The following theorems make it possible to employ logarithms for purposes of computation:

I. *The logarithm of the product of two factors is equal to the sum of the logarithms of the factors.*

II. *The logarithm of a quotient is equal to the logarithm of the dividend minus the logarithm of the divisor.*

III. *The logarithm of the kth power of M is equal to the logarithm of M multiplied by k, where k is any real constant.*

IV. *The logarithm of the positive kth root of M is equal to the logarithm of M divided by k, where k is any real number different from zero.*

In these theorems it is understood that the base is the same throughout. They may be symbolically stated as follows:

$$\text{I. } \log_a MN = \log_a M + \log_a N.$$

$$\text{II. } \log_a \frac{M}{N} = \log_a M - \log_a N.$$

$$\text{III. } \log_a M^k = k \log_a M.$$

$$\text{IV. } \log_a \sqrt[k]{M} = \frac{1}{k} \log_a M.$$

The cologarithm of a number is the negative of its logarithm. Hence it is the logarithm of the reciprocal of the number. The notation colog M is used for the cologarithm of M. We have

$$\text{colog } M = -\log_a M = \log_a \frac{1}{M},$$

and

$$\log_a \frac{M}{N} = \log_a M - \log_a N$$

$$= \log_a M + \text{colog}_a N.$$

Two systems of logarithms are in general use, the *natural* or *Napierian* for which the base is

$$e = 2.71828 \cdots$$

$$= 1 + 1 + \frac{1}{2!} + \frac{1}{3!} + \cdots,$$

and the *common* or *Briggs* for which the base is 10. The common or Briggs logarithms are generally used in logarithmic computation.

3. Common logarithms. The common logarithms of 10, 100, 1000, $\cdots$, which are positive powers of 10, are positive integers; and the logarithms of .1, .01, .001, $\cdots$, which are negative powers of 10, are negative integers. The logarithm of 1, which is the 0th power of 10, is 0. Thus,

$$\log 1 = \log 10^0 = 0;$$

$$\log 10 = \log 10^1 = 1;$$

$$\log 100 = \log 10^2 = 2;$$

$$\log 1000 = \log 10^3 = 3;$$

$$\cdots \cdots \cdots \cdots \cdots$$

$$\log .1 = \log 10^{-1} = -1;$$

$$\log .01 = \log 10^{-2} = -2;$$

$$\log .001 = \log 10^{-3} = -3;$$

$$\cdots \cdots \cdots \cdots \cdots$$

The logarithms of other numbers can be expressed, at least approximately, by the use of decimal fractions having a finite number of digits.

The value of the common logarithm of a number can be separated into two parts, a positive or a negative integer and a positive decimal fraction. Thus, $\log 225 = 2 + .35218$ and $\log .0225 = -2 + .35218,$

when five places of decimals are employed. The integral part of a common logarithm is called the *characteristic* and the fractional part the *mantissa*. The mantissa of a common logarithm may be obtained from tables. The characteristic may be determined by the use of rules which will now be developed.

From the definition of logarithms to the base 10 it follows that the logarithm of any number between

1 and 10 has the characteristic 0,

10 and 100 has the characteristic 1,

100 and 1000 has the characteristic 2,

· ·

.1 and 1 has the characteristic − 1,

.01 and .1 has the characteristic − 2,

.001 and .01 has the characteristic − 3,

· ·

Thus by considering the powers of 10 which are nearest the given number one can readily determine the characteristic. The following may be easily verified:

For numbers greater than 1 the characteristic is positive and is numerically one less than the number of digits to the left of the decimal point.

For numbers less than 1 the characteristic is negative and is numerically one greater than the number of zeros between the decimal point and the first significant figure.

For example, the characteristic of log 235.4 is 2, of log 23.54 is 1, of log 2.354 is 0, of log .2354 is − 1, and of log. 02354 is − 2. For most computational purposes a negative characteristic is generally written as a positive number minus 10. That is, instead of − 1 we write 9 − 10, instead of − 2 we write 8 − 10, etc. Thus, for example,

$$\log 235.4 = 2.37181;$$
$$\log 23.54 = 1.37181;$$
$$\log 2.354 = .37181;$$
$$\log .2354 = 9.37181 - 10;$$
$$\log .02354 = 8.37181 - 10;$$
$$\log .002354 = 7.37181 - 10.$$

EXPLANATION OF TABLES IN PART I

4. Table I. Common logarithms. By means of Table I the common logarithms of numbers containing not more than four significant digits may be determined directly, and those of numbers containing more than four significant figures by interpolation. Likewise the number corresponding to a given logarithm may be determined directly to four places and by interpolation to five places.

The column headed "N" contains numbers ranging from 1 to 10,000; the column headed "log" the mantissas of the logarithms of the numbers in the preceding column; the column headed "d" the differences of adjacent mantissas; and the marginal tables give the proportional parts corresponding to the differences. The decimal points are omitted from the mantissas and only the significant digits of the differences are tabulated.

To find the logarithm of a given number.

Example 1. Find log 726.3 correct to five decimal places.
From Table I, in the column headed "N," we find the number 7263. Opposite this number, in the column headed "log," is .86112, the mantissa of the required logarithm. Since the decimal point of the given number is preceded by three significant digits, the characteristic of the corresponding logarithm is 2. Hence

$$\log 726.3 = 2.86112.$$

Example 2. Find log 32.398 correct to five decimal places.
The given number is not found in the table, so that the required logarithm must be obtained by interpolation. By the method illustrated in Example 1, the values of log 32.39 and log 32.40 may be obtained, and by applying the interpolation principle of § 1 the value of log 32.398 may be determined.
The arithmetical work of interpolation may be greatly facilitated by the use of the tabulated differences and the tables of proportional parts. Opposite the space between 3239 and 3240, in the column headed "d," we find 14, which is the difference of the mantissas of the logarithms of the two numbers. In the table of proportional parts for 14 and to the right of 8 is 11.2. This is the correction which must be added to log 32.39 to give log 32.398. (The decimal point of 11.2 indicates that the digit preceding it should be added to the digit in the fifth decimal place of log 32.39.) We now have

$$\log 32.39 = 1.51041;$$
$$\text{correction for fifth digit } 8 = \qquad 11.2;$$
$$\log 32.398 = 1.51052.$$

Note that only five decimal places are retained in the value of the logarithm.

Example 3. Find log .00238769 correct to five decimal places.
We obtain log .002387 from Table I. The tabular difference corresponding to the space between log 2387 and log 2388 is 18. From the table of proportional parts for 18 the correction for the fifth digit 6 is obtained. The correction for the sixth digit 9 may be obtained by moving the decimal point of the number opposite 9 in the table of proportional parts for 18 one place to the left. As a result of these processes we obtain

$$\log .002387 = 7.37785 - 10;$$
$$\text{correction for fifth digit } 6 = \qquad 10.8;$$
$$\text{correction for sixth digit } 9 = \qquad 1.62;$$
$$\log .00238769 = 7.37797 - 10.$$

To find the number corresponding to a given logarithm.

Example 4. Find the number whose logarithm is 4.51495; that is, find antilog 4.51495.

We find in the column of Table I headed "log" the mantissa of the given logarithm. The corresponding number is 3273. Since the characteristic of the given logarithm is 4, we know that five digits of the required number precede the decimal point. Therefore,

$$\text{antilog } 4.51495 = 32730.$$

Example 5. Find antilog 9.27268 − 10.

The mantissa of the given logarithm is not tabulated in Table I, so that interpolation will be necessary to obtain the required antilogarithm. In the table the mantissa just smaller than .27268 is .27254. This mantissa corresponds to the number 1873. The difference between the given mantissa and the mantissa of 1873 is

$$.27268 - .27254 = .00014,$$

or simply 14, without the decimal point and the zeros. The tabular difference of the two mantissas adjacent to the given mantissa is 23. In the table of proportional parts for 23 the number in the right-hand column most nearly equal to 14 is 13.8, which corresponds to 6. The fifth digit of the required number is 6. Consequently, the number corresponding best to the given mantissa is 18736, when five significant digits are employed. The characteristic 9 − 10 or − 1 indicates that the decimal point immediately precedes the first significant digit. Hence,

$$\text{antilog } 9.27268 - 10 = .18736.$$

5. Table II. Important constants. This table contains the numerical values and logarithms of a number of important constants.

6. Table III. Logarithmic trigonometric functions. This table gives the logarithmic values of the sine, cosine, tangent, and cotangent of angles for each minute from 0° to 90°. For angles between 0° and 45° the number of degrees and the name of the trigonometric function are read at the top of the page and the number of minutes in the left-hand column; for angles between 45° and 90° the corresponding information is found at the bottom of the page and in the right-hand column.

When the angle is given to the nearest second the logarithmic values of the trigonometric function may be interpolated. However, interpolation for the logarithm of the sine, tangent, and cotangent of small angles is not accurate. Similar conditions exist for the logarithmic values of the cosine, tangent, and cotangent of angles between 86° and 90°. Table IV should be used for obtaining the logarithmic sines and tangents of angles between 0° and 4°, and for the logarithmic cosines and cotangents of angles between 86° and 90°.

Tables of proportional parts which may be used in interpolation of logarithmic values of the functions between 4° and 86° are included in Table III.

To find the value of the logarithm of a function of a given angle.

Example 6. Find log sin 12° 26′ 46″.

On the page of the table for 12° we find 26′ in the first column on the left-hand side. From the column headed "*L* sin" we obtain

$$\log \sin 12° 26′ = 9.33305 - 10.$$

(The − 10 of the characteristic of the logarithm is omitted in the table.) The value of log sin 12° 26′ 46″ must be interpolated between log sin 12° 26′ and log sin 12° 27′. In the third column, opposite the space between these two consecutive values, the tabular difference for 1′ or 60″ is seen to be 57. In the table of proportional parts for 57 the correction for 40″ is 38.0 and that for 6″ is 5.7. These two corrections must be added to the value of log sin 12° 26′. We now have

$$\log \sin 12° 26′ = 9.33305 - 10;$$
$$\text{correction for } 40″ = \qquad 38.0;$$
$$\text{correction for } 6″ = \qquad 5.7;$$
$$\log \sin 12° 26′ 46″ = 9.33349 - 10.$$

Example 7. Find log cot 65° 23′ 27″.

On the page of the table with 65° at the bottom we find 23′ in the right-hand column. From the column with "*L* cot" at the bottom we obtain

$$\log \cot 65° 23′ = 9.66104 - 10.$$

The logarithmic value of the cotangent decreases as the angle increases from 0° to 90°, so that the correction for 27″ must be subtracted from log cot 65° 23′. The tabular difference is 33. In the table of proportional parts for this difference the correction for 20″ is 11.0 and that for 7″ is 3.8. The correction for 27″ is the sum of these two corrections, or 14.8, which must be subtracted from the tabular value of log cot 65° 23′. We now have

$$\log \cot 65° 23′ = 9.66104 - 10;$$
$$\text{correction for } 27″ = \qquad 14.8;$$
$$\log \cot 65° 23′ 27″ = 9.66089 - 10.$$

To find the angle which corresponds to a given value of a trigonometric function.

Example 8. If log tan $\alpha = 9.87382 - 10$, find the value of α which lies between 0° and 90°.

The tabulated value of the logarithm of the tangent just smaller than 9.87382 − 10 is 9.87369 − 10, which corresponds to 36° 47′. The difference between 9.87382 − 10 and 9.87369 − 10 is 13. The tabular difference to be used is 27. In the table of proportional parts for 27 we find that the largest value just smaller than 13 is 9, which corresponds to 20″. We subtract 9 from 13 and obtain 4. This difference corresponds to 9″. Hence the correction is approximately 29″. Therefore

$$\alpha = 36° 47′ 29″.$$

Example 9. If $\log \cos \alpha = 9.92517 - 10$, find the value of α which lies between $0°$ and $90°$.

Since the value of the logarithm of the cosine decreases as the angle increases from $0°$ to $90°$ we must find in the column of the table headed "$L \cos$" a value just larger than $9.92517 - 10$. This required value is $9.92522 - 10$, which corresponds to $32° 40'$. The difference between $9.92522 - 10$ and $9.92517 - 10$ is 5. The tabular difference is 8. In the table of proportional parts for 8 we find that the largest value just smaller than 5 is 4, which corresponds to $30''$. We subtract 4 from 5 and obtain 1. This difference corresponds approximately to $8''$. (When two values which differ by unity in the last digit are equally good approximations it is a good practice to choose that value which ends in an even digit.) Hence the correction for a difference of 5 in the value of the logarithm is approximately $38''$. Therefore,

$$\alpha = 32° 40' 38''.$$

7. Table IV. Logarithmic sines and tangents of small angles. This table gives the values of log sin and log tan for each second between $0°$ and $1°$ and for each ten seconds between $1°$ and $4°$, and of log cos and log tan for each second between $89°$ and $90°$ and for each ten seconds between $86°$ and $89°$. In the ranges between $1°$ and $4°$ and between $86°$ and $89°$ the approximate values of the functions may be obtained for each second by interpolation. Average differences and tables of proportional parts are given with the tables.

8. Table V. Natural trigonometric functions for angles which are multiples of $15°$. This table gives the exact values of the six trigonometric functions for integral multiples of $15°$ from $0°$ to $360°$.

9. Table VI. Natural trigonometric functions. This table gives the actual values of the sine, cosine, tangent, and cotangent for each minute from $0°$ to $90°$. Tabular differences for each minute and tables of proportional parts are provided. The method of interpolation is similar to that which has been illustrated for Table III.

10. Other tables. The remaining tables are listed below and need no further explanation.

Table VII. **Four-Place Logarithms.**

Table VIII. **Four-Place Antilogarithms.**

Table IX. **Four-Place Logarithmic Trigonometric Functions.**

Table X. **Four-Place Natural Trigonometric Functions.**

Table XI. **Logarithmic Trigonometric Functions of Angles expressed in Radian Measure.**

Table XII. **Table for Changing from Sexagesimal to Circular Measure of Angles.**

Table XIII. **Natural Trigonometric Functions of Angles Expressed in Radian Measure.**

Table XIV. **Table for Changing from Circular to Sexagesi~~~~~ ~~easure of Angles.**

TABLES I–XIV

TABLE I. COMMON LOGARITHMS

N	Log	N	Log	N	Log	N	Log	N	Log
0	—	50	69897	100	00000	150	17609	200	30103
1	00000	51	70757	101	00432	151	17898	201	30320
2	30103	52	71600	102	00860	152	18184	202	30535
3	47712	53	72428	103	01284	153	18469	203	30750
4	60206	54	73239	104	01703	154	18752	204	30963
5	69897	55	74036	105	02119	155	19033	205	31175
6	77815	56	74819	106	02531	156	19312	206	31387
7	84510	57	75587	107	02938	157	19590	207	31597
8	90309	58	76343	108	03342	158	19866	208	31806
9	95424	59	77085	109	03743	159	20140	209	32015
10	00000	60	77815	110	04139	160	20412	210	32222
11	04139	61	78533	111	04532	161	20683	211	32428
12	07918	62	79239	112	04922	162	20952	212	32634
13	11394	63	79934	113	05308	163	21219	213	32838
14	14613	64	80618	114	05690	164	21484	214	33041
15	17609	65	81291	115	06070	165	21748	215	33244
16	20412	66	81954	116	06446	166	22011	216	33445
17	23045	67	82607	117	06819	167	22272	217	33646
18	25527	68	83251	118	07188	168	22531	218	33846
19	27875	69	83885	119	07555	169	22789	219	34044
20	30103	70	84510	120	07918	170	23045	220	34242
21	32222	71	85126	121	08279	171	23300	221	34439
22	34242	72	85733	122	08636	172	23553	222	34635
23	36173	73	86332	123	08991	173	23805	223	34830
24	38021	74	86923	124	09342	174	24055	224	35025
25	39794	75	87506	125	09691	175	24304	225	35218
26	41497	76	88081	126	10037	176	24551	226	35411
27	43136	77	88649	127	10380	177	24797	227	35603
28	44716	78	89209	128	10721	178	25042	228	35793
29	46240	79	89763	129	11059	179	25285	229	35984
30	47712	80	90309	130	11394	180	25527	230	36173
31	49136	81	90849	131	11727	181	25768	231	36361
32	50515	82	91381	132	12057	182	26007	232	36549
33	51851	83	91908	133	12385	183	26245	233	36736
34	53148	84	92428	134	12710	184	26482	234	36922
35	54407	85	92942	135	13033	185	26717	235	37107
36	55630	86	93450	136	13354	186	26951	236	37291
37	56820	87	93952	137	13672	187	27184	237	37475
38	57978	88	94448	138	13988	188	27416	238	37658
39	59106	89	94939	139	14301	189	27646	239	37840
40	60206	90	95424	140	14613	190	27875	240	38021
41	61278	91	95904	141	14922	191	28103	241	38202
42	62325	92	96379	142	15229	192	28330	242	38382
43	63347	93	96848	143	15534	193	28556	243	38561
44	64345	94	97313	144	15836	194	28780	244	38739
45	65321	95	97772	145	16137	195	29003	245	38917
46	66276	96	98227	146	16435	196	29226	246	39094
47	67210	97	98677	147	16732	197	29447	247	39270
48	68124	98	99123	148	17026	198	29667	248	39445
49	69020	99	99564	149	17319	199	29885	249	39620
50	69897	100	00000	150	17609	200	30103	250	39794

N	Log	N	Log	N	Log	N	Log	N	Log

N	Log	N	Log	N	Log	N	Log	N	Log
250	39794	**300**	47712	**350**	54407	**400**	60206	**450**	65321
251	39967	301	47857	351	54531	401	60314	451	65418
252	40140	302	48001	352	54654	402	60423	452	65514
253	40312	303	48144	353	54777	403	60531	453	65610
254	40483	304	48287	354	54900	404	60638	454	65706
255	40654	305	48430	355	55023	405	60746	455	65801
256	40824	306	48572	356	55145	406	60853	456	65896
257	40993	307	48714	357	55267	407	60959	457	65992
258	41162	308	48855	358	55388	408	61066	458	66087
259	41330	309	48996	359	55509	409	61172	459	66181
260	41497	**310**	49136	**360**	55630	**410**	61278	**460**	66276
261	41664	311	49276	361	55751	411	61384	461	66370
262	41830	312	49415	362	55871	412	61490	462	66464
263	41996	313	49554	363	55991	413	61595	463	66558
264	42160	314	49693	364	56110	414	61700	464	66652
265	42325	315	49831	365	56229	415	61805	465	66745
266	42488	316	49969	366	56348	416	61909	466	66839
267	42651	317	50106	367	56467	417	62014	467	66932
268	42813	318	50243	368	56585	418	62118	468	67025
269	42975	319	50379	369	56703	419	62221	469	67117
270	43136	**320**	50515	**370**	56820	**420**	62325	**470**	67210
271	43297	321	50651	371	56937	421	62428	471	67302
272	43457	322	50786	372	57054	422	62531	472	67394
273	43616	323	50920	373	57171	423	62634	473	67486
274	43775	324	51055	374	57287	424	62737	474	67578
275	43933	325	51188	375	57403	425	62839	475	67669
276	44091	326	51322	376	57519	426	62941	476	67761
277	44248	327	51455	377	57634	427	63043	477	67852
278	44404	328	51587	378	57749	428	63144	478	67943
279	44560	329	51720	379	57864	429	63246	479	68034
280	44716	**330**	51851	**380**	57978	**430**	63347	**480**	68124
281	44871	331	51983	381	58092	431	63448	481	68215
282	45025	332	52114	382	58206	432	63548	482	68305
283	45179	333	52244	383	58320	433	63649	483	68395
284	45332	334	52375	384	58433	434	63749	484	68485
285	45484	335	52504	385	58546	435	63849	485	68574
286	45637	336	52634	386	58659	436	63949	486	68664
287	45788	337	52763	387	58771	437	64048	487	68753
288	45939	338	52892	388	58883	438	64147	488	68842
289	46090	339	53020	389	58995	439	64246	489	68931
290	46240	**340**	53148	**390**	59106	**440**	64345	**490**	69020
291	46389	341	53275	391	59218	441	64444	491	69108
292	46538	342	53403	392	59329	442	64542	492	69197
293	46687	343	53529	393	59439	443	64640	493	69285
294	46835	344	53656	394	59550	444	64738	494	69373
295	46982	345	53782	395	59660	445	64836	495	69461
296	47129	346	53908	396	59770	446	64933	496	69548
297	47276	347	54033	397	59879	447	65031	497	69636
298	47422	348	54158	398	59988	448	65128	498	69723
299	47567	349	54283	399	60097	499	65225	499	69810
300	47712	**350**	54407	**400**	60206	**450**	65321	**500**	69897
N	Log	N	Log	N	Log	N	Log	N	Log

N	Log	N	Log	N	Log	N	Log	N	Log
500	69897	**550**	74036	**600**	77815	**650**	81291	**700**	84510
501	69984	551	74115	601	77887	651	81358	701	84572
502	70070	552	74194	602	77960	652	81425	702	84634
503	70157	553	74273	603	78032	653	81491	703	84696
504	70243	554	74351	604	78104	654	81558	704	84757
505	70329	555	74429	605	78176	655	81624	705	84819
506	70415	556	74507	606	78247	656	81690	706	84880
507	70501	557	74586	607	78319	657	81757	707	84942
508	70586	558	74663	608	78390	658	81823	708	85003
509	70672	559	74741	609	78462	659	81889	709	85065
510	70757	**560**	74819	**610**	78533	**660**	81954	**710**	85126
511	70842	561	74896	611	78604	661	82020	711	85187
512	70927	562	74974	612	78675	662	82086	712	85248
513	71012	563	75051	613	78746	663	82151	713	85309
514	71096	564	75128	614	78817	664	82217	714	85370
515	71181	565	75205	615	78888	665	82282	715	85431
516	71265	566	75282	616	78958	666	82347	716	85491
517	71349	567	75358	617	79029	667	82413	717	85552
518	71433	568	75435	618	79099	668	82478	718	85612
519	71517	569	75511	619	79169	669	82543	719	85673
520	71600	**570**	75587	**620**	79239	**670**	82607	**720**	85733
521	71684	571	75664	621	79309	671	82672	721	85794
522	71767	572	75740	622	79379	672	82737	722	85854
523	71850	573	75815	623	79449	673	82802	723	85914
524	71933	574	75891	624	79518	674	82866	724	85974
525	72016	575	75967	625	79588	675	82930	725	86034
526	72099	576	76042	626	79657	676	82995	726	86094
527	72181	577	76118	627	79727	677	83059	727	86153
528	72263	578	76193	628	79796	678	83123	728	86213
529	72346	579	76268	629	79865	679	83187	729	86273
530	72428	**580**	76343	**630**	79934	**680**	83251	**730**	86332
531	72509	581	76418	631	80003	681	83315	731	86392
532	72591	582	76492	632	80072	682	83378	732	86451
533	72673	583	76567	633	80140	683	83442	733	86510
534	72754	584	76641	634	80209	684	83506	734	86570
535	72835	585	76716	635	80277	685	83569	735	86629
536	72916	586	76790	636	80346	686	83632	736	86688
537	72997	587	76864	637	80414	687	83696	737	86747
538	73078	588	76938	638	80482	688	83759	738	86806
539	73159	589	77012	639	80550	689	83822	739	86864
540	73239	**590**	77085	**640**	80618	**690**	83885	**740**	86923
541	73320	591	77159	641	80686	691	83948	741	86982
542	73400	592	77232	642	80754	692	84011	742	87040
543	73480	593	77305	643	80821	693	84073	743	87099
544	73560	594	77379	644	80889	694	84136	744	87157
545	73640	595	77452	645	80956	695	84198	745	87216
546	73719	596	77525	646	81023	696	84261	746	87274
547	73799	597	77597	647	81090	697	84323	747	87332
548	73878	598	77670	648	81158	698	84386	748	87390
549	73957	599	77743	649	81224	699	84448	749	87448
550	74036	**600**	77815	**650**	81291	**700**	84510	**750**	87506
N	Log	N	Log	N	Log	N	Log	N	Log

N	Log	N	Log	N	Log	N	Log	N	Log
750	87506	**800**	90309	**850**	92942	**900**	95424	**950**	97772
751	87564	801	90363	851	92993	901	95472	951	97818
752	87622	802	90417	852	93044	902	95521	952	97864
753	87679	803	90472	853	93095	903	95569	953	97909
754	87737	804	90526	854	93146	904	95617	954	97955
755	87795	805	90580	855	93197	905	95665	955	98000
756	87852	806	90634	856	93247	906	95713	956	98046
757	87910	807	90687	857	93298	907	95761	957	98091
758	87967	808	90741	858	93349	908	95809	958	98137
759	88024	809	90795	859	93399	909	95856	959	98182
760	88081	**810**	90849	**860**	93450	**910**	95904	**960**	98227
761	88138	811	90902	861	93500	911	95952	961	98272
762	88195	812	90956	862	93551	912	95999	962	98318
763	88252	813	91009	863	93601	913	96047	963	98363
764	88309	814	91062	864	93651	914	96095	964	98408
765	88366	815	91116	865	93702	915	96142	965	98453
766	88423	816	91169	866	93752	916	96190	966	98498
767	88480	817	91222	867	93802	917	96237	967	98543
768	88536	818	91275	868	93852	918	96284	968	98588
769	88593	819	91328	869	93902	919	96332	969	98632
770	88649	**820**	91381	**870**	93952	**920**	96379	**970**	98677
771	88705	821	91434	871	94002	921	96426	971	98722
772	88762	822	91487	872	94052	922	96473	972	98767
773	88818	823	91540	873	94101	923	96520	973	98811
774	88874	824	91593	874	94151	924	96567	974	98856
775	88930	825	91645	875	94201	925	96614	975	98900
776	88986	826	91698	876	94250	926	96661	976	98945
777	89042	827	91751	877	94300	927	96708	977	98989
778	89098	828	91803	878	94349	928	96755	978	99034
779	89154	829	91855	879	94399	929	96802	979	99078
780	89209	**830**	91908	**880**	94448	**930**	96848	**980**	99123
781	89265	831	91960	881	94498	931	96895	981	99167
782	89321	832	92012	882	94547	932	96942	982	99211
783	89376	833	92065	883	94596	933	96988	983	99255
784	89432	834	92117	884	94645	934	97035	984	99300
785	89487	835	92169	885	94694	935	97081	985	99344
786	89542	836	92221	886	94743	936	97128	986	99388
787	89597	837	92273	887	94792	937	97174	987	99432
788	89653	838	92324	888	94841	938	97220	988	99476
789	89708	839	92376	889	94890	939	97267	989	99520
790	89763	**840**	92428	**890**	94939	**940**	97313	**990**	99564
791	89818	841	92480	891	94988	941	97359	991	99607
792	89873	842	92531	892	95036	942	97405	992	99651
793	89927	843	92583	893	95085	943	97451	993	99695
794	89982	844	92634	894	95134	944	97497	994	99739
795	90037	845	92686	895	95182	945	97543	995	99782
796	90091	846	92737	896	95231	946	97589	996	99826
797	90146	847	92788	897	95279	947	97635	997	99870
798	90200	848	92840	898	95328	948	97681	998	99913
799	90255	849	92891	899	95376	949	97727	999	99957
800	90309	**850**	92942	**900**	95424	**950**	97772	**1000**	00000
N	Log	N	Log	N	Log	N	Log	N	Log

N	Log	d	N	Log	d	N	Log	d	N	Log	d	N	Log	d
1000	00000	43	**1050**	02119	41	**1100**	04139	40	**1150**	06070	38	**1200**	07918	36
1001	043	44	1051	160	42	1101	179	39	1151	108	37	1201	954	36
1002	087	43	1052	202	41	1102	218	40	1152	145	38	1202	990	37
1003	130	43	1053	243	41	1103	258	39	1153	183	38	1203	08027	36
1004	173	44	1054	284	41	1104	297	39	1154	221	37	1204	063	36
1005	00217	43	**1055**	02325	41	**1105**	04336	40	**1155**	06258	38	**1205**	08099	36
1006	260	43	1056	366	41	1106	376	39	1156	296	37	1206	135	36
1007	303	43	1057	407	42	1107	415	39	1157	333	38	1207	171	36
1008	346	43	1058	449	41	1108	454	39	1158	371	37	1208	207	36
1009	389	43	1059	490	41	1109	493	39	1159	408	38	1209	243	36
1010	00432	43	**1060**	02531	41	**1110**	04532	39	**1160**	06446	37	**1210**	08279	35
1011	475	43	1061	572	40	1111	571	39	1161	483	38	1211	314	36
1012	518	43	1062	612	41	1112	610	40	1162	521	37	1212	350	36
1013	561	43	1063	653	41	1113	650	39	1163	558	37	1213	386	36
1014	604	43	1064	694	41	1114	689	38	1164	595	38	1214	422	36
1015	00647	42	**1065**	02735	41	**1115**	04727	39	**1165**	06633	37	**1215**	08458	35
1016	689	43	1066	776	40	1116	766	39	1166	670	37	1216	493	36
1017	732	43	1067	816	41	1117	805	39	1167	707	37	1217	529	36
1018	775	42	1068	857	41	1118	844	39	1168	744	37	1218	565	35
1019	817	43	1069	898	40	1119	883	39	1169	781	38	1219	600	36
1020	00860	43	**1070**	02938	41	**1120**	04922	39	**1170**	06819	37	**1220**	08636	36
1021	903	42	1071	979	40	1121	961	38	1171	856	37	1221	672	35
1022	945	43	1072	03019	41	1122	999	39	1172	893	37	1222	707	36
1023	988	42	1073	060	40	1123	05038	39	1173	930	37	1223	743	35
1024	01030	42	1074	100	41	1124	077	38	1174	967	37	1224	778	36
1025	01072	43	**1075**	03141	40	**1125**	05115	39	**1175**	07004	37	**1225**	08814	35
1026	115	42	1076	181	41	1126	154	38	1176	041	37	1226	849	35
1027	157	42	1077	222	40	1127	192	39	1177	078	37	1227	884	36
1028	199	43	1078	262	40	1128	231	38	1178	115	36	1228	920	35
1029	242	42	1079	302	40	1129	269	39	1179	151	37	1229	955	36
1030	01284	42	**1080**	03342	41	**1130**	05308	38	**1180**	07188	37	**1230**	08991	35
1031	326	42	1081	383	40	1131	346	39	1181	225	37	1231	09026	35
1032	368	42	1082	423	40	1132	385	38	1182	262	36	1232	061	35
1033	410	42	1083	463	40	1133	423	38	1183	298	37	1233	096	36
1034	452	42	1084	503	40	1134	461	39	1184	335	37	1234	132	35
1035	01494	42	**1085**	03543	40	**1135**	05500	38	**1185**	07372	36	**1235**	09167	35
1036	536	42	1086	583	40	1136	538	38	1186	408	37	1236	202	35
1037	578	42	1087	623	40	1137	576	38	1187	445	37	1237	237	35
1038	620	42	1088	663	40	1138	614	38	1188	482	36	1238	272	35
1039	662	41	1089	703	40	1139	652	38	1189	518	37	1239	307	35
1040	01703	42	**1090**	03743	39	**1140**	05690	39	**1190**	07555	36	**1240**	09342	35
1041	745	42	1091	782	40	1141	729	38	1191	591	37	1241	377	35
1042	787	41	1092	822	40	1142	767	38	1192	628	36	1242	412	35
1043	828	42	1093	862	40	1143	805	38	1193	664	36	1243	447	35
1044	870	42	1094	902	39	1144	843	38	1194	700	37	1244	482	35
1045	01912	41	**1095**	03941	40	**1145**	05881	37	**1195**	07737	36	**1245**	09517	35
1046	953	42	1096	981	40	1146	918	38	1196	773	36	1246	552	35
1047	995	41	1097	04021	39	1147	956	38	1197	809	37	1247	587	34
1048	02036	42	1098	060	40	1148	994	38	1198	846	36	1248	621	35
1049	078	41	1099	100	39	1149	06032	38	1199	882	36	1249	656	35
1050	02119		**1100**	04139		**1150**	06070		**1200**	07918		**1250**	09691	

P. P.

	44	43
1	4.4	4.3
2	8.8	8.6
3	13.2	12.9
4	17.6	17.2
5	22.0	21.5
6	26.4	25.8
7	30.8	30.1
8	35.2	34.4
9	39.6	38.7

	42	41
1	4.2	4.1
2	8.4	8.2
3	12.6	12.3
4	16.8	16.4
5	21.0	20.5
6	25.2	24.6
7	29.4	28.7
8	33.6	32.8
9	37.8	36.9

	40	39
1	4.0	3.9
2	8.0	7.8
3	12.0	11.7
4	16.0	15.6
5	20.0	19.5
6	24.0	23.4
7	28.0	27.3
8	32.0	31.2
9	36.0	35.1

	38	37
1	3.8	3.7
2	7.6	7.4
3	11.4	11.1
4	15.2	14.8
5	19.0	18.5
6	22.8	22.2
7	26.6	25.9
8	30.4	29.6
9	34.2	33.3

N	Log	d	N	Log	d	N	Log	d	N	Log	d	N	Log	d	P. P.

Proportional Parts (P.P.)

	36	35		34	33		32	31		30	29
1	3.6	3.5		3.4	3.3		3.2	3.1		3.0	2.9
2	7.2	7.0		6.8	6.6		6.4	6.2		6.0	5.8
3	10.8	10.5		10.2	9.9		9.6	9.3		9.0	8.7
4	14.4	14.0		13.6	13.2		12.8	12.4		12.0	11.6
5	18.0	17.5		17.0	16.5		16.0	15.5		15.0	14.5
6	21.6	21.0		20.4	19.8		19.2	18.6		18.0	17.4
7	25.2	24.5		23.8	23.1		22.4	21.7		21.0	20.3
8	28.8	28.0		27.2	26.4		25.6	24.8		24.0	23.2
9	32.4	31.5		30.6	29.7		28.8	27.9		27.0	26.1

N	Log	d	N	Log	d	N	Log	d	N	Log	d	N	Log	d
1250	09691	35	1300	11394	34	1350	13033	33	1400	14613	31	1450	16137	30
1251	726	34	1301	428	33	1351	066	32	1401	644	31	1451	167	30
1252	760	35	1302	461	33	1352	098	32	1402	675	31	1452	197	30
1253	795	35	1303	494	34	1353	130	32	1403	706	31	1453	227	29
1254	830	34	1304	528	33	1354	162	32	1404	737	31	1454	256	30
1255	09864	35	1305	11561	33	1355	13194	32	1405	14768	31	1455	16286	30
1256	899	35	1306	594	34	1356	226	32	1406	799	30	1456	316	30
1257	934	34	1307	628	33	1357	258	32	1407	829	31	1457	346	30
1258	968	35	1308	661	33	1358	290	32	1408	860	31	1458	376	30
1259	10003	34	1309	694	33	1359	322	32	1409	891	31	1459	406	29
1260	10037	35	1310	11727	33	1360	13354	32	1410	14922	31	1460	16435	30
1261	072	34	1311	760	33	1361	386	32	1411	953	30	1461	465	30
1262	106	34	1312	793	33	1362	418	32	1412	983	31	1462	495	30
1263	140	35	1313	826	34	1363	450	31	1413	15014	31	1463	524	30
1264	175	34	1314	860	33	1364	481	32	1414	045	31	1464	554	30
1265	10209	34	1315	11893	33	1365	13513	32	1415	15076	30	1465	16584	29
1266	243	35	1316	926	33	1366	545	32	1416	106	31	1466	613	30
1267	278	34	1317	959	33	1367	577	32	1417	137	31	1467	643	30
1268	312	34	1318	992	32	1368	609	31	1418	168	30	1468	673	29
1269	346	34	1319	12024	33	1369	640	32	1419	198	31	1469	702	30
1270	10380	35	1320	12057	33	1370	13672	32	1420	15229	30	1470	16732	29
1271	415	34	1321	090	33	1371	704	31	1421	259	31	1471	761	30
1272	449	34	1322	123	33	1372	735	32	1422	290	30	1472	791	29
1273	483	34	1323	156	33	1373	767	32	1423	320	31	1473	820	30
1274	517	34	1324	189	33	1374	799	31	1424	351	30	1474	850	29
1275	10551	34	1325	12222	32	1375	13830	32	1425	15381	31	1475	16879	30
1276	585	34	1326	254	33	1376	862	31	1426	412	30	1476	909	29
1277	619	34	1327	287	33	1377	893	32	1427	442	31	1477	938	29
1278	653	34	1328	320	32	1378	925	31	1428	473	30	1478	967	30
1279	687	34	1329	352	33	1379	956	32	1429	503	31	1479	997	29
1280	10721	34	1330	12385	33	1380	13988	31	1430	15534	30	1480	17026	30
1281	755	34	1331	418	32	1381	14019	32	1431	564	30	1481	056	29
1282	789	34	1332	450	33	1382	051	31	1432	594	31	1482	085	29
1283	823	34	1333	483	33	1383	082	32	1433	625	30	1483	114	29
1284	857	33	1334	516	32	1384	114	31	1434	655	30	1484	143	30
1285	10890	34	1335	12548	33	1385	14145	31	1435	15685	30	1485	17173	29
1286	924	34	1336	581	32	1386	176	32	1436	715	31	1486	202	29
1287	958	34	1337	613	33	1387	208	31	1437	746	30	1487	231	29
1288	992	33	1338	646	32	1388	239	31	1438	776	30	1488	260	29
1289	11025	34	1339	678	32	1389	270	31	1439	806	30	1489	289	30
1290	11059	34	1340	12710	33	1390	14301	32	1440	15836	30	1490	17319	29
1291	093	33	1341	743	32	1391	333	31	1441	866	31	1491	348	29
1292	126	34	1342	775	33	1392	364	31	1442	897	30	1492	377	29
1293	160	33	1343	808	32	1393	395	31	1443	927	30	1493	406	29
1294	193	34	1344	840	32	1394	426	31	1444	957	30	1494	435	29
1295	11227	34	1345	12872	33	1395	14457	32	1445	15987	30	1495	17464	29
1296	261	33	1346	905	32	1396	489	31	1446	16017	30	1496	493	29
1297	294	33	1347	937	32	1397	520	31	1447	047	30	1497	522	29
1298	327	34	1348	969	32	1398	551	31	1448	077	30	1498	551	29
1299	361	33	1349	13001	32	1399	582	31	1449	107	30	1499	580	29
1300	11394		1350	13033		1400	14613		1450	16137		1500	17609	

N	Log	d	N	Log	d	N	Log	d	N	Log	d	N	Log	d
1500	17609	29	**1550**	19033	28	**1600**	20412	27	**1650**	21748	27	**1700**	23045	25
1501	638	29	1551	061	28	1601	439	27	1651	775	26	1701	070	26
1502	667	29	1552	089	28	1602	466	27	1652	801	26	1702	096	25
1503	696	29	1553	117	28	1603	493	27	1653	827	26	1703	121	26
1504	725	29	1554	145	28	1604	520	28	1654	854	26	1704	147	25
1505	17754	28	1555	19173	28	1605	20548	27	1655	21880	26	1705	23172	26
1506	782	29	1556	201	28	1606	575	27	1656	906	26	1706	198	25
1507	811	29	1557	229	28	1607	602	27	1657	932	26	1707	223	26
1508	840	29	1558	257	28	1608	629	27	1658	958	27	1708	249	25
1509	869	29	1559	285	27	1609	656	27	1659	985	26	1709	274	26
1510	17898	28	**1560**	19312	28	**1610**	20683	27	**1660**	22011	26	**1710**	23300	25
1511	926	29	1561	340	28	1611	710	27	1661	037	26	1711	325	25
1512	955	29	1562	368	28	1612	737	26	1662	063	26	1712	350	26
1513	984	29	1563	396	28	1613	763	27	1663	089	26	1713	376	25
1514	18013	28	1564	424	27	1614	790	27	1664	115	26	1714	401	25
1515	18041	29	1565	19451	28	1615	20817	27	1665	22141	26	1715	23426	26
1516	070	29	1566	479	28	1616	844	27	1666	167	27	1716	452	25
1517	099	28	1567	507	28	1617	871	27	1667	194	26	1717	477	25
1518	127	29	1568	535	27	1618	898	27	1668	220	26	1718	502	26
1519	156	28	1569	562	28	1619	925	27	1669	246	26	1719	528	25
1520	18184	29	**1570**	19590	28	**1620**	20952	26	**1670**	22272	26	**1720**	23553	25
1521	213	28	1571	618	27	1621	978	27	1671	298	26	1721	578	25
1522	241	29	1572	645	28	1622	21005	27	1672	324	26	1722	603	26
1523	270	28	1573	673	27	1623	032	27	1673	350	26	1723	629	25
1524	298	29	1574	700	28	1624	059	26	1674	376	25	1724	654	25
1525	18327	28	1575	19728	28	1625	21085	27	1675	22401	26	1725	23679	25
1526	355	29	1576	756	27	1626	112	27	1676	427	26	1726	704	25
1527	384	28	1577	783	28	1627	139	26	1677	453	26	1727	729	25
1528	412	29	1578	811	27	1628	165	27	1678	479	26	1728	754	25
1529	441	28	1579	838	28	1629	192	27	1679	505	26	1729	779	26
1530	18469	29	**1580**	19866	27	**1630**	21219	26	**1680**	22531	26	**1730**	23805	25
1531	498	28	1581	893	28	1631	245	27	1681	557	26	1731	830	25
1532	526	28	1582	921	27	1632	272	27	1682	583	25	1732	855	25
1533	554	29	1583	948	28	1633	299	26	1683	608	26	1733	880	25
1534	583	28	1584	976	27	1634	325	27	1684	634	26	1734	905	25
1535	18611	28	1585	20003	27	1635	21352	26	1685	22660	26	1735	23930	25
1536	639	28	1586	030	28	1636	378	27	1686	686	26	1736	955	25
1537	667	29	1587	058	27	1637	405	26	1687	712	25	1737	980	25
1538	696	28	1588	085	27	1638	431	27	1688	737	26	1738	24005	25
1539	724	28	1589	112	28	1639	458	26	1689	763	26	1739	030	25
1540	18752	28	**1590**	20140	27	**1640**	21484	27	**1690**	22789	25	**1740**	24055	25
1541	780	28	1591	167	27	1641	511	26	1691	814	26	1741	080	25
1542	808	29	1592	194	28	1642	537	27	1692	840	26	1742	105	25
1543	837	28	1593	222	27	1643	564	26	1693	866	25	1743	130	25
1544	865	28	1594	249	27	1644	590	27	1694	891	26	1744	155	25
1545	18893	28	1595	20276	27	1645	21617	26	1695	22917	26	1745	24180	24
1546	921	28	1596	303	27	1646	643	26	1696	943	25	1746	204	25
1547	949	28	1597	330	28	1647	669	27	1697	968	26	1747	229	25
1548	977	28	1598	358	27	1648	696	26	1698	994	25	1748	254	25
1549	19005	28	1599	385	27	1649	722	26	1699	23019	26	1749	279	25
1550	19033		**1600**	20412		**1650**	21748		**1700**	23045		**1750**	24304	
N	Log	d	N	Log	d	N	Log	d	N	Log	d	N	Log	d

P. P.

29		28		27		26		25	
1	2.9	1	2.8	1	2.7	1	2.6	1	2.5
2	5.8	2	5.6	2	5.4	2	5.2	2	5.0
3	8.7	3	8.4	3	8.1	3	7.8	3	7.5
4	11.6	4	11.2	4	10.8	4	10.4	4	10.0
5	14.5	5	14.0	5	13.5	5	13.0	5	12.5
6	17.4	6	16.8	6	16.2	6	15.6	6	15.0
7	20.3	7	19.6	7	18.9	7	18.2	7	17.5
8	23.2	8	22.4	8	21.6	8	20.8	8	20.0
9	26.1	9	25.2	9	24.3	9	23.4	9	22.5

P.P. (Proportional Parts)

24	23	22	21
1 2.4	1 2.3	1 2.2	1 2.1
2 4.8	2 4.6	2 4.4	2 4.2
3 7.2	3 6.9	3 6.6	3 6.3
4 9.6	4 9.2	4 8.8	4 8.4
5 12.0	5 11.5	5 11.0	5 10.5
6 14.4	6 13.8	6 13.2	6 12.6
7 16.8	7 16.1	7 15.4	7 14.7
8 19.2	8 18.4	8 17.6	8 16.8
9 21.6	9 20.7	9 19.8	9 18.9

N	Log	d	N	Log	d	N	Log	d	N	Log	d	N	Log	d
1750	24304	25	**1800**	25527	24	**1850**	26717	24	**1900**	27875	23	**1950**	29003	23
1751	329	24	1801	551	24	1851	741	23	1901	898	23	1951	026	22
1752	353	25	1802	575	25	1852	764	24	1902	921	23	1952	048	22
1753	378	25	1803	600	24	1853	788	23	1903	944	23	1953	070	22
1754	403	25	1804	624	24	1854	811	23	1904	967	22	1954	092	23
1755	24428	24	1805	25648	24	1855	26834	24	1905	27989	23	1955	29115	22
1756	452	25	1806	672	24	1856	858	23	1906	28012	23	1956	137	22
1757	477	25	1807	696	24	1857	881	24	1907	035	23	1957	159	22
1758	502	25	1808	720	24	1858	905	23	1908	058	23	1958	181	22
1759	527	24	1809	744	24	1859	928	23	1909	081	22	1959	203	23
1760	24551	25	**1810**	25768	24	**1860**	26951	24	**1910**	28103	23	**1960**	29226	22
1761	576	25	1811	792	24	1861	975	23	1911	126	23	1961	248	22
1762	601	24	1812	816	24	1862	998	23	1912	149	22	1962	270	22
1763	625	25	1813	840	24	1863	27021	24	1913	171	23	1963	292	22
1764	650	24	1814	864	24	1864	045	23	1914	194	23	1964	314	22
1765	24674	25	1815	25888	24	1865	27068	23	1915	28217	23	1965	29336	22
1766	699	25	1816	912	23	1866	091	23	1916	240	22	1966	358	22
1767	724	24	1817	935	24	1867	114	24	1917	262	23	1967	380	23
1768	748	25	1818	959	24	1868	138	23	1918	285	22	1968	403	22
1769	773	24	1819	983	24	1869	161	23	1919	307	23	1969	425	22
1770	24797	25	**1820**	26007	24	**1870**	27184	23	**1920**	28330	23	**1970**	29447	22
1771	822	24	1821	031	24	1871	207	24	1921	353	22	1971	469	22
1772	846	25	1822	055	24	1872	231	23	1922	375	23	1972	491	22
1773	871	24	1823	079	23	1873	254	23	1923	398	23	1973	513	22
1774	895	25	1824	102	24	1874	277	23	1924	421	22	1974	535	22
1775	24920	24	1825	26126	24	1875	27300	23	1925	28443	23	1975	29557	22
1776	944	25	1826	150	24	1876	323	23	1926	466	22	1976	579	22
1777	969	24	1827	174	24	1877	346	24	1927	488	23	1977	601	22
1778	993	25	1828	198	23	1878	370	23	1928	511	22	1978	623	22
1779	25018	24	1829	221	24	1879	393	23	1929	533	23	1979	645	22
1780	25042	24	**1830**	26245	24	**1880**	27416	23	**1930**	28556	22	**1980**	29667	21
1781	066	25	1831	269	24	1881	439	23	1931	578	23	1981	688	22
1782	091	24	1832	293	23	1882	462	23	1932	601	22	1982	710	22
1783	115	24	1833	316	24	1883	485	23	1933	623	23	1983	732	22
1784	139	25	1834	340	24	1884	508	23	1934	646	22	1984	754	22
1785	25164	24	1835	26364	23	1885	27531	23	1935	28668	23	1985	29776	22
1786	188	24	1836	387	24	1886	554	23	1936	691	22	1986	798	22
1787	212	25	1837	411	24	1887	577	23	1937	713	22	1987	820	22
1788	237	24	1838	435	23	1888	600	23	1938	735	23	1988	842	21
1789	261	24	1839	458	24	1889	623	23	1939	758	22	1989	863	22
1790	25285	25	**1840**	26482	23	**1890**	27646	23	**1940**	28780	23	**1990**	29885	22
1791	310	24	1841	505	24	1891	669	23	1941	803	22	1991	907	22
1792	334	24	1842	529	24	1892	692	23	1942	825	22	1992	929	22
1793	358	24	1843	553	23	1893	715	23	1943	847	23	1993	951	22
1794	382	24	1844	576	24	1894	738	23	1944	870	22	1994	973	21
1795	25406	25	1845	26600	23	1895	27761	23	1945	28892	22	1995	29994	22
1796	431	24	1846	623	24	1896	784	23	1946	914	23	1996	30016	22
1797	455	24	1847	647	23	1897	807	23	1947	937	22	1997	038	22
1798	479	24	1848	670	24	1898	830	22	1948	959	22	1998	060	21
1799	503	24	1849	694	23	1899	852	23	1949	981	22	1999	081	22
1800	25527		**1850**	26717		**1900**	27875		**1950**	29003		**2000**	30103	

N	Log	d	N	Log	d	N	Log	d	N	Log	d	N	Log	d
2000	30103	22	**2050**	31175	22	**2100**	32222	21	**2150**	33244	20	**2200**	34242	20
2001	125	21	2051	197	21	2101	243	20	2151	264	20	2201	262	20
2002	146	22	2052	218	22	2102	263	21	2152	284	20	2202	282	19
2003	168	22	2053	239	21	2103	284	21	2153	304	21	2203	301	20
2004	190	21	2054	260	21	2104	305	20	2154	325	20	2204	321	20
2005	30211	22	2055	31281	21	2105	32325	21	2155	33345	20	2205	34341	20
2006	233	22	2056	302	21	2106	346	20	2156	365	20	2206	361	19
2007	255	21	2057	323	22	2107	366	21	2157	385	20	2207	380	20
2008	276	22	2058	345	21	2108	387	21	2158	405	20	2208	400	20
2009	298	22	2059	366	21	2109	408	20	2159	425	20	2209	420	19
2010	30320	21	**2060**	31387	21	**2110**	32428	21	**2160**	33445	20	**2210**	34439	20
2011	341	22	2061	408	21	2111	449	20	2161	465	21	2211	459	20
2012	363	21	2062	429	21	2112	469	21	2162	486	20	2212	479	19
2013	384	22	2063	450	21	2113	490	20	2163	506	20	2213	498	20
2014	406	22	2064	471	21	2114	510	21	2164	526	20	2214	518	19
2015	30428	21	2065	31492	21	2115	32531	21	2165	33546	20	2215	34537	20
2016	449	22	2066	513	21	2116	552	20	2166	566	20	2216	557	20
2017	471	21	2067	534	21	2117	572	21	2167	586	20	2217	577	19
2018	492	22	2068	555	21	2118	593	20	2168	606	20	2218	596	20
2019	514	21	2069	576	21	2119	613	21	2169	626	20	2219	616	19
2020	30535	22	**2070**	31597	21	**2120**	32634	20	**2170**	33646	20	**2220**	34635	20
2021	557	21	2071	618	21	2121	654	21	2171	666	20	2221	655	19
2022	578	22	2072	639	21	2122	675	20	2172	686	20	2222	674	20
2023	600	21	2073	660	21	2123	695	20	2173	706	20	2223	694	19
2024	621	22	2074	681	21	2124	715	21	2174	726	20	2224	713	20
2025	30643	21	2075	31702	21	2125	32736	20	2175	33746	20	2225	34733	20
2026	664	21	2076	723	21	2126	756	21	2176	766	20	2226	753	19
2027	685	22	2077	744	21	2127	777	20	2177	786	20	2227	772	20
2028	707	21	2078	765	20	2128	797	21	2178	806	20	2228	792	19
2029	728	22	2079	785	21	2129	818	20	2179	826	20	2229	811	19
2030	30750	21	**2080**	31806	21	**2130**	32838	20	**2180**	33846	20	**2230**	34830	20
2031	771	21	2081	827	21	2131	858	21	2181	866	19	2231	850	19
2032	792	22	2082	848	21	2132	879	20	2182	885	20	2232	869	20
2033	814	21	2083	869	21	2133	899	20	2183	905	20	2233	889	19
2034	835	21	2084	890	21	2134	919	21	2184	925	20	2234	908	20
2035	30856	22	2085	31911	20	2135	32940	20	2185	33945	20	2235	34928	19
2036	878	21	2086	931	21	2136	960	20	2186	965	20	2236	947	20
2037	899	21	2087	952	21	2137	980	21	2187	985	20	2237	967	19
2038	920	22	2088	973	21	2138	33001	20	2188	34005	20	2238	986	19
2039	942	21	2089	994	21	2139	021	20	2189	025	19	2239	35005	20
2040	30963	21	**2090**	32015	20	**2140**	33041	21	**2190**	34044	20	**2240**	35025	19
2041	984	22	2091	035	21	2141	062	20	2191	064	20	2241	044	20
2042	31006	21	2092	056	21	2142	082	20	2192	084	20	2242	064	19
2043	027	21	2093	077	21	2143	102	20	2193	104	20	2243	083	19
2044	048	21	2094	098	20	2144	122	21	2194	124	19	2244	102	20
2045	31069	22	2095	32118	21	2145	33143	20	2195	34143	20	2245	35122	19
2046	091	21	2096	139	21	2146	163	20	2196	163	20	2246	141	19
2047	112	21	2097	160	21	2147	183	20	2197	183	20	2247	160	20
2048	133	21	2098	181	20	2148	203	21	2198	203	20	2248	180	19
2049	154	21	2099	201	21	2149	224	20	2199	223	19	2249	199	19
2050	31175		**2100**	32222		**2150**	33244		**2200**	34242		**2250**	35218	

P. P.

22		21		20	
1	2.2	1	2.1	1	2.0
2	4.4	2	4.2	2	4.0
3	6.6	3	6.3	3	6.0
4	8.8	4	8.4	4	8.0
5	11.0	5	10.5	5	10.0
6	13.2	6	12.6	6	12.0
7	15.4	7	14.7	7	14.0
8	17.6	8	16.8	8	16.0
9	19.8	9	18.9	9	18.0

P. P.

19	
1	1.9
2	3.8
3	5.7
4	7.6
5	9.5
6	11.4
7	13.3
8	15.2
9	17.1

18	
1	1.8
2	3.6
3	5.4
4	7.2
5	9.0
6	10.8
7	12.6
8	14.4
9	16.2

17	
1	1.7
2	3.4
3	5.1
4	6.8
5	8.5
6	10.2
7	11.9
8	13.6
9	15.3

N	Log	d	N	Log	d	N	Log	d	N	Log	d	N	Log	d
2250	35218	20	**2300**	36173	19	**2350**	37107	18	**2400**	38021	18	**2450**	38917	17
2251	238	19	2301	192	19	2351	125	19	2401	039	18	2451	934	18
2252	257	19	2302	211	18	2352	144	18	2402	057	18	2452	952	18
2253	276	19	2303	229	19	2353	162	19	2403	075	18	2453	970	17
2254	295	20	2304	248	19	2354	181	18	2404	093	19	2454	987	18
2255	35315	19	2305	36267	19	2355	37199	19	2405	38112	18	2455	39005	18
2256	334	19	2306	286	19	2356	218	18	2406	130	18	2456	023	18
2257	353	19	2307	305	19	2357	236	18	2407	148	18	2457	041	17
2258	372	20	2308	324	18	2358	254	19	2408	166	18	2458	058	18
2259	392	19	2309	342	19	2359	273	18	2409	184	18	2459	076	18
2260	35411	19	**2310**	36361	19	**2360**	37291	19	**2410**	38202	18	**2460**	39094	17
2261	430	19	2311	380	19	2361	310	18	2411	220	18	2461	111	18
2262	449	19	2312	399	19	2362	328	18	2412	238	18	2462	129	17
2263	468	20	2313	418	18	2363	346	19	2413	256	18	2463	146	18
2264	488	19	2314	436	19	2364	365	18	2414	274	18	2464	164	18
2265	35507	19	2315	36455	19	2365	37383	18	2415	38292	18	2465	39182	17
2266	526	19	2316	474	19	2366	401	19	2416	310	18	2466	199	18
2267	545	19	2317	493	18	2367	420	18	2417	328	18	2467	217	18
2268	564	19	2318	511	19	2368	438	19	2418	346	18	2468	235	17
2269	583	20	2319	530	19	2369	457	18	2419	364	18	2469	252	18
2270	35603	19	**2320**	36549	19	**2370**	37475	18	**2420**	38382	17	**2470**	39270	17
2271	622	19	2321	568	18	2371	493	18	2421	399	18	2471	287	18
2272	641	19	2322	586	19	2372	511	19	2422	417	18	2472	305	17
2273	660	19	2323	605	19	2373	530	18	2423	435	18	2473	322	18
2274	679	19	2324	624	18	2374	548	18	2424	453	18	2474	340	18
2275	35698	19	2325	36642	19	2375	37566	19	2425	38471	18	2475	39358	17
2276	717	19	2326	661	19	2376	585	18	2426	489	18	2476	375	18
2277	736	19	2327	680	18	2377	603	18	2427	507	18	2477	393	17
2278	755	19	2328	698	19	2378	621	18	2428	525	18	2478	410	18
2279	774	19	2329	717	19	2379	639	19	2429	543	18	2479	428	17
2280	35793	20	**2330**	36736	18	**2380**	37658	18	**2430**	38561	17	**2480**	39445	18
2281	813	19	2331	754	19	2381	676	18	2431	578	18	2481	463	17
2282	832	19	2332	773	18	2382	694	18	2432	596	18	2482	480	18
2283	851	19	2333	791	19	2383	712	19	2433	614	18	2483	498	17
2284	870	19	2334	810	19	2384	731	18	2434	632	18	2484	515	18
2285	35889	19	2335	36829	18	2385	37749	18	2435	38650	18	2485	39533	17
2286	908	19	2336	847	19	2386	767	18	2436	668	18	2486	550	18
2287	927	19	2337	866	18	2387	785	18	2437	686	17	2487	568	17
2288	946	19	2338	884	19	2388	803	19	2438	703	18	2488	585	17
2289	965	19	2339	903	19	2389	822	18	2439	721	18	2489	602	18
2290	35984	19	**2340**	36922	18	**2390**	37840	18	**2440**	38739	18	**2490**	39620	17
2291	36003	18	2341	940	19	2391	858	18	2441	757	18	2491	637	18
2292	021	19	2342	959	18	2392	876	18	2442	775	17	2492	655	17
2293	040	19	2343	977	19	2393	894	18	2443	792	18	2493	672	18
2294	059	19	2344	996	18	2394	912	19	2444	810	18	2494	690	17
2295	36078	19	2345	37014	19	2395	37931	18	2445	38828	18	2495	39707	17
2296	097	19	2346	033	18	2396	949	18	2446	846	17	2496	724	18
2297	116	19	2347	051	19	2397	967	18	2447	863	18	2497	742	17
2298	135	19	2348	070	18	2398	985	18	2448	881	18	2498	759	18
2299	154	19	2349	088	19	2399	38003	18	2449	899	18	2499	777	17
2300	36173		**2350**	37107		**2400**	38021		**2450**	38917		**2500**	39794	

P. P.	N	Log	d	N	Log	d	N	Log	d	N	Log	d	N	Log	d

N	Log	d	N	Log	d	N	Log	d	N	Log	d	N	Log	d	P. P.
2500	39794	17	**2550**	40654	17	**2600**	41497	17	**2650**	42325	16	**2700**	43136	16	
2501	811	18	2551	671	17	2601	514	17	2651	341	16	2701	152	17	
2502	829	17	2552	688	17	2602	531	16	2652	357	17	2702	169	16	
2503	846	17	2553	705	17	2603	547	17	2653	374	16	2703	185	16	
2504	863	18	2554	722	17	2604	564	17	2654	390	16	2704	201	16	
2505	39881	17	2555	40739	17	2605	41581	16	2655	42406	17	2705	43217	16	
2506	898	17	2556	756	17	2606	597	17	2656	423	16	2706	233	16	
2507	915	18	2557	773	18	2607	614	17	2657	439	16	2707	249	16	
2508	933	17	2558	790	17	2608	631	17	2658	455	17	2708	265	16	
2509	950	17	2559	807	17	2609	647	17	2659	472	16	2709	281	16	
2510	39967	18	**2560**	40824	17	**2610**	41664	17	**2660**	42488	16	**2710**	43297	16	
2511	985	17	2561	841	17	2611	681	16	2661	504	17	2711	313	16	
2512	40002	17	2562	858	17	2612	697	17	2662	521	16	2712	329	16	
2513	019	18	2563	875	18	2613	714	17	2663	537	16	2713	345	16	**18**
2514	037	17	2564	892	17	2614	731	16	2664	553	17	2714	361	16	1 | 1.8
2515	40054	17	2565	40909	17	2615	41747	17	2665	42570	16	2715	43377	16	2 | 3.6
2516	071	17	2566	926	17	2616	764	16	2666	586	16	2716	393	16	3 | 5.4
2517	088	18	2567	943	18	2617	780	17	2667	602	17	2717	409	16	4 | 7.2
2518	106	17	2568	960	17	2618	797	17	2668	619	16	2718	425	16	5 | 9.0
2519	123	17	2569	976	17	2619	814	16	2669	635	16	2719	441	16	6 | 10.8
2520	40140	17	**2570**	40993	17	**2620**	41830	17	**2670**	42651	16	**2720**	43457	16	7 | 12.6
2521	157	18	2571	41010	18	2621	847	16	2671	667	17	2721	473	16	8 | 14.4
2522	175	17	2572	027	17	2622	863	17	2672	684	16	2722	489	16	9 | 16.2
2523	192	17	2573	044	17	2623	880	16	2673	700	16	2723	505	16	
2524	209	17	2574	061	17	2624	896	17	2674	716	16	2724	521	16	
2525	40226	17	2575	41078	17	2625	41913	16	2675	42732	17	2725	43537	16	**17**
2526	243	18	2576	095	16	2626	929	17	2676	749	16	2726	553	16	1 | 1.7
2527	261	17	2577	111	17	2627	946	17	2677	765	16	2727	569	15	2 | 3.4
2528	278	17	2578	128	17	2628	963	16	2678	781	16	2728	584	16	3 | 5.1
2529	295	17	2579	145	17	2629	979	17	2679	797	16	2729	600	16	4 | 6.8
2530	40312	17	**2580**	41162	17	**2630**	41996	16	**2680**	42813	17	**2730**	43616	16	5 | 8.5
2531	329	17	2581	179	17	2631	42012	17	2681	830	16	2731	632	16	6 | 10.2
2532	346	18	2582	196	16	2632	029	16	2682	846	16	2732	648	16	7 | 11.9
2533	364	17	2583	212	17	2633	045	17	2683	862	16	2733	664	16	8 | 13.6
2534	381	17	2584	229	17	2634	062	16	2684	878	16	2734	680	16	9 | 15.3
2535	40398	17	2585	41246	17	2635	42078	17	2685	42894	17	2735	43696	16	
2536	415	17	2586	263	17	2636	095	16	2686	911	16	2736	712	15	
2537	432	17	2587	280	16	2637	111	16	2687	927	16	2737	727	16	
2538	449	17	2588	296	17	2638	127	17	2688	943	16	2738	743	16	**16**
2539	466	17	2589	313	17	2639	144	16	2689	959	16	2739	759	16	1 | 1.6
2540	40483	17	**2590**	41330	17	**2640**	42160	17	**2690**	42975	16	**2740**	43775	16	2 | 3.2
2541	500	18	2591	347	16	2641	177	16	2691	991	17	2741	791	16	3 | 4.8
2542	518	17	2592	363	17	2642	193	17	2692	43008	16	2742	807	16	4 | 6.4
2543	535	17	2593	380	17	2643	210	16	2693	024	16	2743	823	15	5 | 8.0
2544	552	17	2594	397	17	2644	226	17	2694	040	16	2744	838	16	6 | 9.6
2545	40569	17	2595	41414	16	2645	42243	16	2695	43056	16	2745	43854	16	7 | 11.2
2546	586	17	2596	430	17	2646	259	16	2696	072	16	2746	870	16	8 | 12.8
2547	603	17	2597	447	17	2647	275	17	2697	088	16	2747	886	16	9 | 14.4
2548	620	17	2598	464	17	2648	292	16	2698	104	16	2748	902	15	
2549	637	17	2599	481	16	2649	308	17	2699	120	17	2749	917	16	
2550	40654		**2600**	41497		**2650**	42325		**2700**	43136		**2750**	43933		
N	Log	d	N	Log	d	N	Log	d	N	Log	d	N	Log	d	P. P.

P. P.

	15
1	1.5
2	3.0
3	4.5
4	6.0
5	7.5
6	9.0
7	10.5
8	12.0
9	13.5

	14
1	1.4
2	2.8
3	4.2
4	5.6
5	7.0
6	8.4
7	9.8
8	11.2
9	12.6

N	Log	d	N	Log	d	N	Log	d	N	Log	d	N	Log	d
2750	43933	16	2800	44716	15	2850	45484	16	2900	46240	15	2950	46982	15
2751	949	16	2801	731	16	2851	500	15	2901	255	15	2951	997	15
2752	965	16	2802	747	15	2852	515	15	2902	270	15	2952	47012	14
2753	981	15	2803	762	16	2853	530	15	2903	285	15	2953	026	15
2754	996	16	2804	778	15	2854	545	16	2904	300	15	2954	041	15
2755	44012	16	2805	44793	16	2855	45561	15	2905	46315	15	2955	47056	14
2756	028	16	2806	809	15	2856	576	15	2906	330	15	2956	070	15
2757	044	15	2807	824	16	2857	591	15	2907	345	14	2957	085	15
2758	059	16	2808	840	15	2858	606	15	2908	359	15	2958	100	14
2759	075	16	2809	855	16	2859	621	16	2909	374	15	2959	114	15
2760	44091	16	2810	44871	15	2860	45637	15	2910	46389	15	2960	47129	15
2761	107	15	2811	886	16	2861	652	15	2911	404	15	2961	144	15
2762	122	16	2812	902	15	2862	667	15	2912	419	15	2962	159	14
2763	138	16	2813	917	15	2863	682	15	2913	434	15	2963	173	15
2764	154	16	2814	932	16	2864	697	15	2914	449	15	2964	188	14
2765	44170	15	2815	44948	15	2865	45712	16	2915	46464	15	2965	47202	15
2766	185	16	2816	963	16	2866	728	15	2916	479	15	2966	217	15
2767	201	16	2817	979	15	2867	743	15	2917	494	15	2967	232	14
2768	217	15	2818	994	16	2868	758	15	2918	509	14	2968	246	15
2769	232	16	2819	45010	15	2869	773	15	2919	523	15	2969	261	15
2770	44248	16	2820	45025	15	2870	45788	15	2920	46538	15	2970	47276	14
2771	264	15	2821	040	16	2871	803	15	2921	553	15	2971	290	15
2772	279	16	2822	056	15	2872	818	16	2922	568	15	2972	305	14
2773	295	16	2823	071	15	2873	834	15	2923	583	15	2973	319	15
2774	311	15	2824	086	16	2874	849	15	2924	598	15	2974	334	15
2775	44326	16	2825	45102	15	2875	45864	15	2925	46613	14	2975	47349	14
2776	342	16	2826	117	16	2876	879	15	2926	627	15	2976	363	15
2777	358	15	2827	133	15	2877	894	15	2927	642	15	2977	378	14
2778	373	16	2828	148	15	2878	909	15	2928	657	15	2978	392	15
2779	389	15	2829	163	16	2879	924	15	2929	672	15	2979	407	15
2780	44404	16	2830	45179	15	2880	45939	15	2930	46687	15	2980	47422	14
2781	420	16	2831	194	15	2881	954	15	2931	702	14	2981	436	15
2782	436	15	2832	209	16	2882	969	15	2932	716	15	2982	451	14
2783	451	16	2833	225	15	2883	984	16	2933	731	15	2983	465	15
2784	467	16	2834	240	15	2884	46000	15	2934	746	15	2984	480	14
2785	44483	15	2835	45255	16	2885	46015	15	2935	46761	15	2985	47494	15
2786	498	16	2836	271	15	2886	030	15	2936	776	14	2986	509	15
2787	514	15	2837	286	15	2887	045	15	2937	790	15	2987	524	14
2788	529	16	2838	301	16	2888	060	15	2938	805	15	2988	538	15
2789	545	15	2839	317	15	2889	075	15	2939	820	15	2989	553	14
2790	44560	16	2840	45332	15	2890	46090	15	2940	46835	15	2990	47567	15
2791	576	16	2841	347	16	2891	105	15	2941	850	14	2991	582	14
2792	592	15	2842	362	16	2892	120	15	2942	864	15	2992	596	15
2793	607	16	2843	378	15	2893	135	15	2943	879	15	2993	611	14
2794	623	15	2844	393	15	2894	150	15	2944	894	15	2994	625	15
2795	44638	16	2845	45408	15	2895	46165	15	2945	46909	14	2995	47640	14
2796	654	15	2846	423	16	2896	180	15	2946	923	15	2996	654	15
2797	669	16	2847	439	15	2897	195	15	2947	938	15	2997	669	14
2798	685	15	2848	454	15	2898	210	15	2948	953	14	2998	683	15
2799	700	16	2849	469	15	2899	225	15	2949	967	15	2999	698	14
2800	44716		2850	45484		2900	46240		2950	46982		3000	47712	

P. P.	N	Log	d	N	Log	d	N	Log	d	N	Log	d	N	Log	d

N	Log	d	N	Log	d	N	Log	d	N	Log	d	N	Log	d	P. P.
3000	47712	15	**3050**	48430	14	**3100**	49136	14	**3150**	49831	14	**3200**	50515	14	
3001	727	14	3051	444	14	3101	150	14	3151	845	14	3201	529	13	
3002	741	15	3052	458	15	3102	164	14	3152	859	13	3202	542	14	
3003	756	14	3053	473	14	3103	178	14	3153	872	14	3203	556	13	
3004	770	14	3054	487	14	3104	192	14	3154	886	14	3204	569	14	
3005	47784	15	3055	48501	14	3105	49206	14	3155	49900	14	3205	50583	13	
3006	799	14	3056	515	15	3106	220	14	3156	914	13	3206	596	14	
3007	813	14	3057	530	14	3107	234	14	3157	927	14	3207	610	13	
3008	828	14	3058	544	14	3108	248	14	3158	941	14	3208	623	14	
3009	842	15	3059	558	14	3109	262	14	3159	955	14	3209	637	14	
3010	47857	14	**3060**	48572	14	**3110**	49276	14	**3160**	49969	13	**3210**	50651	13	
3011	871	14	3061	586	15	3111	290	14	3161	982	14	3211	664	14	
3012	885	15	3062	601	14	3112	304	14	3162	996	14	3212	678	13	
3013	900	14	3063	615	14	3113	318	14	3163	50010	14	3213	691	14	
3014	914	15	3064	629	14	3114	332	14	3164	024	13	3214	705	13	
3015	47929	14	3065	48643	14	3115	49346	14	3165	50037	14	3215	50718	14	**15**
3016	943	15	3066	657	14	3116	360	14	3166	051	14	3216	732	13	1 1.5
3017	958	14	3067	671	14	3117	374	14	3167	065	14	3217	745	14	2 3.0
3018	972	14	3068	686	14	3118	388	14	3168	079	13	3218	759	13	3 4.5
3019	986	15	3069	700	14	3119	402	13	3169	092	14	3219	772	14	4 6.0
3020	48001	14	**3070**	48714	14	**3120**	49415	14	**3170**	50106	14	**3220**	50786	13	5 7.5
3021	015	14	3071	728	14	3121	429	14	3171	120	13	3221	799	14	6 9.0
3022	029	15	3072	742	14	3122	443	14	3172	133	14	3222	813	13	7 10.5
3023	044	14	3073	756	14	3123	457	14	3173	147	14	3223	826	14	8 12.0
3024	058	15	3074	770	15	3124	471	14	3174	161	14	3224	840	13	9 13.5
3025	48073	14	3075	48785	14	3125	49485	14	3175	50174	14	3225	50853	13	
3026	087	14	3076	799	14	3126	499	14	3176	188	14	3226	866	14	
3027	101	15	3077	813	14	3127	513	14	3177	202	13	3227	880	13	
3028	116	14	3078	827	14	3128	527	14	3178	215	14	3228	893	14	
3029	130	14	3079	841	14	3129	541	13	3179	229	14	3229	907	13	
3030	48144	15	**3080**	48855	14	**3130**	49554	14	**3180**	50243	13	**3230**	50920	14	**14**
3031	159	14	3081	869	14	3131	568	14	3181	256	14	3231	934	13	1 1.4
3032	173	14	3082	883	14	3132	582	14	3182	270	14	3232	947	14	2 2.8
3033	187	15	3083	897	15	3133	596	14	3183	284	13	3233	961	13	3 4.2
3034	202	14	3084	911	15	3134	610	14	3184	297	14	3234	974	13	4 5.6
3035	48216	14	3085	48926	14	3135	49624	14	3185	50311	14	3235	50987	14	5 7.0
3036	230	14	3086	940	14	3136	638	13	3186	325	13	3236	51001	13	6 8.4
3037	244	15	3087	954	15	3137	651	14	3187	338	14	3237	014	14	7 9.8
3038	259	14	3088	968	14	3138	665	14	3188	352	13	3238	028	13	8 11.2
3039	273	14	3089	982	14	3139	679	14	3189	365	14	3239	041	14	9 12.6
3040	48287	15	**3090**	48996	14	**3140**	49693	14	**3190**	50379	14	**3240**	51055	13	
3041	302	14	3091	49010	14	3141	707	14	3191	393	13	3241	068	13	
3042	316	14	3092	024	14	3142	721	13	3192	406	14	3242	081	14	
3043	330	14	3093	038	14	3143	734	14	3193	420	13	3243	095	13	
3044	344	15	3094	052	14	3144	748	14	3194	433	14	3244	108	13	
3045	48359	14	3095	49066	14	3145	49762	14	3195	50447	14	3245	51121	14	
3046	373	14	3096	080	14	3146	776	14	3196	461	13	3246	135	13	
3047	387	14	3097	094	14	3147	790	13	3197	474	14	3247	148	14	
3048	401	15	3098	108	15	3148	803	14	3198	488	13	3248	162	13	
3049	416	14	3099	122	14	3149	817	14	3199	501	14	3249	175	13	
3050	48430		**3100**	49136		**3150**	49831		**3200**	50515		**3250**	51188		
N	Log	d	N	Log	d	N	Log	d	N	Log	d	N	Log	d	P. P.

P. P.

Proportional parts for 13:

1	1.3
2	2.6
3	3.9
4	5.2
5	6.5
6	7.8
7	9.1
8	10.4
9	11.7

Proportional parts for 12:

1	1.2
2	2.4
3	3.6
4	4.8
5	6.0
6	7.2
7	8.4
8	9.6
9	10.8

N	Log	d	N	Log	d	N	Log	d	N	Log	d	N	Log	d
3250	51188	14	**3300**	51851	14	**3350**	52504	13	**3400**	53148	13	**3450**	53782	12
3251	202	13	3301	865	13	3351	517	13	3401	161	12	3451	794	13
3252	215	13	3302	878	13	3352	530	13	3402	173	13	3452	807	13
3253	228	14	3303	891	13	3353	543	13	3403	186	13	3453	820	12
3254	242	13	3304	904	13	3354	556	13	3404	199	13	3454	832	13
3255	51255	13	3305	51917	13	3355	52569	13	3405	53212	12	3455	53845	12
3256	268	14	3306	930	13	3356	582	13	3406	224	13	3456	857	13
3257	282	13	3307	943	14	3357	595	13	3407	237	13	3457	870	12
3258	295	13	3308	957	13	3358	608	13	3408	250	13	3458	882	13
3259	308	14	3309	970	13	3359	621	13	3409	263	12	3459	895	13
3260	51322	13	**3310**	51983	13	**3360**	52634	13	**3410**	53275	13	**3460**	53908	12
3261	335	13	3311	996	13	3361	647	13	3411	288	13	3461	920	13
3262	348	14	3312	52009	14	3362	660	13	3412	301	13	3462	933	12
3263	362	13	3313	022	13	3363	673	13	3413	314	12	3463	945	13
3264	375	13	3314	035	13	3364	686	13	3414	326	13	3464	958	12
3265	51388	14	3315	52048	13	3365	52699	12	3415	53339	13	3465	53970	13
3266	402	13	3316	061	14	3366	711	13	3416	352	12	3466	983	12
3267	415	13	3317	075	13	3367	724	13	3417	364	13	3467	995	13
3268	428	13	3318	088	13	3368	737	13	3418	377	13	3468	54008	12
3269	441	14	3319	101	13	3369	750	13	3419	390	13	3469	020	13
3270	51455	13	**3320**	52114	13	**3370**	52763	13	**3420**	53403	12	**3470**	54033	12
3271	468	13	3321	127	13	3371	776	13	3421	415	13	3471	045	13
3272	481	14	3322	140	13	3372	789	13	3422	428	13	3472	058	12
3273	495	13	3323	153	13	3373	802	13	3423	441	12	3473	070	13
3274	508	13	3324	166	13	3374	815	12	3424	453	13	3474	083	12
3275	51521	13	3325	52179	13	3375	52827	13	3425	53466	13	3475	54095	13
3276	534	14	3326	192	13	3376	840	13	3426	479	12	3476	108	12
3277	548	13	3327	205	13	3377	853	13	3427	491	13	3477	120	13
3278	561	13	3328	218	13	3378	866	13	3428	504	13	3478	133	12
3279	574	13	3329	231	13	3379	879	13	3429	517	12	3479	145	13
3280	51587	14	**3330**	52244	13	**3380**	52892	13	**3430**	53529	13	**3480**	54158	12
3281	601	13	3331	257	13	3381	905	12	3431	542	13	3481	170	13
3282	614	13	3332	270	14	3382	917	13	3432	555	12	3482	183	12
3283	627	13	3333	284	13	3383	930	13	3433	567	13	3483	195	13
3284	640	14	3334	297	13	3384	943	13	3434	580	13	3484	208	12
3285	51654	13	3335	52310	13	3385	52956	13	3435	53593	12	3485	54220	13
3286	667	13	3336	323	13	3386	969	13	3436	605	13	3486	233	12
3287	680	13	3337	336	13	3387	982	12	3437	618	13	3487	245	13
3288	693	13	3338	349	13	3388	994	13	3438	631	12	3488	258	12
3289	706	14	3339	362	13	3389	53007	13	3439	643	13	3489	270	13
3290	51720	13	**3340**	52375	13	**3390**	53020	13	**3440**	53656	12	**3490**	54283	12
3291	733	13	3341	388	13	3391	033	13	3441	668	13	3491	295	12
3292	746	13	3342	401	13	3392	046	12	3442	681	13	3492	307	13
3293	759	14	3343	414	13	3393	058	13	3443	694	12	3493	320	12
3294	772	13	3344	427	14	3394	071	13	3444	706	13	3494	332	13
3295	51786	13	3345	52440	13	3395	53084	13	3445	53719	13	3495	54345	12
3296	799	13	3346	453	13	3396	097	13	3446	732	12	3496	357	13
3297	812	13	3347	466	13	3397	110	12	3447	744	13	3497	370	12
3298	825	13	3348	479	13	3398	122	13	3448	757	12	3498	382	12
3299	838	13	3349	492	12	3399	135	13	3449	769	13	3499	394	13
3300	51851		**3350**	52504		**3400**	53148		**3450**	53782		**3500**	54407	

P. P.	N	Log	d	N	Log	d	N	Log	d	N	Log	d	N	Log	d

N	Log	d	N	Log	d	N	Log	d	N	Log	d	N	Log	d	P. P.
3500	54407	12	3550	55023	12	3600	55630	12	3650	56229	12	3700	56820	12	
3501	419	13	3551	035	12	3601	642	12	3651	241	12	3701	832	12	
3502	432	12	3552	047	13	3602	654	12	3652	253	12	3702	844	11	
3503	444	12	3553	060	12	3603	666	12	3653	265	12	3703	855	12	
3504	456	13	3554	072	12	3604	678	13	3654	277	12	3704	867	12	
3505	54469	12	3555	55084	12	3605	55691	12	3655	56289	12	3705	56879	12	
3506	481	13	3556	096	12	3606	703	12	3656	301	11	3706	891	11	
3507	494	12	3557	108	13	3607	715	12	3657	312	12	3707	902	12	
3508	506	12	3558	121	12	3608	727	12	3658	324	12	3708	914	12	
3509	518	13	3559	133	13	3609	739	12	3659	336	12	3709	926	11	
3510	54531	12	3560	55145	12	3610	55751	12	3660	56348	12	3710	56937	12	
3511	543	12	3561	157	12	3611	763	12	3661	360	12	3711	949	12	
3512	555	13	3562	169	13	3612	775	13	3662	372	12	3712	961	11	
3513	568	12	3563	182	12	3613	787	12	3663	384	12	3713	972	12	
3514	580	13	3564	194	12	3614	799	12	3664	396	11	3714	984	12	**13**
3515	54593	12	3565	55206	12	3615	55811	12	3665	56407	12	3715	56996	12	1 | 1.3
3516	605	12	3566	218	12	3616	823	12	3666	419	12	3716	57008	11	2 | 2.6
3517	617	13	3567	230	12	3617	835	12	3667	431	12	3717	019	12	3 | 3.9
3518	630	12	3568	242	13	3618	847	13	3668	443	12	3718	031	12	4 | 5.2
3519	642	12	3569	255	12	3619	859	12	3669	455	12	3719	043	11	5 | 6.5
															6 | 7.8
3520	54654	13	3570	55267	12	3620	55871	12	3670	56467	11	3720	57054	12	7 | 9.1
3521	667	12	3571	279	12	3621	883	12	3671	478	12	3721	066	12	8 | 10.4
3522	679	12	3572	291	12	3622	895	12	3672	490	12	3722	078	11	9 | 11.7
3523	691	13	3573	303	12	3623	907	12	3673	502	12	3723	089	12	
5524	704	12	3574	315	13	3624	919	13	3674	514	12	3724	101	12	
3525	54716	12	3575	55328	12	3625	55931	12	3675	56526	12	3725	57113	11	
3526	728	13	3576	340	12	3626	943	12	3676	538	11	3726	124	12	
3527	741	12	3577	352	12	3627	955	12	3677	549	12	3727	136	12	
3528	753	12	3578	364	12	3628	967	12	3678	561	12	3728	148	11	
3529	765	12	3579	376	12	3629	979	12	3679	573	12	3729	159	12	
3530	54777	13	3580	55388	12	3630	55991	12	3680	56585	12	3730	57171	12	**12**
3531	790	12	3581	400	13	3631	56003	12	3681	597	11	3731	183	11	1 | 1.2
3532	802	12	3582	413	12	3632	015	12	3682	608	12	3732	194	12	2 | 2.4
3533	814	13	3583	425	12	3633	027	11	3683	620	12	3733	206	11	3 | 3.6
3534	827	12	3584	437	12	3634	038	12	3684	632	12	3734	217	12	4 | 4.8
3535	54839	12	3585	55449	12	3635	56050	12	3685	56644	12	3735	57229	12	5 | 6.0
3536	851	13	3586	461	12	3636	•062	12	3686	656	11	3736	241	11	6 | 7.2
3537	864	12	3587	473	12	3637	074	12	3687	667	12	3737	252	12	7 | 8.4
3538	876	12	3588	485	12	3638	086	12	3688	679	12	3738	264	12	8 | 9.6
3539	888	12	3589	497	12	3639	098	12	3689	691	12	3739	276	12	9 | 10.8
3540	54900	13	3590	55509	13	3640	56110	12	3690	56703	11	3740	57287	12	
3541	913	12	3591	522	12	3641	122	12	3691	714	12	3741	299	11	
3542	925	12	3592	534	12	3642	134	12	3692	726	12	3742	310	12	
3543	937	12	3593	546	12	3643	146	12	3693	738	12	3743	322	12	
3544	949	13	3594	558	12	3644	158	12	3694	750	12	3744	334	11	
3545	54962	12	3595	55570	12	3645	56170	12	3695	56761	12	3745	57345	12	
3546	974	12	3596	582	12	3646	182	12	3696	773	12	3746	357	11	
3547	986	12	3597	594	12	3647	194	11	3697	785	12	3747	368	12	
3548	998	13	3598	606	12	3648	205	12	3698	797	11	3748	380	12	
3549	55011	12	3599	618	12	3649	217	12	3699	808	12	3749	392	11	
3550	55023		3600	55630		3650	56229		3700	56820		3750	57403		
N	Log	d	N	Log	d	N	Log	d	N	Log	d	N	Log	d	P. P.

P. P.	N	Log	d	N	Log	d	N	Log	d	N	Log	d	N	Log	d
	3750	57403	12	**3800**	57978	12	**3850**	58546	11	**3900**	59106	12	**3950**	59660	11
	3751	415	11	3801	990	11	3851	557	12	3901	118	11	3951	671	11
	3752	426	12	3802	58001	12	3852	569	11	3902	129	11	3952	682	11
	3753	438	11	3803	013	11	3853	580	11	3903	140	11	3953	693	11
	3754	449	12	3804	024	11	3854	591	11	3904	151	11	3954	704	11
	3755	57461	12	3805	58035	12	3855	58602	12	3905	59162	11	3955	59715	11
	3756	473	11	3806	047	11	3856	614	11	3906	173	11	3956	726	11
	3757	484	12	3807	058	12	3857	625	11	3907	184	11	3957	737	11
	3758	496	11	3808	070	11	3858	636	11	3908	195	12	3958	748	11
	3759	507	12	3809	081	11	3859	647	12	3909	207	11	3959	759	11
	3760	57519	11	**3810**	58092	12	**3860**	58659	11	**3910**	59218	11	**3960**	59770	10
	3761	530	12	3811	104	11	3861	670	11	3911	229	11	3961	780	11
	3762	542	11	3812	115	12	3862	681	11	3912	240	11	3962	791	11
	3763	553	12	3813	127	11	3863	692	12	3913	251	11	3963	802	11
11	3764	565	11	3814	138	11	3864	704	11	3914	262	11	3964	813	11
1 | 1.1	3765	57576	12	3815	58149	12	3865	58715	11	3915	59273	11	3965	59824	11
2 | 2.2	3766	588	12	3816	161	11	3866	726	11	3916	284	11	3966	835	11
3 | 3.3	3767	600	11	3817	172	12	3867	737	12	3917	295	11	3967	846	11
4 | 4.4	3768	611	12	3818	184	11	3868	749	11	3918	306	12	3968	857	11
5 | 5.5	3769	623	11	3819	195	11	3869	760	11	3919	318	11	3969	868	11
6 | 6.6	**3770**	57634	12	**3820**	58206	12	**3870**	58771	11	**3920**	59329	11	**3970**	59879	11
7 | 7.7	3771	646	11	3821	218	11	3871	782	12	3921	340	11	3971	890	11
8 | 8.8	3772	657	12	3822	229	11	3872	794	11	3922	351	11	3972	901	11
9 | 9.9	3773	669	11	3823	240	12	3873	805	11	3923	362	11	3973	912	11
	3774	680	12	3824	252	11	3874	816	11	3924	373	11	3974	923	11
	3775	57692	11	3825	58263	11	3875	58827	11	3925	59384	11	3975	59934	11
	3776	703	12	3826	274	12	3876	838	12	3926	395	11	3976	945	11
	3777	715	11	3827	286	11	3877	850	11	3927	406	11	3977	956	10
	3778	726	12	3828	297	12	3878	861	11	3928	417	11	3978	966	11
10	3779	738	11	3829	309	11	3879	872	11	3929	428	11	3979	977	11
1 | 1.0	**3780**	57749	12	**3830**	58320	11	**3880**	58883	11	**3930**	59439	11	**3980**	59988	11
2 | 2.0	3781	761	11	3831	331	12	3881	894	12	3931	450	11	3981	999	11
3 | 3.0	3782	772	12	3832	343	11	3882	906	11	3932	461	11	3982	60010	11
4 | 4.0	3783	784	11	3833	354	11	3883	917	11	3933	472	11	3983	021	11
5 | 5.0	3784	795	12	3834	365	12	3884	928	11	3934	483	11	3984	032	11
6 | 6.0	3785	57807	11	3835	58377	11	3885	58939	11	3935	59494	12	3985	60043	11
7 | 7.0	3786	818	12	3836	388	11	3886	950	11	3936	506	11	3986	054	11
8 | 8.0	3787	830	11	3837	399	11	3887	961	12	3937	517	11	3987	065	11
9 | 9.0	3788	841	11	3838	410	12	3888	973	11	3938	528	11	3988	076	10
	3789	852	12	3839	422	11	3889	984	11	3939	539	11	3989	086	11
	3790	57864	11	**3840**	58433	11	**3890**	58995	11	**3940**	59550	11	**3990**	60097	11
	3791	875	12	3841	444	12	3891	59006	11	3941	561	11	3991	108	11
	3792	887	11	3842	456	11	3892	017	11	3942	572	11	3992	119	11
	3793	898	12	3843	467	11	3893	028	12	3943	583	11	3993	130	11
	3794	910	11	3844	478	12	3894	040	11	3944	594	11	3994	141	11
	3795	57921	12	3845	58490	11	3895	59051	11	3945	59605	11	3995	60152	11
	3796	933	11	3846	501	11	3896	062	11	3946	616	11	3996	163	10
	3797	944	11	3847	512	12	3897	073	11	3947	627	11	3997	173	11
	3798	955	12	3848	524	11	3898	084	11	3948	638	11	3998	184	11
	3799	967	11	3849	535	11	3899	095	11	3949	649	11	3999	195	11
	3800	57978		**3850**	58546		**3900**	59106		**3950**	59660		**4000**	60206	
P. P.	N	Log	d	N	Log	d	N	Log	d	N	Log	d	N	Log	d

N	Log	d	N	Log	d	N	Log	d	N	Log	d	N	Log	d	P. P.
4000	60206	11	**4050**	60746	10	**4100**	61278	11	**4150**	61805	10	**4200**	62325	10	
4001	217	11	4051	756	11	4101	289	11	4151	815	11	4201	335	11	
4002	228	11	4052	767	11	4102	300	10	4152	826	10	4202	346	10	
4003	239	10	4053	778	10	4103	310	11	4153	836	11	4203	356	10	
4004	249	11	4054	788	11	4104	321	10	4154	847	10	4204	366	11	
4005	60260	11	4055	60799	11	4105	61331	11	4155	61857	11	4205	62377	10	
4006	271	11	4056	810	11	4106	342	10	4156	868	10	4206	387	10	
4007	282	11	4057	821	10	4107	352	11	4157	878	10	4207	397	11	
4008	293	11	4058	831	11	4108	363	11	4158	888	11	4208	408	10	
4009	304	10	4059	842	11	4109	374	10	4159	899	10	4209	418	10	
4010	60314	11	**4060**	60853	10	**4110**	61384	11	**4160**	61909	11	**4210**	62428	11	
4011	325	11	4061	863	11	4111	395	10	4161	920	10	4211	439	10	
4012	336	11	4062	874	11	4112	405	11	4162	930	11	4212	449	10	
4013	347	11	4063	885	10	4113	416	10	4163	941	10	4213	459	10	
4014	358	11	4064	895	11	4114	426	11	4164	951	11	4214	469	11	
4015	60369	10	4065	60906	11	4115	61437	11	4165	61962	10	4215	62480	10	**11**
4016	379	11	4066	917	10	4116	448	10	4166	972	10	4216	490	10	1 | 1.1
4017	390	11	4067	927	11	4117	458	11	4167	982	11	4217	500	11	2 | 2.2
4018	401	11	4068	938	11	4118	469	10	4168	993	10	4218	511	10	3 | 3.3
4019	412	11	4069	949	10	4119	479	11	4169	62003	11	4219	521	10	4 | 4.4
4020	60423	10	**4070**	60959	11	**4120**	61490	10	**4170**	62014	10	**4220**	62531	11	5 | 5.5
4021	433	11	4071	970	11	4121	500	11	4171	024	10	4221	542	10	6 | 6.6
4022	444	11	4072	981	10	4122	511	10	4172	034	11	4222	552	10	7 | 7.7
4023	455	11	4073	991	11	4123	521	11	4173	045	10	4223	562	10	8 | 8.8
4024	466	11	4074	61002	11	4124	532	10	4174	055	11	4224	572	11	9 | 9.9
4025	60477	10	4075	61013	10	4125	61542	11	4175	62066	10	4225	62583	10	
4026	487	11	4076	023	11	4126	553	10	4176	076	10	4226	593	10	
4027	498	11	4077	034	11	4127	563	11	4177	086	11	4227	603	10	
4028	509	11	4078	045	10	4128	574	10	4178	097	10	4228	613	11	
4029	520	11	4079	055	11	4129	584	11	4179	107	11	4229	624	10	
4030	60531	10	**4080**	61066	11	**4130**	61595	11	**4180**	62118	10	**4230**	62634	10	**10**
4031	541	11	4081	077	10	4131	606	10	4181	128	10	4231	644	11	1 | 1.0
4032	552	11	4082	087	11	4132	616	11	4182	138	11	4232	655	10	2 | 2.0
4033	563	11	4083	098	11	4133	627	10	4183	149	10	4233	665	10	3 | 3.0
4034	574	10	4084	109	10	4134	637	11	4184	159	11	4234	675	10	4 | 4.0
4035	60584	11	4085	61119	11	4135	61648	10	4185	62170	10	4235	62685	11	5 | 5.0
4036	595	11	4086	130	10	4136	658	11	4186	180	10	4236	696	10	6 | 6.0
4037	606	11	4087	140	11	4137	669	10	4187	190	11	4237	706	10	7 | 7.0
4038	617	10	4088	151	11	4138	679	11	4188	201	10	4238	716	10	8 | 8.0
4039	627	11	4089	162	10	4139	690	10	4189	211	10	4239	726	11	9 | 9.0
4040	60638	11	**4090**	61172	11	**4140**	61700	11	**4190**	62221	11	**4240**	62737	10	
4041	649	11	4091	183	11	4141	711	10	4191	232	10	4241	747	10	
4042	660	10	4092	194	10	4142	721	10	4192	242	10	4242	757	10	
4043	670	11	4093	204	11	4143	731	11	4193	252	11	4243	767	11	
4044	681	11	4094	215	10	4144	742	10	4194	263	10	4244	778	10	
4045	60692	11	4095	61225	11	4145	61752	11	4195	62273	11	4245	62788	10	
4046	703	10	4096	236	11	4146	763	10	4196	284	10	4246	798	10	
4047	713	11	4097	247	10	4147	773	11	4197	294	10	4247	808	10	
4048	724	11	4098	257	11	4148	784	10	4198	304	11	4248	818	11	
4049	735	11	4099	268	10	4149	794	11	4199	315	10	4249	829	10	
4050	60746		**4100**	61278		**4150**	61805		**4200**	62325		**4250**	62839		
N	Log	d	N	Log	d	N	Log	d	N	Log	d	N	Log	d	P. P.

P. P.	N	Log	d	N	Log	d	N	Log	d	N	Log	d	N	Log	d
	4250	62839	10	**4300**	63347	10	**4350**	63849	10	**4400**	64345	10	**4450**	64836	10
	4251	849	10	4301	357	10	4351	859	10	4401	355	10	4451	846	10
	4252	859	11	4302	367	10	4352	869	10	4402	365	10	4452	856	9
	4253	870	10	4303	377	10	4353	879	10	4403	375	10	4453	865	10
	4254	880	10	4304	387	10	4354	889	10	4404	385	10	4454	875	10
	4255	62890	10	**4305**	63397	10	**4355**	63899	10	**4405**	64395	9	**4455**	64885	10
	4256	900	10	4306	407	10	4356	909	10	4406	404	10	4456	895	9
	4257	910	11	4307	417	11	4357	919	10	4407	414	10	4457	904	10
	4258	921	10	4308	428	10	4358	929	10	4408	424	10	4458	914	10
	4259	931	10	4309	438	10	4359	939	10	4409	434	10	4459	924	9
	4260	62941	10	**4310**	63448	10	**4360**	63949	10	**4410**	64444	10	**4460**	64933	10
	4261	951	10	4311	458	10	4361	959	10	4411	454	10	4461	943	10
	4262	961	11	4312	468	10	4362	969	10	4412	464	9	4462	953	10
	4263	972	10	4313	478	10	4363	979	9	4413	473	10	4463	963	9
	4264	982	10	4214	488	10	4364	988	10	4414	483	10	4464	972	10
	4265	62992	10	**4315**	63498	10	**4365**	63998	10	**4415**	64493	10	**4465**	64982	10
	4266	63002	10	4316	508	10	4366	64008	10	4416	503	10	4466	992	10
	4267	012	10	4317	518	10	4367	018	10	4417	513	10	4467	65002	9
	4268	022	11	4318	528	10	4368	028	10	4418	523	9	4468	011	10
	4269	033	10	4319	538	10	4369	038	10	4419	532	10	4469	021	10
9	**4270**	63043	10	**4320**	63548	10	**4370**	64048	10	**4420**	64542	10	**4470**	65031	9
1 | 0.9	4271	053	10	4321	558	10	4371	058	10	4421	552	10	4471	040	10
2 | 1.8	4272	063	10	4322	568	11	4372	068	10	4422	562	10	4472	050	10
3 | 2.7	4273	073	10	4323	579	10	4373	078	10	4423	572	10	4473	060	10
4 | 3.6	4274	083	11	4324	589	10	4374	088	10	4424	582	9	4474	070	9
5 | 4.5	**4275**	63094	10	**4325**	63599	10	**4375**	64098	10	**4425**	64591	10	**4475**	65079	10
6 | 5.4	4276	104	10	4326	609	10	4376	108	10	4426	601	10	4476	089	10
7 | 6.3	4277	114	10	4327	619	10	4377	118	10	4427	611	10	4477	099	9
8 | 7.2	4278	124	10	4328	629	10	4378	128	9	4428	621	10	4478	108	10
9 | 8.1	4279	134	10	4329	639	10	4379	137	10	4429	631	9	4479	118	10
	4280	63144	11	**4330**	63649	10	**4380**	64147	10	**4430**	64640	10	**4480**	65128	9
	4281	155	10	4331	659	10	4381	157	10	4431	650	10	4481	137	10
	4282	165	10	4332	669	10	4382	167	10	4432	660	10	4482	147	10
	4283	175	10	4333	679	10	4383	177	10	4433	670	10	4483	157	10
	4284	185	10	4334	689	10	4384	187	10	4434	680	9	4484	167	9
	4285	63195	10	**4335**	63699	10	**4385**	64197	10	**4435**	64689	10	**4485**	65176	10
	4286	205	10	4336	709	10	4386	207	10	4436	699	10	4486	186	10
	4287	215	10	4337	719	10	4387	217	10	4437	709	10	4487	196	9
	4288	225	11	4338	729	10	4388	227	10	4438	719	10	4488	205	10
	4289	236	10	4339	739	10	4389	237	9	4439	729	9	4489	215	10
	4290	63246	10	**4340**	63749	10	**4390**	64246	10	**4440**	64738	10	**4490**	65225	9
	4291	256	10	4341	759	10	4391	256	10	4441	748	10	4491	234	10
	4292	266	10	4342	769	10	4392	266	10	4442	758	10	4492	244	10
	4293	276	10	4343	779	10	4393	276	10	4443	768	9	4493	254	10
	4294	286	10	4344	789	10	4394	286	10	4444	777	10	4494	263	9
	4295	63296	10	**4345**	63799	10	**4395**	64296	10	**4445**	64787	10	**4495**	65273	10
	4296	306	11	4346	809	10	4396	306	10	4446	797	10	4496	283	9
	4297	317	10	4347	819	10	4397	316	10	4447	807	9	4497	292	10
	4298	327	10	4348	829	10	4398	326	9	4448	816	10	4498	302	10
	4299	337	10	4349	839	10	4399	335	10	4449	826	10	4499	312	9
	4300	63347		**4350**	63849		**4400**	64345		**4450**	64836		**4500**	65321	
P. P.	N	Log	d	N	Log	d	N	Log	d	N	Log	d	N	Log	d

N	Log	d	N	Log	d	N	Log	d	N	Log	d	N	Log	d	P. P.
4500	65321	10	**4550**	65801	10	**4600**	66276	9	**4650**	66745	10	**4700**	67210	9	
4501	331	10	4551	811	9	4601	285	10	4651	755	9	4701	219	9	
4502	341	9	4552	820	10	4602	295	9	4652	764	9	4702	228	9	
4503	350	10	4553	830	9	4603	304	10	4653	773	10	4703	237	10	
4504	360	9	4554	839	10	4604	314	9	4654	783	9	4704	247	9	
4505	65369	10	4555	65849	9	4605	66323	9	4655	66792	9	4705	67256	9	
4506	379	10	4556	858	10	4606	332	10	4656	801	10	4706	265	9	
4507	389	9	4557	868	9	4607	342	9	4657	811	9	4707	274	10	
4508	398	10	4558	877	10	4608	351	10	4658	820	9	4708	284	9	
4509	408	10	4559	887	9	4609	361	9	4659	829	10	4709	293	9	
4510	65418	9	**4560**	65896	10	**4610**	66370	10	**4660**	66839	9	**4710**	67302	9	
4511	427	10	4561	906	10	4611	380	9	4661	848	9	4711	311	10	
4512	437	10	4562	916	9	4612	389	9	4662	857	10	4712	321	9	
4513	447	9	4563	925	10	4613	398	10	4663	867	9	4713	330	9	
4514	456	10	4564	935	9	4614	408	9	4664	876	9	4714	339	9	**10**
4515	65466	9	4565	65944	10	4615	66417	10	4665	66885	9	4715	67348	9	1 | 1.0
4516	475	10	4566	954	9	4616	427	9	4666	894	10	4716	357	10	2 | 2.0
4517	485	10	4567	963	10	4617	436	9	4667	904	9	4717	367	9	3 | 3.0
4518	495	9	4568	973	9	4618	445	10	4668	913	9	4718	376	9	4 | 4.0
4519	504	10	4569	982	10	4619	455	9	4669	922	10	4719	385	9	5 | 5.0
															6 | 6.0
															7 | 7.0
															8 | 8.0
4520	65514	9	**4570**	65992	9	**4620**	66464	10	**4670**	66932	9	**4720**	67394	9	9 | 9.0
4521	523	10	4571	66001	10	4621	474	9	4671	941	9	4721	403	10	
4522	533	10	4572	011	9	4622	483	9	4672	950	10	4722	413	9	
4523	543	9	4573	020	10	4623	492	10	4673	960	9	4723	422	9	
4524	552	10	4574	030	9	4624	502	9	4674	969	9	4724	431	9	
4525	65562	9	4575	66039	10	4625	66511	10	4675	66978	9	4725	67440	9	
4526	571	10	4576	049	9	4626	521	9	4676	987	10	4726	449	10	
4527	581	10	4577	058	10	4627	530	9	4677	997	9	4727	459	9	
4528	591	9	4578	068	9	4628	539	10	4678	67006	9	4728	468	9	
4529	600	10	4579	077	10	4629	549	9	4679	015	10	4729	477	9	**9**
4530	65610	9	**4580**	66087	9	**4630**	66558	9	**4680**	67025	9	**4730**	67486	9	1 | 0.9
4531	619	10	4581	096	10	4631	567	10	4681	034	9	4731	495	9	2 | 1.8
4532	629	10	4582	106	9	4632	577	9	4682	043	9	4732	504	10	3 | 2.7
4533	639	9	4583	115	9	4633	586	10	4683	052	10	4733	514	9	4 | 3.6
4534	648	10	4584	124	10	4634	596	9	4684	062	9	4734	523	9	5 | 4.5
4535	65658	9	4585	66134	9	4635	66605	9	4685	67071	9	4735	67532	9	6 | 5.4
4536	667	10	4586	143	10	4636	614	10	4686	080	9	4736	541	9	7 | 6.3
4537	677	9	4587	153	9	4637	624	9	4687	089	10	4737	550	10	8 | 7.2
4538	686	10	4588	162	10	4638	633	9	4688	099	9	4738	560	9	9 | 8.1
4539	696	10	4589	172	9	4639	642	10	4689	108	9	4739	569	9	
4540	65706	9	**4590**	66181	10	**4640**	66652	9	**4690**	67117	10	**4740**	67578	9	
4541	715	10	4591	191	9	4641	661	10	4691	127	9	4741	587	9	
4542	725	9	4592	200	10	4642	671	9	4692	136	9	4742	596	9	
4543	734	10	4593	210	9	4643	680	9	4693	145	9	4743	605	9	
4544	744	9	4594	219	10	4644	689	10	4694	154	10	4744	614	10	
4545	65753	10	4595	66229	9	4645	66699	9	4695	67164	9	4745	67624	9	
4546	763	9	4596	238	9	4646	708	9	4696	173	9	4746	633	9	
4547	772	10	4597	247	10	4647	717	10	4697	182	9	4747	642	9	
4548	782	10	4598	257	9	4648	727	9	4698	191	10	4748	651	9	
4549	792	9	4599	266	10	4649	736	9	4699	201	9	4749	660	9	
4550	65801		**4600**	66276		**4650**	66745		**4700**	67210		**4750**	67669		
N	Log	d	N	Log	d	N	Log	d	N	Log	d	N	Log	d	P. P.

P. P.	N	Log	d	N	Log	d	N	Log	d	N	Log	d	N	Log	d
	4750	67669	10	**4800**	68124	9	**4850**	68574	9	**4900**	69020	8	**4950**	69461	8
	4751	679	9	4801	133	9	4851	583	9	4901	028	9	4951	469	9
	4752	688	9	4802	142	9	4852	592	9	4902	037	9	4952	478	9
	4753	697	9	4803	151	9	4853	601	9	4903	046	9	4953	487	9
	4754	706	9	4804	160	9	4854	610	9	4904	055	9	4954	496	8
	4755	67715	9	4805	68169	9	4855	68619	9	4905	69064	9	4955	69504	9
	4756	724	9	4806	178	9	4856	628	9	4906	073	9	4956	513	9
	4757	733	9	4807	187	9	4857	637	9	4907	082	8	4957	522	9
	4758	742	10	4808	196	9	4858	646	9	4908	090	9	4958	531	8
	4759	752	9	4809	205	10	4859	655	9	4909	099	9	4959	539	9
	4760	67761	9	**4810**	68215	9	**4860**	68664	9	**4910**	69108	9	**4960**	69548	9
	4761	770	9	4811	224	9	4861	673	8	4911	117	9	4961	557	9
	4762	779	9	4812	233	9	4862	681	9	4912	126	9	4962	566	8
	4763	788	9	4813	242	9	4863	690	9	4913	135	9	4963	574	9
	4764	797	9	4814	251	9	4864	699	9	4914	144	8	4964	583	9
	4765	67806	9	4815	68260	9	4865	68708	9	4915	69152	9	4965	69592	9
	4766	815	10	4816	269	9	4866	717	9	4916	161	9	4966	601	8
	4767	825	9	4817	278	9	4867	726	9	4917	170	9	4967	609	9
	4768	834	9	4818	287	9	4868	735	9	4918	179	9	4968	618	9
	4769	843	9	4819	296	9	4869	744	9	4919	188	9	4969	627	9
	4770	67852	9	**4820**	68305	9	**4870**	68753	9	**4920**	69197	8	**4970**	69636	8
	4771	861	9	4821	314	9	4871	762	9	4921	205	9	4971	644	9
	4772	870	9	4822	323	9	4872	771	9	4922	214	9	4972	653	9
	4773	879	9	4823	332	9	4873	780	9	4923	223	9	4973	662	9
	4774	888	9	4824	341	9	4874	789	8	4924	232	9	4974	671	8
	4775	67897	9	4825	68350	9	4875	68797	9	4925	69241	8	4975	69679	9
	4776	906	10	4826	359	9	4876	806	9	4926	249	9	4976	688	9
	4777	916	9	4827	368	9	4877	815	9	4927	258	9	4977	697	8
	4778	925	9	4828	377	9	4878	824	9	4928	267	9	4978	705	9
	4779	934	9	4829	386	9	4879	833	9	4929	276	9	4979	714	9
	4780	67943	9	**4830**	68395	9	**4880**	68842	9	**4930**	69285	9	**4980**	69723	9
	4781	952	9	4831	404	9	4881	851	9	4931	294	8	4981	732	8
	4782	961	9	4832	413	9	4882	860	9	4932	302	9	4982	740	9
	4783	970	9	4833	422	9	4883	869	9	4933	311	9	4983	749	9
	4784	979	9	4834	431	9	4884	878	8	4934	320	9	4984	758	9
	4785	67988	9	4835	68440	9	4885	68886	9	4935	69329	9	4985	69767	8
	4786	997	9	4836	449	9	4886	895	9	4936	338	8	4986	775	9
	4787	68006	9	4837	458	9	4887	904	9	4937	346	9	4987	784	9
	4788	015	9	4838	467	9	4888	913	9	4938	355	9	4988	793	8
	4789	024	10	4839	476	9	4889	922	9	4939	364	9	4989	801	9
	4790	68034	9	**4840**	68485	9	**4890**	68931	9	**4940**	69373	8	**4990**	69810	9
	4791	043	9	4841	494	8	4891	940	9	4941	381	9	4991	819	8
	4792	052	9	4842	502	9	4892	949	9	4942	390	9	4992	827	9
	4793	061	9	4843	511	9	4893	958	8	4943	399	9	4993	836	9
	4794	070	9	4844	520	9	4894	966	9	4944	408	9	4994	845	9
	4795	68079	9	4845	68529	9	4895	68975	9	4945	69417	8	4995	69854	8
	4796	088	9	4846	538	9	4896	984	9	4946	425	9	4996	862	9
	4797	097	9	4847	547	9	4897	993	9	4947	434	9	4997	871	9
	4798	106	9	4848	556	9	4898	69002	9	4948	443	9	4998	880	8
	4799	115	9	4849	565	9	4899	011	9	4949	452	9	4999	888	9
	4800	68124		**4850**	68574		**4900**	69020		**4950**	69461		**5000**	69897	

P.P. column (left):

8

1	0.8
2	1.6
3	2.4
4	3.2
5	4.0
6	4.8
7	5.6
8	6.4
9	7.2

30 MATHEMATICAL TABLES AND FORMULAS [I

N	Log	d	N	Log	d	N	Log	d	N	Log	d	N	Log	d
5000	69897	9	**5050**	70329	9	**5100**	70757	9	**5150**	71181	8	**5200**	71600	9
5001	906	8	5051	338	8	5101	766	8	5151	189	9	5201	609	8
5002	914	9	5052	346	9	5102	774	9	5152	198	8	5202	617	8
5003	923	9	5053	355	9	5103	783	8	5153	206	8	5203	625	9
5004	932	8	5054	364	8	5104	791	9	5154	214	9	5204	634	8
5005	69940	9	5055	70372	9	5105	70800	8	5155	71223	8	5205	71642	8
5006	949	9	5056	381	8	5106	808	9	5156	231	9	5206	650	9
5007	958	8	5057	389	9	5107	817	8	5157	240	8	5207	659	8
5008	966	9	5058	398	8	5108	825	9	5158	248	9	5208	667	8
5009	975	9	5059	406	9	5109	834	8	5159	257	8	5209	675	9
5010	69984	8	**5060**	70415	9	**5110**	70842	9	**5160**	71265	8	**5210**	71684	8
5011	992	9	5061	424	8	5111	851	8	5161	273	9	5211	692	8
5012	70001	9	5062	432	9	5112	859	9	5162	282	8	5212	700	9
5013	010	8	5063	441	8	5113	868	8	5163	290	9	5213	709	8
5014	018	9	5064	449	9	5114	876	9	5164	299	8	5214	717	8
5015	70027	9	5065	70458	9	5115	70885	8	5165	71307	8	5215	71725	9
5016	036	8	5066	467	8	5116	893	9	5166	315	9	5216	734	8
5017	044	9	5067	475	9	5117	902	8	5167	324	8	5217	742	8
5018	053	9	5068	484	8	5118	910	9	5168	332	9	5218	750	9
5019	062	8	5069	492	9	5119	919	8	5169	341	8	5219	759	8
5020	70070	9	**5070**	70501	8	**5120**	70927	8	**5170**	71349	8	**5220**	71767	8
5021	079	9	5071	509	9	5121	935	9	5171	357	9	5221	775	9
5022	088	8	5072	518	8	5122	944	8	5172	366	8	5222	784	8
5023	096	9	5073	526	9	5123	952	9	5173	374	9	5223	792	8
5024	105	9	5074	535	9	5124	961	8	5174	383	8	5224	800	9
5025	70114	8	5075	70544	8	5125	70969	9	5175	71391	8	5225	71809	8
5026	122	9	5076	552	9	5126	978	8	5176	399	9	5226	817	8
5027	131	9	5077	561	8	5127	986	9	5177	408	8	5227	825	9
5028	140	8	5078	569	9	5128	995	8	5178	416	9	5228	834	8
5029	148	9	5079	578	8	5129	71003	9	5179	425	8	5229	842	8
5030	70157	8	**5080**	70586	9	**5130**	71012	8	**5180**	71433	8	**5230**	71850	8
5031	165	9	5081	595	8	5131	020	9	5181	441	9	5231	858	9
5032	174	9	5082	603	9	5132	029	8	5182	450	8	5232	867	8
5033	183	8	5083	612	9	5133	037	9	5183	458	8	5233	875	8
5034	191	9	5084	621	8	5134	046	8	5184	466	9	5234	883	9
5035	70200	9	5085	70629	9	5135	71054	9	5185	71475	8	5235	71892	8
5036	209	8	5086	638	8	5136	063	8	5186	483	9	5236	900	8
5037	217	9	5087	646	9	5137	071	8	5187	492	8	5237	908	9
5038	226	8	5088	655	8	5138	079	9	5188	500	8	5238	917	8
5039	234	9	5089	663	9	5139	088	8	5189	508	9	5239	925	8
5040	70243	9	**5090**	70672	8	**5140**	71096	9	**5190**	71517	8	**5240**	71933	8
5041	252	8	5091	680	9	5141	105	8	5191	525	8	5241	941	9
5042	260	9	5092	689	8	5142	113	9	5192	533	9	5242	950	8
5043	269	9	5093	697	9	5143	122	8	5193	542	8	5243	958	8
5044	278	8	5094	706	8	5144	130	9	5194	550	9	5244	966	9
5045	70286	9	5095	70714	9	5145	71139	8	5195	71559	8	5245	71975	8
5046	295	8	5096	723	8	5146	147	8	5196	567	8	5246	983	8
5047	303	9	5097	731	9	5147	155	9	5197	575	9	5247	991	8
5048	312	9	5098	740	9	5148	164	8	5198	584	8	5248	999	9
5049	321	8	5099	749	8	5149	172	9	5199	592	8	5249	72008	8
5050	70329		**5100**	70757		**5150**	71181		**5200**	71600		**5250**	72016	

P. P.

9	
1	0.9
2	1.8
3	2.7
4	3.6
5	4.5
6	5.4
7	6.3
8	7.2
9	8.1

8	
1	0.8
2	1.6
3	2.4
4	3.2
5	4.0
6	4.8
7	5.6
8	6.4
9	7.2

P.P.	N	Log	d	N	Log	d	N	Log	d	N	Log	d	N	Log	d
	5250	72016	8	**5300**	72428	8	**5350**	72835	8	**5400**	73239	8	**5450**	73040	8
	5251	024	8	5301	436	8	5351	843	9	5401	247	8	5451	648	8
	5252	032	9	5302	444	8	5352	852	8	5402	255	8	5452	656	8
	5253	041	8	5303	452	8	5353	860	8	5403	263	9	5453	664	8
	5254	049	8	5304	460	9	5354	868	8	5404	272	8	5454	672	7
	5255	72057	9	5305	72469	8	5355	72876	8	5405	73280	8	5455	73679	8
	5256	066	8	5306	477	8	5356	884	8	5406	288	8	5456	687	8
	5257	074	8	5307	485	8	5357	892	8	5407	296	8	5457	695	8
	5258	082	8	5308	493	8	5358	900	8	5408	304	8	5458	703	8
	5259	090	9	5309	501	8	5359	908	8	5409	312	8	5459	711	8
	5260	72099	8	**5310**	72509	9	**5360**	72916	9	**5410**	73320	8	**5460**	73719	8
	5261	107	8	5311	518	8	5361	925	8	5411	328	8	5461	727	8
	5262	115	8	5312	526	8	5362	933	8	5412	336	8	5462	735	8
	5263	123	9	5313	534	8	5363	941	8	5413	344	8	5463	743	8
	5264	132	8	5314	542	8	5364	949	8	5414	352	8	5464	751	8
	5265	72140	8	5315	72550	8	5365	72957	8	5415	73360	8	5465	73759	8
	5266	148	8	5316	558	9	5366	965	8	5416	368	8	5466	767	8
	5267	156	9	5317	567	8	5367	973	8	5417	376	8	5467	775	8
	5268	165	8	5318	575	8	5368	981	8	5418	384	8	5468	783	8
	5269	173	8	5319	583	8	5369	989	8	5419	392	8	5469	791	8
	5270	72181	8	**5320**	72591	8	**5370**	72997	9	**5420**	73400	8	**5470**	73799	8
	5271	189	9	5321	599	8	5371	73006	8	5421	408	8	5471	807	8
7	5272	198	8	5322	607	9	5372	014	8	5422	416	8	5472	815	8
1 | 0.7	5273	206	8	5323	616	8	5373	022	8	5423	424	8	5473	823	7
2 | 1.4	5274	214	8	5324	624	8	5374	030	8	5424	432	8	5474	830	8
3 | 2.1	5275	72222	8	5325	72632	8	5375	73038	8	5425	73440	8	5475	73838	8
4 | 2.8	5276	230	9	5326	640	8	5376	046	8	5426	448	8	5476	846	8
5 | 3.5	5277	239	8	5327	648	8	5377	054	8	5427	456	8	5477	854	8
6 | 4.2	5278	247	8	5328	656	9	5378	062	8	5428	464	8	5478	862	8
7 | 4.9	5279	255	8	5329	665	8	5379	070	8	5429	472	8	5479	870	8
8 | 5.6	**5280**	72263	9	**5330**	72673	8	**5380**	73078	8	**5430**	73480	8	**5480**	73878	8
9 | 6.3	5281	272	8	5331	681	8	5381	086	8	5431	488	8	5481	886	8
	5282	280	8	5332	689	8	5382	094	8	5432	496	8	5482	894	8
	5283	288	8	5333	697	8	5383	102	9	5433	504	8	5483	902	8
	5284	296	8	5334	705	8	5384	111	8	5434	512	8	5484	910	8
	5285	72304	9	5335	72713	9	5385	73119	8	5435	73520	8	5485	73918	8
	5286	313	8	5336	722	8	5386	127	8	5436	528	8	5486	926	7
	5287	321	8	5337	730	8	5387	135	8	5437	536	8	5487	933	8
	5288	329	8	5338	738	8	5388	143	8	5438	544	8	5488	941	8
	5289	337	9	5339	746	8	5389	151	8	5439	552	8	5489	949	8
	5290	72346	8	**5340**	72754	8	**5390**	73159	8	**5440**	73560	8	**5490**	73957	8
	5291	354	8	5341	762	8	5391	167	8	5441	568	8	5491	965	8
	5292	362	8	5342	770	9	5392	175	8	5442	576	8	5492	973	8
	5293	370	8	5343	779	8	5393	183	8	5443	584	8	5493	981	8
	5294	378	9	5344	787	8	5394	191	8	5444	592	8	5494	989	8
	5295	72387	8	5345	72795	8	5395	73199	8	5445	73600	8	5495	73997	8
	5296	395	8	5346	803	8	5396	207	8	5446	608	8	5496	74005	8
	5297	403	8	5347	811	8	5397	215	8	5447	616	8	5497	013	7
	5298	411	9	5348	819	8	5398	223	8	5448	624	8	5498	020	8
	5299	419		5349	827		5399	231		5449	632		5499	028	8
	5300	72428		**5350**	72835		**5400**	73239		**5450**	73640		**5500**	74036	

P.P.	N	Log	d	N	Log	d	N	Log	d	N	Log	d	N	Log	d

N	Log	d	N	Log	d	N	Log	d	N	Log	d	N	Log	d	P. P.
5500	74036	8	**5550**	74429	8	**5600**	74819	8	**5650**	75205	8	**5700**	75587	8	
5501	044	8	5551	437	8	5601	827	7	5651	213	7	5701	595	8	
5502	052	8	5552	445	8	5602	834	8	5652	220	8	5702	603	7	
5503	060	8	5553	453	8	5603	842	8	5653	228	8	5703	610	8	
5504	068	8	5554	461	7	5604	850	8	5654	236	7	5704	618	8	
5505	74076	8	5555	74468	8	5605	74858	7	5655	75243	8	5705	75626	7	
5506	084	8	5556	476	8	5606	865	8	5656	251	8	5706	633	8	
5507	092	7	5557	484	7	5607	873	8	5657	259	7	5707	641	7	
5508	099	8	5558	492	8	5608	881	8	5658	266	8	5708	648	8	
5509	107	8	5559	500	8	5609	889	7	5659	274	8	5709	656	8	
5510	74115	8	**5560**	74507	8	**5610**	74896	8	**5660**	75282	7	**5710**	75664	7	
5511	123	8	5561	515	8	5611	904	8	5661	289	8	5711	671	8	
5512	131	8	5562	523	8	5612	912	8	5662	297	8	5712	679	7	
5513	139	8	5563	531	8	5613	920	7	5663	305	7	5713	686	8	
5514	147	8	5564	539	8	5614	927	8	5664	312	8	5714	694	8	
5515	74155	7	5565	74547	7	5615	74935	8	5665	75320	8	5715	75702	7	
5516	162	8	5566	554	8	5616	943	7	5666	328	7	5716	709	8	
5517	170	8	5567	562	8	5617	950	8	5667	335	8	5717	717	7	
5518	178	8	5568	570	8	5618	958	8	5668	343	8	5718	724	8	
5519	186	8	5569	578	8	5619	966	8	5669	351	7	5719	732	8	
5520	74194	8	**5570**	74586	7	**5620**	74974	7	**5670**	75358	8	**5720**	75740	7	
5521	202	8	5571	593	8	5621	981	8	5671	366	8	5721	747	8	**8**
5522	210	8	5572	601	8	5622	989	8	5672	374	7	5722	755	7	1 | 0.8
5523	218	7	5573	609	8	5623	997	8	5673	381	8	5723	762	8	2 | 1.6
5524	225	8	5574	617	7	5624	75005	7	5674	389	8	5724	770	8	3 | 2.4
															4 | 3.2
5525	74233	8	5575	74624	8	5625	75012	8	5675	75397	7	5725	75778	7	5 | 4.0
5526	241	8	5576	632	8	5626	020	8	5676	404	8	5726	785	8	6 | 4.8
5527	249	8	5577	640	8	5627	028	7	5677	412	8	5727	793	7	7 | 5.6
5528	257	8	5578	648	8	5628	035	8	5678	420	7	5728	800	8	8 | 6.4
5529	265	8	5579	656	7	5629	043	8	5679	427	8	5729	808	7	9 | 7.2
5530	74273	7	**5580**	74663	8	**5630**	75051	8	**5680**	75435	7	**5730**	75815	8	
5531	280	8	5581	671	8	5631	059	7	5681	442	8	5731	823	8	
5532	288	8	5582	679	8	5632	066	8	5682	450	8	5732	831	7	
5533	296	8	5583	687	8	5633	074	8	5683	458	7	5733	838	8	
5534	304	8	5584	695	7	5634	082	7	5684	465	8	5734	846	7	
5535	74312	8	5585	74702	8	5635	75089	8	5685	75473	8	5735	75853	8	
5536	320	7	5586	710	8	5636	097	8	5686	481	7	5736	861	7	
5537	327	8	5587	718	8	5637	105	8	5687	488	8	5737	868	8	
5538	335	8	5588	726	7	5638	113	7	5688	496	8	5738	876	8	
5539	343	8	5589	733	8	5639	120	8	5689	504	7	5739	884	7	
5540	74351	8	**5590**	74741	8	**5640**	75128	8	**5690**	75511	8	**5740**	75891	8	
5541	359	8	5591	749	8	5641	136	7	5691	519	7	5741	899	7	
5542	367	7	5592	757	7	5642	143	8	5692	526	8	5742	906	8	
5543	374	8	5593	764	8	5643	151	8	5693	534	8	5743	914	7	
5544	382	8	5594	772	8	5644	159	7	5694	542	7	5744	921	8	
5545	74390	8	5595	74780	8	5645	75166	8	5695	75549	8	5745	75929	8	
5546	398	8	5596	788	8	5646	174	8	5696	557	8	5746	937	7	
5547	406	8	5597	796	7	5647	182	7	5697	565	7	5747	944	8	
5548	414	7	5598	803	8	5648	189	8	5698	572	8	5748	952	7	
5549	421	8	5599	811	8	5649	197	8	5699	580	7	5749	959	7	
5550	74429		**5600**	74819		**5650**	75205		**5700**	75587		**5750**	75967		
N	Log	d	N	Log	d	N	Log	d	N	Log	d	N	Log	d	P. P.

P. P.	N	Log	d	N	Log	d	N	Log	d	N	Log	d	N	Log	d
	5750	75967	7	**5800**	76343	7	**5850**	76716	7	**5900**	77085	8	**5950**	77452	7
	5751	974	8	5801	350	8	5851	723	7	5901	093	7	5951	459	7
	5752	982	7	5802	358	7	5852	730	8	5902	100	7	5952	466	8
	5753	989	8	5803	365	8	5853	738	7	5903	107	8	5953	474	7
	5754	997	8	5804	373	7	5854	745	8	5904	115	7	5954	481	7
	5755	76005	7	5805	76380	8	5855	76753	7	5905	77122	7	5955	77488	7
	5756	012	8	5806	388	7	5856	760	8	5906	129	8	5956	495	8
	5757	020	7	5807	395	8	5857	768	7	5907	137	7	5957	503	7
	5758	027	8	5808	403	7	5858	775	7	5908	144	7	5958	510	7
	5759	035	7	5809	410	8	5859	782	8	5909	151	8	5959	517	8
	5760	76042	8	**5810**	76418	7	**5860**	76790	7	**5910**	77159	7	**5960**	77525	7
	5761	050	7	5811	425	8	5861	797	8	5911	166	7	5961	532	7
	5762	057	8	5812	433	7	5862	805	7	5912	173	8	5962	539	7
	5763	065	7	5813	440	8	5863	812	7	5913	181	7	5963	546	8
	5764	072	8	5814	448	7	5864	819	8	5914	188	7	5964	554	7
	5765	76080	7	5815	76455	7	5865	76827	7	5915	77195	8	5965	77561	7
	5766	087	8	5816	462	8	5866	834	8	5916	203	7	5966	568	8
	5767	095	7	5817	470	7	5867	842	7	5917	210	7	5967	576	7
	5768	103	8	5818	477	8	5868	849	7	5918	217	8	5968	583	7
	5769	110	8	5819	485	7	5869	856	8	5919	225	7	5969	590	7
7	**5770**	76118	7	**5820**	76492	8	**5870**	76864	7	**5920**	77232	8	**5970**	77597	8
1 0.7	5771	125	8	5821	500	7	5871	871	8	5921	240	7	5971	605	7
2 1.4	5772	133	7	5822	507	8	5872	879	7	5922	247	7	5972	612	7
3 2.1	5773	140	8	5823	515	7	5873	886	7	5923	254	8	5973	619	8
4 2.8	5774	148	7	5824	522	8	5874	893	8	5924	262	7	5974	627	7
5 3.5	5775	76155	8	5825	76530	7	5875	76901	7	5925	77269	7	5975	77634	7
6 4.2	5776	163	7	5826	537	8	5876	908	8	5926	276	7	5976	641	7
7 4.9	5777	170	8	5827	545	7	5877	916	7	5927	283	8	5977	648	8
8 5.6	5778	178	7	5828	552	7	5878	923	7	5928	291	7	5978	656	7
9 6.3	5779	185	8	5829	559	8	5879	930	8	5929	298	7	5979	663	7
	5780	76193	7	**5830**	76567	7	**5880**	76938	7	**5930**	77305	8	**5980**	77670	7
	5781	200	8	5831	574	8	5881	945	8	5931	313	7	5981	677	8
	5782	208	7	5832	582	7	5882	953	7	5932	320	7	5982	685	7
	5783	215	8	5833	589	8	5883	960	7	5933	327	8	5983	692	7
	5784	223	7	5834	597	7	5884	967	8	5934	335	7	5984	699	7
	5785	76230	8	5835	76604	8	5885	76975	7	5935	77342	7	5985	77706	8
	5786	238	7	5836	612	7	5886	982	7	5936	349	8	5986	714	7
	5787	245	8	5837	619	7	5887	989	8	5937	357	7	5987	721	7
	5788	253	7	5838	626	8	5888	997	7	5938	364	7	5988	728	7
	5789	260	8	5839	634	7	5889	77004	8	5939	371	8	5989	735	8
	5790	76268	7	**5840**	76641	8	**5890**	77012	7	**5940**	77379	7	**5990**	77743	7
	5791	275	8	5841	649	7	5891	019	7	5941	386	7	5991	750	7
	5792	283	7	5842	656	8	5892	026	8	5942	393	8	5992	757	7
	5793	290	8	5843	664	7	5893	034	7	5943	401	7	5993	764	8
	5794	298	7	5844	671	7	5894	041	7	5944	408	7	5994	772	7
	5795	76305	8	5845	76678	8	5895	77048	8	5945	77415	7	5995	77779	7
	5796	313	7	5846	686	7	5896	056	7	5946	422	8	5996	786	7
	5797	320	8	5847	693	8	5897	063	7	5947	430	7	5997	793	8
	5798	328	8	5848	701	7	5898	070	8	5948	437	7	5998	801	7
	5799	335	8	5849	708	8	5899	078	7	5949	444	8	5999	808	7
	5800	76343		**5850**	76716		**5900**	77085		**5950**	77452		**6000**	77815	
P. P.	N	Log	d	N	Log	d	N	Log	d	N	Log	d	N	Log	d

N	Log	d	N	Log	d	N	Log	d	N	Log	d	N	Log	d	P. P.
6000	77815	7	**6050**	78176	7	**6100**	78533	7	**6150**	78888	7	**6200**	79239	7	
6001	822	8	6051	183	7	6101	540	7	6151	895	7	6201	246	7	
6002	830	7	6052	190	7	6102	547	7	6152	902	7	6202	253	7	
6003	837	7	6053	197	7	6103	554	7	6153	909	7	6203	260	7	
6004	844	7	6054	204	7	6104	561	8	6154	916	7	6204	267	7	
6005	77851	8	6055	78211	8	6105	78569	7	6155	78923	7	6205	79274	7	
6006	859	7	6056	219	7	6106	576	7	6156	930	7	6206	281	7	
6007	866	7	6057	226	7	6107	583	7	6157	937	7	6207	288	7	
6008	873	7	6058	233	7	6108	590	7	6158	944	7	6208	295	7	
6009	880	7	6059	240	7	6109	597	7	6159	951	7	6209	302	7	
6010	77887	8	**6060**	78247	7	**6110**	78604	7	**6160**	78958	7	**6210**	79309	7	
6011	895	7	6061	254	8	6111	611	7	6161	965	7	6211	316	7	
6012	902	7	6062	262	7	6112	618	7	6162	972	7	6212	323	7	
6013	909	7	6063	269	7	6113	625	8	6163	979	7	6213	330	7	
6014	916	8	6064	276	7	6114	633	7	6164	986	7	6214	337	7	
6015	77924	7	6065	78283	7	6115	78640	7	6165	78993	7	6215	79344	7	**8**
6016	931	7	6066	290	7	6116	647	7	6166	79000	7	6216	351	7	1 | 0.8
6017	938	7	6067	297	8	6117	654	7	6167	007	7	6217	358	7	2 | 1.6
6018	945	7	6068	305	7	6118	661	7	6168	014	7	6218	365	7	3 | 2.4 4 | 3.2
6019	952	8	6069	312	7	6119	668	7	6169	021	8	6219	372	7	5 | 4.0 6 | 4.8
6020	77960	7	**6070**	78319	7	**6120**	78675	7	**6170**	79029	7	**6220**	79379	7	7 | 5.6 8 | 6.4
6021	967	7	6071	326	7	6121	682	7	6171	036	7	6221	386	7	9 | 7.2
6022	974	7	6072	333	7	6122	689	7	6172	043	7	6222	393	7	
6023	981	7	6073	340	7	6123	696	8	6173	050	7	6223	400	7	
6024	988	8	6074	347	8	6124	704	7	6174	057	7	6224	407	7	
6025	77996	7	6075	78355	7	6125	78711	7	6175	79064	7	6225	79414	7	
6026	78003	7	6076	362	7	6126	718	7	6176	071	7	6226	421	7	
6027	010	7	6077	369	7	6127	725	7	6177	078	7	6227	428	7	
6028	017	8	6078	376	7	6128	732	7	6178	085	7	6228	435	7	
6029	025	7	6079	383	7	6129	739	7	6179	092	7	6229	442	7	**7**
6030	78032	7	**6080**	78390	8	**6130**	78746	7	**6180**	79099	7	**6230**	79449	7	1 | 0.7
6031	039	7	6081	398	7	6131	753	7	6181	106	7	6231	456	7	2 | 1.4
6032	046	7	6082	405	7	6132	760	7	6182	113	7	6232	463	7	3 | 2.1 4 | 2.8
6033	053	8	6083	412	7	6133	767	7	6183	120	7	6233	470	7	5 | 3.5 6 | 4.2
6034	061	7	6084	419	7	6134	774	7	6184	127	7	6234	477	7	7 | 4.9 8 | 5.6
6035	78068	7	6085	78426	7	6135	78781	8	6185	79134	7	6235	79484	7	9 | 6.3
6036	075	7	6086	433	7	6136	789	7	6186	141	7	6236	491	7	
6037	082	7	6087	440	7	6137	796	7	6187	148	7	6237	498	7	
6038	089	8	6088	447	8	6138	803	7	6188	155	7	6238	505	7	
6039	097	7	6089	455	7	6139	810	7	6189	162	7	6239	511	6, 7	
6040	78104	7	**6090**	78462	7	**6140**	78817	7	**6190**	79169	7	**6240**	79518	7	
6041	111	7	6091	469	7	6141	824	7	6191	176	7	6241	525	7	
6042	118	7	6092	476	7	6142	831	7	6192	183	7	6242	532	7	
6043	125	7	6093	483	7	6143	838	7	6193	.190	7	6243	539	7	
6044	132	8	6094	490	7	6144	845	7	6194	197	7	6244	546	7	
6045	78140	7	6095	78497	7	6145	78852	7	6195	79204	7	6245	79553	7	
6046	147	7	6096	504	8	6146	859	7	6196	211	7	6246	560	7	
6047	154	7	6097	512	7	6147	866	7	6197	218	7	6247	567	7	
6048	161	7	6098	519	7	6148	873	7	6198	225	7	6248	574	7	
6049	168	8	6099	526	7	6149	880	8	6199	232	8	6249	581	7	
6050	78176		**6100**	78533		**6150**	78888		**6200**	79239		**6250**	79588		
N	Log	d	N	Log	d	N	Log	d	N	Log	d	N	Log	d	P. P.

P. P.	N	Log	d	N	Log	d	N	Log	d	N	Log	d	N	Log	d
	6250	79588	7	**6300**	79934	7	**6350**	80277	7	**6400**	80618	7	**6450**	80956	7
	6251	595	7	6301	941	7	6351	284	7	6401	625	7	6451	963	6
	6252	602	7	6302	948	7	6352	291	7	6402	632	6	6452	969	7
	6253	609	7	6303	955	7	6353	298	7	6403	638	7	6453	976	7
	6254	616	7	6304	962	7	6354	305	7	6404	645	7	6454	983	7
	6255	79623	7	6305	79969	6	6355	80312	6	6405	80652	7	6455	80990	6
	6256	630	7	6306	975	7	6356	318	7	6406	659	6	6456	996	7
	6257	637	7	6307	982	7	6357	325	7	6407	665	7	6457	81003	7
	6258	644	6	6308	989	7	6358	332	7	6408	672	7	6458	010	7
	6259	650	7	6309	996	7	6359	339	7	6409	679	7	6459	017	6
	6260	79657	7	**6310**	80003	7	**6360**	80346	7	**6410**	80686	7	**6460**	81023	7
	6261	664	7	6311	010	7	6361	353	6	6411	693	6	6461	030	7
	6262	671	7	6312	017	7	6362	359	7	6412	699	7	6462	037	6
	6263	678	7	6313	024	6	6363	366	7	6413	706	7	6463	043	7
	6264	685	7	6314	030	7	6364	373	7	6414	713	7	6464	050	7
	6265	79692	7	6315	80037	7	6365	80380	7	6415	80720	6	6465	81057	7
	6266	699	7	6316	044	7	6366	387	6	6416	726	7	6466	064	6
	6267	706	7	6317	051	7	6367	393	7	6417	733	7	6467	070	7
	6268	713	7	6318	058	7	6368	400	7	6418	740	7	6468	077	7
	6269	720	7	6319	065	7	6369	407	7	6419	747	7	6469	084	6
6	**6270**	79727	7	**6320**	80072	7	**6370**	80414	7	**6420**	80754	6	**6470**	81090	7
1 | 0.6	6271	734	7	6321	079	6	6371	421	7	6421	760	7	6471	097	7
2 | 1.2	6272	741	7	6322	085	7	6372	428	6	6422	767	7	6472	104	7
3 | 1.8	6273	748	6	6323	092	7	6373	434	7	6423	774	7	6473	111	6
4 | 2.4	6274	754	7	6324	099	7	6374	441	7	6424	781	6	6474	117	7
5 | 3.0															
6 | 3.6	6275	79761	7	6325	80106	7	6375	80448	7	6425	80787	7	6475	81124	7
7 | 4.2	6276	768	7	6326	113	7	6376	455	7	6426	794	7	6476	131	6
8 | 4.8	6277	775	7	6327	120	7	6377	462	6	6427	801	7	6477	137	7
9 | 5.4	6278	782	7	6328	127	7	6378	468	7	6428	808	6	6478	144	7
	6279	789	7	6329	134	6	6379	475	7	6429	814	7	6479	151	7
	6280	79796	7	**6330**	80140	7	**6380**	80482	7	**6430**	80821	7	**6480**	81158	6
	6281	803	7	6331	147	7	6381	489	7	6431	828	7	6481	164	7
	6282	810	7	6332	154	7	6382	496	6	6432	835	6	6482	171	7
	6283	817	7	6333	161	7	6383	502	7	6433	841	7	6483	178	6
	6284	824	7	6334	168	7	6384	509	7	6434	848	7	6484	184	7
	6285	79831	6	6335	80175	7	6385	80516	7	6435	80855	7	6485	81191	7
	6286	837	7	6336	182	6	6386	523	7	6436	862	6	6486	198	6
	6287	844	7	6337	188	7	6387	530	6	6437	868	7	6487	204	7
	6288	851	7	6338	195	7	6388	536	7	6438	875	7	6488	211	7
	6289	858	7	6339	202	7	6389	543	7	6439	882	7	6489	218	6
	6290	79865	7	**6340**	80209	7	**6390**	80550	7	**6440**	80889	6	**6490**	81224	7
	6291	872	7	6341	216	7	6391	557	7	6441	895	7	6491	231	7
	6292	879	7	6342	223	6	6392	564	6	6442	902	7	6492	238	7
	6293	886	7	6343	229	7	6393	570	7	6443	909	7	6493	245	6
	6294	893	7	6344	236	7	6394	577	7	6444	916	6	6494	251	7
	6295	79900	6	6345	80243	7	6395	80584	7	6445	80922	7	6495	81258	7
	6296	906	7	6346	250	7	6396	591	7	6446	929	7	6496	265	6
	6297	913	7	6347	257	7	6397	598	6	6447	936	7	6497	271	7
	6298	920	7	6348	264	7	6398	604	7	6448	943	6	6498	278	7
	6299	927	7	6349	271	6	6399	611	7	6449	949	7	6499	285	6
	6300	79934		**6350**	80277		**6400**	80618		**6450**	80956		**6500**	81291	
P. P.	N	Log	d	N	Log	d	N	Log	d	N	Log	d	N	Log	d

N	Log	d	N	Log	d	N	Log	d	N	Log	d	N	Log	d
6500	81291	7	6550	81624	7	6600	81954	7	6650	82282	7	6700	82607	7
6501	298	7	6551	631	6	6601	961	7	6651	289	6	6701	614	6
6502	305	6	6552	637	7	6602	968	6	6652	295	7	6702	620	7
6503	311	7	6553	644	7	6603	974	7	6653	302	6	6703	627	6
6504	318	7	6554	651	6	6604	981	6	6654	308	7	6704	633	7
6505	81325	6	6555	81657	7	6605	81987	7	6655	82315	6	6705	82640	6
6506	331	7	6556	664	7	6606	994	6	6656	321	7	6706	646	7
6507	338	7	6557	671	6	6607	82000	6	6657	328	7	6707	653	6
6508	345	6	6558	677	7	6608	007	7	6658	334	6	6708	659	7
6509	351	7	6559	684	6	6609	014	6	6659	341	7	6709	666	6
6510	81358	7	6560	81690	7	6610	82020	7	6660	82347	7	6710	82672	7
6511	365	6	6561	697	7	6611	027	6	6661	354	6	6711	679	6
6512	371	7	6562	704	6	6612	033	7	6662	360	7	6712	685	7
6513	378	7	6563	710	7	6613	040	6	6663	367	6	6713	692	6
6514	385	6	6564	717	6	6614	046	7	6664	373	7	6714	698	7
6515	81391	7	6565	81723	7	6615	82053	7	6665	82380	7	6715	82705	6
6516	398	7	6566	730	7	6616	060	6	6666	387	6	6716	711	7
6517	405	6	6567	737	6	6617	066	7	6667	393	7	6717	718	6
6518	411	7	6568	743	7	6618	073	6	6668	400	6	6718	724	6
6519	418	7	6569	750	7	6619	079	7	6669	406	7	6719	730	7
6520	81425	6	6570	81757	6	6620	82086	6	6670	82413	6	6720	82737	6
6521	431	7	6571	763	7	6621	092	7	6671	419	7	6721	743	7
6522	438	7	6572	770	6	6622	099	6	6672	426	6	6722	750	6
6523	445	6	6573	776	7	6623	105	7	6673	432	7	6723	756	7
6524	451	7	6574	783	7	6624	112	7	6674	439	6	6724	763	6
6525	81458	7	6575	81790	6	6625	82119	6	6675	82445	7	6725	82769	7
6526	465	6	6576	796	7	6626	125	7	6676	452	6	6726	776	6
6527	471	7	6577	803	6	6627	132	6	6677	458	7	6727	782	7
6528	478	7	6578	809	7	6628	138	7	6678	465	6	6728	789	6
6529	485	6	6579	816	7	6629	145	6	6679	471	7	6729	795	7
6530	81491	7	6580	81823	6	6630	82151	7	6680	82478	6	6730	82802	6
6531	498	7	6581	829	7	6631	158	6	6681	484	7	6731	808	6
6532	505	6	6582	836	6	6632	164	7	6682	491	6	6732	814	7
6533	511	7	6583	842	7	6633	171	7	6683	497	7	6733	821	6
6534	518	7	6584	849	7	6634	178	6	6684	504	6	6734	827	7
6535	81525	6	6585	81856	6	6635	82184	7	6685	82510	7	6735	82834	6
6536	531	7	6586	862	7	6636	191	6	6686	517	6	6736	840	7
6537	538	6	6587	869	6	6637	197	7	6687	523	7	6737	847	6
6538	544	7	6588	875	7	6638	204	6	6688	530	6	6738	853	7
6539	551	7	6589	882	7	6639	210	7	6689	536	7	6739	860	6
6540	81558	6	6590	81889	6	6640	82217	6	6690	82543	6	6740	82866	6
6541	564	7	6591	895	7	6641	223	7	6691	549	7	6741	872	7
6542	571	7	6592	902	6	6642	230	6	6692	556	6	6742	879	6
6543	578	6	6593	908	7	6643	236	7	6693	562	7	6743	885	7
6544	584	7	6594	915	6	6644	243	6	6694	569	6	6744	892	6
6545	81591	7	6595	81921	7	6645	82249	7	6695	82575	7	6745	82898	7
6546	598	6	6596	928	7	6646	256	7	6696	582	6	6746	905	6
6547	604	7	6597	935	6	6647	263	6	6697	588	7	6747	911	7
6548	611	6	6598	941	7	6648	269	6	6698	595	6	6748	918	6
6549	617	7	6599	948	6	6649	276	6	6699	601	6	6749	924	6
6550	81624		6600	81954		6650	82282		6700	82607		6750	82930	

P. P.

	7
1	0.7
2	1.4
3	2.1
4	2.8
5	3.5
6	4.2
7	4.9
8	5.6
9	6.3

Proportional parts (left margin):

6

1	0.6
2	1.2
3	1.8
4	2.4
5	3.0
6	3.6
7	4.2
8	4.8
9	5.4

P.P.	N	Log	d	N	Log	d	N	Log	d	N	Log	d	N	Log	d
	6750	82930	7	**6800**	83251	6	**6850**	83569	6	**6900**	83885	6	**6950**	84198	7
	6751	937	6	6801	257	7	6851	575	7	6901	891	6	6951	205	6
	6752	943	7	6802	264	6	6852	582	6	6902	897	7	6952	211	6
	6753	950	6	6803	270	6	6853	588	6	6903	904	6	6953	217	6
	6754	956	7	6804	276	7	6854	594	7	6904	910	6	6954	223	7
	6755	82963	6	6805	83283	6	6855	83601	6	6905	83916	7	6955	84230	6
	6756	969	6	6806	289	7	6856	607	6	6906	923	6	6956	236	6
	6757	975	7	6807	296	6	6857	613	7	6907	929	6	6957	242	6
	6758	982	6	6808	302	6	6858	620	6	6908	935	7	6958	248	7
	6759	988	7	6809	308	7	6859	626	6	6909	942	6	6959	255	6
	6760	82995	6	**6810**	83315	6	**6860**	83632	7	**6910**	83948	6	**6960**	84261	6
	6761	83001	7	6811	321	6	6861	639	6	6911	954	6	6961	267	6
	6762	008	6	6812	327	7	6862	645	6	6912	960	7	6962	273	7
	6763	014	6	6813	334	6	6863	651	7	6913	967	6	6963	280	6
	6764	020	7	6814	340	7	6864	658	6	6914	973	6	6964	286	6
	6765	83027	6	6815	83347	6	6865	83664	6	6915	83979	6	6965	84292	6
	6766	033	7	6816	353	6	6866	670	7	6916	985	7	6966	298	7
	6767	040	6	6817	359	7	6867	677	6	6917	992	6	6967	305	6
	6768	046	6	6818	366	6	6868	683	6	6918	998	6	6968	311	6
	6769	052	7	6819	372	6	6869	689	7	6919	84004	7	6969	317	6
	6770	83059	6	**6820**	83378	7	**6870**	83696	6	**6920**	84011	6	**6970**	84323	7
	6771	065	7	6821	385	6	6871	702	6	6921	017	6	6971	330	6
	6772	072	6	6822	391	7	6872	708	7	6922	023	6	6972	336	6
	6773	078	7	6823	398	6	6873	715	6	6923	029	7	6973	342	6
	6774	085	6	6824	404	6	6874	721	6	6924	036	6	6974	348	6
	6775	83091	6	6825	83410	7	6875	83727	7	6925	84042	6	6975	84354	7
	6776	097	7	6826	417	6	6876	734	6	6926	048	7	6976	361	6
	6777	104	6	6827	423	6	6877	740	6	6927	055	6	6977	367	6
	6778	110	7	6828	429	7	6878	746	7	6928	061	6	6978	373	6
	6779	117	6	6829	436	6	6879	753	6	6929	067	6	6979	379	7
	6780	83123	6	**6830**	83442	6	**6880**	83759	6	**6930**	84073	7	**6980**	84386	6
	6781	129	7	6831	448	7	6881	765	6	6931	080	6	6981	392	6
	6782	136	6	6832	455	6	6882	771	7	6932	086	6	6982	398	6
	6783	142	7	6833	461	6	6883	778	6	6933	092	6	6983	404	6
	6784	149	6	6834	467	7	6884	784	6	6934	098	7	6984	410	7
	6785	83155	6	6835	83474	6	6885	83790	7	6935	84105	6	6985	84417	6
	6786	161	7	6836	480	7	6886	797	6	6936	111	6	6986	423	6
	6787	168	6	6837	487	6	6887	803	6	6937	117	6	6987	429	6
	6788	174	7	6838	493	6	6888	809	7	6938	123	7	6988	435	7
	6789	181	6	6839	499	7	6889	816	6	6939	130	6	6989	442	6
	6790	83187	6	**6840**	83506	6	**6890**	83822	6	**6940**	84136	6	**6990**	84448	6
	6791	193	7	6841	512	6	6891	828	7	6941	142	6	6991	454	6
	6792	200	6	6842	518	7	6892	835	6	6942	148	7	6992	460	6
	6793	206	7	6843	525	6	6893	841	6	6943	155	6	6993	466	7
	6794	213	6	6844	531	6	6894	847	6	6944	161	6	6994	473	6
	6795	83219	6	6845	83537	7	6895	83853	7	6945	84167	6	6995	84479	6
	6796	225	7	6846	544	6	6896	860	6	6946	173	7	6996	485	6
	6797	232	6	6847	550	6	6897	866	6	6947	180	6	6997	491	6
	6798	238	7	6848	556	7	6898	872	7	6948	186	6	6998	497	7
	6799	245	6	6849	563	6	6899	879	6	6949	192	6	6999	504	6
	6800	83251		**6850**	83569		**6900**	83885		**6950**	84198		**7000**	84510	
P.P.	N	Log	d	N	Log	d	N	Log	d	N	Log	d	N	Log	d

N	Log	d	N	Log	d	N	Log	d	N	Log	d	N	Log	d
7000	84510	6	7050	84819	6	7100	85126	6	7150	85431	6	7200	85733	6
7001	516	6	7051	825	6	7101	132	6	7151	437	6	7201	739	6
7002	522	6	7052	831	6	7102	138	6	7152	443	6	7202	745	6
7003	528	7	7053	837	7	7103	144	6	7153	449	6	7203	751	6
7004	535	6	7054	844	6	7104	150	6	7154	455	6	7204	757	6
7005	84541	6	7055	84850	6	7105	85156	7	7155	85461	6	7205	85763	6
7006	547	6	7056	856	6	7106	163	6	7156	467	6	7206	769	6
7007	553	6	7057	862	6	7107	169	6	7157	473	6	7207	775	6
7008	559	7	7058	868	6	7108	175	6	7158	479	6	7208	781	7
7009	566	6	7059	874	6	7109	181	6	7159	485	6	7209	788	6
7010	84572	6	7060	84880	7	7110	85187	6	7160	85491	6	7210	85794	6
7011	578	6	7061	887	6	7111	193	6	7161	497	6	7211	800	6
7012	584	6	7062	893	6	7112	199	6	7162	503	6	7212	806	6
7013	590	7	7063	899	6	7113	205	6	7163	509	7	7213	812	6
7014	597	6	7064	905	6	7114	211	6	7164	516	6	7214	818	6
7015	84603	6	7065	84911	6	7115	85217	7	7165	85522	6	7215	85824	6
7016	609	6	7066	917	7	7116	224	6	7166	528	6	7216	830	6
7017	615	6	7067	924	6	7117	230	6	7167	534	6	7217	836	6
7018	621	7	7068	930	6	7118	236	6	7168	540	6	7218	842	6
7019	628	6	7069	936	6	7119	242	6	7169	546	6	7219	848	6
7020	84634	6	7070	84942	6	7120	85248	6	7170	85552	6	7220	85854	6
7021	640	6	7071	948	6	7121	254	6	7171	558	6	7221	860	6
7022	646	6	7072	954	6	7122	260	6	7172	564	6	7222	866	6
7023	652	6	7073	960	7	7123	266	6	7173	570	6	7223	872	6
7024	658	7	7074	967	6	7124	272	6	7174	576	6	7224	878	6
7025	84665	6	7075	84973	6	7125	85278	7	7175	85582	6	7225	85884	6
7026	671	6	7076	979	6	7126	285	6	7176	588	6	7226	890	6
7027	677	6	7077	985	6	7127	291	6	7177	594	6	7227	896	6
7028	683	6	7078	991	6	7128	297	6	7178	600	6	7228	902	6
7029	689	7	7079	997	6	7129	303	6	7179	606	6	7229	908	6
7030	84696	6	7080	85003	6	7130	85309	6	7180	85612	6	7230	85914	6
7031	702	6	7081	009	7	7131	315	6	7181	618	7	7231	920	6
7032	708	6	7082	016	6	7132	321	6	7182	625	6	7232	926	6
7033	714	6	7083	022	6	7133	327	6	7183	631	6	7233	932	6
7034	720	6	7084	028	6	7134	333	6	7184	637	6	7234	938	6
7035	84726	7	7085	85034	6	7135	85339	6	7185	85643	6	7235	85944	6
7036	733	6	7086	040	6	7136	345	7	7186	649	6	7236	950	6
7037	739	6	7087	046	6	7137	352	6	7187	655	6	7237	956	6
7038	745	6	7088	052	6	7138	358	6	7188	661	6	7238	962	6
7039	751	6	7089	058	7	7139	364	6	7189	667	6	7239	968	6
7040	84757	6	7090	85065	6	7140	85370	6	7190	85673	6	7240	85974	6
7041	763	7	7091	071	6	7141	376	6	7191	679	6	7241	980	6
7042	770	6	7092	077	6	7142	382	6	7192	685	6	7242	986	6
7043	776	6	7093	083	6	7143	388	6	7193	691	6	7243	992	6
7044	782	6	7094	089	6	7144	394	6	7194	697	6	7244	998	6
7045	84788	6	7095	85095	6	7145	85400	6	7195	85703	6	7245	86004	6
7046	794	6	7096	101	6	7146	406	6	7196	709	6	7246	010	6
7047	800	7	7097	107	7	7147	412	6	7197	715	6	7247	016	6
7048	807	6	7098	114	6	7148	418	7	7198	721	6	7248	022	6
7049	813	6	7099	120	6	7149	425	6	7199	727	6	7249	028	6
7050	84819		7100	85126		7150	85431		7200	85733		7250	86034	
N	Log	d	N	Log	d	N	Log	d	N	Log	d	N	Log	d

P. P.

7	
1	0.7
2	1.4
3	2.1
4	2.8
5	3.5
6	4.2
7	4.9
8	5.6
9	6.3

6	
1	0.6
2	1.2
3	1.8
4	2.4
5	3.0
6	3.6
7	4.2
8	4.8
9	5.4

P. P.	N	Log	d	N	Log	d	N	Log	d	N	Log	d	N	Log	d
	7250	86034	6	**7300**	86332	6	**7350**	86629	6	**7400**	86923	6	**7450**	87216	5
	7251	040	6	7301	338	6	7351	635	6	7401	929	6	7451	221	6
	7252	046	6	7302	344	6	7352	641	5	7402	935	6	7452	227	6
	7253	052	6	7303	350	6	7353	646	6	7403	941	6	7453	233	6
	7254	058	6	7304	356	6	7354	652	6	7404	947	6	7454	239	6
	7255	86064	6	7305	86362	6	7355	86658	6	7405	86953	5	7455	87245	6
	7256	070	6	7306	368	6	7356	664	6	7406	958	6	7456	251	5
	7257	076	6	7307	374	6	7357	670	6	7407	964	6	7457	256	6
	7258	082	6	7308	380	6	7358	676	6	7408	970	6	7458	262	6
	7259	088	6	7309	386	6	7359	682	6	7409	976	6	7459	268	6
	7260	86094	6	**7310**	86392	6	**7360**	86688	6	**7410**	86982	6	**7460**	87274	6
	7261	100	6	7311	398	6	7361	694	6	7411	988	6	7461	280	6
	7262	106	6	7312	404	6	7362	700	5	7412	994	6	7462	286	5
	7263	112	6	7313	410	5	7363	705	6	7413	999	6	7463	291	6
	7264	118	6	7314	415	6	7364	711	6	7414	87005	6	7464	297	6
	7265	86124	6	7315	86421	6	7365	86717	6	7415	87011	6	7465	87303	6
	7266	130	6	7316	427	6	7366	723	6	7416	017	6	7466	309	6
	7267	136	5	7317	433	6	7367	729	6	7417	023	6	7467	315	5
	7268	141	6	7318	439	6	7368	735	6	7418	029	6	7468	320	6
	7269	147	6	7319	445	6	7369	741	6	7419	035	5	7469	326	6
	7270	86153	6	**7320**	86451	6	**7370**	86747	6	**7420**	87040	6	**7470**	87332	6
	7271	159	6	7321	457	6	7371	753	6	7421	046	6	7471	338	6
	7272	165	6	7322	463	6	7372	759	5	7422	052	6	7472	344	5
	7273	171	6	7323	469	6	7373	764	6	7423	058	6	7473	349	6
	7274	177	6	7324	475	6	7374	770	6	7424	064	6	7474	355	6
	7275	86183	6	7325	86481	6	7375	86776	6	7425	87070	5	7475	87361	6
	7276	189	6	7326	487	6	7376	782	6	7426	075	6	7476	367	6
	7277	195	6	7327	493	6	7377	788	6	7427	081	6	7477	373	6
	7278	201	6	7328	499	5	7378	794	6	7428	087	6	7478	379	5
	7279	207	6	7329	504	6	7379	800	6	7429	093	6	7479	384	6
	7280	86213	6	**7330**	86510	6	**7380**	86806	6	**7430**	87099	6	**7480**	87390	6
	7281	219	6	7331	516	6	7381	812	5	7431	105	6	7481	396	6
	7282	225	6	7332	522	6	7382	817	6	7432	111	5	7482	402	6
	7283	231	6	7333	528	6	7383	823	6	7433	116	6	7483	408	5
	7284	237	6	7334	534	6	7384	829	6	7434	122	6	7484	413	6
	7285	86243	6	7335	86540	6	7385	86835	6	7435	87128	6	7485	87419	6
	7286	249	6	7336	546	6	7386	841	6	7436	134	6	7486	425	6
	7287	255	6	7337	552	6	7387	847	6	7437	140	6	7487	431	6
	7288	261	6	7338	558	6	7388	853	6	7438	146	5	7488	437	5
	7289	267	6	7339	564	6	7389	859	5	7439	151	6	7489	442	6
	7290	86273	6	**7340**	86570	6	**7390**	86864	6	**7440**	87157	6	**7490**	87448	6
	7291	279	6	7341	576	5	7391	870	6	7441	163	6	7491	454	6
	7292	285	6	7342	581	6	7392	876	6	7442	169	6	7492	460	6
	7293	291	6	7343	587	6	7393	882	6	7443	175	6	7493	466	5
	7294	297	6	7344	593	6	7394	888	6	7444	181	5	7494	471	6
	7295	86303	5	7345	86599	6	7395	86894	6	7445	87186	6	7495	87477	6
	7296	308	6	7346	605	6	7396	900	6	7446	192	6	7496	483	6
	7297	314	6	7347	611	6	7397	906	5	7447	198	6	7497	489	6
	7298	320	6	7348	617	6	7398	911	6	7448	204	6	7498	495	6
	7299	326	6	7349	623	6	7399	917	6	7449	210	6	7499	500	6
	7300	86332		**7350**	86629		**7400**	86923		**7450**	87216		**7500**	87506	
P. P.	N	Log	d	N	Log	d	N	Log	d	N	Log	d	N	Log	d

P. P.

5

1	0.5
2	1.0
3	1.5
4	2.0
5	2.5
6	3.0
7	3.5
8	4.0
9	4.5

N	Log	d	N	Log	d	N	Log	d	N	Log	d	N	Log	d	P. P.
7500	87506	6	**7550**	87795	5	**7600**	88081	6	**7650**	88366	6	**7700**	88649	6	
7501	512	6	7551	800	6	7601	087	6	7651	372	5	7701	655	5	
7502	518	5	7552	806	6	7602	093	5	7652	377	6	7702	660	6	
7503	523	6	7553	812	6	7603	098	6	7653	383	6	7703	666	6	
7504	529	6	7554	818	5	7604	104	6	7654	389	6	7704	672	5	
7505	87535	6	**7555**	87823	6	**7605**	88110	6	**7655**	88395	5	**7705**	88677	6	
7506	541	6	7556	829	6	7606	116	5	7656	400	6	7706	683	6	
7507	547	5	7557	835	6	7607	121	6	7657	406	6	7707	689	5	
7508	552	6	7558	841	5	7608	127	6	7658	412	5	7708	694	6	
7509	558	6	7559	846	6	7609	133	5	7659	417	6	7709	700	5	
7510	87564	6	**7560**	87852	6	**7610**	88138	6	**7660**	88423	6	**7710**	88705	6	
7511	570	6	7561	858	6	7611	144	6	7661	429	5	7711	711	6	
7512	576	5	7562	864	5	7612	150	6	7662	434	6	7712	717	5	
7513	581	6	7563	869	6	7613	156	5	7663	440	6	7713	722	6	
7514	587	6	7564	875	6	7614	161	6	7664	446	5	7714	728	6	
7515	87593	6	**7565**	87881	6	**7615**	88167	6	**7665**	88451	6	**7715**	88734	5	
7516	599	5	7566	887	5	7616	173	5	7666	457	6	7716	739	6	
7517	604	6	7567	892	6	7617	178	6	7667	463	5	7717	745	5	
7518	610	6	7568	898	6	7618	184	6	7668	468	6	7718	750	6	
7519	616	6	7569	904	6	7619	190	5	7669	474	6	7719	756	6	
7520	87622	6	**7570**	87910	5	**7620**	88195	6	**7670**	88480	5	**7720**	88762	5	**6**
7521	628	5	7571	915	6	7621	201	6	7671	485	6	7721	767	6	1 0.6
7522	633	6	7572	921	6	7622	207	6	7672	491	6	7722	773	6	2 1.2
7523	639	6	7573	927	6	7623	213	5	7673	497	5	7723	779	5	3 1.8
7524	645	6	7574	933	5	7624	218	6	7674	502	6	7724	784	6	4 2.4
7525	87651	5	**7575**	87938	6	**7625**	88224	6	**7675**	88508	5	**7725**	88790	5	5 3.0
7526	656	6	7576	944	6	7626	230	5	7676	513	6	7726	795	6	6 3.6
7527	662	6	7577	950	5	7627	235	6	7677	519	6	7727	801	6	7 4.2
7528	668	6	7578	955	6	7628	241	6	7678	525	5	7728	807	5	8 4.8
7529	674	5	7579	961	6	7629	247	5	7679	530	6	7729	812	6	9 5.4
7530	87679	6	**7580**	87967	6	**7630**	88252	6	**7680**	88536	6	**7730**	88818	6	
7531	685	6	7581	973	5	7631	258	6	7681	542	5	7731	824	5	
7532	691	6	7582	978	6	7632	264	6	7682	547	6	7732	829	6	
7533	697	6	7583	984	6	7633	270	5	7683	553	6	7733	835	5	
7534	703	5	7584	990	6	7634	275	6	7684	559	5	7734	840	6	
7535	87708	6	**7585**	87996	5	**7635**	88281	6	**7685**	88564	6	**7735**	88846	6	
7536	714	6	7586	88001	6	7636	287	5	7686	570	6	7736	852	5	
7537	720	6	7587	007	6	7637	292	6	7687	576	5	7737	857	6	
7538	726	5	7588	013	5	7638	298	6	7688	581	6	7738	863	5	
7539	731	6	7589	018	6	7639	304	5	7689	587	6	7739	868	6	
7540	87737	6	**7590**	88024	6	**7640**	88309	6	**7690**	88593	5	**7740**	88874	6	
7541	743	6	7591	030	6	7641	315	6	7691	598	6	7741	880	5	
7542	749	5	7592	036	5	7642	321	5	7692	604	6	7742	885	6	
7543	754	6	7593	041	6	7643	326	6	7693	610	5	7743	891	6	
7544	760	6	7594	047	6	7644	332	6	7694	615	6	7744	897	5	
7545	87766	6	**7595**	88053	5	**7645**	88338	5	**7695**	88621	6	**7745**	88902	6	
7546	772	5	7596	058	6	7646	343	6	7696	627	5	7746	908	5	
7547	777	6	7597	064	6	7647	349	6	7697	632	6	7747	913	6	
7548	783	6	7598	070	6	7648	355	5	7698	638	5	7748	919	6	
7549	789	6	7599	076	5	7649	360	6	7699	643	6	7749	925	5	
7550	87795		**7600**	88081		**7650**	88366		**7700**	88649		**7750**	88930		
N	Log	d	N	Log	d	N	Log	d	N	Log	d	N	Log	d	P. P.

COMMON LOGARITHMS

P.P.

	5
1	0.5
2	1.0
3	1.5
4	2.0
5	2.5
6	3.0
7	3.5
8	4.0
9	4.5

N	Log	d	N	Log	d	N	Log	d	N	Log	d	N	Log	d
7750	88930	6	7800	89209	6	7850	89487	5	7900	89763	5	7950	90037	5
7751	936	5	7801	215	6	7851	492	6	7901	768	6	7951	042	6
7752	941	6	7802	221	5	7852	498	6	7902	774	5	7952	048	5
7753	947	6	7803	226	6	7853	504	5	7903	779	6	7953	053	6
7754	953	5	7804	232	5	7854	509	6	7904	785	5	7954	059	5
7755	88958	6	7805	89237	6	7855	89515	5	7905	89790	6	7955	90064	5
7756	964	5	7806	243	5	7856	520	6	7906	796	5	7956	069	6
7757	969	6	7807	248	6	7857	526	5	7907	801	6	7957	075	5
7758	975	6	7808	254	6	7858	531	6	7908	807	5	7958	080	6
7759	981	5	7809	260	5	7859	537	5	7909	812	6	7959	086	5
7760	88986	6	7810	89265	6	7860	89542	6	7910	89818	5	7960	90091	6
7761	992	5	7811	271	5	7861	548	5	7911	823	6	7961	097	5
7762	997	6	7812	276	6	7862	553	6	7912	829	5	7962	102	6
7763	89003	6	7813	282	5	7863	559	5	7913	834	6	7963	108	5
7764	009	5	7814	287	6	7864	564	6	7914	840	5	7964	113	6
7765	89014	6	7815	89293	5	7865	89570	5	7915	89845	6	7965	90119	5
7766	020	5	7816	298	6	7866	575	6	7916	851	5	7966	124	5
7767	025	6	7817	304	6	7867	581	5	7917	856	6	7967	129	6
7768	031	6	7818	310	5	7868	586	6	7918	862	5	7968	135	5
7769	037	5	7819	315	6	7869	592	5	7919	867	6	7969	140	6
7770	89042	6	7820	89321	5	7870	89597	6	7920	89873	5	7970	90146	5
7771	048	5	7821	326	6	7871	603	6	7921	878	5	7971	151	6
7772	053	6	7822	332	5	7872	609	5	7922	883	6	7972	157	5
7773	059	5	7823	337	6	7873	614	6	7923	889	5	7973	162	6
7774	064	6	7824	343	5	7874	620	5	7924	894	6	7974	168	5
7775	89070	6	7825	89348	6	7875	89625	6	7925	89900	5	7975	90173	6
7776	076	5	7826	354	6	7876	631	5	7926	905	6	7976	179	5
7777	081	6	7827	360	5	7877	636	6	7927	911	5	7977	184	5
7778	087	5	7828	365	6	7878	642	5	7928	916	6	7978	189	6
7779	092	6	7829	371	5	7879	647	6	7929	922	6	7979	195	5
7780	89098	6	7830	89376	6	7880	89653	5	7930	89927	6	7980	90200	6
7781	104	5	7831	382	5	7881	658	6	7931	933	5	7981	206	5
7782	109	6	7832	387	6	7882	664	5	7932	938	6	7982	211	6
7783	115	5	7833	393	5	7883	669	6	7933	944	5	7983	217	5
7784	120	6	7834	398	6	7884	675	5	7934	949	6	7984	222	5
7785	89126	5	7835	89404	5	7885	89680	6	7935	89955	5	7985	90227	6
7786	131	6	7836	409	6	7886	686	5	7936	960	6	7986	233	5
7787	137	6	7837	415	6	7887	691	6	7937	966	5	7987	238	6
7788	143	5	7838	421	5	7888	697	5	7938	971	6	7988	244	5
7789	148	6	7839	426	6	7889	702	6	7939	977	5	7989	249	6
7790	89154	5	7840	89432	5	7890	89708	5	7940	89982	6	7990	90255	5
7791	159	6	7841	437	6	7891	713	6	7941	988	5	7991	260	6
7792	165	5	7842	443	5	7892	719	5	7942	993	5	7992	266	5
7793	170	6	7843	448	6	7893	724	6	7943	998	6	7993	271	5
7794	176	6	7844	454	5	7894	730	5	7944	90004	5	7994	276	6
7795	89182	5	7845	89459	6	7895	89735	6	7945	90009	6	7995	90282	5
7796	187	6	7846	465	5	7896	741	5	7946	015	5	7996	287	6
7797	193	5	7847	470	6	7897	746	6	7947	020	6	7997	293	5
7798	198	6	7848	476	5	7898	752	5	7948	026	5	7998	298	6
7799	204	5	7849	481	6	7899	757	6	7949	031	6	7999	304	5
7800	89209		7850	89487		7900	89763		7950	90037		8000	90309	

P.P. | N | Log | d | N | Log | d | N | Log | d | N | Log | d | N | Log | d

N	Log	d	N	Log	d	N	Log	d	N	Log	d	N	Log	d	P.P.
8000	90309	5	8050	90580	5	8100	90849	5	8150	91116	5	8200	91381	6	
8001	314	6	8051	585	5	8101	854	5	8151	121	5	8201	387	5	
8002	320	5	8052	590	6	8102	859	5	8152	126	6	8202	392	5	
8003	325	6	8053	596	5	8103	865	6	8153	132	5	8203	397	6	
8004	331	5	8054	601	6	8104	870	5	8154	137	5	8204	403	5	
8005	90336	6	8055	90607	5	8105	90875	5	8155	91142	6	8205	91408	5	
8006	342	5	8056	612	5	8106	881	6	8156	148	5	8206	413	5	
8007	347	5	8057	617	6	8107	886	5	8157	153	5	8207	418	6	
8008	352	6	8058	623	5	8108	891	5	8158	158	6	8208	424	5	
8009	358	5	8059	628	6	8109	897	6	8159	164	5	8209	429	5	
8010	90363	6	8060	90634	5	8110	90902	5	8160	91169	5	8210	91434	6	
8011	369	5	8061	639	5	8111	907	5	8161	174	6	8211	440	5	
8012	374	6	8062	644	6	8112	913	6	8162	180	5	8212	445	5	
8013	380	5	8063	650	5	8113	918	5	8163	185	5	8213	450	5	
8014	385	5	8064	655	5	8114	924	6	8164	190	6	8214	455	6	
8015	90390	6	8065	90660	6	8115	90929	5	8165	91196	5	8215	91461	5	
8016	396	5	8066	666	5	8116	934	5	8166	201	5	8216	466	5	
8017	401	6	8067	671	6	8117	940	6	8167	206	6	8217	471	6	
8018	407	5	8068	677	5	8118	945	5	8168	212	5	8218	477	5	
8019	412	5	8069	682	5	8119	950	5	8169	217	5	8219	482	5	
8020	90417	6	8070	90687	6	8120	90956	6	8170	91222	6	8220	91487	5	**6**
8021	423	5	8071	693	5	8121	961	5	8171	228	5	8221	492	6	1 0.6
8022	428	6	8072	698	5	8122	966	5	8172	233	5	8222	498	5	2 1.2
8023	434	5	8073	703	6	8123	972	6	8173	238	5	8223	503	5	3 1.8
8024	439	6	8074	709	5	8124	977	5	8174	243	6	8224	508	6	4 2.4
8025	90445	5	8075	90714	6	8125	90982	5	8175	91249	5	8225	91514	5	5 3.0
8026	450	5	8076	720	5	8126	988	5	8176	254	5	8226	519	5	6 3.6
8027	455	6	8077	725	5	8127	993	5	8177	259	6	8227	524	5	7 4.2
8028	461	5	8078	730	6	8128	998	6	8178	265	5	8228	529	6	8 4.8
8029	466	6	8079	736	5	8129	91004	5	8179	270	5	8229	535	5	9 5.4
8030	90472	5	8080	90741	6	8130	91009	5	8180	91275	6	8230	91540	5	
8031	477	5	8081	747	5	8131	014	6	8181	281	5	8231	545	6	
8032	482	6	8082	752	5	8132	020	5	8182	286	5	8232	551	5	
8033	488	5	8083	757	6	8133	025	5	8183	291	6	8233	556	5	
8034	493	6	8084	763	5	8134	030	6	8184	297	5	8234	561	5	
8035	90499	5	8085	90768	5	8135	91036	5	8185	91302	5	8235	91566	6	
8036	504	5	8086	773	6	8136	041	5	8186	307	5	8236	572	5	
8037	509	6	8087	779	5	8137	046	6	8187	312	6	8237	577	5	
8038	515	5	8088	784	5	8138	052	5	8188	318	5	8238	582	5	
8039	520	6	8089	789	6	8139	057	5	8189	323	5	8239	587	6	
8040	90526	5	8090	90795	5	8140	91062	6	8190	91328	6	8240	91593	5	
8041	531	5	8091	800	6	8141	068	5	8191	334	5	8241	598	5	
8042	536	6	8092	806	5	8142	073	5	8192	339	5	8242	603	6	
8043	542	5	8093	811	5	8143	078	6	8193	344	6	8243	609	5	
8044	547	6	8094	816	6	8144	084	5	8194	350	5	8244	614	5	
8045	90553	5	8095	90822	5	8145	91089	5	8195	91355	5	8245	91619	5	
8046	558	5	8096	827	5	8146	094	6	8196	360	5	8246	624	6	
8047	563	6	8097	832	6	8147	100	5	8197	365	6	8247	630	5	
8048	569	5	8098	838	5	8148	105	5	8198	371	5	8248	635	5	
8049	574	6	8099	843	6	8149	110	6	8199	376	5	8249	640	5	
8050	90580		8100	90849		8150	91116		8200	91381		8250	91645		
N	Log	d	N	Log	d	N	Log	d	N	Log	d	N	Log	d	P.P.

P.P.	N	Log	d	N	Log	d	N	Log	d	N	Log	d	N	Log	d
	8250	91645	6	**8300**	91908	5	**8350**	92169	5	**8400**	92428	5	**8450**	92686	5
	8251	651	5	8301	913	5	8351	174	5	8401	433	5	8451	691	5
	8252	656	5	8302	918	6	8352	179	5	8402	438	5	8452	696	5
	8253	661	5	8303	924	5	8353	184	5	8403	443	6	8453	701	5
	8254	666	6	8304	929	5	8354	189	6	8404	449	5	8454	706	5
	8255	91672	5	8305	91934	5	8355	92195	5	8405	92454	5	8455	92711	5
	8256	677	5	8306	939	5	8356	200	5	8406	459	5	8456	716	6
	8257	682	5	8307	944	6	8357	205	5	8407	464	5	8457	722	5
	8258	687	6	8308	950	5	8358	210	5	8408	469	5	8458	727	5
	8259	693	5	8309	955	5	8359	215	6	8409	474	6	8459	732	5
	8260	91698	5	**8310**	91960	5	**8360**	92221	5	**8410**	92480	5	**8460**	92737	5
	8261	703	6	8311	965	6	8361	226	5	8411	485	5	8461	742	5
	8262	709	5	8312	971	5	8362	231	5	8412	490	5	8462	747	5
	8263	714	5	8313	976	5	8363	236	5	8413	495	5	8463	752	6
	8264	719	5	8314	981	5	8364	241	6	8414	500	5	8464	758	5
	8265	91724	6	8315	91986	5	8365	92247	5	8415	92505	6	8465	92763	5
	8266	730	5	8316	991	6	8366	252	5	8416	511	5	8466	768	5
	8267	735	5	8317	997	5	8367	257	5	8417	516	5	8467	773	5
	8268	740	5	8318	92002	5	8368	262	5	8418	521	5	8468	778	5
	8269	745	6	8319	007	5	8369	267	6	8419	526	5	8469	783	5
	8270	91751	5	**8320**	92012	6	**8370**	92273	5	**8420**	92531	5	**8470**	92788	5
5	8271	756	5	8321	018	5	8371	278	5	8421	536	6	8471	793	6
1 0.5	8272	761	5	8322	023	5	8372	283	5	8422	542	5	8472	799	5
2 1.0	8273	766	6	8323	028	6	8373	288	5	8423	547	5	8473	804	5
3 1.5	8274	772	5	8324	033	5	8374	293	5	8424	552	5	8474	809	5
4 2.0	8275	91777	5	8325	92038	6	8375	92298	6	8425	92557	5	8475	92814	5
5 2.5	8276	782	5	8326	044	5	8376	304	5	8426	562	5	8476	819	5
6 3.0	8277	787	6	8327	049	5	8377	309	5	8427	567	5	8477	824	5
7 3.5	8278	793	5	8328	054	5	8378	314	5	8428	572	6	8478	829	6
8 4.0	8279	798	5	8329	059	6	8379	319	5	8429	578	5	8479	834	6
9 4.5	**8280**	91803	5	**8330**	92065	5	**8380**	92324	6	**8430**	92583	5	**8480**	92840	5
	8281	808	6	8331	070	6	8381	330	5	8431	588	5	8481	845	5
	8282	814	5	8332	075	5	8382	335	5	8432	593	5	8482	850	5
	8283	819	5	8333	080	5	8383	340	5	8433	598	5	8483	855	5
	8284	824	5	8334	085	6	8384	345	5	8434	603	6	8484	860	6
	8285	91829	5	8335	92091	5	8385	92350	5	8435	92609	5	8485	92865	5
	8286	834	6	8336	096	5	8386	355	6	8436	614	5	8486	870	5
	8287	840	5	8337	101	5	8387	361	5	8437	619	5	8487	875	6
	8288	845	5	8338	106	5	8388	366	5	8438	624	5	8488	881	5
	8289	850	5	8339	111	6	8389	371	5	8439	629	5	8489	886	5
	8290	91855	6	**8340**	92117	5	**8390**	92376	5	**8440**	92634	5	**8490**	92891	5
	8291	861	5	8341	122	5	8391	381	6	8441	639	6	8491	896	5
	8292	866	5	8342	127	5	8392	387	5	8442	645	5	8492	901	5
	8293	871	5	8343	132	5	8393	392	5	8443	650	5	8493	906	5
	8294	876	6	8344	137	6	8394	397	5	8444	655	5	8494	911	5
	8295	91882	5	8345	92143	5	8395	92402	5	8445	92660	5	8495	92916	5
	8296	887	5	8346	148	5	8396	407	5	8446	665	5	8496	921	6
	8297	892	5	8347	153	5	8397	412	6	8447	670	6	8497	927	5
	8298	897	6	8348	158	5	8398	418	5	8448	675	5	8498	932	5
	8299	903	5	8349	163	6	8399	423	5	8449	681	5	8499	937	5
	8300	91908		**8350**	92169		**8400**	92428		**8450**	92686		**8500**	92942	
P. P.	N	Log	d	N	Log	d	N	Log	d	N	Log	d	N	Log	d

N	Log	d	N	Log	d	N	Log	d	N	Log	d	N	Log	d	P.P.
8500	92942	5	**8550**	93197	5	**8600**	93450	5	**8650**	93702	5	**8700**	93952	5	
8501	947	5	8551	202	5	8601	455	5	8651	707	5	8701	957	5	
8502	952	5	8552	207	5	8602	460	5	8652	712	5	8702	962	5	
8503	957	5	8553	212	5	8603	465	5	8653	717	5	8703	967	5	
8504	962	5	8554	217	5	8604	470	5	8654	722	5	8704	972	5	
8505	92967	6	8555	93222	5	8605	93475	5	8655	93727	5	8705	93977	5	
8506	973	5	8556	227	5	8606	480	5	8656	732	5	8706	982	5	
8507	978	5	8557	232	5	8607	485	5	8657	737	5	8707	987	5	
8508	983	5	8558	237	5	8608	490	5	8658	742	5	8708	992	5	
8509	988	5	8559	242	5	8609	495	5	8659	747	5	8709	997	5	
8510	92993	5	**8560**	93247	5	**8610**	93500	5	**8660**	93752	5	**8710**	94002	5	
8511	998	5	8561	252	6	8611	505	5	8661	757	5	8711	007	5	
8512	93003	5	8562	258	5	8612	510	5	8662	762	5	8712	012	5	
8513	008	5	8563	263	5	8613	515	5	8663	767	5	8713	017	5	
8514	013	5	8564	268	5	8614	520	6	8664	772	5	8714	022	5	
8515	93018	6	8565	93273	5	8615	93526	5	8665	93777	5	8715	94027	5	
8516	024	5	8566	278	5	8616	531	5	8666	782	5	8716	032	5	
8517	029	5	8567	283	5	8617	536	5	8667	787	5	8717	037	5	
8518	034	5	8568	288	5	8618	541	5	8668	792	5	8718	042	5	
8519	039	5	8569	293	5	8619	546	5	8669	797	5	8719	047	5	
8520	93044	5	**8570**	93298	5	**8620**	93551	5	**8670**	93802	5	**8720**	94052	5	
8521	049	5	8571	303	5	8621	556	5	8671	807	5	8721	057	5	
8522	054	5	8572	308	5	8622	561	5	8672	812	5	8722	062	5	
8523	059	5	8573	313	5	8623	566	5	8673	817	5	8723	067	5	
8524	064	5	8574	318	5	8624	571	5	8674	822	5	8724	072	5	
8525	93069	6	8575	93323	5	8625	93576	5	8675	93827	5	8725	94077	5	
8526	075	5	8576	328	6	8626	581	5	8676	832	5	8726	082	4	
8527	080	5	8577	334	5	8627	586	5	8677	837	5	8727	086	5	
8528	085	5	8578	339	5	8628	591	5	8678	842	5	8728	091	5	
8529	090	5	8579	344	5	8629	596	5	8679	847	5	8729	096	5	
8530	93095	5	**8580**	93349	5	**8630**	93601	5	**8680**	93852	5	**8730**	94101	5	
8531	100	5	8581	354	5	8631	606	5	8681	857	5	8731	106	5	
8532	105	5	8582	359	5	8632	611	5	8682	862	5	8732	111	5	
8533	110	5	8583	364	5	8633	616	5	8683	867	5	8733	116	5	
8534	115	5	8584	369	5	8634	621	5	8684	872	5	8734	121	5	
8535	93120	5	8585	93374	5	8635	93626	5	8685	93877	5	8735	94126	5	
8536	125	6	8586	379	5	8636	631	5	8686	882	5	8736	131	5	
8537	131	5	8587	384	5	8637	636	5	8687	887	5	8737	136	5	
8538	136	5	8588	389	5	8638	641	5	8688	892	5	8738	141	5	
8539	141	5	8589	394	5	8639	646	5	8689	897	5	8739	146	5	
8540	93146	5	**8590**	93399	5	**8640**	93651	5	**8690**	93902	5	**8740**	94151	5	
8541	151	5	8591	404	5	8641	656	5	8691	907	5	8741	156	5	
8542	156	5	8592	409	5	8642	661	5	8692	912	5	8742	161	5	
8543	161	5	8593	414	6	8643	666	5	8693	917	5	8743	166	5	
8544	166	5	8594	420	5	8644	671	5	8694	922	5	8744	171	5	
8545	93171	5	8595	93425	5	8645	93676	6	8695	93927	5	8745	94176	5	
8546	176	5	8596	430	5	8646	682	5	8696	932	5	8746	181	5	
8547	181	5	8597	435	5	8647	687	5	8697	937	5	8747	186	5	
8548	186	6	8598	440	5	8648	692	5	8698	942	5	8748	191	5	
8549	192	5	8599	445	5	8649	697	5	8699	947	5	8749	196	5	
8550	93197		**8600**	93450		**8650**	93702		**8700**	93952		**8750**	94201		
N	Log	d	N	Log	d	N	Log	d	N	Log	d	N	Log	d	P.P.

P. P.

	6
1	0.6
2	1.2
3	1.8
4	2.4
5	3.0
6	3.6
7	4.2
8	4.8
9	5.4

	5
1	0.5
2	1.0
3	1.5
4	2.0
5	2.5
6	3.0
7	3.5
8	4.0
9	4.5

P. P.	N	Log	d	N	Log	d	N	Log	d	N	Log	d	N	Log	d
	8750	94201	5	8800	94448	5	8850	94694	5	8900	94939	5	8950	95182	5
	8751	206	5	8801	453	5	8851	699	5	8901	944	5	8951	187	5
	8752	211	5	8802	458	5	8852	704	5	8902	949	5	8952	192	5
	8753	216	5	8803	463	5	8853	709	5	8903	954	5	8953	197	5
	8754	221	5	8804	468	5	8854	714	5	8904	959	4	8954	202	5
	8755	94226	5	8805	94473	5	8855	94719	5	8905	94963	5	8955	95207	4
	8756	231	5	8806	478	5	8856	724	5	8906	968	5	8956	211	5
	8757	236	4	8807	483	5	8857	729	5	8907	973	5	8957	216	5
	8758	240	5	8808	488	5	8858	734	4	8908	978	5	8958	221	5
	8759	245	5	8809	493	5	8859	738	5	8909	983	5	8959	226	5
	8760	94250	5	8810	94498	5	8860	94743	5	8910	94988	5	8960	95231	5
	8761	255	5	8811	503	4	8861	748	5	8911	993	5	8961	236	4
	8762	260	5	8812	507	5	8862	753	5	8912	998	5	8962	240	5
	8763	265	5	8813	512	5	8863	758	5	8913	95002	5	8963	245	5
	8764	270	5	8814	517	5	8864	763	5	8914	007	5	8964	250	5
	8765	94275	5	8815	94522	5	8865	94768	5	8915	95012	5	8965	95255	5
	8766	280	5	8816	527	5	8866	773	5	8916	017	5	8966	260	5
	8767	285	5	8817	532	5	8867	778	5	8917	022	5	8967	265	5
	8768	290	5	8818	537	5	8868	783	4	8918	027	5	8968	270	4
	8769	295	5	8819	542	5	8869	787	5	8919	032	4	8969	274	5
	8770	94300	5	8820	94547	5	8870	94792	5	8920	95036	5	8970	95279	5
	8771	305	5	8821	552	5	8871	797	5	8921	041	5	8971	284	5
4	8772	310	5	8822	557	5	8872	802	5	8922	046	5	8972	289	5
1 0.4	8773	315	5	8823	562	5	8873	807	5	8923	051	5	8973	294	5
2 0.8	8774	320	5	8824	567	4	8874	812	5	8924	056	5	8974	299	4
3 1.2	8775	94325	5	8825	94571	5	8875	94817	5	8925	95061	5	8975	95303	5
4 1.6	8776	330	5	8826	576	5	8876	822	5	8926	066	5	8976	308	5
5 2.0	8777	335	5	8827	581	5	8877	827	5	8927	071	4	8977	313	5
6 2.4	8778	340	5	8828	586	5	8878	832	4	8928	075	5	8978	318	5
7 2.8	8779	345	4	8829	591	5	8879	836	5	8929	080	5	8979	323	5
8 3.2	8780	94349	5	8830	94596	5	8880	94841	5	8930	95085	5	8980	95328	4
9 3.6	8781	354	5	8831	601	5	8881	846	5	8931	090	5	8981	332	5
	8782	359	5	8832	606	5	8882	851	5	8932	095	5	8982	337	5
	8783	364	5	8833	611	5	8883	856	5	8933	100	5	8983	342	5
	8784	369	5	8834	616	5	8884	861	5	8934	105	4	8984	347	5
	8785	94374	5	8835	94621	5	8885	94866	5	8935	95109	5	8985	95352	5
	8786	379	5	8836	626	4	8886	871	5	8936	114	5	8986	357	4
	8787	384	5	8837	630	5	8887	876	4	8937	119	5	8987	361	5
	8788	389	5	8838	635	5	8888	880	5	8938	124	5	8988	366	5
	8789	394	5	8839	640	5	8889	885	5	8939	129	5	8989	371	5
	8790	94399	5	8840	94645	5	8890	94890	5	8940	95134	5	8990	95376	5
	8791	404	5	8841	650	5	8891	895	5	8941	139	4	8991	381	5
	8792	409	5	8842	655	5	8892	900	5	8942	143	5	8992	386	4
	8793	414	5	8843	660	5	8893	905	5	8943	148	5	8993	390	5
	8794	419	5	8844	665	5	8894	910	5	8944	153	5	8994	395	5
	8795	94424	5	8845	94670	5	8895	94915	4	8945	95158	5	8995	95400	5
	8796	429	4	8846	675	5	8896	919	5	8946	163	5	8996	405	5
	8797	433	5	8847	680	5	8897	924	5	8947	168	5	8997	410	5
	8798	438	5	8848	685	4	8898	929	5	8948	173	4	8998	415	4
	8799	443	5	8849	689	5	8899	934	5	8949	177	5	8999	419	5
	8800	94448		8850	94694		8900	94939		8950	95182		9000	95424	
P. P.	N	Log	d	N	Log	d	N	Log	d	N	Log	d	N	Log	d

N	Log	d	N	Log	d	N	Log	d	N	Log	d	N	Log	d	P. P.
9000	95424	5	9050	95665	5	9100	95904	5	9150	96142	5	9200	96379	5	
9001	429	5	9051	670	4	9101	909	5	9151	147	5	9201	384	4	
9002	434	5	9052	674	5	9102	914	4	9152	152	4	9202	388	5	
9003	439	5	9053	679	5	9103	918	5	9153	156	5	9203	393	5	
9004	444	4	9054	684	5	9104	923	5	9154	161	5	9204	398	4	
9005	95448	5	9055	95689	5	9105*	95928	5	9155	96166	5	9205	96402	5	
9006	453	5	9056	694	4	9106	933	5	9156	171	4	9206	407	5	
9007	458	5	9057	698	5	9107	938	4	9157	175	5	9207	412	5	
9008	463	5	9058	703	5	9108	942	5	9158	180	5	9208	417	4	
9009	468	4	9059	708	5	9109	947	5	9159	185	5	9209	421	5	
9010	95472	5	9060	95713	5	9110	95952	5	9160	96190	4	9210	96426	5	
9011	477	5	9061	718	4	9111	957	4	9161	194	5	9211	431	4	
9012	482	5	9062	722	5	9112	961	5	9162	199	5	9212	435	5	
9013	487	5	9063	727	5	9113	966	5	9163	204	5	9213	440	5	
9014	492	5	9064	732	5	9114	971	5	9164	209	4	9214	445	5	
9015	95497	4	9065	95737	5	9115	95976	4	9165	96213	5	9215	96450	4	
9016	501	5	9066	742	4	9116	980	5	9166	218	5	9216	454	5	
9017	506	5	9067	746	5	9117	985	5	9167	223	4	9217	459	5	
9018	511	5	9068	751	5	9118	990	5	9168	227	5	9218	464	4	
9019	516	5	9069	756	5	9119	995	4	9169	232	5	9219	468	5	
9020	95521	4	9070	95761	5	9120	95999	5	9170	96237	5	9220	96473	5	**5**
9021	525	5	9071	766	4	9121	96004	5	9171	242	4	9221	478	5	1 | 0.5
9022	530	5	9072	770	5	9122	009	5	9172	246	5	9222	483	4	2 | 1.0
9023	535	5	9073	775	5	9123	014	5	9173	251	5	9223	487	5	3 | 1.5
9024	540	5	9074	780	5	9124	019	4	9174	256	5	9224	492	5	4 | 2.0
															5 | 2.5
9025	95545	5	9075	95785	4	9125	96023	5	9175	96261	4	9225	96497	4	6 | 3.0
9026	550	4	9076	789	5	9126	028	5	9176	265	5	9226	501	5	7 | 3.5
9027	554	5	9077	794	5	9127	033	5	9177	270	5	9227	506	5	8 | 4.0
9028	559	5	9078	799	5	9128	038	4	9178	275	5	9228	511	4	9 | 4.5
9029	564	5	9079	804	5	9129	042	5	9179	280	4	9229	515	5	
9030	95569	5	9080	95809	4	9130	96047	5	9180	96284	5	9230	96520	5	
9031	574	4	9081	813	5	9131	052	5	9181	289	5	9231	525	5	
9032	578	5	9082	818	5	9132	057	4	9182	294	4	9232	530	4	
9033	583	5	9083	823	5	9133	061	5	9183	298	5	9233	534	5	
9034	588	5	9084	828	4	9134	066	5	9184	303	5	9234	539	5	
9035	95593	5	9085	95832	5	9135	96071	5	9185	96308	5	9235	96544	4	
9036	598	4	9086	837	5	9136	076	4	9186	313	4	9236	548	5	
9037	602	5	9087	842	5	9137	080	5	9187	317	5	9237	553	5	
9038	607	5	9088	847	5	9138	085	5	9188	322	5	9238	558	4	
9039	612	5	9089	852	4	9139	090	5	9189	327	5	9239	562	5	
9040	95617	5	9090	95856	5	9140	96095	4	9190	96332	4	9240	96567	5	
9041	622	4	9091	861	5	9141	099	5	9191	336	5	9241	572	5	
9042	626	5	9092	866	5	9142	104	5	9192	341	5	9242	577	4	
9043	631	5	9093	871	4	9143	109	5	9193	346	4	9243	581	5	
9044	636	5	9094	875	5	9144	114	4	9194	350	5	9244	586	5	
9045	95641	5	9095	95880	5	9145	96118	5	9195	96355	5	9245	96591	4	
9046	646	4	9096	885	5	9146	123	5	9196	360	5	9246	595	5	
9047	650	5	9097	890	5	9147	128	5	9197	365	4	9247	600	5	
9048	655	5	9098	895	4	9148	133	4	9198	369	5	9248	605	4	
9049	660	5	9099	899	5	9149	137	5	9199	374	5	9249	609	5	
9050	95665		9100	95904		9150	96142		9200	96379		9250	96614		
N	Log	d	N	Log	d	N	Log	d	N	Log	d	N	Log	d	P. P.

P. P.	N	Log	d	N	Log	d	N	Log	d	N	Log	d	N	Log	d
	9250	96614	5	**9300**	96848	5	**9350**	97081	5	**9400**	97313	4	**9450**	97543	5
	9251	619	5	9301	853	5	9351	086	4	9401	317	5	9451	548	4
	9252	624	4	9302	858	4	9352	090	5	9402	322	5	9452	552	5
	9253	628	5	9303	862	5	9353	095	5	9403	327	4	9453	557	5
	9254	633	5	9304	867	5	9354	100	4	9404	331		9454	562	4
	9255	96638	4	9305	96872	4	9355	97104	5	9405	97336	4	9455	97566	5
	9256	642	5	9306	876	5	9356	109	5	9406	340	5	9456	571	4
	9257	647	5	9307	881	5	9357	114	4	9407	345	5	9457	575	5
	9258	652	4	9308	886	4	9358	118	5	9408	350	4	9458	580	5
	9259	656	5	9309	890	5	9359	123	5	9409	354	5	9459	585	4
	9260	96661	5	**9310**	96895	5	**9360**	97128	4	**9410**	97359	5	**9460**	97589	5
	9261	666	4	9311	900	4	9361	132	5	9411	364	4	9461	594	4
	9262	670	5	9312	904	5	9362	137	5	9412	368	5	9462	598	5
	9263	675	5	9313	909	5	9363	142	4	9413	373	4	9463	603	4
	9264	680	5	9314	914	4	9364	146	5	9414	377	5	9464	607	5
	9265	96685	4	9315	96918	5	9365	97151	4	9415	97382	5	9465	97612	5
	9266	689	5	9316	923	5	9366	155	5	9416	387	4	9466	617	4
	9267	694	5	9317	928	4	9367	160	5	9417	391	5	9467	621	5
	9268	699	4	9318	932	5	9368	165	4	9418	396	4	9468	626	4
	9269	703	5	9319	937	5	9369	169	5	9419	400	5	9469	630	5
4	**9270**	96708	5	**9320**	96942	4	**9370**	97174	5	**9420**	97405	5	**9470**	97635	5
1 0.4	9271	713	4	9321	946	5	9371	179	4	9421	410	4	9471	640	4
2 0.8	9272	717	5	9322	951	5	9372	183	5	9422	414	5	9472	644	5
3 1.2	9273	722	5	9323	956	4	9373	188	4	9423	419	5	9473	649	4
4 1.6	9274	727	4	9324	960	5	9374	192	5	9424	424	4	9474	653	5
5 2.0	9275	96731	5	9325	96965	5	9375	97197	5	9425	97428	5	9475	97658	5
6 2.4	9276	736	5	9326	970	4	9376	202	4	9426	433	4	9476	663	4
7 2.8	9277	741	4	9327	974	5	9377	206	5	9427	437	5	9477	667	5
8 3.2	9278	745	5	9328	979	5	9378	211	5	9428	442	5	9478	672	4
9 3.6	9279	750	5	9329	984	4	9379	216	4	9429	447	4	9479	676	5
	9280	96755	4	**9330**	96988	5	**9380**	97220	5	**9430**	97451	5	**9480**	97681	4
	9281	759	5	9331	993	4	9381	225	5	9431	456	4	9481	685	5
	9282	764	5	9332	997	5	9382	230	4	9432	460	5	9482	690	5
	9283	769	5	9333	97002	5	9383	234	5	9433	465	5	9483	695	4
	9284	774	4	9334	007	4	9384	239	4	9434	470	4	9484	699	5
	9285	96778	5	9335	97011	5	9385	97243	5	9435	97474	5	9485	97704	4
	9286	783	5	9336	016	5	9386	248	5	9436	479	4	9486	708	5
	9287	788	4	9337	021	4	9387	253	4	9437	483	5	9487	713	4
	9288	792	5	9338	025	5	9388	257	5	9438	488	5	9488	717	5
	9289	797	5	9339	030	5	9389	262	5	9439	493	4	9489	722	5
	9290	96802	4	**9340**	97035	4	**9390**	97267	4	**9440**	97497	5	**9490**	97727	4
	9291	806	5	9341	039	5	9391	271	5	9441	502	4	9491	731	5
	9292	811	5	9342	044	5	9392	276	4	9442	506	5	9492	736	4
	9293	816	4	9343	049	4	9393	280	5	9443	511	5	9493	740	5
	9294	820	5	9344	053	5	9394	285	5	9444	516	4	9494	745	4
	9295	96825	5	9345	97058	5	9395	97290	4	9445	97520	5	9495	97749	5
	9296	830	4	9346	063	4	9396	294	5	9446	525	4	9496	754	5
	9297	834	5	9347	067	5	9397	299	5	9447	529	5	9497	759	4
	9298	839	5	9348	072	5	9398	304	4	9448	534	5	9498	763	5
	9299	844	4	9349	077	4	9399	308	5	9449	539	4	9499	768	4
	9300	96848		**9350**	97081		**9400**	97313		**9450**	97543		**9500**	97772	
P. P.	N	Log	d	N	Log	d	N	Log	d	N	Log	d	N	Log	d

N	Log	d	N	Log	d	N	Log	d	N	Log	d	N	Log	d
9500	97772	5	**9550**	98000	5	**9600**	98227	5	**9650**	98453	4	**9700**	98677	5
9501	777	5	9551	005	5	9601	232	4	9651	457	5	9701	682	4
9502	782	4	9552	009	4	9602	236	5	9652	462	4	9702	686	5
9503	786	5	9553	014	5	9603	241	4	9653	466	5	9703	691	4
9504	791	4	9554	019	4	9604	245	5	9654	471	4	9704	695	5
9505	97795	5	9555	98023	5	9605	98250	4	9655	98475	5	9705	98700	4
9506	800	4	9556	028	4	9606	254	5	9656	480	4	9706	704	5
9507	804	5	9557	032	5	9607	259	4	9657	484	5	9707	709	4
9508	809	4	9558	037	4	9608	263	5	9658	489	4	9708	713	4
9509	813	5	9559	041	5	9609	268	4	9659	493	5	9709	717	5
9510	97818	5	**9560**	98046	4	**9610**	98272	5	**9660**	98498	4	**9710**	98722	4
9511	823	4	9561	050	5	9611	277	4	9661	502	5	9711	726	5
9512	827	5	9562	055	4	9612	281	5	9662	507	4	9712	731	4
9513	832	4	8563	059	5	9613	286	4	9663	511	5	9713	735	5
9514	836	5	9564	064	4	9614	290	5	9664	516	4	9714	740	4
9515	97841	4	9565	98068	5	9615	98295	4	9665	98520	5	9715	98744	5
9516	845	5	9566	073	5	9616	299	5	9666	525	4	9716	749	4
9517	850	5	9567	078	4	9617	304	4	9667	529	5	9717	753	5
9518	855	4	9568	082	5	9618	308	5	9668	534	4	9718	758	4
9519	859	5	9569	087	4	9619	313	5	9669	538	5	9719	762	5
9520	97864	4	**9570**	98091	5	**9620**	98318	4	**9670**	98543	4	**9720**	98767	4
9521	868	5	9571	096	4	9621	322	5	9671	547	5	9721	771	5
9522	873	4	9572	100	5	9622	327	4	9672	552	4	9722	776	4
9523	877	4	9573	105	4	9623	331	5	9673	556	5	9723	780	4
9524	882	5	9574	109	5	9624	336	4	9674	561	4	9724	784	5
9525	97886	5	9575	98114	4	9625	98340	5	9675	98565	5	9725	98789	4
9526	891	5	9576	118	5	9626	345	4	9676	570	4	9726	793	5
9527	896	4	9577	123	4	9627	349	5	9677	574	5	9727	798	4
9528	900	5	9578	127	5	9628	354	4	9678	579	4	9728	802	5
9529	905	4	9579	132	5	9629	358	5	9679	583	5	9729	807	4
9530	97909	5	**9580**	98137	4	**9630**	98363	4	**9680**	98588	4	**9730**	98811	5
9531	914	4	9581	141	5	9631	367	5	9681	592	5	9731	816	4
9532	918	5	9582	146	4	9632	372	4	9682	597	4	9732	820	5
9533	923	5	9583	150	5	9633	376	5	9683	601	4	9733	825	4
9534	928	4	9584	155	4	9634	381	4	9684	605	5	9734	829	5
9535	97932	5	9585	98159	5	9635	98385	5	9685	98610	4	9735	98834	4
9536	937	4	9586	164	4	9636	390	4	9686	614	5	9736	838	5
9537	941	5	9587	168	5	9637	394	5	9687	619	4	9737	843	4
9538	946	4	9588	173	4	9638	399	4	9688	623	5	9738	847	4
9539	950	5	9589	177	5	9639	403	5	9689	628	4	9739	851	5
9540	97955	4	**9590**	98182	4	**9640**	98408	4	**9690**	98632	5	**9740**	98856	4
9541	959	5	9591	186	5	9641	412	5	9691	637	4	9741	860	5
9542	964	4	9592	191	4	9642	417	4	9692	641	5	9742	865	4
9543	968	5	9593	195	5	9643	421	5	9693	646	4	9743	869	5
9544	973	5	9594	200	4	9644	426	4	9694	650	5	9744	874	4
9545	97978	4	9595	98204	5	9645	98430	5	9695	98655	4	9745	98878	5
9546	982	5	9596	209	5	9646	435	4	9696	659	5	9746	883	4
9547	987	4	9597	214	4	9647	439	5	9697	664	4	9747	887	5
9548	991	5	9598	218	5	9648	444	4	9698	668	5	9748	892	4
9549	996	4	9599	223	4	9649	448	5	9699	673	4	9749	896	4
9550	98000		**9600**	98227		**9650**	98453		**9700**	98677		**9750**	98900	

P. P.

	5
1	0.5
2	1.0
3	1.5
4	2.0
5	2.5
6	3.0
7	3.5
8	4.0
9	4.5

P. P. table (left margin):

	4
1	0.4
2	0.8
3	1.2
4	1.6
5	2.0
6	2.4
7	2.8
8	3.2
9	3.6

N	Log	d	N	Log	d	N	Log	d	N	Log	d	N	Log	d
9750	98900	5	**9800**	99123	4	**9850**	99344	4	**9900**	99564	4	**9950**	99782	5
9751	905	4	9801	127	4	9851	348	4	9901	568	4	9951	787	4
9752	909	5	9802	131	5	9852	352	5	9902	572	5	9952	791	4
9753	914	4	9803	136	4	9853	357	4	9903	577	4	9953	795	5
9754	918	5	9804	140	5	9854	361	5	9904	581	4	9954	800	4
9755	98923	4	9805	99145	4	9855	99366	4	9905	99585	5	9955	99804	4
9756	927	5	9806	149	5	9856	370	4	9906	590	4	9956	808	5
9757	932	4	9807	154	4	9857	374	5	9907	594	5	9957	813	4
9758	936	5	9808	158	4	9858	379	4	9908	599	4	9958	817	5
9759	941	4	9809	162	5	9859	383	5	9909	603	4	9959	822	4
9760	98945	4	**9810**	99167	4	**9860**	99388	4	**9910**	99607	5	**9960**	99826	4
9761	949	5	9811	171	5	9861	392	4	9911	612	4	9961	830	5
9762	954	4	9812	176	4	9862	396	5	9912	616	5	9962	835	4
9763	958	5	9813	180	5	9863	401	4	9913	621	4	9963	839	4
9764	963	4	9814	185	4	9864	405	5	9914	625	4	9964	843	5
9765	98967	5	9815	99189	4	9865	99410	4	9915	99629	5	9965	99848	4
9766	972	4	9816	193	5	9866	414	5	9916	634	4	9966	852	4
9767	976	5	9817	198	4	9867	419	4	9917	638	4	9967	856	5
9768	981	4	9818	202	5	9868	423	4	9918	642	5	9968	861	4
9769	985	4	9819	207	4	9869	427	5	9919	647	4	9969	865	5
9770	98989	5	**9820**	99211	5	**9870**	99432	4	**9920**	99651	5	**9970**	99870	4
9771	994	4	9821	216	4	9871	436	5	9921	656	4	9971	874	4
9772	998	4	9822	220	4	9872	441	4	9922	660	4	9972	878	5
9773	99003	5	9823	224	5	9873	445	4	9923	664	5	9973	883	4
9774	007	4	9824	229	4	9874	449	5	9924	669	4	9974	887	4
9775	99012	4	9825	99233	5	9875	99454	4	9925	99673	4	9975	99891	5
9776	016	5	9826	238	4	9876	458	5	9926	677	5	9976	896	4
9777	021	4	9827	242	5	9877	463	4	9927	682	4	9977	900	4
9778	025	4	9828	247	4	9878	467	4	9928	686	5	9978	904	5
9779	029	5	9829	251	4	9879	471	5	9929	691	4	9979	909	4
9780	99034	4	**9830**	99255	5	**9880**	99476	4	**9930**	99695	4	**9980**	99913	4
9781	038	5	9831	260	4	9881	480	4	9931	699	5	9981	917	5
9782	043	4	9832	264	5	9882	484	5	9932	704	4	9982	922	4
9783	047	5	9833	269	4	9883	489	4	9933	708	4	9983	926	4
9784	052	4	9834	273	4	9884	493	5	9934	712	5	9984	930	5
9785	99056	5	9835	99277	5	9885	99498	4	9935	99717	4	9985	99935	4
9786	061	4	9836	282	4	9886	502	4	9936	721	5	9986	939	5
9787	065	4	9837	286	5	9887	506	5	9937	726	4	9987	944	4
9788	069	5	9838	291	4	9888	511	4	9938	730	4	9988	948	4
9789	074	4	9839	295	5	9889	515	5	9939	734	5	9989	952	5
9790	99078	5	**9840**	99300	4	**9890**	99520	4	**9940**	99739	4	**9990**	99957	4
9791	083	4	9841	304	4	9891	524	4	9941	743	4	9991	961	4
9792	087	5	9842	308	5	9892	528	5	9942	747	5	9992	965	5
9793	092	4	9843	313	4	9893	533	4	9943	752	4	9993	970	4
9794	096	4	9844	317	5	9894	537	5	9944	756	4	9994	974	4
9795	99100	5	9845	99322	4	9895	99542	4	9945	99760	5	9995	99978	5
9796	105	4	9846	326	4	9896	546	4	9946	765	4	9996	983	4
9797	109	5	9847	330	5	9897	550	5	9947	769	5	9997	987	4
9798	114	4	9848	335	4	9898	555	4	9948	774	4	9998	991	5
9799	118	5	9849	339	5	9899	559	4	9949	778	4	9999	996	4
9800	99123		**9850**	99344		**9900**	99564		**9950**	99782		**10000**	00000	

TABLE II. IMPORTANT CONSTANTS

Number	Log	Number	Log
$\pi = 3.14159265$	0.4971499		
$2\pi = 6.28318531$	0.7981799	$\pi^2 = 9.86960440$	0.9942997
$4\pi = 12.56637061$	1.0992099	$\dfrac{1}{\pi^2} = 0.10132118$	9.0057003–10
$\dfrac{\pi}{2} = 1.57079633$	0.1961199	$\sqrt{\pi} = 1.77245385$	0.2485749
$\dfrac{\pi}{3} = 1.04719755$	0.0200286	$\dfrac{1}{\sqrt{\pi}} = 0.56418958$	9.7514251–10
$\dfrac{4\pi}{3} = 4.18879020$	0.6220886	$\sqrt{\dfrac{3}{\pi}} = 0.97720502$	9.9899857–10
$\dfrac{\pi}{4} = 0.78539816$	9.8950899–10	$\sqrt{\dfrac{4}{\pi}} = 1.12837917$	0.0524551
$\dfrac{\pi}{6} = 0.52359878$	9.7189986–10	$\sqrt[3]{\pi} = 1.46459189$	0.1657166
$\dfrac{1}{\pi} = 0.31830989$	9.5028501–10	$\dfrac{1}{\sqrt[3]{\pi}} = 0.68278406$	9.8342834–10
$\dfrac{1}{2\pi} = 0.15915494$	9.2018201–10	$\sqrt[3]{\pi^2} = 2.14502940$	0.3314332
$\dfrac{3}{\pi} = 0.95492966$	9.9799714–10	$\sqrt[3]{\dfrac{3}{4\pi}} = 0.62035049$	9.7926371–10
$\dfrac{4}{\pi} = 1.27323954$	0.1049101	$\sqrt[3]{\dfrac{\pi}{6}} = 0.80599598$	9.9063329–10

		Log
$e =$ Naperian Base.................... $=$	2.71828183	0.43429448
$M = \log_{10} e$.......................... $=$	0.43429448	9.63778431–10
$1 \div M = \log_e 10$...................... $=$	2.30258509	0.36221569
$180 \div \pi =$ degrees in 1 radian......... $=$	57.2957795	1.75812263
$\pi \div 180 =$ radians in 1°.............. $=$	0.01745329	8.24187737–10
$\pi \div 10800 =$ radians in 1′............. $=$	0.0002908882	6.46372612–10
$\pi \div 648000 =$ radians in 1″............ $=$	0.000004848136811095	4.68557487–10
sin 1″............................... $=$	0.000004848136811076	4.68557487–10
tan 1″............................... $=$	0.000004848136811133	4.68557487–10
centimeters in 1 ft..................... $=$	30.480	1.4840150
feet in 1 cm......................... $=$	0.032808	8.5159850–10
inches in 1 m........................ $=$	39.37	1.5951654
pounds in 1 kg....................... $=$	2.20462	0.3433340
kilograms in 1 lb..................... $=$	0.453593	9.6566660–10

$$\pi = 3.14159 \quad 26535 \quad 89793 \quad 23846 \quad 26433 \quad 83280$$
$$e = 2.71828 \quad 18284 \quad 59045 \quad 23536 \quad 02874 \quad 71353$$
$$M = 0.43429 \quad 44819 \quad 03251 \quad 82765 \quad 11289 \quad 18917$$
$$1 \div M = 2.30258 \quad 50929 \quad 94045 \quad 68401 \quad 79914 \quad 54684$$
$$\log_{10} \pi = 0.49714 \quad 98726 \quad 94133 \quad 85435 \quad 12682 \quad 88291$$
$$\log_e \pi = 1.14472 \quad 98858 \quad 49400 \quad 17414 \quad 34273 \quad 51353$$

TABLE III. LOGARITHMIC TRIGONOMETRIC FUNCTIONS

	0°						
'	**L Sin**	**d**	**L Tan**	**c d**	**L Cot**	**L Cos**	
0	————		————		————	0.00 000	60
1	6.46 373		6.46 373		3.53 627	000	59
2	.76 476	30103	.76 476	30103	.23 524	000	58
3	.94 085	17609	.94 085	17609	.05 915	000	57
4	7.06 579	12494	7.06 579	12494	2.93 421	000	56
		9691		9691			
5	7.16 270		7.16 270		2.83 730	0.00 000	55
6	.24 188	7918	.24 188	7918	.75 812	000	54
7	.30 882	6694	.30 882	6694	.69 118	000	53
8	.36 682	5800	.36 682	5800	.63 318	000	52
9	.41 797	5115	.41 797	5115	.58 203	000	51
		4576		4576			
10	7.46 373		7.46 373		2.53 627	0.00 000	50
11	.50 512	4139	.50 512	4139	.49 488	000	49
12	.54 291	3779	.54 291	3779	.45 709	000	48
13	.57 767	3476	.57 767	3476	.42 233	000	47
14	.60 985	3218	.60 986	3219	.39 014	000	46
		2997		2996			
15	7.63 982		7.63 982		2.36 018	0.00 000	45
16	.66 784	2802	.66 785	2803	.33 215	000	44
17	.69 417	2633	.69 418	2633	.30 582	9.99 999	43
18	.71 900	2483	.71 900	2482	.28 100	999	42
19	.74 248	2348	.74 248	2348	.25 752	999	41
		2227		2228			
20	7.76 475		7.76 476		2.23 524	9.99 999	40
21	.78 594	2119	.78 595	2119	.21 405	999	39
22	.80 615	2021	.80 615	2020	.19 385	999	38
23	.82 545	1930	.82 546	1931	.17 454	999	37
24	.84 393	1848	.84 394	1848	.15 606	999	36
		1773		1773			
25	7.86 166		7.86 167		2.13 833	9.99 999	35
26	.87 870	1704	.87 871	1704	.12 129	999	34
27	.89 509	1639	.89 510	1639	.10 490	999	33
28	.91 088	1579	.91 089	1579	.08 911	999	32
29	.92 612	1524	.92 613	1524	.07 387	998	31
		1472		1473			
30	7.94 084		7.94 086		2.05 914	9.99 998	30
31	.95 508	1424	.95 510	1424	.04 490	998	29
32	.96 887	1379	.96 889	1379	.03 111	998	28
33	.98 223	1336	.98 225	1336	.01 775	998	27
34	.99 520	1297	.99 522	1297	.00 478	998	26
		1259		1259			
35	8.00 779		8.00 781		1.99 219	9.99 998	25
36	.02 002	1223	.02 004	1223	.97 996	998	24
37	.03 192	1190	.03 194	1190	.96 806	997	23
38	.04 350	1158	.04 353	1159	.95 647	997	22
39	.05 478	1128	.05 481	1128	.94 519	997	21
		1100		1100			
40	8.06 578		8.06 581		1.93 419	9.99 997	20
41	.07 650	1072	.07 653	1072	.92347	997	19
42	.08 696	1046	.08 700	1047	.91 300	997	18
43	.09 718	1022	.09 722	1022	.90 278	997	17
44	.10 717	999	.10 720	998	.89 280	996	16
		976		976			
45	8.11 693		8.11 696		1.88 304	9.99 996	15
46	.12 647	954	.12 651	955	.87 349	996	14
47	.13 581	934	.13 585	934	.86 415	996	13
48	.14 495	914	.14 500	915	.85 500	996	12
49	.15 391	896	.15 395	895	.84 605	996	11
		877		878			
50	8.16 268		8.16 273		1.83 727	9.99 995	10
51	.17 128	860	.17 133	860	.82 867	995	9
52	.17 971	843	.17 976	843	.82 024	995	8
53	.18 798	827	.18 804	828	.81 196	995	7
54	.19 610	812	.19 616	812	.80 384	995	6
		797		797			
55	8.20 407		8.20 413		1.79 587	9.99 994	5
56	.21 189	782	.21 195	782	.78 805	994	4
57	.21 958	769	.21 964	769	.78 036	994	3
58	.22 713	755	.22 720	756	.77 280	994	2
59	.23 456	743	.23 462	742	.76 538	994	1
		730		730			
60	8.24 186		8.24 192		1.75 808	9.99 993	0
	L Cos	**d**	**L Cot**	**c d**	**L Tan**	**L Sin**	**'**
	89°						

For more accurate values of L Sin and L Cos for interpolated values of the angle use Table IV.

1°

′	L Sin	d	L Tan	c d	L Cot	L Cos	
0	8.24 186	717	8.24 192	718	1.75 808	9.99 993	60
1	.24 903	706	.24 910	706	.75 090	993	59
2	.25 609	695	.25 616	696	.74 384	993	58
3	.26 304	684	.26 312	684	.73 688	993	57
4	.26 988	673	.26 996	673	.73 004	992	56
5	8.27 661	663	8.27 669	663	1.72 331	9.99 992	55
6	.28 324	653	.28 332	654	.71 668	992	54
7	.28 977	644	.28 986	643	.71 014	992	53
8	.29 621	634	.29 629	634	.70 371	992	52
9	.30 255	624	.30 263	625	.69 737	991	51
10	8.30 879	616	8.30 888	617	1.69 112	9.99 991	50
11	.31 495	608	.31 505	607	.68 495	991	49
12	.32 103	599	.32 112	599	.67 888	990	48
13	.32 702	590	.32 711	591	.67 289	990	47
14	.33 292	583	.33 302	584	.66 698	990	46
15	8.33 875	575	8.33 886	575	1.66 114	9.99 990	45
16	.34 450	568	.34 461	568	.65 539	989	44
17	.35 018	560	.35 029	561	.64 971	989	43
18	.35 578	553	.35 590	553	.64 410	989	42
19	.36 131	547	.36 143	546	.63 857	989	41
20	8.36 678	539	8.36 689	540	1.63 311	9.99 988	40
21	.37 217	533	.37 229	533	.62 771	988	39
22	.37 750	526	.37 762	527	.62 238	988	38
23	.38 276	520	.38 289	520	.61 711	987	37
24	.38 796	514	.38 809	514	.61 191	987	36
25	8.39 310	508	8.39 323	509	1.60 677	9.99 987	35
26	.39 818	502	.39 832	502	.60 168	986	34
27	.40 320	496	.40 334	496	.59 666	986	33
28	.40 816	491	.40 830	491	.59 170	986	32
29	.41 307	485	.41 321	486	.58 679	985	31
30	8.41 792	480	8.41 807	480	1.58 193	9.99 985	30
31	.42 272	474	.42 287	475	.57 713	985	29
32	.42 746	470	.42 762	470	.57 238	984	28
33	.43 216	464	.43 232	464	.56 768	984	27
34	.43 680	459	.43 696	460	.56 304	984	26
35	8.44 139	455	8.44 156	455	1.55 844	9.99 983	25
36	.44 594	450	.44 611	450	.55 389	983	24
37	.45 044	445	.45 061	446	.54 939	983	23
38	.45 489	441	.45 507	441	.54 493	982	22
39	.45 930	436	.45 948	437	.54 052	982	21
40	8.46 366	433	8.46 385	432	1.53 615	9.99 982	20
41	.46 799	427	.46 817	428	.53 183	981	19
42	.47 226	424	.47 245	424	.52 755	981	18
43	.47 650	419	.47 669	420	.52 331	981	17
44	.48 069	416	.48 089	416	.51 911	980	16
45	8.48 485	411	8.48 505	412	1.51 495	9.99 980	15
46	.48 896	408	.48 917	408	.51 083	979	14
47	.49 304	404	.49 325	404	.50 675	979	13
48	.49 708	400	.49 729	401	.50 271	979	12
49	.50 108	396	.50 130	397	.49 870	978	11
50	8.50 504	393	8.50 527	393	1.49 473	9.99 978	10
51	.50 897	390	.50 920	390	.49 080	977	9
52	.51 287	386	.51 310	386	.48 690	977	8
53	.51 673	382	.51 696	383	.48 304	977	7
54	.52 055	379	.52 079	380	.47 921	976	6
55	8.52 434	376	8.52 459	376	1.47 541	9.99 976	5
56	.52 810	373	.52 835	373	.47 165	975	4
57	.53 183	369	.53 208	370	.46 792	975	3
58	.53 552	367	.53 578	367	.46 422	974	2
59	.53 919	363	.53 945	363	.46 055	974	1
60	8.54 282		8.54 308		1.45 692	9.99 974	0
	L Cos	d	L Cot	c d	L Tan	L Sin	′

88°

For more accurate values of L Sin and L Cos for interpolated values of the angle use Table IV.

2°							
′	L Sin	d	L Tan	c d	L Cot	L Cos	
0	8.54 282	360	8.54 308	361	1.45 692	9.99 974	**60**
1	642	357	669	358	331	973	59
2	999	355	.55 027	355	.44 973	973	58
3	.55 354	351	382	352	618	972	57
4	705	349	734	349	266	972	56
5	8.56 054	346	8.56 083	346	1.43 917	9.99 971	**55**
6	400	343	429	344	571	971	54
7	743	341	773	341	227	970	53
8	.57 084	337	.57 114	338	.42 886	970	52
9	421	336	452	336	548	969	51
10	8.57 757	332	8.57 788	333	1.42 212	9.99 969	**50**
11	.58 089	330	.58 121	330	.41 879	968	49
12	419	328	451	328	549	968	48
13	747	325	779	326	221	967	47
14	.59 072	323	.59 105	323	.40 895	967	46
15	8.59 395	320	8.59 428	321	1.40 572	9.99 967	**45**
16	715	318	749	319	251	966	44
17	.60 033	316	.60 068	316	.39 932	966	43
18	349	313	384	314	616	965	42
19	662	311	698	311	302	964	41
20	8.60 973	309	8.61 009	310	1.38 991	9.99 964	**40**
21	.61 282	307	319	307	681	963	39
22	589	305	626	305	374	963	38
23	894	302	931	303	069	962	37
24	.62 196	301	.62 234	301	.37 766	962	36
25	8.62 497	298	8.62 535	299	1.37 465	9.99 961	**35**
26	795	296	834	297	166	961	34
27	.63 091	294	.63 131	295	.36 869	960	33
28	385	293	426	292	574	960	32
29	678	290	718	291	282	959	31
30	8.63 968	288	8.64 009	289	1.35 991	9.99 959	**30**
31	.64 256	287	298	287	702	958	29
32	543	284	585	285	415	958	28
33	827	283	870	284	130	957	27
34	.65 110	281	.65 154	281	.34 846	956	26
35	8.65 391	279	8.65 435	280	1.34 565	9.99 956	**25**
36	670	277	715	278	285	955	24
37	947	276	993	276	007	955	23
38	.66 223	274	.66 269	274	.33 731	954	22
39	497	272	543	273	457	954	21
40	8.66 769	270	8.66 816	271	1.33 184	9.99 953	**20**
41	.67 039	269	.67 087	269	.32 913	952	19
42	308	267	356	268	644	952	18
43	575	266	624	266	376	951	17
44	841	263	890	264	110	951	16
45	8.68 104	263	8.68 154	263	1.31 846	9.99 950	**15**
46	367	260	417	261	583	949	14
47	627	259	678	260	322	949	13
48	886	258	938	258	062	948	12
49	.69 144	256	.69 196	257	.30 804	948	11
50	8.69 400	254	8.69 453	255	1.30 547	9.99 947	**10**
51	654	253	708	254	292	946	9
52	907	252	962	252	038	946	8
53	.70 159	250	.70 214	251	.29 786	945	7
54	409	249	465	249	535	944	6
55	8.70 658	247	8.70 714	248	1.29 286	9.99 944	**5**
56	905	246	962	246	038	943	4
57	.71 151	244	.71 208	245	.28 792	942	3
58	395	243	453	244	547	942	2
59	638	242	697	243	303	941	1
60	8.71 880		8.71 940		1.28 060	9.99 940	**0**
	L Cos	d	L Cot	c d	L Tan	L Sin	′
87°							

For more accurate values of L Sin and L Cos for interpolated values of the angle use Table IV.

3°

′	L Sin	d	L Tan	c d	L Cot	L Cos	′
0	8.71 880	240	8.71 940	241	1.28 060	9.99 940	60
1	.72 120	239	.72 181	239	.27 819	940	59
2	359	238	420	239	580	939	58
3	597	237	659	237	341	938	57
4	834	235	896	236	104	938	56
5	8.73 069	234	8.73 132	234	1.26 868	9.99 937	55
6	303	232	366	234	634	936	54
7	535	232	600	232	400	936	53
8	767	230	832	231	168	935	52
9	997	229	.74 063	229	.25 937	934	51
10	8.74 226	228	8.74 292	229	1.25 708	9.99 934	50
11	454	226	521	227	479	933	49
12	680	226	748	226	252	932	48
13	906	224	974	225	026	932	47
14	.75 130	223	.75 199	224	.24 801	931	46
15	8.75 353	222	8.75 423	222	1.24 577	9.99 930	45
16	575	220	645	222	355	929	44
17	795	220	867	220	133	929	43
18	.76 015	219	.76 087	219	.23 913	928	42
19	234	217	306	219	694	927	41
20	8.76 451	216	8.76 525	217	1.23 475	9.99 926	40
21	667	216	742	216	258	926	39
22	883	214	958	215	042	925	38
23	.77 097	213	.77 173	214	.22 827	924	37
24	310	212	387	213	613	923	36
25	8.77 522	211	8.77 600	211	1.22 400	9.99 923	35
26	733	210	811	211	189	922	34
27	943	209	.78 022	210	.21 978	921	33
28	.78 152	208	232	209	768	920	32
29	360	208	441	208	559	920	31
30	8.78 568	206	8.78 649	206	1.21 351	9.99 919	30
31	774	205	855	206	145	918	29
32	979	204	.79 061	205	.20 939	917	28
33	.79 183	203	266	204	734	917	27
34	386	202	470	203	530	916	26
35	8.79 588	201	8.79 673	202	1.20 327	9.99 915	25
36	789	201	875	201	125	914	24
37	990	199	.80 076	201	.19 924	913	23
38	.80 189	199	277	199	723	913	22
39	388	197	476	198	524	912	21
40	8.80 585	197	8.80 674	198	1.19 326	9.99 911	20
41	782	196	872	196	128	910	19
42	978	195	.81 068	196	.18 932	909	18
43	.81 173	194	264	195	736	909	17
44	367	193	459	194	541	908	16
45	8.81 560	192	8.81 653	193	1.18 347	9.99 907	15
46	752	192	846	192	154	906	14
47	944	190	.82 038	192	.17 962	905	13
48	.82 134	190	230	190	770	904	12
49	324	189	420	190	580	904	11
50	8.82 513	188	8.82 610	189	1.17 390	9.99 903	10
51	701	187	799	188	201	902	9
52	888	187	987	188	013	901	8
53	.83 075	186	.83 175	186	.16 825	900	7
54	261	185	361	186	639	899	6
55	8.83 446	184	8.83 547	185	1.16 453	9.99 898	5
56	630	183	732	184	268	898	4
57	813	183	916	184	084	897	3
58	996	181	.84 100	182	.15 900	896	2
59	.84 177	181	282	182	718	895	1
60	8.84 358		8.84 464		1.15 536	9.99 894	0
	L Cos	d	L Cot	c d	L Tan	L Sin	′

P. P.

For more accurate values of L Sin and L Cos for interpolated values of the angle use Table IV.

	182	181	179	178	177
1	3.0	3.0	3.0	3.0	3.0
2	6.1	6.0	6.0	5.9	5.9
3	9.1	9.0	9.0	8.9	8.8
4	12.1	12.1	11.9	11.9	11.8
5	15.2	15.1	14.9	14.8	14.8
6	18.2	18.1	17.9	17.8	17.7
7	21.2	21.1	20.9	20.8	20.6
8	24.3	24.1	23.9	23.7	23.6
9	27.3	27.2	26.8	26.7	26.6
10	30.3	30.2	29.8	29.7	29.5
20	60.7	60.3	59.7	59.3	59.0
30	91.0	90.5	89.5	89.0	88.5
40	121.3	120.7	119.3	118.7	118.0
50	151.7	150.8	149.2	148.3	147.5

	176	175	174	173	172
1	2.9	2.9	2.9	2.9	2.9
2	5.9	5.8	5.8	5.8	5.7
3	8.8	8.8	8.7	8.6	8.6
4	11.7	11.7	11.6	11.5	11.5
5	14.7	14.6	14.5	14.4	14.3
6	17.6	17.5	17.4	17.3	17.2
7	20.5	20.4	20.3	20.2	20.1
8	23.5	23.3	23.2	23.1	22.9
9	26.4	26.2	26.1	26.0	25.8
10	29.3	29.2	29.0	28.8	28.7
20	58.7	58.3	58.0	57.7	57.3
30	88.0	87.5	87.0	86.5	86.0
40	117.3	116.7	116.0	115.3	114.7
50	146.7	145.8	145.0	144.2	143.3

	171	170	169	168	167
1	2.8	2.8	2.8	2.8	2.8
2	5.7	5.7	5.6	5.6	5.6
3	8.6	8.5	8.4	8.4	8.4
4	11.4	11.3	11.3	11.2	11.1
5	14.2	14.2	14.1	14.0	13.9
6	17.1	17.0	16.9	16.8	16.7
7	20.0	19.8	19.7	19.6	19.5
8	22.8	22.7	22.5	22.4	22.3
9	25.6	25.5	25.4	25.2	25.0
10	28.5	28.3	28.2	28.0	27.8
20	57.0	56.7	56.3	56.0	55.7
30	85.5	85.0	84.5	84.0	83.5
40	114.0	113.3	112.7	112.0	111.3
50	142.5	141.7	140.8	140.0	139.2

	166	165	164	163	162
1	2.8	2.8	2.7	2.7	2.7
2	5.5	5.5	5.5	5.4	5.4
3	8.3	8.2	8.2	8.2	8.1
4	11.1	11.0	10.9	10.9	10.8
5	13.8	13.8	13.7	13.6	13.5
6	16.6	16.5	16.4	16.3	16.2
7	19.4	19.2	19.1	19.0	18.9
8	22.1	22.0	21.9	21.7	21.6
9	24.9	24.8	24.6	24.4	24.3
10	27.7	27.5	27.3	27.2	27.0
20	55.3	55.0	54.7	54.3	54.0
30	83.0	82.5	82.0	81.5	81.0
40	110.7	110.0	109.3	108.7	108.0
50	138.3	137.5	136.7	135.8	135.0

86°

4°

| P. P. | | | | | | | ' | L Sin | d | L Tan | c d | L Cot | L Cos | |

P. P.

	161	160	159	158
1	2.7	2.7	2.6	2.6
2	5.4	5.3	5.3	5.3
3	8.0	8.0	8.0	7.9
4	10.7	10.7	10.6	10.5
5	13.4	13.3	13.2	13.2
6	16.1	16.0	15.9	15.8
7	18.8	18.7	18.6	18.4
8	21.5	21.3	21.2	21.1
9	24.2	24.0	23.8	23.7
10	26.8	26.7	26.5	26.3
20	53.7	53.3	53.0	52.7
30	80.5	80.0	79.5	79.0
40	107.3	106.7	106.0	105.3
50	134.2	133.3	132.5	131.7

	157	156	155	154
1	2.6	2.6	2.6	2.6
2	5.2	5.2	5.2	5.1
3	7.8	7.8	7.8	7.7
4	10.5	10.4	10.3	10.3
5	13.1	13.0	12.9	12.8
6	15.7	15.6	15.5	15.4
7	18.3	18.2	18.1	18.0
8	20.9	20.8	20.7	20.5
9	23.6	23.4	23.2	23.1
10	26.2	26.0	25.8	25.7
20	52.3	52.0	51.7	51.3
30	78.5	78.0	77.5	77.0
40	104.7	104.0	103.3	102.7
50	130.8	130.0	129.2	128.3

	153	152	151
1	2.6	2.5	2.5
2	5.1	5.1	5.0
3	7.6	7.6	7.6
4	10.2	10.1	10.1
5	12.8	12.7	12.6
6	15.3	15.2	15.1
7	17.8	17.7	17.6
8	20.4	20.3	20.1
9	23.0	22.8	22.6
10	25.5	25.3	25.2
20	51.0	50.7	50.3
30	76.5	76.0	75.5
40	102.0	101.3	100.7
50	127.5	126.7	125.8

	150	149	148
1	2.5	2.5	2.5
2	5.0	5.0	4.9
3	7.5	7.4	7.4
4	10.0	9.9	9.9
5	12.5	12.4	12.3
6	15.0	14.9	14.8
7	17.5	17.4	17.3
8	20.0	19.9	19.7
9	22.5	22.4	22.2
10	25.0	24.8	24.7
20	50.0	49.7	49.3
30	75.0	74.5	74.0
40	100.0	99.3	98.7
50	125.0	124.2	123.3

	147	146	145
1	2.4	2.4	2.4
2	4.9	4.9	4.8
3	7.4	7.3	7.2
4	9.8	9.7	9.7
5	12.2	12.2	12.1
6	14.7	14.6	14.5
7	17.2	17.0	16.9
8	19.6	19.5	19.3
9	22.0	21.9	21.8
10	24.5	24.3	24.2
20	49.0	48.7	48.3
30	73.5	73.0	72.5
40	98.0	97.3	96.7
50	122.5	121.7	120.8

'	L Sin	d	L Tan	c d	L Cot	L Cos	
0	8.84 358	181	8.84 464	182	1.15 536	9.99 894	60
1	539		646		354	893	59
2	718	179	826	180	174	892	58
3	897	179	.85 006	180	.14 994	891	57
4	.85 075	178	185	179	815	891	56
		177		178			
5	8.85 252	177	8.85 363	177	1.14 637	9.99 890	55
6	429	176	540	177	460	889	54
7	605	176	717	176	283	888	53
8	780	175	893	176	107	887	52
9	955	175	.86 069	174	.13 931	886	51
		173					
10	8.86 128	173	8.86 243	174	1.13 757	9.99 885	50
11	301	173	417	174	583	884	49
12	474	171	591	172	409	883	48
13	645	171	763	172	237	882	47
14	816	171	935	171	065	881	46
15	8.86 987	169	8.87 106	171	1.12 894	9.99 880	45
16	.87 156	169	277	170	723	879	44
17	325	169	447	169	553	879	43
18	494	167	616	169	384	878	42
19	661	168	785	168	215	877	41
20	8.87 829	166	8.87 953	167	1.12 047	9.99 876	40
21	995	166	.88 120	167	.11 880	875	39
22	.88 161	165	287	166	713	874	38
23	326	164	453	165	547	873	37
24	490	164	618	165	382	872	36
25	8.88 654	163	8.88 783	165	1.11 217	9.99 871	35
26	817	163	948	163	052	870	34
27	980	163	.89 111	163	.10 889	869	33
28	.89 142	162	274	163	726	868	32
29	304	160	437	161	563	867	31
30	8.89 464	161	8.89 598	162	1.10 402	9.99 866	30
31	625	159	760	160	240	865	29
32	784	159	920	160	080	864	28
33	943	159	.90 080	160	.09 920	863	27
34	.90 102	158	240	160	760	862	26
35	8.90 260	157	8.90 399	158	1.09 601	9.99 861	25
36	417	157	557	158	443	860	24
37	574	156	715	157	285	859	23
38	730	155	872	157	128	858	22
39	885	155	.91 029	156	.08 971	857	21
40	8.91 040	155	8.91 185	155	1.08 815	9.99 856	20
41	195	154	340	155	660	855	19
42	349	153	495	155	505	854	18
43	502	153	650	153	350	853	17
44	655	152	803	154	197	852	16
45	8.91 807	152	8.91 957	153	1.08 043	9.99 851	15
46	959	151	.92 110	152	.07 890	850	14
47	.92 110	151	262	152	738	848	13
48	261	150	414	151	586	847	12
49	411	150	565	151	435	846	11
50	8.92 561	149	8.92 716	150	1.07 284	9.99 845	10
51	710	149	866	150	134	844	9
52	859	148	.93 016	149	.06 984	843	8
53	.93 007	147	165	148	835	842	7
54	154	147	313	149	687	841	6
55	8.93 301	147	8.93 462	147	1.06 538	9.99 840	5
56	448	146	609	147	391	839	4
57	594	146	756	147	244	838	3
58	740	145	903	146	097	837	2
59	885	145	.94 049	146	.05 951	836	1
60	8.94 030		8.94 195		1.05 805	9.99 834	0

| P. P. | | | | | | | ' | L Cos | d | L Cot | c d | L Tan | L Sin | ' |

85°

5°

'	L Sin	d	L Tan	c d	L Cot	L Cos	'
0	8.94 030	144	8.94 195	145	1.05 805	9.99 834	60
1	174	143	340	145	660	833	59
2	317	144	485	145	515	832	58
3	461	142	630	143	370	831	57
4	603	143	773	144	227	830	56
5	8.94 746	141	8.94 917	143	1.05 083	9.99 829	55
6	887	142	.95 060	142	.04 940	828	54
7	.95 029	141	202	142	798	827	53
8	170	140	344	142	656	825	52
9	310	140	486	141	514	824	51
10	8.95 450	139	8.95 627	140	1.04 373	9.99 823	50
11	589	139	767	141	233	822	49
12	728	139	908	139	092	821	48
13	867	138	.96 047	140	.03 953	820	47
14	.96 005	138	187	138	813	819	46
15	8.96 143	137	8.96 325	139	1.03 675	9.99 817	45
16	280	137	464	138	536	816	44
17	417	136	602	137	398	815	43
18	553	136	739	138	261	814	42
19	689	136	877	136	123	813	41
20	8.96 825	135	8.97 013	137	1.02 987	9.99 812	40
21	960	135	150	135	850	810	39
22	.97 095	134	285	136	715	809	38
23	229	134	421	135	579	808	37
24	363	133	556	135	444	807	36
25	8.97 496	133	8.97 691	134	1.02 309	9.99 806	35
26	629	133	825	134	175	804	34
27	762	132	959	133	041	803	33
28	894	132	.98 092	133	.01 908	802	32
29	.98 026	131	225	133	775	801	31
30	8.98 157	131	8.98 358	132	1.01 642	9.99 800	30
31	288	131	490	132	510	798	29
32	419	130	622	131	378	797	28
33	549	130	753	131	247	796	27
34	679	129	884	131	116	795	26
35	8.98 808	129	8.99 015	130	1.00 985	9.99 793	25
36	937	129	145	130	855	792	24
37	.99 066	128	275	130	725	791	23
38	194	128	405	129	595	790	22
39	322	128	534	128	466	788	21
40	8.99 450	127	8.99 662	129	1.00 338	9.99 787	20
41	577	127	791	128	209	786	19
42	704	126	919	127	081	785	18
43	830	126	.00 046	128	0.99 954	783	17
44	956	126	174	127	826	782	16
45	9.00 082	125	9.00 301	126	0.99 699	9.99 781	15
46	207	125	427	126	573	780	14
47	332	124	553	126	447	778	13
48	456	125	679	126	321	777	12
49	581	123	805	125	195	776	11
50	9.00 704	124	9.00 930	125	0.99 070	9.99 775	10
51	828	123	.01 055	124	.98 945	773	9
52	951	123	179	124	821	772	8
53	.01 074	122	303	124	697	771	7
54	196	122	427	123	573	769	6
55	9.01 318	122	9.01 550	123	0.98 450	9.99 768	5
56	440	121	673	123	327	767	4
57	561	121	796	122	204	765	3
58	682	121	918	122	082	764	2
59	803	120	.02 040	122	.97 960	763	1
60	9.01 923		9.02 162		0.97 838	9.99 761	0
	L Cos	d	L Cot	c d	L Tan	L Sin	'

84°

P. P.

	145	144	143	142	141
1	2.4	2.4	2.4	2.4	2.4
2	4.8	4.8	4.8	4.7	4.7
3	7.2	7.2	7.2	7.1	7.0
4	9.7	9.6	9.5	9.5	9.4
5	12.1	12.0	11.9	11.8	11.8
6	14.5	14.4	14.3	14.2	14.1
7	16.9	16.8	16.7	16.6	16.4
8	19.3	19.2	19.1	18.9	18.8
9	21.8	21.6	21.4	21.3	21.2
10	24.2	24.0	23.8	23.7	23.5
20	48.3	48.0	47.7	47.3	47.0
30	72.5	72.0	71.5	71.0	70.5
40	96.7	96.0	95.3	94.7	94.0
50	120.8	120.0	119.2	118.3	117.5

	140	139	138	137	136
1	2.3	2.3	2.3	2.3	2.3
2	4.7	4.6	4.6	4.6	4.5
3	7.0	7.0	6.9	6.8	6.8
4	9.3	9.3	9.2	9.1	9.1
5	11.7	11.6	11.5	11.4	11.3
6	14.0	13.9	13.8	13.7	13.6
7	16.3	16.2	16.1	16.0	15.9
8	18.7	18.5	18.4	18.3	18.1
9	21.0	20.8	20.7	20.6	20.4
10	23.3	23.2	23.0	22.8	22.7
20	46.7	46.3	46.0	45.7	45.3
30	70.0	69.5	69.0	68.5	68.0
40	93.3	92.7	92.0	91.3	90.7
50	116.7	115.8	115.0	114.2	113.3

	135	134	133	132	131
1	2.2	2.2	2.2	2.2	2.2
2	4.5	4.5	4.4	4.4	4.4
3	6.8	6.7	6.6	6.6	6.6
4	9.0	8.9	8.9	8.8	8.7
5	11.2	11.2	11.1	11.0	10.9
6	13.5	13.4	13.3	13.2	13.1
7	15.8	15.6	15.5	15.4	15.3
8	18.0	17.9	17.7	17.6	17.5
9	20.2	20.1	20.0	19.8	19.6
10	22.5	22.3	22.2	22.0	21.8
20	45.0	44.7	44.3	44.0	43.7
30	67.5	67.0	66.5	66.0	65.5
40	90.0	89.3	88.7	88.0	87.3
50	112.5	111.7	110.8	110.0	109.2

	130	129	128	127	126
1	2.2	2.2	2.1	2.1	2.1
2	4.3	4.3	4.3	4.2	4.2
3	6.5	6.4	6.4	6.4	6.3
4	8.7	8.6	8.5	8.5	8.4
5	10.8	10.8	10.7	10.6	10.5
6	13.0	12.9	12.8	12.7	12.6
7	15.2	15.0	14.9	14.8	14.7
8	17.3	17.2	17.1	16.9	16.8
9	19.5	19.4	19.2	19.0	18.9
10	21.7	21.5	21.3	21.2	21.0
20	43.3	43.0	42.7	42.3	42.0
30	65.0	64.5	64.0	63.5	63.0
40	86.7	86.0	85.3	84.7	84.0
50	108.3	107.5	106.7	105.8	105.0

	125	124	123	122
1	2.1	2.1	2.0	2.0
2	4.2	4.1	4.1	4.1
3	6.2	6.2	6.2	6.1
4	8.3	8.3	8.2	8.1
5	10.4	10.3	10.2	10.2
6	12.5	12.4	12.3	12.2
7	14.6	14.5	14.4	14.2
8	16.7	16.5	16.4	16.3
9	18.8	18.6	18.4	18.3
10	20.8	20.7	20.5	20.3
20	41.7	41.3	41.0	40.7
30	62.5	62.0	61.5	61.0
40	83.3	82.7	82.0	81.3
50	104.2	103.3	102.5	101.7

P. P.

6°

P. P.

	121	120	119	118
1	2.0	2.0	2.0	2.0
2	4.0	4.0	4.0	3.9
3	6.0	6.0	6.0	5.9
4	8.1	8.0	7.9	7.9
5	10.1	10.0	9.9	9.8
6	12.1	12.0	11.9	11.8
7	14.1	14.0	13.9	13.8
8	16.1	16.0	15.9	15.7
9	18.2	18.0	17.8	17.7
10	20.2	20.0	19.8	19.7
20	40.3	40.0	39.7	39.3
30	60.5	60.0	59.5	59.0
40	80.7	80.0	79.3	78.7
50	100.8	100.0	99.2	98.3

	117	116	115	114
1	1.9	1.9	1.9	1.9
2	3.9	3.9	3.8	3.8
3	5.8	5.8	5.8	5.7
4	7.8	7.7	7.7	7.6
5	9.8	9.7	9.6	9.5
6	11.7	11.6	11.5	11.4
7	13.6	13.5	13.4	13.3
8	15.6	15.5	15.3	15.2
9	17.6	17.4	17.2	7.1
10	19.5	19.3	19.2	19.0
20	39.0	38.7	38.3	38.0
30	58.5	58.0	57.5	57.0
40	78.0	77.3	76.7	76.0
50	97.5	96.7	95.8	95.0

	113	112	111	110
1	1.9	1.9	1.8	1.8
2	3.8	3.7	3.7	3.7
3	5.6	5.6	5.6	5.5
4	7.5	7.5	7.4	7.3
5	9.4	9.3	9.2	9.2
6	11.3	11.2	11.1	11.0
7	13.2	13.1	13.0	12.8
8	15.1	14.9	14.8	14.7
9	17.0	16.8	16.6	16.5
10	18.8	18.7	18.5	18.3
20	37.7	37.3	37.0	36.7
30	56.5	56.0	55.5	55.0
40	75.3	74.7	74.0	73.3
50	94.2	93.3	92.5	91.7

	109	108	107	106
1	1.8	1.8	1.8	1.8
2	3.6	3.6	3.6	3.5
3	5.4	5.4	5.4	5.3
4	7.3	7.2	7.1	7.1
5	9.1	9.0	8.9	8.8
6	10.9	10.8	10.7	10.6
7	12.7	12.6	12.5	12.4
8	14.5	14.4	14.3	14.1
9	16.4	16.2	16.0	15.9
10	18.2	18.0	17.8	17.7
20	36.3	36.0	35.7	35.3
30	54.5	54.0	53.5	53.0
40	72.7	72.0	71.3	70.7
50	90.8	90.0	89.2	88.3

	105	104	103
1	1.8	1.7	1.7
2	3.5	3.5	3.4
3	5.2	5.2	5.2
4	7.0	6.9	6.9
5	8.8	8.7	8.6
6	10.5	10.4	10.3
7	12.2	12.1	12.0
8	14.0	13.9	13.7
9	15.8	15.6	15.4
10	17.5	17.3	17.2
20	35.0	34.7	34.3
30	52.5	52.0	51.5
40	70.0	69.3	68.7
50	87.5	86.7	85.8

'	L Sin	d	L Tan	c d	L Cot	L Cos	
0	9.01 923		9.02 162		0.97 838	9.99 761	60
1	.02 043	120	283	121	717	760	59
2	163	120	404	121	596	759	58
3	283	120	525	121	475	757	57
4	402	119	645	120	355	756	56
5	9.02 520	118	9.02 766	121	0.97 234	9.99 755	55
6	639	119	885	119	115	753	54
7	757	118	.03 005	120	.96 995	752	53
8	874	117	124	119	876	751	52
9	992	118	242	118	758	749	51
10	9.03 109	117	9.03 361	119	0.96 639	9.99 748	50
11	226	117	479	118	521	747	49
12	342	116	597	118	403	745	48
13	458	116	714	117	286	744	47
14	574	116	832	118	168	742	46
15	9.03 690	116	9.03 948	116	0.96 052	9.99 741	45
16	805	115	.04 065	117	.95 935	740	44
17	920	115	181	116	819	738	43
18	.04 034	114	297	116	703	737	42
19	149	115	413	116	587	736	41
20	9.04 262	113	9.04 528	115	0.95 472	9.99 734	40
21	376	114	643	115	357	733	39
22	490	114	768	115	242	731	38
23	603	113	873	115	127	730	37
24	715	112	987	114	013	728	36
25	9.04 828	113	9.05 101	114	0.94 899	9.99 727	35
26	940	112	214	113	786	726	34
27	.05 052	112	328	114	672	724	33
28	164	112	441	113	559	723	32
29	275	111	553	112	447	721	31
30	9.05 386	111	9.05 666	113	0.94 334	9.99 720	30
31	497	111	778	112	222	718	29
32	607	110	890	112	110	717	28
33	717	110	.06 002	112	.93 998	716	27
34	827	110	113	111	887	714	26
35	9.05 937	110	9.06 224	111	0.93 776	9.99 713	25
36	.06 046	109	335	111	665	711	24
37	155	109	445	110	555	710	23
38	264	109	556	111	444	708	22
39	372	108	666	110	334	707	21
40	9.06 481	109	9.06 775	109	0.93 225	9.99 705	20
41	589	108	885	110	115	704	19
42	696	107	994	109	006	702	18
43	804	108	.07 103	109	.92 897	701	17
44	911	107	211	108	789	699	16
45	9.07 018	107	9.07 320	109	0.92 680	9.99 698	15
46	124	106	428	108	572	696	14
47	231	107	536	108	464	695	13
48	337	106	643	107	357	693	12
49	442	105	751	108	249	692	11
50	9.07 548	106	9.07 858	107	0.92 142	9.99 690	10
51	653	105	964	106	036	689	9
52	758	105	.08 071	107	.91 929	687	8
53	863	105	177	106	823	686	7
54	968	105	283	106	717	684	6
55	9.08 072	104	9.08 389	106	0.91 611	9.99 683	5
56	176	104	495	106	505	681	4
57	280	104	600	105	400	680	3
58	383	103	705	105	295	678	2
59	486	103	810	105	190	677	1
60	9.08 589	103	9.08 914	104	0.91 086	9.99 675	0

P. P.	L Cos	d	L Cot	c d	L Tan	L Sin	'

83°

7°

′	L Sin	d	L Tan	c d	L Cot	L Cos	
0	9.08 589		9.08 914		0.91 086	9.99 675	60
1	692	103	9.09 019	105	.90 981	674	59
2	795	103	123	104	877	672	58
3	897	102	227	104	773	670	57
4	999	102	330	103	670	669	56
		102		104			
5	9.09 101		9.09 434		0.90 566	9.99 667	55
6	202	101	537	103	463	666	54
7	304	102	640	103	360	664	53
8	405	101	742	102	258	663	52
9	506	101	845	103	155	661	51
		100		102			
10	9.09 606		9.09 947		0.90 053	9.99 659	50
11	707	101	.10 049	102	.89 951	658	49
12	807	100	150	101	850	656	48
13	907	100	252	102	748	655	47
14	9.10 006	99	353	101	647	653	46
		100		101			
15	9.10 106		9.10 454		0.89 546	9.99 651	45
16	205	99	555	101	445	650	44
17	304	99	656	101	344	648	43
18	402	98	756	100	244	647	42
19	501	99	856	100	144	645	41
		98		100			
20	9.10 599		9.10 956		0.89 044	9.99 643	40
21	697	98	.11 056	100	.88 944	642	39
22	795	98	155	99	845	640	38
23	893	98	254	99	746	638	37
24	990	97	353	99	647	637	36
		97		99			
25	9.11 087		9.11 452		0.88 548	9.99 635	35
26	184	97	551	99	449	633	34
27	281	97	649	98	351	632	33
28	377	96	747	98	253	630	32
29	474	97	845	98	155	629	31
		96		98			
30	9.11 570		9.11 943		0.88 057	9.99 627	30
31	666	96	.12 040	97	.87 960	625	29
32	761	95	138	98	862	624	28
33	857	96	235	97	765	622	27
34	952	95	332	97	668	620	26
		95		96			
35	9.12 047		9.12 428		0.87 572	9.99 618	25
36	142	95	525	97	475	617	24
37	236	94	621	96	379	615	23
38	331	95	717	96	283	613	22
39	425	94	813	96	187	612	21
		94		96			
40	9.12 519		9.12 909		0.87 091	9.99 610	20
41	612	93	.13 004	95	.86 996	608	19
42	706	94	099	95	901	607	18
43	799	93	194	95	806	605	17
44	892	93	289	95	711	603	16
		93		95			
45	9.12 985		9.13 384		0.86 616	9.99 601	15
46	.13 078	93	478	94	522	600	14
47	171	93	573	95	427	598	13
48	263	92	667	94	333	596	12
49	355	92	761	94	239	595	11
		92		93			
50	9.13 447		9.13 854		0.86 146	9.99 593	10
51	539	92	948	94	052	591	9
52	630	91	.14 041	93	.85 959	589	8
53	722	92	134	93	866	588	7
54	813	91	227	93	773	586	6
		91		93			
55	9.13 904		9.14 320		0.85 680	9.99 584	5
56	994	90	412	92	588	582	4
57	.14 085	91	504	92	496	581	3
58	175	90	597	93	403	579	2
59	266	91	688	91	312	577	1
		90		92			
60	9.14 356		9.14 780		0.85 220	9.99 575	0
	L Cos	d	L Cot	c d	L Tan	L Sin	′

82°

P. P.

	105	104	103	102
1	1.8	1.7	1.7	1.7
2	3.5	3.5	3.4	3.4
3	5.2	5.2	5.2	5.1
4	7.0	6.9	6.9	6.8
5	8.8	8.7	8.6	8.5
6	10.5	10.4	10.3	10.2
7	12.2	12.1	12.0	11.9
8	14.0	13.9	13.7	13.6
9	15.8	15.6	15.4	15.3
10	17.5	17.3	17.2	17.0
20	35.0	34.7	34.3	34.0
30	52.5	52.0	51.5	51.0
40	70.0	69.3	68.7	68.0
50	87.5	86.7	85.8	85.0

	101	100	99	98
1	1.7	1.7	1.6	1.6
2	3.4	3.3	3.3	3.3
3	5.0	5.0	5.0	4.9
4	6.7	6.7	6.6	6.5
5	8.4	8.3	8.2	8.2
6	10.1	10.0	9.9	9.8
7	11.8	11.7	11.6	11.4
8	13.5	13.3	13.2	13.1
9	15.2	15.0	14.8	14.7
10	16.8	16.7	16.5	16.3
20	33.7	33.3	33.0	32.7
30	50.5	50.0	49.5	49.0
40	67.3	66.7	66.0	65.3
50	84.2	83.3	82.5	81.7

	97	96	95	94
1	1.6	1.6	1.6	1.6
2	3.2	3.2	3.2	3.1
3	4.8	4.8	4.8	4.7
4	6.5	6.4	6.3	6.3
5	8.1	8.0	7.9	7.8
6	9.7	9.6	9.5	9.4
7	11.3	11.2	11.1	11.0
8	12.9	12.8	12.7	12.5
9	14.6	14.4	14.2	14.1
10	16.2	16.0	15.8	15.7
20	32.3	32.0	31.7	31.3
30	48.5	48.0	47.5	47.0
40	64.7	64.0	63.3	62.7
50	80.8	80.0	79.2	78.3

	93	92	91	90
1	1.6	1.5	1.5	1.5
2	3.1	3.1	3.0	3.0
3	4.6	4.6	4.6	4.5
4	6.2	6.1	6.1	6.0
5	7.8	7.7	7.6	7.5
6	9.3	9.2	9.1	9.0
7	10.8	10.7	10.6	10.5
8	12.4	12.3	12.1	12.0
9	14.0	13.8	13.6	13.5
10	15.5	15.3	15.2	15.0
20	31.0	30.7	30.3	30.0
30	46.5	46.0	45.5	45.0
40	62.0	61.3	60.7	60.0
50	77.5	76.7	75.8	75.0

P. P.

8°							
P. P.	'	L Sin	d	L Tan	c d	L Cot	L Cos	

P. P.	'	L Sin	d	L Tan	c d	L Cot	L Cos	
	0	9.14 356	89	9.14 780	92	0.85 220	9.99 575	**60**
	1	445	90	872	91	128	574	59
	2	535	89	963	91	037	572	58
	3	624	90	.15 054	91	.84 946	570	57
	4	714	89	145	91	855	568	56
	5	9.14 803	88	9.15 236	91	0.84 764	9.99 566	**55**
	6	891	89	327	90	673	565	54
	7	980	89	417	91	583	563	53
	8	.15 069	88	508	90	492	561	52
	9	157	88	598	90	402	559	51
	10	9.15 245	88	9.15 688	89	0.84 312	9.99 557	**50**
	11	333	88	777	90	223	556	49
	12	421	87	867	89	133	554	48
	13	508	88	956	90	044	552	47
	14	596	87	.16 046	89	.83 954	550	46
	15	9.15 683	87	9.16 135	89	0.83 865	9.99 548	**45**
	16	770	87	224	88	776	546	44
	17	857	87	312	89	688	545	43
	18	944	86	401	88	599	543	42
	19	.16 030	86	489	88	511	541	41
	20	9.16 116	87	9.16 577	88	0.83 423	9.99 539	**40**
	21	203	86	665	88	335	537	39
	22	289	85	753	88	247	535	38
	23	374	86	841	87	159	533	37
	24	460	85	928	88	072	532	36
	25	9.16 545	86	0.17 016	87	0.82 984	9.99 530	**35**
	26	631	85	103	87	897	528	34
	27	716	85	190	87	810	526	33
	28	801	85	277	86	723	524	32
	29	886	84	363	87	637	522	31
	30	9.16 970	85	9.17 450	86	0.82 550	9.99 520	**30**
	31	.17 055	84	536	86	464	518	29
	32	139	84	622	86	378	517	28
	33	223	84	708	86	292	515	27
	34	307	84	794	86	206	513	26
	35	9.17 391	83	9.17 880	85	0.82 120	9.99 511	**25**
	36	474	84	965	86	035	509	24
	37	558	83	.18 051	85	.81 949	507	23
	38	641	83	136	85	864	505	22
	39	724	83	221	85	779	503	21
	40	9.17 807	83	9.18 306	85	0.81 694	9.99 501	**20**
	41	890	83	391	84	609	499	19
	42	973	82	475	85	525	497	18
	43	.18 055	82	560	84	440	495	17
	44	137	83	644	84	356	494	16
	45	9.18 220	82	9.18 728	84	0.81 272	9.99 492	**15**
	46	302	81	812	84	188	490	14
	47	383	82	896	83	104	488	13
	48	465	82	979	84	021	486	12
	49	547	81	.19 063	83	.80 937	484	11
	50	9.18 628	81	9.19 146	83	0.80 854	9.99 482	**10**
	51	709	81	229	83	771	480	9
	52	790	81	312	83	688	478	8
	53	871	81	395	83	605	476	7
	54	952	81	478	83	522	474	6
	55	9.19 033	80	9.19 561	82	0.80 439	9.99 472	**5**
	56	113	80	643	82	357	470	4
	57	193	80	725	82	275	468	3
	58	273	80	807	82	193	466	2
	59	353	80	889	82	111	464	1
	60	9.19 433		9.19 971		0.80 029	9.99 462	**0**

P. P. tables:

	89	88	87
1	1.5	1.5	1.4
2	3.0	2.9	2.9
3	4.4	4.4	4.4
4	5.9	5.9	5.8
5	7.4	7.3	7.2
6	8.9	8.8	8.7
7	10.4	10.3	10.2
8	11.9	11.7	11.6
9	13.4	13.2	13.0
10	14.8	14.7	14.5
20	29.7	29.3	29.0
30	44.5	44.0	43.5
40	59.3	58.7	58.0
50	74.2	73.3	72.5

	86	85	84
1	1.4	1.4	1.4
2	2.9	2.8	2.8
3	4.3	4.2	4.2
4	5.7	5.7	5.6
5	7.2	7.1	7.0
6	8.6	8.5	8.4
7	10.0	9.9	9.8
8	11.5	11.3	11.2
9	12.9	12.8	12.6
10	14.3	14.2	14.0
20	28.7	28.3	28.0
30	43.0	42.5	42.0
40	57.3	56.7	56.0
50	71.7	70.8	70.0

	83	82	81	80
1	1.4	1.4	1.4	1.3
2	2.8	2.7	2.7	2.7
3	4.2	4.1	4.0	4.0
4	5.5	5.5	5.4	5.3
5	6.9	6.8	6.8	6.7
6	8.3	8.2	8.1	8.0
7	9.7	9.6	9.4	9.3
8	11.1	10.9	10.8	10.7
9	12.4	12.3	12.2	12.0
10	13.8	13.7	13.5	13.3
20	27.7	27.3	27.0	26.7
30	41.5	41.0	40.5	40.0
40	55.3	54.7	54.0	53.3
50	69.2	68.3	67.5	66.7

P. P.		L Cos	d	L Cot	c d	L Tan	L Sin	'

| 81° | | | | | | | | |

9°

'	L Sin	d	L Tan	c d	L Cot	L Cos	
0	9.19 433	80	9.19 971	82	0.80 029	9.99 462	60
1	513	79	9.20 053	82	.79 947	460	59
2	592	80	134	81	866	458	58
3	672	79	216	82	784	456	57
4	751	79	297	81	703	454	56
5	9.19 830	79	9.20 378	81	0.79 622	9.99 452	55
6	909	79	459	81	541	450	54
7	988	79	540	81	460	448	53
8	.20 067	79	621	81	379	446	52
9	145	78	701	80	299	444	51
10	9.20 223	79	9.20 782	81	0.79 218	9.99 442	50
11	302	78	862	80	138	440	49
12	380	78	942	80	058	438	48
13	458	78	.21 022	80	.78 978	436	47
14	535	77	102	80	898	434	46
15	9.20 613	78	9.21 182	80	0.78 818	9.99 432	45
16	691	77	261	79	739	429	44
17	768	77	341	80	659	427	43
18	845	77	420	79	580	425	42
19	922	77	499	79	501	423	41
20	9.20 999	77	9.21 578	79	0.78 422	9.99 421	40
21	.21 076	77	657	79	343	419	39
22	153	76	736	79	264	417	38
23	229	77	814	78	186	415	37
24	306	76	893	79	107	413	36
25	9.21 382	76	9.21 971	78	0.78 029	9.99 411	35
26	458	76	.22 049	78	.77 951	409	34
27	534	76	127	78	873	407	33
28	610	75	205	78	795	404	32
29	685	76	283	78	717	402	31
30	9.21 761	75	9.22 361	77	0.77 639	9.99 400	30
31	836	76	438	78	562	398	29
32	912	75	516	77	484	396	28
33	987	75	593	77	407	394	27
34	.22 062	75	670	77	330	392	26
35	9.22 137	74	9.22 747	77	0.77 253	9.99 390	25
36	211	75	824	77	176	388	24
37	286	75	901	76	099	385	23
38	361	74	977	77	023	383	22
39	435	74	.23 054	76	.76 946	381	21
40	9.22 509	74	9.23 130	76	0.76 870	9.99 379	20
41	583	74	206	77	794	377	19
42	657	74	283	76	717	375	18
43	731	74	359	76	641	372	17
44	805	73	435	75	565	370	16
45	9.22 878	74	9.23 510	76	0.76 490	9.99 368	15
46	952	73	586	75	414	366	14
47	.23 025	73	661	76	339	364	13
48	098	73	737	75	263	362	12
49	171	73	812	75	188	359	11
50	9.23 244	73	9.23 887	75	0.76 113	9.99 357	10
51	317	73	962	75	038	355	9
52	390	72	.24 037	75	.75 963	353	8
53	462	73	112	74	888	351	7
54	535	72	186	75	814	348	6
55	9.23 607	72	9.24 261	74	0.75 739	9.99 346	5
56	679	73	335	75	665	344	4
57	752	71	410	74	590	342	3
58	823	72	484	74	516	340	2
59	895	72	558	74	442	337	1
60	9.23 967		9.24 632		0.75 368	9.99 335	0
	L Cos	d	L Cot	c d	L Tan	L Sin	'

P. P.

	82	81	80	79
1	1.4	1.4	1.3	1.3
2	2.7	2.7	2.7	2.6
3	4.1	4.0	4.0	4.0
4	5.5	5.4	5.3	5.3
5	6.8	6.8	6.7	6.6
6	8.2	8.1	8.0	7.9
7	9.6	9.4	9.3	9.2
8	10.9	10.8	10.7	10.5
9	12.3	12.2	12.0	11.8
10	13.7	13.5	13.3	13.2
20	27.3	27.0	26.7	26.3
30	41.0	40.5	40.0	39.5
40	54.7	54.0	53.3	52.7
50	68.3	67.5	66.7	65.8

	78	77	76	75
1	1.3	1.3	1.3	1.2
2	2.6	2.6	2.5	2.5
3	3.9	3.8	3.8	3.8
4	5.2	5.1	5.1	5.0
5	6.5	6.4	6.3	6.2
6	7.8	7.7	7.6	7.5
7	9.1	9.0	8.9	8.8
8	10.4	10.3	10.1	10.0
9	11.7	11.6	11.4	11.2
10	13.0	12.8	12.7	12.5
20	26.0	25.7	25.3	25.0
30	39.0	38.5	38.0	37.5
40	52.0	51.3	50.7	50.0
50	65.0	64.2	63.3	62.5

	74	73	72	71
1	1.2	1.2	1.2	1.2
2	2.5	2.4	2.4	2.4
3	3.7	3.6	3.6	3.6
4	4.9	4.9	4.8	4.7
5	6.2	6.1	6.0	5.9
6	7.4	7.3	7.2	7.1
7	8.6	8.5	8.4	8.3
8	9.9	9.7	9.6	9.5
9	11.1	11.0	10.8	10.6
10	12.3	12.2	12.0	11.8
20	24.7	24.3	24.0	23.7
30	37.0	36.5	36.0	35.5
40	49.3	48.7	48.0	47.3
50	61.7	60.8	60.0	59.2

P. P.

80°

10°

P. P.	′	L Sin	d	L Tan	c d	L Cot	L Cos	d	
	0	9.23 967	72	9.24 632	74	0.75 368	9.99 335	2	**60**
	1	.24 039	71	706	73	294	333	2	59
	2	110	71	779	74	221	331	2	58
	3	181	72	853	73	147	328	3	57
	4	253	71	926	74	074	326	2	56
	5	9.24 324	71	9.25 000	73	0.75 000	9.99 324	2	**55**
	6	395	71	073	73	.74 927	322	3	54
	7	466	70	146	73	854	319	2	53
	8	536	71	219	73	781	317	2	52
	9	607	70	292	73	708	315	2	51
	10	9.24 677	71	9.25 365	72	0.74 635	9.99 313	3	**50**
	11	748	70	437	73	563	310	2	49
	12	818	70	510	72	490	308	2	48
	13	888	70	582	73	418	306	2	47
	14	958	70	655	72	345	304	3	46
	15	9.25 028	70	9.25 727	72	0.74 273	9.99 301	2	**45**
	16	098	70	799	72	201	299	2	44
	17	168	69	871	72	129	297	3	43
	18	237	70	943	72	057	294	2	42
	19	307	69	.26 015	71	.73 985	292	2	41
	20	9.25 376	69	9.26 086	72	0.73 914	9.99 290	2	**40**
	21	445	69	158	71	842	288	3	39
	22	514	69	229	72	771	285	2	38
	23	583	69	301	71	699	283	2	37
	24	652	69	372	71	628	281	3	36
	25	9.25 721	69	9.26 443	71	0.73 557	9.99 278	2	**35**
	26	790	68	514	71	486	276	2	34
	27	858	69	585	70	415	274	3	33
	28	927	68	655	71	345	271	2	32
	29	995	68	726	71	274	269	2	31
	30	9.26 063	68	9.26 797	70	0.73 203	9.99 267	3	**30**
	31	131	68	867	70	133	264	2	29
	32	199	68	937	71	063	262	2	28
	33	267	68	.27 008	70	.72 992	260	3	27
	34	335	68	078	70	922	257	2	26
	35	9.26 403	67	9.27 148	70	0.72 852	9.99 255	3	**25**
	36	470	68	218	70	782	252	2	24
	37	538	67	288	69	712	250	2	23
	38	605	67	357	70	643	248	3	22
	39	672	67	427	69	573	245	2	21
	40	9.26 739	67	9.27 496	70	0.72 504	9.99 243	2	**20**
	41	806	67	566	69	434	241	3	19
	42	873	67	635	69	365	238	2	18
	43	940	67	704	69	296	236	3	17
	44	.27 007	66	773	69	227	233	2	16
	45	9.27 073	67	9.27 842	69	0.72 158	9.99 231	2	**15**
	46	140	66	911	69	089	229	3	14
	47	206	67	980	69	020	226	2	13
	48	273	66	.28 049	68	.71 951	224	3	12
	49	339	66	117	69	883	221	2	11
	50	9.27 405	66	9.28 186	68	0.71 814	9.99 219	2	**10**
	51	471	66	254	69	746	217	3	9
	52	537	65	323	68	677	214	2	8
	53	602	66	391	68	609	212	3	7
	54	668	66	459	68	541	209	2	6
	55	9.27 734	65	9.28 527	68	0.71 473	9.99 207	3	**5**
	56	799	65	595	67	405	204	2	4
	57	864	66	662	68	338	202	2	3
	58	930	65	730	68	270	200	3	2
	59	995	65	798	67	202	197	2	1
	60	9.28 060		9.28 865		0.71 135	9.99 195		**0**
P. P.		L Cos	d	L Cot	c d	L Tan	L Sin	d	′

79°

P. P.

	70	69	68	67
1	1.2	1.2	1.1	1.1
2	2.3	2.3	2.3	2.2
3	3.5	3.4	3.4	3.4
4	4.7	4.6	4.5	4.5
5	5.8	5.8	5.7	5.6
6	7.0	6.9	6.8	6.7
7	8.2	8.0	7.9	7.8
8	9.3	9.2	9.1	8.9
9	10.5	10.4	10.2	10.0
10	11.7	11.5	11.3	11.2
20	23.3	23.0	22.7	22.3
30	35.0	34.5	34.0	33.5
40	46.7	46.0	45.3	44.7
50	58.3	57.5	56.7	55.8

	66	65	3	2
1	1.1	1.1	0.0	0.0
2	2.2	2.2	0.1	0.1
3	3.3	3.2	0.2	0.1
4	4.4	4.3	0.2	0.1
5	5.5	5.4	0.2	0.2
6	6.6	6.5	0.3	0.2
7	7.7	7.6	0.4	0.2
8	8.8	8.7	0.4	0.3
9	9.9	9.8	0.4	0.3
10	11.0	10.8	0.5	0.3
20	22.0	21.7	1.0	0.7
30	33.0	32.5	1.5	1.0
40	44.0	43.3	2.0	1.3
50	55.0	54.2	2.5	1.7

11°

'	L Sin	d	L Tan	c d	L Cot	L Cos	d	
0	9.28 060	65	9.28 865	68	0.71 135	9.99 195	3	60
1	125	65	933	67	067	192	2	59
2	190	64	.29 000	67	000	190	3	58
3	254	65	067	67	.70 933	187	2	57
4	319	65	134	67	866	185	3	56
5	9.28 384	64	9.29 201	67	0.70 799	9.99 182	2	55
6	448	64	268	67	732	180	3	54
7	512	65	335	67	665	177	2	53
8	577	64	402	66	598	175	3	52
9	641	64	468	67	532	172	2	51
10	9.28 705	64	9.29 535	66	0.70 465	9.99 170	3	50
11	769	64	601	67	399	167	2	49
12	833	63	668	66	332	165	3	48
13	896	64	734	66	266	162	2	47
14	960	64	800	66	200	160	3	46
15	9.29 024	63	9.29 866	66	0.70 134	9.99 157	2	45
16	087	63	932	66	068	155	3	44
17	150	64	998	66	002	152	2	43
18	214	63	.30 064	66	.69 936	150	3	42
19	277	63	130	65	870	147	2	41
20	9.29 340	63	9.30 195	66	0.69 805	9.99 145	3	40
21	403	63	261	65	739	142	2	39
22	466	63	326	65	674	140	3	38
23	529	62	391	66	609	137	2	37
24	591	63	457	65	543	135	3	36
25	9.29 654	62	9.30 522	65	0.69 478	9.99 132	2	35
26	716	63	587	65	413	130	3	34
27	779	62	652	65	348	127	3	33
28	841	62	717	65	283	124	2	32
29	903	63	782	64	218	122	3	31
30	9.29 966	62	9.30 846	65	0.69 154	9.99 119	2	30
31	.30 028	62	911	64	089	117	3	29
32	090	61	975	65	025	114	2	28
33	151	62	.31 040	64	.68 960	112	3	27
34	213	62	104	64	896	109	3	26
35	9.30 275	61	9.31 168	65	0.68 832	9.99 106	2	25
36	336	62	233	64	767	104	3	24
37	398	61	297	64	703	101	2	23
38	459	62	361	64	639	099	3	22
39	521	61	425	64	575	096	3	21
40	9.30 582	61	9.31 489	63	0.68 511	9.99 093	2	20
41	643	61	552	64	448	091	3	19
42	704	61	616	63	384	088	2	18
43	765	61	679	64	321	086	3	17
44	826	61	743	63	257	083	3	16
45	9.30 887	60	9.31 806	64	0.68 194	9.99 080	3	15
46	947	61	870	63	130	078	3	14
47	.31 008	60	933	63	067	075	3	13
48	068	61	996	63	004	072	2	12
49	129	60	.32 059	63	.67 941	070	3	11
50	9.31 189	61	9.32 122	63	0.67 878	9.99 067	3	10
51	250	60	185	63	815	064	2	9
52	310	60	248	63	752	062	3	8
53	370	60	311	62	689	059	3	7
54	430	60	373	63	627	056	2	6
55	9.31 490	59	9.32 436	62	0.67 564	9.99 054	3	5
56	549	60	498	63	502	051	3	4
57	609	60	561	62	439	048	2	3
58	669	59	623	62	377	046	3	2
59	728	60	685	62	315	043	3	1
60	9.31 788		9.32 747		0.67 253	9.99 040		0

| | L Cos | d | L Cot | c d | L Tan | L Sin | d | ' |

P. P.

	68	67	66	65
1	1.1	1.1	1.1	1.1
2	2.3	2.2	2.2	2.2
3	3.4	3.4	3.3	3.2
4	4.5	4.5	4.4	4.3
5	5.7	5.6	5.5	5.4
6	6.8	6.7	6.6	6.5
7	7.9	7.8	7.7	7.6
8	9.1	8.9	8.8	8.7
9	10.2	10.0	9.9	9.8
10	11.3	11.2	11.0	10.8
20	22.7	22.3	22.0	21.7
30	34.0	33.5	33.0	32.5
40	45.3	44.7	44.0	43.3
50	56.7	55.8	55.0	54.2

	64	63	62	61
1	1.1	1.0	1.0	1.0
2	2.1	2.1	2.1	2.0
3	3.2	3.2	3.1	3.0
4	4.3	4.2	4.1	4.1
5	5.3	5.2	5.2	5.1
6	6.4	6.3	6.2	6.1
7	7.5	7.4	7.2	7.1
8	8.5	8.4	8.3	8.1
9	9.6	9.4	9.3	9.2
10	10.7	10.5	10.3	10.2
20	21.3	21.0	20.7	20.3
30	32.0	31.5	31.0	30.5
40	42.7	42.0	41.3	40.7
50	53.3	52.5	51.7	50.8

P. P.

78°

12°

′	L Sin	d	L Tan	c d	L Cot	L Cos	d	
0	9.31 788	59	9.32 747	63	0.67 253	9.99 040	2	60
1	847	60	810	62	190	038	3	59
2	907	59	872	61	128	035	3	58
3	966	59	933	62	067	032	3	57
4	.32 025	59	995	62	005	030	2	56
5	9.32 084	59	9.33 057	62	0.66 943	9.99 027	3	55
6	143	59	119	61	881	024	3	54
7	202	59	180	62	820	022	2	53
8	261	58	242	61	758	019	3	52
9	319	59	303	62	697	016	3	51
10	9.32 378	59	9.33 365	61	0.66 635	9.99 013	2	50
11	437	58	426	61	574	011	3	49
12	495	58	487	61	513	008	3	48
13	553	59	548	61	452	005	3	47
14	612	58	609	61	391	002	2	46
15	9.32 670	58	9.33 670	61	0.66 330	9.99 000	3	45
16	728	58	731	61	269	.98 997	3	44
17	786	58	792	61	208	994	3	43
18	844	58	853	60	147	991	2	42
19	902	58	913	61	087	989	3	41
20	9.32 960	58	9.33 974	60	0.66 026	9.98 986	3	40
21	.33 018	57	.34 034	61	.65 966	983	3	39
22	075	58	095	60	905	980	2	38
23	133	57	155	60	845	978	3	37
24	190	58	215	61	785	975	3	36
25	9.33 248	57	9.34 276	60	0.65 724	9.98 972	3	35
26	305	57	336	60	664	969	2	34
27	362	58	396	60	604	967	3	33
28	420	57	456	60	544	964	3	32
29	477	57	516	60	484	961	3	31
30	9.33 534	57	9.34 576	59	0.65 424	9.98 958	3	30
31	591	56	635	60	365	955	2	29
32	647	57	695	60	305	953	3	28
33	704	57	755	59	245	950	3	27
34	761	57	814	60	186	947	3	26
35	9.33 818	56	9.34 874	59	0.65 126	9.98 944	3	25
36	874	57	933	59	067	941	3	24
37	931	56	992	59	008	938	3	23
38	987	56	.35 051	60	.64 949	936	3	22
39	.34 043	57	111	59	889	933	3	21
40	9.34 100	56	9.35 170	59	0.64 830	9.98 930	3	20
41	156	56	229	59	771	927	3	19
42	212	56	288	59	712	924	3	18
43	268	56	347	58	653	921	2	17
44	324	56	405	59	595	919	3	16
45	9.34 380	56	9.35 464	59	0.64 536	9.98 916	3	15
46	436	55	523	58	477	913	3	14
47	491	56	581	59	419	910	3	13
48	547	55	640	58	360	907	3	12
49	602	56	698	59	302	904	3	11
50	9.34 658	55	9.35 757	58	0.64 243	9.98 901	3	10
51	713	56	815	58	185	898	2	9
52	769	55	873	58	127	896	3	8
53	824	55	931	58	069	893	3	7
54	879	55	989	58	011	890	3	6
55	9.34 934	55	9.36 047	58	0.63 953	9.98 887	3	5
56	989	55	105	58	895	884	3	4
57	.35 044	55	163	58	837	881	3	3
58	099	55	221	58	779	878	3	2
59	154	55	279	57	721	875	3	1
60	9.35 209		9.36 336		0.63 664	9.98 872		0

P. P.

	60	59	58	57
1	1.0	1.0	1.0	1.0
2	2.0	2.0	1.9	1.9
3	3.0	3.0	2.9	2.8
4	4.0	3.9	3.9	3.8
5	5.0	4.9	4.8	4.8
6	6.0	5.9	5.8	5.7
7	7.0	6.9	6.8	6.6
8	8.0	7.9	7.7	7.6
9	9.0	8.8	8.7	8.6
10	10.0	9.8	9.7	9.5
20	20.0	19.7	19.3	19.0
30	30.0	29.5	29.0	28.5
40	40.0	39.3	38.7	38.0
50	50.0	49.2	48.3	47.5

	56	55	3	2
1	0.9	0.9	0.0	0.0
2	1.9	1.8	0.1	0.1
3	2.8	2.8	0.2	0.1
4	3.7	3.7	0.2	0.1
5	4.7	4.6	0.2	0.2
6	5.6	5.5	0.3	0.2
7	6.5	6.4	0.4	0.2
8	7.5	7.3	0.4	0.3
9	8.4	8.2	0.4	0.3
10	9.3	9.2	0.5	0.3
20	18.7	18.3	1.0	0.7
30	28.0	27.5	1.5	1.0
40	37.3	36.7	2.0	1.3
50	46.7	45.8	2.5	1.7

P. P.	L Cos	d	L Cot	c d	L Tan	L Sin	d	′

77°

13°

'	L Sin	d	L Tan	c d	L Cot	L Cos	d		P. P.
0	9.35 209	54	9.36 336	58	0.63 664	9.98 872	3	60	
1	263	55	394	58	606	869	2	59	
2	318	55	452	57	548	867	3	58	
3	373	54	509	57	491	864	3	57	
4	427	54	566	58	434	861	3	56	
5	9.35 481	55	9.36 624	57	0.63 376	9.98 858	3	55	
6	536	54	681	57	319	855	3	54	
7	590	54	738	57	262	852	3	53	
8	644	54	795	57	205	849	3	52	
9	698	54	852	57	148	846	3	51	
10	9.35 752	54	9.36 909	57	0.63 091	9.98 843	3	50	
11	806	54	966	57	034	840	3	49	
12	860	54	.37 023	57	.62 977	837	3	48	
13	914	54	080	57	920	834	3	47	
14	968	54	137	56	863	831	3	46	
15	9.36 022	53	9.37 193	57	0.62 807	9.98 828	3	45	
16	075	54	250	56	750	825	3	44	58　57　56　55
17	129	53	306	57	694	822	3	43	1　1.0　1.0　0.9　0.9
18	182	54	363	56	637	819	3	42	2　1.9　1.9　1.9　1.8
19	236	53	419	57	581	816	3	41	3　2.9　2.8　2.8　2.8
20	9.36 289	53	9.37 476	56	0.62 524	9.98 813	3	40	4　3.9　3.8　3.7　3.7
21	342	53	532	56	468	810	3	39	5　4.8　4.8　4.7　4.6
22	395	54	588	56	412	807	3	38	6　5.8　5.7　5.6　5.5
23	449	53	644	56	356	804	3	37	7　6.8　6.6　6.5　6.4
24	502	53	700	56	300	801	3	36	8　7.7　7.6　7.5　7.3
25	9.36 555	53	9.37 756	56	0.62 244	9.98 798	3	35	9　8.7　8.6　8.4　8.2
26	608	52	812	56	188	795	3	34	10　9.7　9.5　9.3　9.2
27	660	53	868	56	132	792	3	33	20　19.3　19.0　18.7　18.3
28	713	53	924	56	076	789	3	32	30　29.0　28.5　28.0　27.5
29	766	53	980	55	020	786	3	31	40　38.7　38.0　37.3　36.7
30	9.36 819	52	9.38 035	56	0.61 965	9.98 783	3	30	50　48.3　47.5　46.7　45.8
31	871	53	091	56	909	780	3	29	
32	924	52	147	55	853	777	3	28	
33	976	52	202	55	798	774	3	27	54　53　52　51
34	.37 028	53	257	56	743	771	3	26	1　0.9　0.9　0.9　0.8
35	9.37 081	52	9.38 313	55	0.61 687	9.98 768	3	25	2　1.8　1.8　1.7　1.7
36	133	52	368	55	632	765	3	24	3　2.7　2.6　2.6　2.6
37	185	52	423	56	577	762	3	23	4　3.6　3.5　3.5　3.4
38	237	52	479	55	521	759	3	22	5　4.5　4.4　4.3　4.2
39	289	52	534	55	466	756	3	21	6　5.4　5.3　5.2　5.1
40	9.37 341	52	9.38 589	55	0.61 411	9.98 753	3	20	7　6.3　6.2　6.1　6.0
41	393	52	644	55	356	750	4	19	8　7.2　7.1　6.9　6.8
42	445	52	699	55	301	746	3	18	9　8.1　8.0　7.8　7.6
43	497	52	754	54	246	743	3	17	10　9.0　8.8　8.7　8.5
44	549	51	808	55	192	740	3	16	20　18.0　17.7　17.3　17.0
45	9.37 600	52	9.38 863	55	0.61 137	9.98 737	3	15	30　27.0　26.5　26.0　25.5
46	652	51	918	54	082	734	3	14	40　36.0　35.3　34.7　34.0
47	703	52	972	55	028	731	3	13	50　45.0　44.2　43.3　42.5
48	755	51	.39 027	55	.60 973	728	3	12	
49	806	52	082	54	918	725	3	11	
50	9.37 858	51	9.39 136	54	0.60 864	9.98 722	3	10	
51	909	51	190	55	810	719	4	9	
52	960	51	245	54	755	715	3	8	
53	.38 011	51	299	54	701	712	3	7	
54	062	51	353	54	647	709	3	6	
55	9.38 113	51	9.39 407	54	0.60 593	9.98 706	3	5	
56	164	51	461	54	539	703	3	4	
57	215	51	515	54	485	700	3	3	
58	266	51	569	54	431	697	3	2	
59	317	51	623	54	377	694	4	1	
60	9.38 368		9.39 677		0.60 323	9.98 690		0	
	L Cos	d	L Cot	c d	L Tan	L Sin	d	'	P. P.

76°

		14°							
P. P.		′	L Sin	d	L Tan	c d	L Cot	L Cos	d	
		0	9.38 368	50	9.39 677	54	0.60 323	9.98 690	3	**60**
		1	418	50	731	54	269	687	3	59
		2	469	51	785	54	215	684	3	58
		3	519	50	838	53	162	681	3	57
		4	570	51	892	54	108	678	3	56
				50		53			3	
		5	9.38 620	50	9.39 945	54	0.60 055	9.98 675	4	**55**
		6	670	51	999	54	001	671	3	54
		7	721	50	.40 052	53	.59 948	668	3	53
		8	771	50	106	54	894	665	3	52
		9	821	50	159	53	841	662	3	51
		10	9.38 871	50	9.40 212	54	0.59 788	9.98 659	3	**50**
		11	921	50	266	53	734	656	4	49
		12	971	50	319	53	681	652	3	48
		13	.39 021	50	372	53	628	649	3	47
		14	071	50	425	53	575	646	3	46
		15	9.39 121	49	9.40 478	53	0.59 522	9.98 643	3	**45**
		16	170	50	531	53	469	640	4	44
		17	220	50	584	52	416	636	3	43
		18	270	49	636	53	364	633	3	42
		19	319	50	689	53	311	630	3	41
		20	9.39 369	49	9.40 742	53	0.59 258	9.98 627	4	**40**
		21	418	49	795	52	205	623	3	39
		22	467	50	847	53	153	620	3	38
		23	517	49	900	52	100	617	3	37
		24	566	49	952	53	048	614	4	36
		25	9.39 615	49	9.41 005	52	0.58 995	9.98 610	3	**35**
		26	664	49	057	52	943	607	3	34
		27	713	49	109	52	891	604	3	33
		28	762	49	161	53	839	601	4	32
		29	811	49	214	52	786	597	3	31
		30	9.39 860	49	9.41 266	52	0.58 734	9.98 594	3	**30**
		31	909	49	318	52	682	591	3	29
		32	958	48	370	52	630	588	4	28
		33	.40 006	49	422	52	578	584	3	27
		34	055	48	474	52	526	581	3	26
		35	9.40 103	49	9.41 526	52	0.58 474	9.98 578	4	**25**
		36	152	48	578	51	422	574	3	24
		37	200	49	629	52	371	571	3	23
		38	249	48	681	52	319	568	3	22
		39	297	49	733	51	267	565	4	21
		40	9.40 346	48	9.41 784	52	0.58 216	9.98 561	3	**20**
		41	394	48	836	51	164	558	3	19
		42	442	48	887	52	113	555	4	18
		43	490	48	939	51	061	551	3	17
		44	538	48	990	51	010	548	3	16
		45	9.40 586	48	9.42 041	52	0.57 959	9.98 545	4	**15**
		46	634	48	093	51	907	541	3	14
		47	682	48	144	51	856	538	3	13
		48	730	48	195	51	805	535	4	12
		49	778	47	246	51	754	531	3	11
		50	9.40 825	48	9.42 297	51	0.57 703	9.98 528	3	**10**
		51	873	48	348	51	652	525	4	9
		52	921	47	399	51	601	521	3	8
		53	968	48	450	51	550	518	3	7
		54	.41 016	47	501	51	499	515	4	6
		55	9.41 063	48	9.42 552	51	0.57 448	9.98 511	3	**5**
		56	111	47	603	50	397	508	3	4
		57	158	47	653	51	347	505	4	3
		58	205	47	704	51	296	501	3	2
		59	252	48	755	50	245	498	4	1
		60	9.41 300		9.42 805		0.57 195	9.98 494		**0**
P. P.			L Cos	d	L Cot	c d	L Tan	L Sin	d	′
				75°						

P. P. side tables:

	50	49	48	47
1	0.8	0.8	0.8	0.8
2	1.7	1.6	1.6	1.6
3	2.5	2.4	2.4	2.4
4	3.3	3.3	3.2	3.1
5	4.2	4.1	4.0	3.9
6	5.0	4.9	4.8	4.7
7	5.8	5.7	5.6	5.5
8	6.7	6.5	6.4	6.3
9	7.5	7.4	7.2	7.0
10	8.3	8.2	8.0	7.8
20	16.7	16.3	16.0	15.7
30	25.0	24.5	24.0	23.5
40	33.3	32.7	32.0	31.3
50	41.7	40.8	40.0	39.2

	4	3	2
1	0.1	0.0	0.0
2	0.1	0.1	0.1
3	0.2	0.2	0.1
4	0.3	0.2	0.1
5	0.3	0.2	0.2
6	0.4	0.3	0.2
7	0.5	0.4	0.2
8	0.5	0.4	0.3
9	0.6	0.4	0.3
10	0.7	0.5	0.3
20	1.3	1.0	0.7
30	2.0	1.5	1.0
40	2.7	2.0	1.3
50	3.3	2.5	1.7

15°

′	L Sin	d	L Tan	c d	L Cot	L Cos	d		P. P.
0	9.41 300	47	9.42 805	51	0.57 195	9.98 494	3	60	
1	347	47	856	50	144	491	3	59	
2	394	47	906	51	094	488	4	58	
3	441	47	957	50	043	484	3	57	
4	488	47	.43 007	50	.56 993	481	4	56	
5	9.41 535	47	9.43 057	51	0.56 943	9.98 477	3	55	
6	582	46	108	50	892	474	3	54	
7	628	47	158	50	842	471	4	53	
8	675	47	208	50	792	467	3	52	
9	722	46	258	50	742	464	4	51	
10	9.41 768	47	9.43 308	50	0.56 692	9.98 460	3	50	
11	815	46	358	50	642	457	4	49	
12	861	47	408	50	592	453	3	48	
13	908	46	458	50	542	450	3	47	
14	954	47	508	50	492	447	4	46	
15	9.42 001	46	9.43 558	49	0.56 442	9.98 443	3	45	
16	047	46	607	50	393	440	4	44	
17	093	47	657	50	343	436	4	43	
18	140	46	707	50	293	433	4	42	**51 50 49**
19	186	46	756	49	244	429	3	41	1 0.8 0.8 0.8
20	9.42 232	46	9.43 806	50	0.56 194	9.98 426	4	40	2 1.7 1.7 1.6
21	278	46	855	49	145	422	4	39	3 2.6 2.5 2.4
22	324	46	905	50	095	419	3	38	4 3.4 3.3 3.3
23	370	46	954	49	046	415	4	37	5 4.2 4.2 4.1
24	416	45	.44 004	50	.55 996	412	3	36	6 5.1 5.0 4.9
25	9.42 461	46	9.44 053	49	0.55 947	9.98 409	4	35	7 6.0 5.8 5.7
26	507	46	102	49	898	405	3	34	8 6.8 6.7 6.5
27	553	46	151	50	849	402	4	33	9 7.6 7.5 7.4
28	599	45	201	49	799	398	3	32	10 8.5 8.3 8.2
29	644	46	250	49	750	395	4	31	20 17.0 16.7 16.3
30	9.42 690	45	9.44 299	49	0.55 701	9.98 391	3	30	30 25.5 25.0 24.5
31	735	46	348	49	652	388	4	29	40 34.0 33.3 32.7
32	781	45	397	49	603	384	3	28	50 42.5 41.7 40.8
33	826	46	446	49	554	381	4	27	
34	872	45	495	49	505	377	4	26	
35	9.42 917	45	9.44 544	48	0.55 456	9.98 373	3	25	**48 47 46**
36	962	46	592	49	408	370	4	24	1 0.8 0.8 0.8
37	.43 008	45	641	49	359	366	3	23	2 1.6 1.6 1.5
38	053	45	690	48	310	363	4	22	3 2.4 2.4 2.3
39	098	45	738	49	262	359	3	21	4 3.2 3.1 3.1
40	9.43 143	45	9.44 787	49	0.55 213	9.98 356	4	20	5 4.0 3.9 3.8
41	188	45	836	48	164	352	3	19	6 4.8 4.7 4.6
42	233	45	884	49	116	349	4	18	7 5.6 5.5 5.4
43	278	45	933	48	067	345	3	17	8 6.4 6.3 6.1
44	323	44	981	48	019	342	4	16	9 7.2 7.0 6.9
45	9.43 367	45	9.45 029	49	0.54 971	9.98 338	4	15	10 8.0 7.8 7.7
46	412	45	078	48	922	334	3	14	20 16.0 15.7 15.3
47	457	45	126	48	874	331	4	13	30 24.0 23.5 23.0
48	502	44	174	48	826	327	4	12	40 32.0 31.3 30.7
49	546	45	222	49	778	324	4	11	50 40.0 39.2 38.3
50	9.43 591	44	9.45 271	48	0.54 729	9.98 320	3	10	
51	635	45	319	48	681	317	4	9	
52	680	44	367	48	633	313	4	8	
53	724	45	415	48	585	309	3	7	
54	769	44	463	48	537	306	4	6	
55	9.43 813	44	9.45 511	48	0.54 489	9.98 302	3	5	
56	857	44	559	47	441	299	4	4	
57	901	45	606	48	394	295	4	3	
58	946	44	654	48	346	291	3	2	
59	990	44	702	48	298	288	4	1	
60	9.44 034		9.45 750		0.54 250	9.98 284		0	

	L Cos	d	L Cot	c d	L Tan	L Sin	d	′	P. P.

74°

16°								
P. P.	′	L Sin	d	L Tan	c d	L Cot	L Cos	d	

′	L Sin	d	L Tan	c d	L Cot	L Cos	d	
0	9.44 034		9.45 750		0.54 250	9.98 284		**60**
1	078	44	797	47	203	281	3	59
2	122	44	845	48	155	277	4	58
3	166	44	892	47	108	273	4	57
4	210	44	940	48	060	270	3	56
		43		47			4	
5	9.44 253	44	9.45 987	48	0.54 013	9.98 266	4	**55**
6	297	44	.46 035	47	.53 965	262	4	54
7	341	44	082	48	918	259	3	53
8	385	43	130	47	870	255	4	52
9	428	44	177	47	823	251	3	51
10	9.44 472	44	9.46 224	47	0.53 776	9.98 248	4	**50**
11	516	43	271	48	729	244	4	49
12	559	43	319	47	681	240	3	48
13	602	44	366	47	634	237	4	47
14	646	43	413	47	587	233	4	46
15	9.44 689	44	9.46 460	47	0.53 540	9.98 229	3	**45**
16	733	43	507	47	493	226	4	44
17	776	43	554	47	446	222	4	43
18	819	43	601	47	399	218	3	42
19	862	43	648	46	352	215	4	41
20	9.44 905	43	9.46 694	47	0.53 306	9.98 211	4	**40**
21	948	44	741	47	259	207	3	39
22	992	43	788	47	212	204	4	38
23	.45 035	42	835	46	165	200	4	37
24	077	43	881	47	119	196	4	36
25	9.45 120	43	9.46 928	47	0.53 072	9.98 192	3	**35**
26	163	43	975	46	025	189	4	34
27	206	43	.47 021	47	.52 979	185	4	33
28	249	43	068	46	932	181	4	32
29	292	42	114	46	886	177	3	31
30	9.45 334	43	9.47 160	47	0.52 840	9.98 174	4	**30**
31	377	42	207	46	793	170	4	29
32	419	43	253	46	747	166	4	28
33	462	42	299	47	701	162	3	27
34	504	43	346	46	654	159	4	26
35	9.45 547	42	9.47 392	46	0.52 608	9.98 155	4	**25**
36	589	43	438	46	562	151	4	24
37	632	42	484	46	516	147	3	23
38	674	42	530	46	470	144	4	22
39	716	42	576	46	424	140	4	21
40	9.45 758	43	9.47 622	46	0.52 378	9.98 136	4	**20**
41	801	42	668	46	332	132	3	19
42	843	42	714	46	286	129	4	18
43	885	42	760	46	240	125	4	17
44	927	42	806	46	194	121	4	16
45	9.45 969	42	9.47 852	45	0.52 148	9.98 117	4	**15**
46	.46 011	42	897	46	103	113	3	14
47	053	42	943	46	057	110	4	13
48	095	41	989	46	011	106	4	12
49	136	42	.48 035	45	.51 965	102	4	11
50	9.46 178	42	9.48 080	46	0.51 920	9.98 098	4	**10**
51	220	42	126	45	874	094	4	9
52	262	41	171	46	829	090	3	8
53	303	42	217	45	783	087	4	7
54	345	41	262	45	738	083	4	6
55	9.46 386	42	9.48 307	46	0.51 693	9.98 079	4	**5**
56	428	41	353	45	647	075	4	4
57	469	42	398	45	602	071	4	3
58	511	41	443	46	557	067	4	2
59	552	42	489	45	511	063	3	1
60	9.46 594		9.48 534		0.51 466	9.98 060		**0**

| **P. P.** | | L Cos | d | L Cot | c d | L Tan | L Sin | d | ′ |

73°

P. P. left table:

	45	**44**	**43**
1	0.8	0.7	0.7
2	1.5	1.5	1.4
3	2.2	2.2	2.2
4	3.0	2.9	2.9
5	3.8	3.7	3.6
6	4.5	4.4	4.3
7	5.2	5.1	5.0
8	6.0	5.9	5.7
9	6.8	6.6	6.4
10	7.5	7.3	7.2
20	15.0	14.7	14.3
30	22.5	22.0	21.5
40	30.0	29.3	28.7
50	37.5	36.7	35.8

	42	**41**	**4**	**3**
1	0.7	0.7	0.1	0.0
2	1.4	1.4	0.1	0.1
3	2.1	2.0	0.2	0.2
4	2.8	2.7	0.3	0.2
5	3.5	3.4	0.3	0.2
6	4.2	4.1	0.4	0.3
7	4.9	4.8	0.5	0.4
8	5.6	5.5	0.5	0.4
9	6.3	6.2	0.6	0.4
10	7.0	6.8	0.7	0.5
20	14.0	13.7	1.3	1.0
30	21.0	20.5	2.0	1.5
40	28.0	27.3	2.7	2.0
50	35.0	34.2	3.3	2.5

17°

′	L Sin	d	L Tan	c d	L Cot	L Cos	d		P. P.
0	9.46 594	41	9.48 534	45	0.51 466	9.98 060	4	60	
1	635	41	579	45	421	056	4	59	
2	676	41	624	45	376	052	4	58	
3	717	41	669	45	331	048	4	57	
4	758	42	714	45	286	044	4	56	
5	9.46 800	41	9.48 759	45	0.51 241	9.98 040	4	55	
6	841	41	804	45	196	036	4	54	
7	882	41	849	45	151	032	4	53	
8	923	41	894	45	106	029	3	52	
9	964	41	939	45	061	025	4	51	
10	9.47 005	40	9.48 984	45	0.51 016	9.98 021	4	50	
11	045	41	.49 029	44	.50 971	017	4	49	
12	086	41	073	45	927	013	4	48	
13	127	41	118	45	882	009	4	47	
14	168	41	163	44	837	005	4	46	
15	9.47 209	40	9.49 207	45	0.50 793	9.98 001	4	45	
16	249	41	252	44	748	.97 997	4	44	
17	290	40	296	45	704	993	4	43	
18	330	41	341	44	659	989	3	42	
19	371	40	385	45	615	986	4	41	
20	9.47 411	41	9.49 430	44	0.50 570	9.97 982	4	40	
21	452	40	474	45	526	978	4	39	
22	492	41	519	44	481	974	4	38	
23	533	40	563	44	437	970	4	37	
24	573	40	607	45	393	966	4	36	
25	9.47 613	41	9.49 652	44	0.50 348	9.97 962	4	35	
26	654	40	696	44	304	958	4	34	
27	694	40	740	44	260	954	4	33	
28	734	40	784	44	216	950	4	32	
29	774	40	828	44	172	946	4	31	
30	9.47 814	40	9.49 872	44	0.50 128	9.97 942	4	30	
31	854	40	916	44	084	938	4	29	
32	894	40	960	44	040	934	4	28	
33	934	40	.50 004	44	.49 996	930	4	27	
34	974	40	048	44	952	926	4	26	
35	9.48 014	40	9.50 092	44	0.49 908	9.97 922	4	25	
36	054	40	136	44	864	918	4	24	
37	094	39	180	43	820	914	4	23	
38	133	40	223	44	777	910	4	22	
39	173	40	267	44	733	906	4	21	
40	9.48 213	39	9.50 311	44	0.49 689	9.97 902	4	20	
41	252	40	355	43	645	898	4	19	
42	292	40	398	44	602	894	4	18	
43	332	39	442	43	558	890	4	17	
44	371	40	485	44	515	886	4	16	
45	9.48 411	39	9.50 529	43	0.49 471	9.97 882	4	15	
46	450	40	572	44	428	878	4	14	
47	490	39	616	43	384	874	4	13	
48	529	39	659	44	341	870	4	12	
49	568	39	703	43	297	866	5	11	
50	9.48 607	40	9.50 746	43	0.49 254	9.97 861	4	10	
51	647	39	789	44	211	857	4	9	
52	686	39	833	43	167	853	4	8	
53	725	39	876	43	124	849	4	7	
54	764	39	919	43	081	845	4	6	
55	9.48 803	39	9.50 962	43	0.49 038	9.97 841	4	5	
56	842	39	.51 005	43	.48 995	837	4	4	
57	881	39	048	44	952	833	4	3	
58	920	39	092	43	908	829	4	2	
59	959	39	135	43	865	825	4	1	
60	9.48 998		9.51 178		0.48 822	9.97 821		0	
	L Cos	d	L Cot	c d	L Tan	L Sin	d	′	P. P.

P. P.

	45	44	43
1	0.8	0.7	0.7
2	1.5	1.5	1.4
3	2.2	2.2	2.2
4	3.0	2.9	2.9
5	3.8	3.7	3.6
6	4.5	4.4	4.3
7	5.2	5.1	5.0
8	6.0	5.9	5.7
9	6.8	6.6	6.4
10	7.5	7.3	7.2
20	15.0	14.7	14.3
30	22.5	22.0	21.5
40	30.0	29.3	28.7
50	37.5	36.7	35.8

	42	41	40
1	0.7	0.7	0.7
2	1.4	1.4	1.3
3	2.1	2.0	2.0
4	2.8	2.7	2.7
5	3.5	3.4	3.3
6	4.2	4.1	4.0
7	4.9	4.8	4.7
8	5.6	5.5	5.3
9	6.3	6.2	6.0
10	7.0	6.8	6.7
20	14.0	13.7	13.3
30	21.0	20.5	20.0
40	28.0	27.3	26.7
50	35.0	34.2	33.3

72°

18°

	39	38	37
1	0.6	0.6	0.6
2	1.3	1.3	1.2
3	2.0	1.9	1.8
4	2.6	2.5	2.5
5	3.2	3.2	3.1
6	3.9	3.8	3.7
7	4.6	4.4	4.3
8	5.2	5.1	4.9
9	5.8	5.7	5.6
10	6.5	6.3	6.2
20	13.0	12.7	12.3
30	19.5	19.0	18.5
40	26.0	25.3	24.7
50	32.5	31.7	30.8

	5	4	3
1	0.1	0.1	0.0
2	0.2	0.1	0.1
3	0.2	0.2	0.2
4	0.3	0.3	0.2
5	0.4	0.3	0.2
6	0.5	0.4	0.3
7	0.6	0.5	0.4
8	0.7	0.5	0.4
9	0.8	0.6	0.4
10	0.8	0.7	0.5
20	1.7	1.3	1.0
30	2.5	2.0	1.5
40	3.3	2.7	2.0
50	4.2	3.3	2.5

'	L Sin	d	L Tan	c d	L Cot	L Cos	d	
0	9.48 998	39	9.51 178	43	0.48 822	9.97 821	4	60
1	.49 037	39	221	43	779	817	5	59
2	076	39	264	42	736	812	4	58
3	115	38	306	43	694	808	4	57
4	153	39	349	43	651	804	4	56
5	9.49 192	39	9.51 392	43	0.48 608	9.97 800	4	55
6	231	38	435	43	565	796	4	54
7	269	39	478	42	522	792	4	53
8	308	39	520	43	480	788	4	52
9	347	38	563	43	437	784	5	51
10	9.49 385	39	9.51 606	42	0.48 394	9.97 779	4	50
11	424	38	648	43	352	775	4	49
12	462	38	691	43	309	771	4	48
13	500	39	734	42	266	767	4	47
14	539	38	776	43	224	763	4	46
15	9.49 577	38	9.51 819	42	0.48 181	9.97 759	5	45
16	615	39	861	42	139	754	4	44
17	654	38	903	43	097	750	4	43
18	692	38	946	42	054	746	4	42
19	730	38	988	43	012	742	4	41
20	9.49 768	38	9.52 031	42	0.47 969	9.97 738	4	40
21	806	38	073	42	927	734	5	39
22	844	38	115	42	885	729	4	38
23	882	38	157	43	843	725	4	37
24	920	38	200	42	800	721	4	36
25	9.49 958	38	9.52 242	42	0.47 758	9.97 717	4	35
26	996	38	284	42	716	713	5	34
27	.50 034	38	326	42	674	708	4	33
28	072	38	368	42	632	704	4	32
29	110	38	410	42	590	700	4	31
30	9.50 148	37	9.52 452	42	0.47 548	9.97 696	5	30
31	185	38	494	42	506	691	4	29
32	223	38	536	42	464	687	4	28
33	261	37	578	42	422	683	4	27
34	298	38	620	41	380	679	5	26
35	9.50 336	38	9.52 661	42	0.47 339	9.97 674	4	25
36	374	37	703	42	297	670	4	24
37	411	38	745	42	255	666	4	23
38	449	37	787	42	213	662	5	22
39	486	37	829	41	171	657	4	21
40	9.50 523	38	9.52 870	42	0.47 130	9.97 653	4	20
41	561	37	912	41	088	649	4	19
42	598	37	953	42	047	645	5	18
43	635	38	995	42	005	640	4	17
44	673	37	.53 037	41	.46 963	636	4	16
45	9.50 710	37	9.53 078	42	0.46 922	9.97 632	4	15
46	747	37	120	41	880	628	5	14
47	784	37	161	41	839	623	4	13
48	821	37	202	42	798	619	4	12
49	858	38	244	41	756	615	5	11
50	9.50 896	37	9.53 285	42	0.46 715	9.97 610	4	10
51	933	37	327	41	673	606	4	9
52	970	37	368	41	632	602	5	8
53	.51 007	36	409	41	591	597	4	7
54	043	37	450	42	550	593	4	6
55	9.51 080	37	9.53 492	41	0.46 508	9.97 589	5	5
56	117	37	533	41	467	584	4	4
57	154	37	574	41	426	580	4	3
58	191	36	615	41	385	576	5	2
59	227	37	656	41	344	571	4	1
60	9.51 264		9.53 697		0.46 303	9.97 567		0

| P. P. | | | L Cos | d | L Cot | c d | L Tan | L Sin | d | ' |

71°

19°

'	L Sin	d	L Tan	c d	L Cot	L Cos	d	
0	9.51 264	37	9.53 697	41	0.46 303	9.97 567	4	60
1	301	37	738	41	262	563	5	59
2	338	37	779	41	221	558	4	58
3	374	36	820	41	180	554	4	57
4	411	37	861	41	139	550	5	56
		36		41			5	
5	9.51 447	37	9.53 902	41	0.46 098	9.97 545	4	55
6	484	36	943	41	057	541	5	54
7	520	37	984	41	016	536	4	53
8	557	36	.54 025	40	.45 975	532	4	52
9	593	36	065	41	935	528	5	51
10	9.51 629	37	9.54 106	41	0.45 894	9.97 523	4	50
11	666	36	147	40	853	519	4	49
12	702	36	187	41	813	515	5	48
13	738	36	228	41	772	510	4	47
14	774	37	269	40	731	506	5	46
15	9.51 811	36	9.54 309	41	0.45 691	9.97 501	4	45
16	847	36	350	40	650	497	5	44
17	883	36	390	41	610	492	4	43
18	919	36	431	41	569	488	4	42
19	955	36	471	41	529	484	5	41
20	9.51 991	36	9.54 512	40	0.45 488	9.97 479	4	40
21	.52 027	36	552	41	448	475	5	39
22	063	36	593	40	407	470	4	38
23	099	36	633	40	367	466	5	37
24	135	36	673	41	327	461	4	36
25	9.52 171	36	9.54 714	40	0.45 286	9.97 457	4	35
26	207	35	764	40	246	453	5	34
27	242	36	794	41	206	448	4	33
28	278	36	835	40	165	444	5	32
29	314	36	875	40	125	439	4	31
30	9.52 350	35	9.54 915	40	0.45 085	9.97 435	5	30
31	385	36	955	40	045	430	4	29
32	421	35	995	40	005	426	5	28
33	456	36	.55 035	40	.44 965	421	4	27
34	492	35	075	40	925	417	5	26
35	9.52 527	36	9.55 115	40	0.44 885	9.97 412	4	25
36	563	35	155	40	845	408	5	24
37	598	36	195	40	805	403	4	23
38	634	35	235	40	765	399	5	22
39	669	36	275	40	725	394	4	21
40	9.52 705	35	9.55 315	40	0.44 685	9.97 390	5	20
41	740	35	355	40	645	385	4	19
42	775	36	395	39	605	381	5	18
43	811	35	434	40	566	376	4	17
44	846	35	474	40	526	372	5	16
45	9.52 881	35	9.55 514	40	0.44 486	9.97 367	4	15
46	916	35	554	39	446	363	5	14
47	951	35	593	40	407	358	5	13
48	986	35	633	40	367	353	4	12
49	.53 021	35	673	39	327	349	5	11
50	9.53 056	36	9.55 712	40	0.44 288	9.97 344	4	10
51	092	34	752	39	248	340	5	9
52	126	35	791	40	209	335	4	8
53	161	35	831	39	169	331	5	7
54	196	35	870	40	130	326	4	6
55	9.53 231	35	9.55 910	39	0.44 090	9.97 322	5	5
56	266	35	949	40	051	317	5	4
57	301	35	989	39	011	312	5	3
58	336	34	.56 028	39	.43 972	308	4	2
59	370	35	067	40	933	303	5	1
60	9.53 405		9.56 107		0.43 893	9.97 299		0

| | L Cos | d | L Cot | c d | L Tan | L Sin | d | ' |

70°

P. P.

	41	40	39
1	0.7	0.7	0.6
2	1.4	1.3	1.3
3	2.0	2.0	2.0
4	2.7	2.7	2.6
5	3.4	3.3	3.2
6	4.1	4.0	3.9
7	4.8	4.7	4.6
8	5.5	5.3	5.2
9	6.2	6.0	5.8
10	6.8	6.7	6.5
20	13.7	13.3	13.0
30	20.5	20.0	19.5
40	27.3	26.7	26.0
50	34.2	33.3	32.5

	38	37	36
1	0.6	0.6	0.6
2	1.3	1.2	1.2
3	1.9	1.8	1.8
4	2.5	2.5	2.4
5	3.2	3.1	3.0
6	3.8	3.7	3.6
7	4.4	4.3	4.2
8	5.1	4.9	4.8
9	5.7	5.6	5.4
10	6.3	6.2	6.0
20	12.7	12.3	12.0
30	19.0	18.5	18.0
40	25.3	24.7	24.0
50	31.7	30.8	30.0

20°

P. P.

35	34	33
1	0.6	0.6	0.6
2	1.2	1.1	1.1
3	1.8	1.7	1.6
4	2.3	2.3	2.2
5	2.9	2.8	2.8
6	3.5	3.4	3.3
7	4.1	4.0	3.8
8	4.7	4.5	4.4
9	5.2	5.1	5.0
10	5.8	5.7	5.5
20	11.7	11.3	11.0
30	17.5	17.0	16.5
40	23.3	22.7	22.0
50	29.2	28.3	27.5

5	4
1	0.1	0.1
2	0.2	0.1
3	0.2	0.2
4	0.3	0.3
5	0.4	0.3
6	0.5	0.4
7	0.6	0.5
8	0.7	0.5
9	0.8	0.6
10	0.8	0.7
20	1.7	1.3
30	2.5	2.0
40	3.3	2.7
50	4.2	3.3

′	L Sin	d	L Tan	c d	L Cot	L Cos	d	′
0	9.53 405	35	9.56 107	39	0.43 893	9.97 299	5	**60**
1	440	35	146	39	854	294	5	59
2	475	34	185	39	815	289	4	58
3	509	35	224	40	776	285	5	57
4	544	34	264	39	736	280	4	56
5	9.53 578	35	9.56 303	39	0.43 697	9.97 276	5	**55**
6	613	34	342	39	658	271	5	54
7	647	35	381	39	619	266	4	53
8	682	34	420	39	580	262	5	52
9	716	35	459	39	541	257	5	51
10	9.53 751	34	9.56 498	39	0.43 502	9.97 252	4	**50**
11	785	34	537	39	463	248	5	49
12	819	35	576	39	424	243	5	48
13	854	34	615	39	385	238	4	47
14	888	34	654	39	346	234	5	46
15	9.53 922	35	9.56 693	39	0.43 307	9.97 229	5	**45**
16	957	34	732	39	268	224	4	44
17	991	34	771	39	229	220	5	43
18	.54 025	34	810	39	190	215	5	42
19	059	34	849	38	151	210	4	41
20	9.54 093	34	9.56 887	39	0.43 113	9.97 206	5	**40**
21	127	34	926	39	074	201	5	39
22	161	34	965	39	035	196	4	38
23	195	34	.57 004	38	.42 996	192	5	37
24	229	34	042	39	958	187	5	36
25	9.54 263	34	9.57 081	39	0.42 919	9.97 182	4	**35**
26	297	34	120	38	880	178	5	34
27	331	34	158	39	842	173	5	33
28	365	34	197	38	803	168	5	32
29	399	34	235	39	765	163	4	31
30	9.54 433	33	9.57 274	38	0.42 726	9.97 159	5	**30**
31	466	34	312	39	688	154	5	29
32	500	34	351	38	649	149	4	28
33	534	33	389	39	611	145	5	27
34	567	34	428	38	572	140	5	26
35	9.54 601	34	9.57 466	38	0.42 534	9.97 135	5	**25**
36	635	33	504	39	496	130	4	24
37	668	34	543	38	457	126	5	23
38	702	33	581	38	419	121	5	22
39	735	34	619	39	381	116	5	21
40	9.54 769	33	9.57 658	38	0.42 342	9.97 111	4	**20**
41	802	34	696	38	304	107	5	19
42	836	33	734	38	266	102	5	18
43	869	34	772	38	228	097	5	17
44	903	33	810	39	190	092	5	16
45	9.54 936	33	9.57 849	38	0.42 151	9.97 087	4	**15**
46	969	34	887	38	113	083	5	14
47	.55 003	33	925	38	075	078	5	13
48	036	33	963	38	037	073	5	12
49	069	33	.58 001	38	.41 999	068	5	11
50	9.55 102	34	9.58 039	38	0.41 961	9.97 063	4	**10**
51	136	33	077	38	923	059	5	9
52	169	33	115	38	885	054	5	8
53	202	33	153	38	847	049	5	7
54	235	33	191	38	809	044	5	6
55	9.55 268	33	9.58 229	38	0.41 771	9.97 039	4	**5**
56	301	33	267	37	733	035	5	4
57	334	33	304	38	696	030	5	3
58	367	33	342	38	658	025	5	2
59	400	33	380	38	620	020	5	1
60	9.55 433		9.58 418		0.41 582	9.97 015		**0**

| P. P. | | L Cos | d | L Cot | c d | L Tan | L Sin | d | ′ |

69°

21°

′	L Sin	d	L Tan	c d	L Cot	L Cos	d	′
0	9.55 433	33	9.58 418	37	0.41 582	9.97 015	5	60
1	466	33	455	38	545	010	5	59
2	499	33	493	38	507	005	4	58
3	532	32	531	38	469	001	5	57
4	564	33	569	37	431	.96 996	5	56
5	9.55 597	33	9.58 606	38	0.41 394	9.96 991	5	55
6	630	33	644	37	356	986	5	54
7	663	32	681	38	319	981	5	53
8	695	33	719	38	281	976	5	52
9	728	33	757	37	243	971	5	51
10	9.55 761	32	9.58 794	38	0.41 206	9.96 966	4	50
11	793	33	832	37	168	962	5	49
12	826	32	869	38	131	957	5	48
13	858	33	907	37	093	952	5	47
14	891	32	944	37	056	947	5	46
15	9.55 923	33	9.58 981	38	0.41 019	9.96 942	5	45
16	956	32	.59 019	37	.40 981	937	5	44
17	988	33	056	38	944	932	5	43
18	.56 021	32	094	37	906	927	5	42
19	053	32	131	37	869	922	5	41
20	9.56 085	33	9.59 168	37	0.40 832	9.96 917	5	40
21	118	32	205	38	795	912	5	39
22	150	32	243	37	757	907	4	38
23	182	33	280	37	720	903	5	37
24	215	32	317	37	683	898	5	36
25	9.56 247	32	9.59 354	37	0.40 646	9.96 893	5	35
26	279	32	391	38	609	888	5	34
27	311	32	429	37	571	883	5	33
28	343	32	466	37	534	878	5	32
29	375	33	503	37	497	873	5	31
30	9.56 408	32	9.59 540	37	0.40 460	9.96 868	5	30
31	440	32	577	37	423	863	5	29
32	472	32	614	37	386	858	5	28
33	504	32	651	37	349	853	5	27
34	536	32	688	37	312	848	5	26
35	9.56 568	31	9.59 725	37	0.40 275	9.96 843	5	25
36	599	32	762	37	238	838	5	24
37	631	32	799	36	201	833	5	23
38	663	32	835	37	165	828	5	22
39	695	32	872	37	128	823	5	21
40	9.56 727	32	9.59 909	37	0.40 091	9.96 818	5	20
41	759	31	946	37	054	813	5	19
42	790	32	983	36	017	808	5	18
43	822	32	.60 019	37	.39 981	803	5	17
44	854	32	056	37	944	798	5	16
45	9.56 886	31	9.60 093	37	0.39 907	9.96 793	5	15
46	917	32	130	36	870	788	5	14
47	949	31	166	37	834	783	5	13
48	980	32	203	37	797	778	6	12
49	.57 012	32	240	36	760	772	5	11
50	9.57 044	31	9.60 276	37	0.39 724	9.96 767	5	10
51	075	32	313	36	687	762	5	9
52	107	31	349	37	651	757	5	8
53	138	31	386	36	614	752	5	7
54	169	32	422	37	578	747	5	6
55	9.57 201	31	9.60 459	36	0.39 541	9.96 742	5	5
56	232	32	495	37	505	737	5	4
57	264	31	532	36	468	732	5	3
58	295	31	568	37	432	727	5	2
59	326	32	605	36	395	722	5	1
60	9.57 358		9.60 641		0.39 359	9.96 717		0
	L Cos	d	L Cot	c d	L Tan	L Sin	d	′

P. P.

	38	37	36
1	0.6	0.6	0.6
2	1.3	1.2	1.2
3	1.9	1.8	1.8
4	2.5	2.5	2.4
5	3.2	3.1	3.0
6	3.8	3.7	3.6
7	4.4	4.3	4.2
8	5.1	4.9	4.8
9	5.7	5.6	5.4
10	6.3	6.2	6.0
20	12.7	12.3	12.0
30	19.0	18.5	18.0
40	25.3	24.7	24.0
50	31.7	30.8	30.0

	35	33	32
1	0.6	0.6	0.5
2	1.2	1.1	1.1
3	1.8	1.6	1.6
4	2.3	2.2	2.1
5	2.9	2.8	2.7
6	3.5	3.3	3.2
7	4.1	3.8	3.7
8	4.7	4.4	4.3
9	5.2	5.0	4.8
10	5.8	5.5	5.3
20	11.7	11.0	10.7
30	17.5	16.5	16.0
40	23.3	22.0	21.3
50	29.2	27.5	26.7

22°

P. P.

31	30	29
1	0.5	0.5	0.5
2	1.0	1.0	1.0
3	1.6	1.5	1.4
4	2.1	2.0	1.9
5	2.6	2.5	2.4
6	3.1	3.0	2.9
7	3.6	3.5	3.4
8	4.1	4.0	3.9
9	4.6	4.5	4.4
10	5.2	5.0	4.8
20	10.3	10.0	9.7
30	15.5	15.0	14.5
40	20.7	20.0	19.3
50	25.8	25.0	24.2

6	5	4
1	0.1	0.1	0.1
2	0.2	0.2	0.1
3	0.3	0.2	0.2
4	0.4	0.3	0.3
5	0.5	0.4	0.3
6	0.6	0.5	0.4
7	0.7	0.6	0.5
8	0.8	0.7	0.5
9	0.9	0.8	0.6
10	1.0	0.8	0.7
20	2.0	1.7	1.3
30	3.0	2.5	2.0
40	4.0	3.3	2.7
50	5.0	4.2	3.3

′	L Sin	d	L Tan	c d	L Cot	L Cos	d	
0	9.57 358	31	9.60 641	36	0.39 359	9.96 717	6	60
1	389	31	677	37	323	711	5	59
2	420	31	714	36	286	706	5	58
3	451	31	750	36	250	701	5	57
4	482	32	786	37	214	696	5	56
5	9.57 514	31	9.60 823	36	0.39 177	9.96 691	5	55
6	545	31	859	36	141	686	5	54
7	576	31	895	36	105	681	5	53
8	607	31	931	36	069	676	5	52
9	638	31	967	37	033	670	5	51
10	9.57 669	31	9.61 004	36	0.38 996	9.96 665	5	50
11	700	31	040	36	960	660	5	49
12	731	31	076	36	924	655	5	48
13	762	31	112	36	888	650	5	47
14	793	31	148	36	852	645	5	46
15	9.57 824	31	9.61 184	36	0.38 816	9.96 640	6	45
16	855	30	220	36	780	634	5	44
17	885	31	256	36	744	629	5	43
18	916	31	292	36	708	624	5	42
19	947	31	328	36	672	619	5	41
20	9.57 978	30	9.61 364	36	0.38 636	9.96 614	6	40
21	.58 008	31	400	36	600	608	5	39
22	039	31	436	36	564	603	5	38
23	070	31	472	36	528	598	5	37
24	101	30	508	36	492	593	5	36
25	9.58 131	31	9.61 544	35	0.38 456	9.96 588	6	35
26	162	30	579	36	421	582	5	34
27	192	31	615	36	385	577	5	33
28	223	30	651	36	349	572	5	32
29	253	31	687	35	313	567	5	31
30	9.58 284	30	9.61 722	36	0.38 278	9.96 562	6	30
31	314	31	758	36	242	556	5	29
32	345	30	794	36	206	551	5	28
33	375	31	830	35	170	546	5	27
34	406	30	865	36	135	541	6	26
35	9.58 436	31	9.61 901	35	0.38 099	9.96 535	5	25
36	467	30	936	36	064	530	5	24
37	497	30	972	36	028	525	5	23
38	527	30	.62 008	35	.37 992	520	6	22
39	557	31	043	36	957	514	5	21
40	9.58 588	30	9.62 079	35	0.37 921	9.96 509	5	20
41	618	30	114	36	886	504	6	19
42	648	30	150	35	850	498	5	18
43	678	31	185	36	815	493	5	17
44	709	30	221	35	779	488	5	16
45	9.58 739	30	9.62 256	36	0.37 744	9.96 483	6	15
46	769	30	292	35	708	477	5	14
47	799	30	327	35	673	472	5	13
48	829	30	362	36	638	467	6	12
49	859	30	398	35	602	461	5	11
50	9.58 889	30	9.62 433	35	0.37 567	9.96 456	5	10
51	919	30	468	36	532	451	6	9
52	949	30	504	35	496	445	5	8
53	979	30	539	35	461	440	5	7
54	.59 009	30	574	35	426	435	6	6
55	9.59 039	30	9.62 609	36	0.37 391	9.96 429	5	5
56	069	29	645	35	355	424	5	4
57	098	30	680	35	320	419	6	3
58	128	30	715	35	285	413	5	2
59	158	30	750	35	250	408	5	1
60	9.59 188		9.62 785		0.37 215	9.96 403		0

| **P. P.** | | L Cos | d | L Cot | c d | L Tan | L Sin | d | ′ |

67°

23°

'	L Sin	d	L Tan	c d	L Cot	L Cos	d	'
0	9.59 188	30	9.62 785	35	0.37 215	9.96 403	6	60
1	218	29	820	35	180	397	5	59
2	247	30	855	35	145	392	5	58
3	277	30	890	36	110	387	6	57
4	307	29	926	35	074	381	5	56
5	9.59 336	30	9.62 961	35	0.37 039	9.96 376	6	55
6	366	30	996	35	004	370	5	54
7	396	29	.63 031	35	.36 969	365	5	53
8	425	30	066	35	934	360	6	52
9	455	29	101	34	899	354	5	51
10	9.59 484	30	9.63 135	35	0.36 865	9.96 349	6	50
11	514	29	170	35	830	343	5	49
12	543	30	205	35	795	338	5	48
13	573	29	240	35	760	333	6	47
14	602	30	275	35	725	327	5	46
15	9.59 632	29	9.63 310	35	0.36 690	9.96 322	6	45
16	661	29	345	34	655	316	6	44
17	690	30	379	35	621	311	6	43
18	720	29	414	35	586	305	6	42
19	749	29	449	35	551	300	6	41
20	9.59 778	30	9.63 484	35	0.36 516	9.96 294	5	40
21	808	29	519	34	481	289	5	39
22	837	29	553	35	447	284	6	38
23	866	29	588	34	412	278	5	37
24	895	29	623	34	377	273	6	36
25	9.59 924	30	9.63 657	35	0.36 343	9.96 267	6	35
26	954	29	692	34	308	262	5	34
27	983	29	726	35	274	256	5	33
28	.60 012	29	761	34	239	251	6	32
29	041	29	796	35	204	245	5	31
30	9.60 070	29	9.63 830	35	0.36 170	9.96 240	6	30
31	099	29	865	34	135	234	5	29
32	128	29	899	35	101	229	6	28
33	157	29	934	34	066	223	5	27
34	186	29	968	35	032	218	6	26
35	9.60 215	29	9.64 003	34	0.35 997	9.96 212	6	25
36	244	29	037	35	963	207	5	24
37	273	29	072	34	928	201	6	23
38	302	29	106	34	894	196	5	22
39	331	28	140	35	860	190	5	21
40	9.60 359	29	9.64 175	34	0.35 825	9.96 185	6	20
41	388	29	209	34	791	179	5	19
42	417	29	243	35	757	174	6	18
43	446	28	278	34	722	168	6	17
44	474	29	312	34	688	162	5	16
45	9.60 503	29	9.64 346	35	0.35 654	9.96 157	6	15
46	532	29	381	34	619	151	5	14
47	561	28	415	34	585	146	6	13
48	589	29	449	34	551	140	5	12
49	618	28	483	34	517	135	6	11
50	9.60 646	29	9.64 517	35	0.35 483	9.96 129	6	10
51	675	29	552	34	448	123	5	9
52	704	28	586	34	414	118	6	8
53	732	29	620	34	380	112	5	7
54	761	28	654	34	346	107	6	6
55	9.60 789	29	9.64 688	34	0.35 312	9.96 101	6	5
56	818	28	722	34	278	095	5	4
57	846	29	756	34	244	090	6	3
58	875	28	790	34	210	084	6	2
59	903	28	824	34	176	079	5	1
60	9.60 931		9.64 858		0.35 142	9.96 073		0

L Cos	d	L Cot	c d	L Tan	L Sin	d	'

P. P.

	36	35	34
1	0.6	0.6	0.6
2	1.2	1.2	1.1
3	1.8	1.8	1.7
4	2.4	2.3	2.3
5	3.0	2.9	2.8
6	3.6	3.5	3.4
7	4.2	4.1	4.0
8	4.8	4.7	4.5
9	5.4	5.2	5.1
10	6.0	5.8	5.7
20	12.0	11.7	11.3
30	18.0	17.5	17.0
40	24.0	23.3	22.7
50	30.0	29.2	28.3

	33	30
1	0.6	0.5
2	1.1	1.0
3	1.6	1.5
4	2.2	2.0
5	2.8	2.5
6	3.3	3.0
7	3.8	3.5
8	4.4	4.0
9	5.0	4.5
10	5.5	5.0
20	11.0	10.0
30	16.5	15.0
40	22.0	20.0
50	27.5	25.0

66°

24°

P. P.	'	L Sin	d	L Tan	c d	L Cot	L Cos	d	
	0	9.60 931	29	9.64 858	34	0.35 142	9.96 073	6	**60**
	1	960	28	892	34	108	067	5	59
	2	988	28	926	34	074	062	6	58
	3	.61 016	29	960	34	040	056	6	57
	4	045	28	994	34	006	050	5	56
	5	9.61 073	28	9.65 028	34	0.34 972	9.96 045	6	**55**
	6	101	28	062	34	938	039	5	54
	7	129	29	096	34	904	034	6	53
	8	158	28	130	34	870	028	6	52
	9	186	28	164	33	836	022	5	51
	10	9.61 214	28	9.65 197	34	0.34 803	9.96 017	6	**50**
	11	242	28	231	34	769	011	6	49
	12	270	28	265	34	735	005	5	48
	13	298	28	299	34	701	000	6	47
	14	326	28	333	33	667	.95 994	6	46
29 28 27	**15**	9.61 354	28	9.65 366	34	0.34 634	9.95 988	6	**45**
1 | 0.5 0.5 0.4	16	382	29	400	34	600	982	5	44
2 | 1.0 0.9 0.9	17	411	27	434	33	566	977	6	43
3 | 1.4 1.4 1.4	18	438	28	467	34	533	971	6	42
4 | 1.9 1.9 1.8	19	466	28	501	34	499	965	5	41
5 | 2.4 2.3 2.2	**20**	9.61 494	28	9.65 535	33	0.34 465	9.95 960	6	**40**
6 | 2.9 2.8 2.7	21	522	28	568	34	432	954	6	39
7 | 3.4 3.3 3.2	22	550	28	602	34	398	948	6	38
8 | 3.9 3.7 3.6	23	578	28	636	33	364	942	5	37
9 | 4.4 4.2 4.0	24	606	28	669	34	331	937	6	36
10 | 4.8 4.7 4.5	**25**	9.61 634	28	9.65 703	33	0.34 297	9.95 931	6	**35**
20 | 9.7 9.3 9.0	26	662	27	736	34	264	925	5	34
30 | 14.5 14.0 13.5	27	689	28	770	33	230	920	6	33
40 | 19.3 18.7 18.0	28	717	28	803	34	197	914	6	32
50 | 24.2 23.3 22.5	29	745	28	837	33	163	908	6	31
	30	9.61 773	27	9.65 870	34	0.34 130	9.95 902	5	**30**
	31	800	28	904	33	096	897	6	29
	32	828	28	937	34	063	891	6	28
	33	856	27	971	33	029	885	6	27
	34	883	28	.66 004	34	.33 996	879	6	26
6 5	**35**	9.61 911	28	9.66 038	33	0.33 962	9.95 873	5	**25**
1 | 0.1 0.1	36	939	27	071	33	929	868	6	24
2 | 0.2 0.2	37	966	28	104	34	896	862	6	23
3 | 0.3 0.2	38	994	27	138	33	862	856	6	22
4 | 0.4 0.3	39	.62 021	28	171	33	829	850	6	21
5 | 0.5 0.4	**40**	9.62 049	27	9.66 204	34	0.33 796	9.95 844	5	**20**
6 | 0.6 0.5	41	076	28	238	33	762	839	6	19
7 | 0.7 0.6	42	104	27	271	33	729	833	6	18
8 | 0.8 0.7	43	131	28	304	33	696	827	6	17
9 | 0.9 0.8	44	159	27	337	34	663	821	6	16
10 | 1.0 0.8	**45**	9.62 186	28	9.66 371	33	0.33 629	9.95 815	5	**15**
20 | 2.0 1.7	46	214	27	404	33	596	810	6	14
30 | 3.0 2.5	47	241	27	437	33	563	804	6	13
40 | 4.0 3.3	48	268	28	470	33	530	798	6	12
50 | 5.0 4.2	49	296	27	503	34	497	792	6	11
	50	9.62 323	27	9.66 537	33	0.33 463	9.95 786	6	**10**
	51	350	27	570	33	430	780	5	9
	52	377	28	603	33	397	775	6	8
	53	405	27	636	33	364	769	6	7
	54	432	27	669	33	331	763	6	6
	55	9.62 459	27	9.66 702	33	0.33 298	9.95 757	6	**5**
	56	486	27	735	33	265	751	6	4
	57	513	28	768	33	232	745	6	3
	58	541	27	801	33	199	739	6	2
	59	568	27	834	33	166	733	5	1
	60	9.62 595		9.66 867		0.33 133	9.95 728		**0**
P. P.		L Cos	d	L Cot	c d	L Tan	L Sin	d	'

65°

25°

'	L Sin	d	L Tan	c d	L Cot	L Cos	d	'
0	9.62 595	27	9.66 867	33	0.33 133	9.95 728	6	60
1	622	27	900	33	100	722	6	59
2	649	27	933	33	067	716	6	58
3	676	27	966	33	034	710	6	57
4	703	27	999	33	001	704	6	56
5	9.62 730	27	9.67 032	33	0.32 968	9.95 698	6	55
6	757	27	065	33	935	692	6	54
7	784	27	098	33	902	686	6	53
8	811	27	131	32	869	680	6	52
9	838	27	163	33	837	674	6	51
10	9.62 865	27	9.67 196	33	0.32 804	9.95 668	5	50
11	892	26	229	33	771	663	6	49
12	918	27	262	33	738	657	6	48
13	945	27	295	32	705	651	6	47
14	972	27	327	33	673	645	6	46
15	9.62 999	27	9.67 360	33	0.32 640	9.95 639	6	45
16	.63 026	26	393	33	607	633	6	44
17	052	27	426	32	574	627	6	43
18	079	27	458	33	542	621	6	42
19	106	27	491	33	509	615	6	41
20	9.63 133	26	9.67 524	32	0.32 476	9.95 609	6	40
21	159	27	556	33	444	603	6	39
22	186	27	589	33	411	597	6	38
23	213	26	622	32	378	591	6	37
24	239	27	654	33	346	585	6	36
25	9.63 266	26	9.67 687	32	0.32 313	9.95 579	6	35
26	292	27	719	33	281	573	6	34
27	319	26	752	33	248	567	6	33
28	345	27	785	32	215	561	6	32
29	372	26	817	33	183	555	6	31
30	9.63 398	27	9.67 850	32	0.32 150	9.95 549	6	30
31	425	26	882	33	118	543	6	29
32	451	27	915	32	085	537	6	28
33	478	26	947	33	053	531	6	27
34	504	27	980	32	020	525	6	26
35	9.63 531	26	9.68 012	32	0.31 988	9.95 519	6	25
36	557	26	044	33	956	513	6	24
37	583	27	077	32	923	507	7	23
38	610	26	109	33	891	500	6	22
39	636	26	142	32	858	494	6	21
40	9.63 662	27	9.68 174	32	0.31 826	9.95 488	6	20
41	689	26	206	33	794	482	6	19
42	715	26	239	32	761	476	6	18
43	741	26	271	32	729	470	6	17
44	767	27	303	33	697	464	6	16
45	9.63 794	26	9.68 336	32	0.31 664	9.95 458	6	15
46	820	26	368	32	632	452	6	14
47	846	26	400	32	600	446	6	13
48	872	26	432	33	568	440	6	12
49	898	26	465	32	535	434	7	11
50	9.63 924	26	9.68 497	32	0.31 503	9.95 427	6	10
51	950	26	529	32	471	421	6	9
52	976	26	561	32	439	415	6	8
53	.64 002	26	593	33	407	409	6	7
54	028	26	626	32	374	403	6	6
55	9.64 054	26	9.68 658	32	0.31 342	9.95 397	6	5
56	080	26	690	32	310	391	7	4
57	106	26	722	32	278	384	6	3
58	132	26	754	32	246	378	6	2
59	158	26	786	32	214	372	6	1
60	9.64 184		9.68 818		0.31 182	9.95 366		0

	L Cos	d	L Cot	c d	L Tan	L Sin	d	'

P. P.

	33	32	31
1	0.6	0.5	0.5
2	1.1	1.1	1.0
3	1.6	1.6	1.6
4	2.2	2.1	2.1
5	2.8	2.7	2.6
6	3.3	3.2	3.1
7	3.8	3.7	3.6
8	4.4	4.3	4.1
9	5.0	4.8	4.6
10	5.5	5.3	5.2
20	11.0	10.7	10.3
30	16.5	16.0	15.5
40	22.0	21.3	20.7
50	27.5	26.7	25.8

	27	26
1	0.4	0.4
2	0.9	0.9
3	1.4	1.3
4	1.8	1.7
5	2.2	2.2
6	2.7	2.6
7	3.2	3.0
8	3.6	3.5
9	4.0	3.9
10	4.5	4.3
20	9.0	8.7
30	13.5	13.0
40	18.0	17.3
50	22.5	21.7

64°

26°

P. P.	'	L Sin	d	L Tan	c d	L Cot	L Cos	d	
	0	9.64 184	26	9.68 818	32	0.31 182	9.95 366	6	**60**
	1	210	26	850	32	150	360	6	59
	2	236	26	882	32	118	354	6	58
	3	262	26	914	32	086	348	7	57
	4	288	25	946	32	054	341	6	56
	5	9.64 313	26	9.68 978	32	0.31 022	9.95 335	6	**55**
	6	339	26	.69 010	32	.30 990	329	6	54
	7	365	26	042	32	958	323	6	53
	8	391	26	074	32	926	317	7	52
	9	417	25	106	32	894	310	6	51
	10	9.64 442	26	9.69 138	32	0.30 862	9.95 304	6	**50**
	11	468	26	170	32	830	298	6	49
	12	494	25	202	32	798	292	6	48
	13	519	26	234	32	766	286	7	47
	14	545	26	266	32	734	279	6	46
25 24	**15**	9.64 571	25	9.69 298	31	0.30 702	9.95 273	6	**45**
1 0.4 0.4	16	596	26	329	32	671	267	6	44
2 0.8 0.8	17	622	25	361	32	639	261	7	43
3 1.2 1.2	18	647	26	393	32	607	254	6	42
4 1.7 1.6	19	673	25	425	32	575	248	6	41
5 2.1 2.0	**20**	9.64 698	26	9.69 457	31	0.30 543	9.95 242	6	**40**
6 2.5 2.4	21	724	25	488	32	512	236	7	39
7 2.9 2.8	22	749	26	520	32	480	229	6	38
8 3.3 3.2	23	775	25	552	32	448	223	6	37
9 3.8 3.6	24	800	26	584	31	416	217	6	36
10 4.2 4.0	**25**	9.64 826	25	9.69 615	32	0.30 385	9.95 211	7	**35**
20 8.3 8.0	26	851	26	647	32	353	204	6	34
30 12.5 12.0	27	877	25	679	31	321	198	6	33
40 16.7 16.0	28	902	25	710	32	290	192	7	32
50 20.8 20.0	29	927	26	742	32	258	185	6	31
	30	9.64 953	25	9.69 774	31	0.30 226	9.95 179	6	**30**
	31	978	25	805	32	195	173	6	29
	32	.65 003	26	837	31	163	167	7	28
	33	029	25	868	32	132	160	6	27
	34	054	25	900	32	100	154	6	26
7 6 5	**35**	9.65 079	25	9.69 932	31	0.30 068	9.95 148	7	**25**
1 0.1 0.1 0.1	36	104	26	963	32	037	141	6	24
2 0.2 0.2 0.2	37	130	25	995	31	005	135	6	23
3 0.4 0.3 0.2	38	155	25	.70 026	32	.29 974	129	7	22
4 0.5 0.4 0.3	39	180	25	058	31	942	122	6	21
5 0.6 0.5 0.4	**40**	9.65 205	25	9.70 089	32	0.29 911	9.95 116	6	**20**
6 0.7 0.6 0.5	41	230	25	121	31	879	110	7	19
7 0.8 0.7 0.6	42	255	26	152	32	848	103	6	18
8 0.9 0.8 0.7	43	281	25	184	31	816	097	7	17
9 1.0 0.9 0.8	44	306	25	215	32	785	090	6	16
10 1.2 1.0 0.8	**45**	9.65 331	25	9.70 247	31	0.29 753	9.95 084	6	**15**
20 2.3 2.0 1.7	46	356	25	278	31	722	078	7	14
30 3.5 3.0 2.5	47	381	25	309	32	691	071	6	13
40 4.7 4.0 3.3	48	406	25	341	31	659	065	6	12
50 5.8 5.0 4.2	49	431	25	372	32	628	059	7	11
	50	9.65 456	25	9.70 404	31	0.29 596	9.95 052	6	**10**
	51	481	25	435	31	565	046	7	9
	52	506	25	466	32	534	039	6	8
	53	531	25	498	31	502	033	6	7
	54	556	24	529	31	471	027	7	6
	55	9.65 580	25	9.70 560	32	0.29 440	9.95 020	6	**5**
	56	605	25	592	31	408	014	7	4
	57	630	25	623	31	377	007	6	3
	58	655	25	654	31	346	001	6	2
	59	680	25	685	32	315	.94 995	7	1
	60	9.65 705		9.70 717		0.29 283	9.94 988		**0**
P. P.		L Cos	d	L Cot	c d	L Tan	L Sin	d	'

63°

27°

'	L Sin	d	L Tan	c d	L Cot	L Cos	d	'
0	9.65 705	24	9.70 717	31	0.29 283	9.94 988	6	60
1	729	25	748	31	252	982	7	59
2	754	25	779	31	221	975	6	58
3	779	25	810	31	190	969	7	57
4	804	24	841	32	159	962	6	56
5	9.65 828	25	9.70 873	31	0.29 127	9.94 956	7	55
6	853	25	904	31	096	949	6	54
7	878	24	935	31	065	943	7	53
8	902	25	966	31	034	936	6	52
9	927	25	997	31	003	930	7	51
10	9.65 952	24	9.71 028	31	0.28 972	9.94 923	6	50
11	976	25	059	31	941	917	6	49
12	.66 001	24	090	31	910	911	7	48
13	025	25	121	32	879	904	6	47
14	050	25	153	31	847	898	7	46
15	9.66 075	24	9.71 184	31	0.28 816	9.94 891	6	45
16	099	25	215	31	785	885	7	44
17	124	24	246	31	754	878	7	43
18	148	25	277	31	723	871	6	42
19	173	24	308	31	692	865	7	41
20	9.66 197	24	9.71 339	31	0.28 661	9.94 858	6	40
21	221	25	370	31	630	852	7	39
22	246	25	401	30	599	845	6	38
23	270	25	431	31	569	839	7	37
24	295	24	462	31	538	832	6	36
25	9.66 319	24	9.71 493	31	0.28 507	9.94 826	7	35
26	343	25	524	31	476	819	6	34
27	368	24	555	31	445	813	7	33
28	392	24	586	31	414	806	7	32
29	416	25	617	31	383	799	6	31
30	9.66 441	24	9.71 648	31	0.28 352	9.94 793	7	30
31	465	24	679	30	321	786	6	29
32	489	24	709	31	291	780	7	28
33	513	24	740	31	260	773	6	27
34	537	25	771	31	229	767	7	26
35	9.66 562	24	9.71 802	31	0.28 198	9.94 760	7	25
36	586	24	833	30	167	753	6	24
37	610	24	863	31	137	747	7	23
38	634	24	894	31	106	740	6	22
39	658	24	925	30	075	734	7	21
40	9.66 682	24	9.71 955	31	0.28 045	9.94 727	7	20
41	706	25	986	31	014	720	6	19
42	731	24	.72 017	31	.27 983	714	7	18
43	755	24	048	30	952	707	6	17
44	779	24	078	31	922	700	6	16
45	9.66 803	24	9.72 109	31	0.27 891	9.94 694	7	15
46	827	24	140	30	860	687	7	14
47	851	24	170	31	830	680	6	13
48	875	24	201	30	799	674	7	12
49	899	23	231	31	769	667	7	11
50	9.66 922	24	9.72 262	31	0.27 738	9.94 660	6	10
51	946	24	293	30	707	654	7	9
52	970	24	323	31	677	647	7	8
53	994	24	354	30	646	640	6	7
54	.67 018	24	384	31	616	634	7	6
55	9.67 042	24	9.72 415	30	0.27 585	9.94 627	7	5
56	066	24	445	31	555	620	6	4
57	090	23	476	30	524	614	7	3
58	113	24	506	31	494	607	7	2
59	137	24	537	30	463	600	7	1
60	9.67 161		9.72 567		0.27 433	9.94 593		0

	L Cos	d	L Cot	c d	L Tan	L Sin	d	'

62°

P. P.

	32	31	30
1	0.5	0.5	0.5
2	1.1	1.0	1.0
3	1.6	1.6	1.5
4	2.1	2.1	2.0
5	2.7	2.6	2.5
6	3.2	3.1	3.0
7	3.7	3.6	3.5
8	4.3	4.1	4.0
9	4.8	4.6	4.5
10	5.3	5.2	5.0
20	10.7	10.3	10.0
30	16.0	15.5	15.0
40	21.3	20.7	20.0
50	26.7	25.8	25.0

	25	24	23
1	0.4	0.4	0.4
2	0.8	0.8	0.8
3	1.2	1.2	1.2
4	1.7	1.6	1.5
5	2.1	2.0	1.9
6	2.5	2.4	2.3
7	2.9	2.8	2.7
8	3.3	3.2	3.1
9	3.8	3.6	3.4
10	4.2	4.0	3.8
20	8.3	8.0	7.7
30	12.5	12.0	11.5
40	16.7	16.0	15.3
50	20.8	20.0	19.2

28°

′	L Sin	d	L Tan	c d	L Cot	L Cos	d	
0	9.67 161	24	9.72 567	31	0.27 433	9.94 593	6	**60**
1	185	23	598	30	402	587	7	59
2	208	24	628	31	372	580	7	58
3	232	24	659	30	341	573	6	57
4	256	24	689	31	311	567	7	56
5	9.67 280	23	9.72 720	30	0.27 280	9.94 560	7	**55**
6	303	24	750	30	250	553	7	54
7	327	23	780	31	220	546	6	53
8	350	24	811	30	189	540	7	52
9	374	24	841	31	159	533	7	51
10	9.67 398	23	9.72 872	30	0.27 128	9.94 526	7	**50**
11	421	24	902	30	098	519	6	49
12	445	23	932	31	068	513	7	48
13	468	24	963	30	037	506	7	47
14	492	23	993	30	007	499	7	46
15	9.67 515	24	9.73 023	31	0.26 977	9.94 492	7	**45**
16	539	23	054	30	946	485	6	44
17	562	24	084	30	916	479	7	43
18	586	23	114	30	886	472	7	42
19	609	24	144	31	856	465	7	41
20	9.67 633	23	9.73 175	30	0.26 825	9.94 458	7	**40**
21	656	24	205	30	795	451	6	39
22	680	23	235	30	765	445	7	38
23	703	23	265	30	735	438	7	37
24	726	24	295	31	705	431	7	36
25	9.67 750	23	0.73 326	30	0.26 674	9.94 424	7	**35**
26	773	23	356	30	644	417	7	34
27	796	24	386	30	614	410	6	33
28	820	23	416	30	584	404	7	32
29	843	23	446	30	554	397	7	31
30	9.67 866	24	9.73 476	31	0.26 524	9.94 390	7	**30**
31	890	23	507	30	493	383	7	29
32	913	23	537	30	463	376	7	28
33	936	23	567	30	433	369	7	27
34	959	23	597	30	403	362	7	26
35	9.67 982	24	9.73 627	30	0.26 373	9.94 355	6	**25**
36	.68 006	23	657	30	343	349	7	24
37	029	23	687	30	313	342	7	23
38	052	23	717	30	283	335	7	22
39	075	23	747	30	253	328	7	21
40	9.68 098	23	9.73 777	30	0.26 223	9.94 321	7	**20**
41	121	23	807	30	193	314	7	19
42	144	23	837	30	163	307	7	18
43	167	23	867	30	133	300	7	17
44	190	23	897	30	103	293	7	16
45	9.68 213	24	9.73 927	30	0.26 073	9.94 286	7	**15**
46	237	23	957	30	043	279	6	14
47	260	23	987	30	013	273	7	13
48	283	22	.74 017	30	.25 983	266	7	12
49	305	23	047	30	953	259	7	11
50	9.68 328	23	9.74 077	30	0.25 923	9.94 252	7	**10**
51	351	23	107	30	893	245	7	9
52	374	23	137	29	863	238	7	8
53	397	23	166	30	834	231	7	7
54	420	23	196	30	804	224	7	6
55	9.68 443	23	9.74 226	30	0.25 774	9.94 217	7	**5**
56	466	23	256	30	744	210	7	4
57	489	23	286	30	714	203	7	3
58	512	22	316	29	684	196	7	2
59	534	23	345	30	655	189	7	1
60	9.68 557		9.74 375		0.25 625	9.94 182		**0**
	L Cos	d	L Cot	c d	L Tan	L Sin	d	′

61°

P. P.

	22	8	7
1	0.4	0.1	0.1
2	0.7	0.3	0.2
3	1.1	0.4	0.4
4	1.5	0.5	0.5
5	1.8	0.7	0.6
6	2.2	0.8	0.7
7	2.6	0.9	0.8
8	2.9	1.1	0.9
9	3.3	1.2	1.0
10	3.7	1.3	1.2
20	7.3	2.7	2.3
30	11.0	4.0	3.5
40	14.7	5.3	4.7
50	18.3	6.7	5.8

P. P.

29°

'	L Sin	d	L Tan	c d	L Cot	L Cos	d	'
0	9.68 557	23	9.74 375	30	0.25 625	9.94 182	7	60
1	580	23	405	30	595	175	7	59
2	603	22	435	30	565	168	7	58
3	625	23	465	29	535	161	7	57
4	648	23	494	30	506	154	7	56
5	9.68 671	23	9.74 524	30	0.25 476	9.94 147	7	55
6	694	23	554	29	446	140	7	54
7	716	22	583	30	417	133	7	53
8	739	23	613	30	387	126	7	52
9	762	22	643	30	357	119	7	51
10	9.68 784	23	9.74 673	29	0.25 327	9.94 112	7	50
11	807	22	702	30	298	105	7	49
12	829	23	732	30	268	098	8	48
13	852	23	762	29	238	090	7	47
14	875	22	791	30	209	083	7	46
15	9.68 897	23	9.74 821	30	0.25 179	9.94 076	7	45
16	920	22	851	29	149	069	7	44
17	942	23	880	30	120	062	7	43
18	965	22	910	29	090	055	7	42
19	987	23	939	30	061	048	7	41
20	9.69 010	22	9.74 969	29	0.25 031	9.94 041	7	40
21	032	22	998	30	002	034	7	39
22	055	22	.75 028	30	.24 972	027	7	38
23	077	23	058	29	942	020	8	37
24	100	22	087	30	913	012	7	36
25	9.69 122	22	9.75 117	29	0.24 883	9.94 005	7	35
26	144	23	146	30	854	.93 998	7	34
27	167	22	176	29	824	991	7	33
28	189	23	205	30	795	984	7	32
29	212	22	235	29	765	977	7	31
30	9.69 234	22	9.75 264	30	0.24 736	9.93 970	7	30
31	256	23	294	29	706	963	8	29
32	279	22	323	30	677	955	7	28
33	301	22	353	29	647	948	7	27
34	323	22	382	29	618	941	7	26
35	9.69 345	23	9.75 411	30	0.24 589	9.93 934	7	25
36	368	22	441	29	559	927	7	24
37	390	22	470	30	530	920	8	23
38	412	22	500	29	500	912	7	22
39	434	22	529	29	471	905	7	21
40	9.69 456	23	9.75 558	30	0.24 442	9.93 898	7	20
41	479	22	588	29	412	891	7	19
42	501	22	617	30	383	884	7	18
43	523	22	647	29	353	876	8	17
44	545	22	676	29	324	869	7	16
45	9.69 567	22	9.75 705	30	0.24 295	9.93 862	7	15
46	589	22	735	29	265	855	8	14
47	611	22	764	29	236	847	7	13
48	633	22	793	29	207	840	7	12
49	655	22	822	30	178	833	7	11
50	9.69 677	22	9.75 852	29	0.24 148	9.93 826	7	10
51	699	22	881	29	119	819	8	9
52	721	22	910	29	090	811	7	8
53	743	22	939	30	061	804	7	7
54	765	22	969	29	031	797	8	6
55	9.69 787	22	9.75 998	29	0.24 002	9.93 789	7	5
56	809	22	.76 027	29	.23 973	782	7	4
57	831	22	056	30	944	775	7	3
58	853	22	086	29	914	768	8	2
59	875	22	115	29	885	760	7	1
60	9.69 897		9.76 144		0.23 856	9.93 753		0
	L Cos	d	L Cot	c d	L Tan	L Sin	d	'

60°

P. P.

	30	29	28
1	0.5	0.5	0.5
2	1.0	1.0	0.9
3	1.5	1.4	1.4
4	2.0	1.9	1.9
5	2.5	2.4	2.3
6	3.0	2.9	2.8
7	3.5	3.4	3.3
8	4.0	3.9	3.7
9	4.5	4.4	4.2
10	5.0	4.8	4.7
20	10.0	9.7	9.3
30	15.0	14.5	14.0
40	20.0	19.3	18.7
50	25.0	24.2	23.3

	23	22	21
1	0.4	0.4	0.4
2	0.8	0.7	0.7
3	1.2	1.1	1.0
4	1.5	1.5	1.4
5	1.9	1.8	1.8
6	2.3	2.2	2.1
7	2.7	2.6	2.4
8	3.1	2.9	2.8
9	3.4	3.3	3.2
10	3.8	3.7	3.5
20	7.7	7.3	7.0
30	11.5	11.0	10.5
40	15.3	14.7	14.0
50	19.2	18.3	17.5

30°

P. P.	'	L Sin	d	L Tan	c d	L Cot	L Cos	d	
	0	9.69 897	22	9.76 144	29	0.23 856	9.93 753	7	**60**
	1	919	22	173	29	827	746	8	59
	2	941	22	202	29	798	738	7	58
	3	963	21	231	30	769	731	7	57
	4	984	22	261	29	739	724	7	56
	5	9.70 006	22	9.76 290	29	0.23 710	9.93 717	8	**55**
	6	028	22	319	29	681	709	7	54
	7	050	22	348	29	652	702	7	53
	8	072	21	377	29	623	695	8	52
	9	093	22	406	29	594	687	7	51
	10	9.70 115	22	9.76 435	29	0.23 565	9.93 680	7	**50**
	11	137	22	464	29	536	673	8	49
	12	159	21	493	29	507	665	7	48
	13	180	22	522	29	478	658	8	47
	14	202	22	551	29	449	650	7	46
	15	9.70 224	21	9.76 580	29	0.23 420	9.93 643	7	**45**
	16	245	22	609	30	391	636	8	44
	17	267	21	639	29	361	628	7	43
	18	288	22	668	29	332	621	7	42
	19	310	22	697	28	303	614	8	41
	20	9.70 332	21	9.76 725	29	0.23 275	9.93 606	7	**40**
	21	353	22	754	29	246	599	8	39
	22	375	21	783	29	217	591	7	38
	23	396	22	812	29	188	584	7	37
	24	418	21	841	29	159	577	8	36
	25	9.70 439	22	9.76 870	29	0.23 130	9.93 569	7	**35**
	26	461	21	899	29	101	562	8	34
	27	482	22	928	29	072	554	7	33
	28	504	21	957	29	043	547	8	32
	29	525	22	986	29	014	539	7	31
	30	9.70 547	21	9.77 015	29	0.22 985	9.93 532	7	**30**
	31	568	22	044	29	956	525	8	29
	32	590	21	073	28	927	517	7	28
	33	611	22	101	29	899	510	8	27
	34	633	21	130	29	870	502	7	26
	35	9.70 654	21	9.77 159	29	0.22 841	9.93 495	8	**25**
	36	675	22	188	29	812	487	7	24
	37	697	21	217	29	783	480	8	23
	38	718	21	246	28	754	472	7	22
	39	739	22	274	29	726	465	8	21
	40	9.70 761	21	9.77 303	29	0.22 697	9.93 457	7	**20**
	41	782	21	332	29	668	450	8	19
	42	803	21	361	29	639	442	7	18
	43	824	22	390	28	610	435	8	17
	44	846	21	418	29	582	427	7	16
	45	9.70 867	21	9.77 447	29	0.22 553	9.93 420	8	**15**
	46	888	21	476	29	524	412	7	14
	47	909	22	505	28	495	405	8	13
	48	931	21	533	29	467	397	7	12
	49	952	21	562	29	438	390	8	11
	50	9.70 973	21	9.77 591	28	0.22 409	9.93 382	7	**10**
	51	994	21	619	29	381	375	8	9
	52	.71 015	21	648	29	352	367	7	8
	53	036	22	677	29	323	360	8	7
	54	058	21	706	28	294	352	8	6
	55	9.71 079	21	9.77 734	29	0.22 266	9.93 344	7	**5**
	56	100	21	763	28	237	337	8	4
	57	121	21	791	29	209	329	7	3
	58	142	21	820	29	180	322	8	2
	59	163	21	849	28	151	314	7	1
	60	9.71 184		9.77 877		0.22 123	9.93 307		**0**
P. P.		L Cos	d	L Cot	c d	L Tan	L Sin	d	'

59°

P. P.

	8	7
1	0.1	0.1
2	0.3	0.2
3	0.4	0.4
4	0.5	0.5
5	0.7	0.6
6	0.8	0.7
7	0.9	0.8
8	1.1	0.9
9	1.2	1.0
10	1.3	1.2
20	2.7	2.3
30	4.0	3.5
40	5.3	4.7
50	6.7	5.8

31°

′	L Sin	d	L Tan	c d	L Cot	L Cos	d	′		P. P.	
0	9.71 184	21	9.77 877	29	0.22 123	9.93 307	8	60			
1	205	21	906	29	094	299	8	59			
2	226	21	935	28	065	291	7	58			
3	247	21	963	29	037	284	8	57			
4	268	21	992	28	008	276	7	56			
5	9.71 289	21	9.78 020	29	0.21 980	9.93 269	8	55			
6	310	21	049	28	951	261	8	54			
7	331	21	077	29	923	253	7	53			
8	352	21	106	29	894	246	8	52			
9	373	20	135	28	865	238	8	51			
10	9.71 393	21	9.78 163	29	0.21 837	9.93 230	7	50			
11	414	21	192	28	808	223	8	49			
12	435	21	220	29	780	215	8	48			
13	456	21	249	28	751	207	7	47			
14	477	21	277	29	723	200	8	46			
15	9.71 498	21	9.78 306	28	0.21 694	9.93 192	8	45		**29** **28** **27**	
16	519	20	334	29	666	184	7	44	1	0.5 0.5 0.4	
17	539	21	363	28	637	177	8	43	2	1.0 0.9 0.9	
18	560	21	391	28	609	169	8	42	3	1.4 1.4 1.4	
19	581	21	419	29	581	161	7	41	4	1.9 1.9 1.8	
									5	2.4 2.3 2.2	
									6	2.9 2.8 2.7	
20	9.71 602	20	9.78 448	28	0.21 552	9.93 154	8	40	7	3.4 3.3 3.2	
21	622	21	476	29	524	146	8	39	8	3.9 3.7 3.6	
22	643	21	505	28	495	138	7	38	9	4.4 4.2 4.0	
23	664	21	533	29	467	131	8	37	10	4.8 4.7 4.5	
24	685	20	562	28	438	123	8	36	20	9.7 9.3 9.0	
									30	14.5 14.0 13.5	
									40	19.3 18.7 18.0	
25	9.71 705	21	9.78 590	28	0.21 410	9.93 115	7	35	50	24.2 23.3 22.5	
26	726	21	618	29	382	108	8	34			
27	747	20	647	28	353	100	8	33			
28	767	21	675	29	325	092	8	32			
29	788	21	704	28	296	084	7	31			
30	9.71 809	20	9.78 732	28	0.21 268	9.93 077	8	30			
31	829	21	760	29	240	069	8	29			
32	850	21	789	28	211	061	8	28		**21** **20** **19**	
33	870	21	817	28	183	053	7	27	1	0.4 0.3 0.3	
34	891	20	845	29	155	046	8	26	2	0.7 0.7 0.6	
									3	1.0 1.0 1.0	
									4	1.4 1.3 1.3	
35	9.71 911	21	9.78 874	28	0.21 126	9.93 038	8	25	5	1.8 1.7 1.6	
36	932	20	902	28	098	030	8	24	6	2.1 2.0 1.9	
37	952	21	930	29	070	022	8	23	7	2.4 2.3 2.2	
38	973	21	959	28	041	014	7	22	8	2.8 2.7 2.5	
39	994	20	987	28	013	007	8	21	9	3.2 3.0 2.8	
									10	3.5 3.3 3.2	
									20	7.0 6.7 6.3	
40	9.72 014	20	9.79 015	28	0.20 985	9.92 999	8	20	30	10.5 10.0 9.5	
41	034	21	043	29	957	991	8	19	40	14.0 13.3 12.7	
42	055	20	072	28	928	983	8	18	50	17.5 16.7 15.8	
43	075	21	100	28	900	976	8	17			
44	096	20	128	28	872	968	8	16			
45	9.72 116	21	9.79 156	29	0.20 844	9.92 960	8	15			
46	137	20	185	28	815	952	8	14			
47	157	20	213	28	787	944	8	13			
48	177	21	241	28	759	936	8	12			
49	198	20	269	28	731	929	8	11			
50	9.72 218	20	9.79 297	29	0.20 703	9.92 921	8	10			
51	238	21	326	28	674	913	8	9			
52	259	20	354	28	646	905	8	8			
53	279	20	382	28	618	897	8	7			
54	299	21	410	28	590	889	8	6			
55	9.72 320	20	9.79 438	28	0.20 562	9.92 881	7	5			
56	340	20	466	29	534	874	8	4			
57	360	21	495	28	505	866	8	3			
58	381	20	523	28	477	858	8	2			
59	401	20	551	28	449	850	8	1			
60	9.72 421		9.79 579		0.20 421	9.92 842		0			
	L Cos	d	L Cot	c d	L Tan	L Sin	d	′		P. P.	

58°

32°

P. P.	′	L Sin	d	L Tan	c d	L Cot	L Cos	d	′
	0	9.72 421	20	9.79 579	28	0.20 421	9.92 842	8	60
	1	441	20	607	28	393	834	8	59
	2	461	20	635	28	365	826	8	58
	3	482	21	663	28	337	818	8	57
	4	502	20	691	28	309	810	7	56
	5	9.72 522	20	9.79 719	28	0.20 281	9.92 803	8	55
	6	542	20	747	29	253	795	8	54
	7	562	20	776	28	224	787	8	53
	8	582	20	804	28	196	779	8	52
	9	602	20	832	28	168	771	8	51
	10	9.72 622	21	9.79 860	28	0.20 140	9.92 763	8	50
	11	643	20	888	28	112	755	8	49
	12	663	20	916	28	084	747	8	48
	13	683	20	944	28	056	739	8	47
	14	703	20	972	28	028	731	8	46
	15	9.72 723	20	9.80 000	28	0.20 000	9.92 723	8	45
	16	743	20	028	28	.19 972	715	8	44
	17	763	20	056	28	944	707	8	43
	18	783	20	084	28	916	699	8	42
	19	803	20	112	28	888	691	8	41
	20	9.72 823	20	9.80 140	28	0.19 860	9.92 683	8	40
	21	843	20	168	27	832	675	8	39
	22	863	20	195	28	805	667	8	38
	23	883	19	223	28	777	659	8	37
	24	902	20	251	28	749	651	8	36
	25	9.72 922	20	9.80 279	28	0.19 721	9.92 643	8	35
	26	942	20	307	28	693	635	8	34
	27	962	20	335	28	665	627	8	33
	28	982	20	363	28	637	619	8	32
	29	.73 002	20	391	28	609	611	8	31
	30	9.73 022	19	9.80 419	28	0.19 581	9.92 603	8	30
	31	041	20	447	27	553	595	8	29
	32	061	20	474	28	526	587	8	28
	33	081	20	502	28	498	579	8	27
	34	101	20	530	28	470	571	8	26
	35	9.73 121	19	9.80 558	28	0.19 442	9.92 563	8	25
	36	140	20	586	28	414	555	9	24
	37	160	20	614	28	386	546	8	23
	38	180	20	642	27	358	538	8	22
	39	200	19	669	28	331	530	8	21
	40	9.73 219	20	9.80 697	28	0.19 303	9.92 522	8	20
	41	239	20	725	28	275	514	8	19
	42	259	19	753	28	247	506	8	18
	43	278	20	781	27	219	498	8	17
	44	298	20	808	28	192	490	8	16
	45	9.73 318	19	9.80 836	28	0.19 164	9.92 482	9	15
	46	337	20	864	28	136	473	8	14
	47	357	20	892	27	108	465	8	13
	48	377	19	919	28	081	457	8	12
	49	396	20	947	28	053	449	8	11
	50	9.73 416	19	9.80 975	28	0.19 025	9.92 441	8	10
	51	435	20	.81 003	27	.18 997	433	8	9
	52	455	19	030	28	970	425	9	8
	53	474	20	058	28	942	416	8	7
	54	494	19	086	27	914	408	8	6
	55	9.73 513	20	9.81 113	28	0.18 887	9.92 400	8	5
	56	533	19	141	28	859	392	8	4
	57	552	20	169	27	831	384	8	3
	58	572	19	196	28	804	376	9	2
	59	591	20	224	28	776	367	8	1
	60	9.73 611		9.81 252		0.18 748	9.92 359		0

P. P. (left margin)

	9	8	7
1	0.2	0.1	0.1
2	0.3	0.3	0.2
3	0.4	0.4	0.4
4	0.6	0.5	0.5
5	0.8	0.7	0.6
6	0.9	0.8	0.7
7	1.0	0.9	0.8
8	1.2	1.1	0.9
9	1.4	1.2	1.0
10	1.5	1.3	1.2
20	3.0	2.7	2.3
30	4.5	4.0	3.5
40	6.0	5.3	4.7
50	7.5	6.7	5.8

P. P.		L Cos	d	L Cot	c d	L Tan	L Sin	d	′

57°

33°

'	L Sin	d	L Tan	c d	L Cot	L Cos	d	'	P. P.
0	9.73 611	19	9.81 252	27	0.18 748	9.92 359	8	60	
1	630	20	279	28	721	351	8	59	
2	650	19	307	28	693	343	8	58	
3	669	20	335	27	665	335	9	57	
4	689	19	362	28	638	326	8	56	
5	9.73 708	19	9.81 390	28	0.18 610	9.92 318	8	55	
6	727	20	418	27	582	310	8	54	
7	747	19	445	28	555	302	9	53	
8	766	19	473	27	527	293	8	52	
9	785	20	500	28	500	285	8	51	
10	9.73 805	19	9.81 528	28	0.18 472	9.92 277	8	50	
11	824	19	556	27	444	269	9	49	
12	843	20	583	28	417	260	8	48	
13	863	19	611	27	389	252	8	47	
14	882	19	638	28	362	244	9	46	
15	9.73 901	20	9.81 666	27	0.18 334	9.92 235	8	45	
16	921	19	693	28	307	227	8	44	
17	940	19	721	27	279	219	8	43	
18	959	19	748	28	252	211	9	42	
19	978	19	776	27	224	202	8	41	
20	9.73 997	20	9.81 803	28	0.18 197	9.92 194	8	40	
21	.74 017	19	831	27	169	186	9	39	
22	036	19	858	28	142	177	8	38	
23	055	19	886	27	114	169	8	37	
24	074	19	913	28	087	161	9	36	
25	9.74 093	20	9.81 941	27	0.18 059	9.92 152	8	35	
26	113	19	968	28	032	144	8	34	
27	132	19	996	27	004	136	9	33	
28	151	19	.82 023	28	.17 977	127	8	32	
29	170	19	051	27	949	119	8	31	
30	9.74 189	19	9.82 078	28	0.17 922	9.92 111	9	30	
31	208	19	106	27	894	102	8	29	
32	227	19	133	28	867	094	8	28	
33	246	19	161	27	839	086	8	27	
34	265	19	188	27	812	077	8	26	
35	9.74 284	19	9.82 215	28	0.17 785	9.92 069	9	25	
36	303	19	243	27	757	060	8	24	
37	322	19	270	28	730	052	8	23	
38	341	19	298	27	702	044	9	22	
39	360	19	325	27	675	035	8	21	
40	9.74 379	19	9.82 352	28	0.17 648	9.92 027	9	20	
41	398	19	380	27	620	018	8	19	
42	417	19	407	28	593	010	8	18	
43	436	19	435	27	565	002	8	17	
44	455	19	462	27	538	.91 993	8	16	
45	9.74 474	19	9.82 489	28	0.17 511	9.91 985	9	15	
46	493	19	517	27	483	976	8	14	
47	512	19	544	27	456	968	8	13	
48	531	18	571	28	429	959	8	12	
49	549	19	599	27	401	951	9	11	
50	9.74 568	19	9.82 626	27	0.17 374	9.91 942	8	10	
51	587	19	653	28	347	934	9	9	
52	606	19	681	27	319	925	8	8	
53	625	19	708	27	292	917	9	7	
54	644	18	735	27	265	908	8	6	
55	9.74 662	19	9.82 762	28	0.17 238	9.91 900	9	5	
56	681	19	790	27	210	891	8	4	
57	700	19	817	27	183	883	9	3	
58	719	18	844	27	156	874	8	2	
59	737	19	871	28	129	866	9	1	
60	9.74 756		9.82 899		0.17 101	9.91 857		0	
	L Cos	d	L Cot	c d	L Tan	L Sin	d	'	P. P.

P. P.

	28	27	26
1	0.5	0.4	0.4
2	0.9	0.9	0.9
3	1.4	1.4	1.3
4	1.9	1.8	1.7
5	2.3	2.2	2.2
6	2.8	2.7	2.6
7	3.3	3.2	3.0
8	3.7	3.6	3.5
9	4.2	4.0	3.9
10	4.7	4.5	4.3
20	9.3	9.0	8.7
30	14.0	13.5	13.0
40	18.7	18.0	17.3
50	23.3	22.5	21.7

	20	19	18
1	0.3	0.3	0.3
2	0.7	0.6	0.6
3	1.0	1.0	0.9
4	1.3	1.3	1.2
5	1.7	1.6	1.5
6	2.0	1.9	1.8
7	2.3	2.2	2.1
8	2.7	2.5	2.4
9	3.0	2.8	2.7
10	3.3	3.2	3.0
20	6.7	6.3	6.0
30	10.0	9.5	9.0
40	13.3	12.7	12.0
50	16.7	15.8	15.0

56°

34°

P. P.	′	L Sin	d	L Tan	c d	L Cot	L Cos	d	
	0	9.74 756	19	9.82 899	27	0.17 101	9.91 857	8	**60**
	1	775	19	926	27	074	849	9	59
	2	794	18	953	27	047	840	8	58
	3	812	19	980	28	020	832	8	57
	4	831	19	.83 008	27	.16 992	823	9	56
	5	9.74 850	18	9.83 035	27	0.16 965	9.91 815	9	**55**
	6	868	19	062	27	938	806	8	54
	7	887	19	089	28	911	798	9	53
	8	906	18	117	27	883	789	8	52
	9	924	19	144	27	856	781	9	51
	10	9.74 943	18	9.83 171	27	0.16 829	9.91 772	9	**50**
	11	961	19	198	27	802	763	8	49
	12	980	19	225	27	775	755	9	48
	13	999	18	252	28	748	746	8	47
	14	.75 017	19	280	27	720	738	9	46
	15	9.75 036	18	9.83 307	27	0.16 693	9.91 729	9	**45**
	16	054	19	334	27	666	720	8	44
	17	073	18	361	27	639	712	9	43
	18	091	19	388	27	612	703	8	42
	19	110	18	415	27	585	695	9	41
	20	.9.75 128	19	9.83 442	28	0.16 558	9.91 686	9	**40**
	21	147	18	470	27	530	677	8	39
	22	165	19	497	27	503	669	9	38
	23	184	18	524	27	476	660	9	37
	24	202	19	551	27	449	651	8	36
9 8	**25**	9.75 221	18	9.83 578	27	0.16 422	9.91 643	9	**35**
1 | 0.2 0.1	26	239	19	605	27	395	634	9	34
2 | 0.3 0.3	27	258	18	632	27	368	625	8	33
3 | 0.4 0.4	28	276	18	659	27	341	617	9	32
4 | 0.6 0.5	29	294	19	686	27	314	608	9	31
5 | 0.8 0.7	**30**	9.75 313	18	9.83 713	27	0.16 287	9.91 599	8	**30**
6 | 0.9 0.8	31	331	19	740	28	260	591	9	29
7 | 1.0 0.9	32	350	18	768	27	232	582	9	28
8 | 1.2 1.1	33	368	18	795	27	205	573	8	27
9 | 1.4 1.2	34	386	19	822	27	178	565	9	26
10 | 1.5 1.3	**35**	9.75 405	18	9.83 849	27	0.16 151	9.91 556	9	**25**
20 | 3.0 2.7	36	423	18	876	27	124	547	9	24
30 | 4.5 4.0	37	441	18	903	27	097	538	8	23
40 | 6.0 5.3	38	459	19	930	27	070	530	9	22
50 | 7.5 6.7	39	478	18	957	27	043	521	9	21
	40	9.75 496	18	9.83 984	27	0.16 016	9.91 512	8	**20**
	41	514	19	.84 011	27	.15 989	504	9	19
	42	533	18	038	27	962	495	9	18
	43	551	18	065	27	935	486	9	17
	44	569	18	092	27	908	477	8	16
	45	.9.75 587	18	9.84 119	27	0.15 881	9.91 469	9	**15**
	46	605	19	146	27	854	460	9	14
	47	624	18	173	27	827	451	9	13
	48	642	18	200	27	800	442	9	12
	49	660	18	227	27	773	433	8	11
	50	9.75 678	18	9.84 254	26	0.15 746	9.91 425	9	**10**
	51	696	18	280	27	720	416	9	9
	52	714	19	307	27	693	407	9	8
	53	733	18	334	27	666	398	9	7
	54	751	18	361	27	639	389	8	6
	55	9.75 769	18	9.84 388	27	0.15 612	9.91 381	9	**5**
	56	787	18	415	27	585	372	9	4
	57	805	18	442	27	558	363	9	3
	58	823	18	469	27	531	354	9	2
	59	841	18	496	27	504	345	9	1
	60	9.75 859		9.84 523		0.15 477	9.91 336		**0**
P. P.		L Cos	d	L Cot	c d	L Tan	L Sin	d	′

55°

35°

′	L Sin	d	L Tan	c d	L Cot	L Cos	d	
0	9.75 859	18	9.84 523	27	0.15 477	9.91 336	8	60
1	877	18	550	27	450	328	8	59
2	895	18	576	26	424	319	9	58
3	913	18	603	27	397	310	9	57
4	931	18	630	27	370	301	9	56
5	9.75 949	18	9.84 657	27	0.15 343	9.91 292	9	55
6	967	18	684	27	316	283	9	54
7	985	18	711	27	289	274	9	53
8	.76 003	18	738	26	262	266	8	52
9	021	18	764	27	236	257	9	51
10	9.76 039	18	9.84 791	27	0.15 209	9.91 248	9	50
11	057	18	818	27	182	239	9	49
12	075	18	845	27	155	230	9	48
13	093	18	872	27	128	221	9	47
14	111	18	899	26	101	212	9	46
15	9.76 129	17	9.84 925	27	0.15 075	9.91 203	9	45
16	146	18	952	27	048	194	9	44
17	164	18	979	27	021	185	9	43
18	182	18	.85 006	27	.14 994	176	9	42
19	200	18	033	26	967	167	9	41
20	9.76 218	18	9.85 059	27	0.14 941	9.91 158	9	40
21	236	17	086	27	914	149	8	39
22	253	18	113	27	887	141	9	38
23	271	18	140	26	860	132	9	37
24	289	18	166	27	834	123	9	36
25	9.76 307	17	9.85 193	27	0.14 807	9.91 114	9	35
26	324	18	220	27	780	105	9	34
27	342	18	247	26	753	096	9	33
28	360	18	273	27	727	087	9	32
29	378	17	300	27	700	078	9	31
30	9.76 395	18	9.85 327	27	0.14 673	9.91 069	9	30
31	413	18	354	26	646	060	9	29
32	431	17	380	27	620	051	9	28
33	448	18	407	27	593	042	9	27
34	466	18	434	26	566	033	10	26
35	9.76 484	17	9.85 460	27	0.14 540	9.91 023	9	25
36	501	18	487	27	513	014	9	24
37	519	18	514	26	486	005	9	23
38	537	17	540	27	460	.90 996	9	22
39	554	18	567	27	433	987	9	21
40	9.76 572	18	9.85 594	26	0.14 406	9.90 978	9	20
41	590	17	620	27	380	969	9	19
42	607	18	647	27	353	960	9	18
43	625	17	674	26	326	951	9	17
44	642	18	700	27	300	942	9	16
45	9.76 660	17	9.85 727	27	0.14 273	9.90 933	9	15
46	677	18	754	26	246	924	9	14
47	695	17	780	27	220	915	9	13
48	712	18	807	27	193	906	10	12
49	730	17	834	26	166	896	9	11
50	9.76 747	18	9.85 860	27	0.14 140	9.90 887	9	10
51	765	17	887	26	113	878	9	9
52	782	18	913	27	087	869	9	8
53	800	17	940	27	060	860	9	7
54	817	18	967	26	033	851	9	6
55	9.76 835	17	9.85 993	27	0.14 007	9.90 842	10	5
56	852	18	.86 020	26	.13 980	832	9	4
57	870	18	046	27	954	823	9	3
58	887	17	073	27	927	814	9	2
59	904	17	100	26	900	805	9	1
60	9.76 922		9.86 126		0.13 874	9.90 796		0
	L Cos	d	L Cot	c d	L Tan	L Sin	d	′

P. P.

	27	26
1	0.4	0.4
2	0.9	0.9
3	1.4	1.3
4	1.8	1.7
5	2.2	2.2
6	2.7	2.6
7	3.2	3.0
8	3.6	3.5
9	4.0	3.9
10	4.5	4.3
20	9.0	8.7
30	13.5	13.0
40	18.0	17.3
50	22.5	21.7

	18	17	16
1	0.3	0.3	0.3
2	0.6	0.6	0.5
3	0.9	0.8	0.8
4	1.2	1.1	1.1
5	1.5	1.4	1.3
6	1.8	1.7	1.6
7	2.1	2.0	1.9
8	2.4	2.3	2.1
9	2.7	2.6	2.4
10	3.0	2.8	2.7
20	6.0	5.7	5.3
30	9.0	8.5	8.0
40	12.0	11.3	10.7
50	15.0	14.2	13.3

54°

36°									
P. P.	′	L Sin	d	L Tan	c d	L Cot	L Cos	d	
	0	9.76 922		9.86 126		0.13 874	9.90 796		**60**
	1	939	17	153	27	847	787	9	59
	2	957	18	179	26	821	777	10	58
	3	974	17	206	27	794	768	9	57
	4	991	17	232	26	768	759	9	56
			18		27			9	
	5	9.77 009	17	9.86 259	26	0.13 741	9.90 750	9	**55**
	6	026	17	285	27	715	741	10	54
	7	043	18	312	26	688	731	9	53
	8	061	17	338	27	662	722	9	52
	9	078	17	365	27	635	713	9	51
	10	9.77 095	17	9.86 392	26	0.13 608	9.90 704	10	**50**
	11	112	18	418	27	582	694	9	49
	12	130	17	445	26	555	685	9	48
	13	147	17	471	27	529	676	9	47
	14	164	17	498	26	502	667	10	46
	15	9.77 181	18	9.86 524	27	0.13 476	9.90 657	9	**45**
	16	199	17	551	26	449	648	9	44
	17	216	17	577	26	423	639	9	43
	18	233	17	603	27	397	630	10	42
	19	250	18	630	26	370	620	9	41
	20	9.77 268	17	9.86 656	27	0.13 344	9.90 611	9	**40**
	21	285	17	683	26	317	602	10	39
	22	302	17	709	27	291	592	9	38
	23	319	17	736	26	264	583	9	37
	24	336	17	762	27	238	574		36
10 9 8	**25**	9.77 353	17	9.86 789	26	0.13 211	9.90 665	10	**35**
1 0.2 0.2 0.1	26	370	17	815	27	185	555	9	34
2 0.3 0.3 0.3	27	387	18	842	26	158	546	9	33
3 0.5 0.4 0.4	28	405	17	868	26	132	537	10	32
4 0.7 0.6 0.5	29	422	17	894	27	106	527	9	31
5 0.8 0.8 0.7	**30**	9.77 439	17	9.86 921	26	0.13 079	9.90 518	9	**30**
6 1.0 0.9 0.8	31	456	17	947	27	053	509	10	29
7 1.2 1.0 0.9	32	473	17	974	26	026	499	9	28
8 1.3 1.2 1.1	33	490	17	.87 000	27	000	490	10	27
9 1.5 1.4 1.2	34	507	17	027	26	.12 973	480	10	26
10 1.7 1.5 1.3	**35**	9.77 524	17	9.87 053	26	0.12 947	9.90 471	9	**25**
20 3.3 3.0 2.7	36	541	17	079	27	921	462	10	24
30 5.0 4.5 4.0	37	558	17	106	26	894	452	9	23
40 6.7 6.0 5.3	38	575	17	132	26	868	443	10	22
50 8.3 7.5 6.7	39	592	17	158	27	842	434	10	21
	40	9.77 609	17	9.87 185	26	0.12 815	9.90 424	9	**20**
	41	626	17	211	27	789	415	10	19
	42	643	17	238	26	762	405	9	18
	43	660	17	264	26	736	396	10	17
	44	677	17	290	27	710	386	9	16
	45	9.77 694	17	9.87 317	26	0.12 683	9.90 377	9	**15**
	46	711	17	343	26	657	368	10	14
	47	728	16	369	27	631	358	9	13
	48	744	17	396	26	604	349	10	12
	49	761	17	422	26	578	339	9	11
	50	9.77 778	17	9.87 448	27	0.12 552	9.90 330	10	**10**
	51	795	17	475	26	525	320	9	9
	52	812	17	501	26	499	311	10	8
	53	829	17	527	27	473	301	10	7
	54	846	16	554	26	446	292	10	6
	55	9.77 862	17	9.87 580	26	0.12 420	9.90 282	9	**5**
	56	879	17	606	27	394	273	10	4
	57	896	17	633	26	367	263	9	3
	58	913	17	659	26	341	254	10	2
	59	930	16	685	26	315	244	9	1
	60	9.77 946		9.87 711		0.12 289	9.90 235		**0**
P. P.		L Cos	d	L Cot	c d	L Tan	L Sin	d	′

53°									

37°

'	L Sin	d	L Tan	c d	L Cot	L Cos	d	'
0	9.77 946	17	9.87 711	27	0.12 289	9.90 235	10	60
1	963	17	738	26	262	225	9	59
2	980	17	764	26	236	216	10	58
3	997	16	790	27	210	206	9	57
4	.78 013	17	817	26	183	197	10	56
5	9.78 030	17	9.87 843	26	0.12 157	9.90 187	9	55
6	047	16	869	26	131	178	10	54
7	063	17	895	27	105	168	9	53
8	080	17	922	26	078	159	10	52
9	097	16	948	26	052	149	10	51
10	9.78 113	17	9.87 974	26	0.12 026	9.90 139	9	50
11	130	17	.88 000	27	000	130	9	49
12	147	16	027	26	.11 973	120	9	48
13	163	17	053	26	947	111	10	47
14	180	17	079	26	921	101	10	46
15	9.78 197	16	9.88 105	26	0.11 895	9.90 091	9	45
16	213	17	131	27	869	082	10	44
17	230	16	158	26	842	072	9	43
18	246	17	184	26	816	063	10	42
19	263	17	210	26	790	053	10	41
20	9.78 280	16	9.88 236	26	0.11 764	9.90 043	9	40
21	296	17	262	27	738	034	10	39
22	313	16	289	26	711	024	10	38
23	329	17	315	26	685	014	9	37
24	346	16	341	26	659	005	10	36
25	9.78 362	17	9.88 367	26	0.11 633	9.89 995	10	35
26	379	16	393	27	607	985	9	34
27	395	17	420	26	580	976	10	33
28	412	16	446	26	554	966	10	32
29	428	17	472	26	528	956	9	31
30	9.78 445	16	9.88 498	26	0.11 502	9.89 947	10	30
31	461	17	524	26	476	937	10	29
32	478	16	550	27	450	927	9	28
33	494	16	577	26	423	918	10	27
34	510	17	603	26	397	908	10	26
35	9.78 527	16	9.88 629	26	0.11 371	9.89 898	10	25
36	543	17	655	26	345	888	9	24
37	560	16	681	26	319	879	10	23
38	576	16	707	26	293	869	10	22
39	592	17	733	26	267	859	10	21
40	9.78 609	16	9.88 759	27	0.11 241	9.89 849	9	20
41	625	17	786	26	214	840	10	19
42	642	16	812	26	188	830	10	18
43	658	16	838	26	162	820	10	17
44	674	17	864	26	136	810	9	16
45	9.78 691	16	9.88 890	26	0.11 110	9.89 801	10	15
46	707	16	916	26	084	791	10	14
47	723	16	942	26	058	781	10	13
48	739	17	968	26	032	771	10	12
49	756	16	994	26	006	761	9	11
50	9.78 772	16	9.89 020	26	0.10 980	9.89 752	10	10
51	788	17	046	27	954	742	10	9
52	805	16	073	26	927	732	10	8
53	821	16	099	26	901	722	10	7
54	837	16	125	26	875	712	10	6
55	9.78 853	16	9.89 151	26	0.10 849	9.89 702	9	5
56	869	17	177	26	823	693	10	4
57	886	16	203	26	797	683	10	3
58	902	16	229	26	771	673	10	2
59	918	16	255	26	745	663	10	1
60	9.78 934		9.89 281		0.10 719	9.89 653		0
	L Cos	d	L Cot	c d	L Tan	L Sin	d	'

52°

P. P.

	27	26	25
1	0.4	0.4	0.4
2	0.9	0.9	0.8
3	1.4	1.3	1.2
4	1.8	1.7	1.7
5	2.2	2.2	2.1
6	2.7	2.6	2.5
7	3.2	3.0	2.9
8	3.6	3.5	3.3
9	4.0	3.9	3.8
10	4.5	4.3	4.2
20	9.0	8.7	8.3
30	13.5	13.0	12.5
40	18.0	17.3	16.7
50	22.5	21.7	20.8

	17	16	15
1	0.3	0.3	0.2
2	0.6	0.5	0.5
3	0.8	0.8	0.8
4	1.1	1.1	1.0
5	1.4	1.3	1.2
6	1.7	1.6	1.5
7	2.0	1.9	1.8
8	2.3	2.1	2.0
9	2.6	2.4	2.2
10	2.8	2.7	2.5
20	5.7	5.3	5.0
30	8.5	8.0	7.5
40	11.3	10.7	10.0
50	14.2	13.3	12.5

38°

P. P.	′	L Sin	d	L Tan	c d	L Cot	L Cos	d	
	0	9.78 934	16	9.89 281	26	0.10 719	9.89 653	10	60
	1	950	17	307	26	693	643	10	59
	2	967	16	333	26	667	633	9	58
	3	983	16	359	26	641	624	10	57
	4	999	16	385	26	615	614	10	56
	5	9.79 015	16	9.89 411	26	0.10 589	9.89 604	10	55
	6	031	16	437	26	563	594	10	54
	7	047	16	463	26	537	584	10	53
	8	063	16	489	26	511	574	10	52
	9	079	16	515	26	485	564	10	51
	10	9.79 095	16	9.89 541	26	0.10 459	9.89 554	10	50
	11	111	17	567	26	433	544	10	49
	12	128	16	593	26	407	534	10	48
	13	144	16	619	26	381	524	10	47
	14	160	16	645	26	355	514	10	46
	15	9.79 176	16	9.89 671	26	0.10 329	9.89 504	9	45
	16	192	16	697	26	303	495	10	44
	17	208	16	723	26	277	485	10	43
	18	224	16	749	26	251	475	10	42
	19	240	16	775	26	225	465	10	41
	20	9.79 256	16	9.89 801	26	0.10 199	9.89 455	10	40
	21	272	16	827	26	173	445	10	39
	22	288	16	853	26	147	435	10	38
	23	304	15	879	26	121	425	10	37
	24	319	16	905	26	095	415	10	36
	25	9.79 335	16	9.89 931	26	0.10 069	9.89 405	10	35
	26	351	16	957	26	043	395	10	34
	27	367	16	983	26	017	385	10	33
	28	383	16	.90 009	26	.09 991	375	11	32
	29	399	16	035	26	965	364	10	31
	30	9.79 415	16	9.90 061	25	0.09 939	9.89 354	10	30
	31	431	16	086	26	914	344	10	29
	32	447	16	112	26	888	334	10	28
	33	463	15	138	26	862	324	10	27
	34	478	16	164	26	836	314	10	26
	35	9.79 494	16	9.90 190	26	0.09 810	9.89 304	10	25
	36	510	16	216	26	784	294	10	24
	37	526	16	242	26	758	284	10	23
	38	542	16	268	26	732	274	10	22
	39	558	15	294	26	706	264	10	21
	40	9.79 573	16	9.90 320	26	0.09 680	9.89 254	10	20
	41	589	16	346	25	654	244	11	19
	42	605	16	371	26	629	233	10	18
	43	621	15	397	26	603	223	10	17
	44	636	16	423	26	577	213	10	16
	45	9.79 652	16	9.90 449	26	0.09 551	9.89 203	10	15
	46	668	16	475	26	525	193	10	14
	47	684	15	501	26	499	183	10	13
	48	699	16	527	26	473	173	11	12
	49	715	16	553	25	447	162	10	11
	50	9.79 731	15	9.90 578	26	0.09 422	9.89 152	10	10
	51	746	16	604	26	396	142	10	9
	52	762	16	630	26	370	132	10	8
	53	778	15	656	26	344	122	10	7
	54	793	16	682	26	318	112	11	6
	55	9.79 809	16	9.90 708	26	0.09 292	9.89 101	10	5
	56	825	15	734	25	266	091	10	4
	57	840	16	759	26	241	081	10	3
	58	856	16	785	26	215	071	11	2
	59	872	15	811	26	189	060	10	1
	60	9.79 887		9.90 837		0.09 163	9.89 050		0

| P. P. | | L Cos | d | L Cot | c d | L Tan | L Sin | d | ′ |

P. P.

	11	10	9
1	0.2	0.2	0.2
2	0.4	0.3	0.3
3	0.6	0.5	0.4
4	0.7	0.7	0.6
5	0.9	0.8	0.8
6	1.1	1.0	0.9
7	1.3	1.2	1.0
8	1.5	1.3	1.2
9	1.6	1.5	1.4
10	1.8	1.7	1.5
20	3.7	3.3	3.0
30	5.5	5.0	4.5
40	7.3	6.7	6.0
50	9.2	8.3	7.5

51°

39°

'	L Sin	d	L Tan	c d	L Cot	L Cos	d		P. P.
0	9.79 887	16	9.90 837	26	0.09 163	9.89 050	10	**60**	
1	903	15	863	26	137	040	10	59	
2	918	16	889	25	111	030	10	58	
3	934	16	914	26	086	020	11	57	
4	950	15	940	26	060	009	10	56	
5	9.79 965	16	9.90 966	26	0.09 034	9.88 999	10	**55**	
6	981	15	992	26	008	989	11	54	
7	996	16	.91 018	25	.08 982	978	10	53	
8	.80 012	15	043	26	957	968	10	52	
9	027	16	069	26	931	958	10	51	
10	9.80 043	15	9.91 095	26	0.08 905	9.88 948	11	**50**	
11	058	16	121	26	879	937	10	49	
12	074	15	147	25	853	927	10	48	
13	089	16	172	26	828	917	11	47	
14	105	15	198	26	802	906	10	46	
15	9.80 120	16	9.91 224	26	0.08 776	9.88 896	10	**45**	
16	136	15	250	26	750	886	11	44	**26** **25**
17	151	15	276	25	724	875	10	43	1 0.4 0.4
18	166	16	301	26	699	865	10	42	2 0.9 0.8
19	182	15	327	26	673	855	11	41	3 1.3 1.2
20	9.80 197	16	9.91 353	26	0.08 647	9.88 844	10	**40**	4 1.7 1.7 5 2.2 2.1
21	213	15	379	25	621	834	10	39	6 2.6 2.5
22	228	16	404	26	596	824	10	38	7 3.0 2.9
23	244	15	430	26	570	813	11	37	8 3.5 3.3
24	259	15	456	26	544	803	10	36	9 3.9 3.8 10 4.3 4.2
25	9.80 274	16	9.91 482	25	0.08 518	9.88 793	11	**35**	20 8.7 8.3
26	290	15	507	26	493	782	10	34	30 13.0 12.5
27	305	15	533	26	467	772	11	33	40 17.3 16.7
28	320	16	559	26	441	761	10	32	50 21.7 20.8
29	336	15	585	25	415	751	10	31	
30	9.80 351	15	9.91 610	26	0.08 390	9.88 741	11	**30**	
31	366	16	636	26	364	730	10	29	
32	382	15	662	26	338	720	11	28	
33	397	15	688	25	312	709	10	27	
34	412	16	713	26	287	699	11	26	**16** **15** **14**
35	9.80 428	15	9.91 739	26	0.08 261	9.88 688	10	**25**	1 0.3 0.2 0.2
36	443	15	765	26	235	678	10	24	2 0.5 0.5 0.5
37	458	15	791	25	209	668	11	23	3 0.8 0.8 0.7
38	473	16	816	26	184	657	10	22	4 1.1 1.0 0.9
39	489	15	842	26	158	647	11	21	5 1.3 1.2 1.2 6 1.6 1.5 1.4
40	9.80 504	15	9.91 868	25	0.08 132	9.88 636	10	**20**	7 1.9 1.8 1.6
41	519	15	893	26	107	626	11	19	8 2.1 2.0 1.9
42	534	16	919	26	081	615	10	18	9 2.4 2.2 2.1
43	550	15	945	26	055	605	11	17	10 2.7 2.5 2.3
44	565	15	971	25	029	594	10	16	20 5.3 5.0 4.7 30 8.0 7.5 7.0
45	9.80 580	15	9.91 996	26	0.08 004	9.88 584	11	**15**	40 10.7 10.0 9.3
46	595	15	.92 022	26	.07 978	573	10	14	50 13.3 12.5 11.7
47	610	15	048	25	952	563	11	13	
48	625	16	073	26	927	552	10	12	
49	641	15	099	26	901	542	11	11	
50	9.80 656	15	9.92 125	25	0.07 875	9.88 531	10	**10**	
51	671	15	150	26	850	521	11	9	
52	686	15	176	26	824	510	11	8	
53	701	15	202	25	798	499	10	7	
54	716	15	227	26	773	489	11	6	
55	9.80 731	15	9.92 253	26	0.07 747	9.88 478	10	**5**	
56	746	16	279	25	721	468	11	4	
57	762	15	304	26	696	457	10	3	
58	777	15	330	26	670	447	11	2	
59	792	15	356	25	644	436	11	1	
60	9.80 807		9.92 381		0.07 619	9.88 425		**0**	
	L Cos	d	L Cot	c d	L Tan	L Sin	d	'	P. P.

50°

40°

P. P.

11	10
1	0.2	0.2
2	0.4	0.3
3	0.6	0.5
4	0.7	0.7
5	0.9	0.8
6	1.1	1.0
7	1.3	1.2
8	1.5	1.3
9	1.6	1.5
10	1.8	1.7
20	3.7	3.3
30	5.5	5.0
40	7.3	6.7
50	9.2	8.3

′	L Sin	d	L Tan	c d	L Cot	L Cos	d	
0	9.80 807	15	9.92 381	26	0.07 619	9.88 425	10	60
1	822	15	407	26	593	415	11	59
2	837	15	433	26	567	404	10	58
3	852	15	458	25	542	394	11	57
4	867	15	484	26	516	383	11	56
5	9.80 882	15	9.92 510	25	0.07 490	9.88 372	10	55
6	897	15	535	26	465	362	11	54
7	912	15	561	26	439	351	11	53
8	927	15	587	25	413	340	10	52
9	942	15	612	26	388	330	11	51
10	9.80 957	15	9.92 638	25	0.07 362	9.88 319	11	50
11	972	15	663	26	337	308	10	49
12	987	15	689	26	311	298	11	48
13	.81 002	15	715	25	285	287	11	47
14	017	15	740	26	260	276	10	46
15	9.81 032	15	9.92 766	26	0.07 234	9.88 266	11	45
16	047	14	792	25	208	255	11	44
17	061	15	817	26	183	244	10	43
18	076	15	843	25	157	234	11	42
19	091	15	868	26	132	223	11	41
20	9.81 106	15	9.92 894	26	0.07 106	9.88 212	11	40
21	121	15	920	25	080	201	10	39
22	136	15	945	26	055	191	11	38
23	151	15	971	25	029	180	11	37
24	166	14	996	26	004	169	11	36
25	9.81 180	15	9.93 022	26	0.06 978	9.88 158	10	35
26	195	15	048	25	952	148	11	34
27	210	15	073	26	927	137	11	33
28	225	15	099	25	901	126	11	32
29	240	14	124	26	876	115	10	31
30	9.81 254	15	9.93 150	25	0.06 850	9.88 105	11	30
31	269	15	175	26	825	094	11	29
32	284	15	201	26	799	083	11	28
33	299	15	227	25	773	072	11	27
34	314	14	252	26	748	061	10	26
35	9.81 328	15	9.93 278	25	0.06 722	9.88 051	11	25
36	343	15	303	26	697	040	11	24
37	358	14	329	25	671	029	11	23
38	372	15	354	26	646	018	11	22
39	387	15	380	26	620	007	11	21
40	9.81 402	15	9.93 406	25	0.06 594	9.87 996	11	20
41	417	14	431	26	569	985	10	19
42	431	15	457	25	543	975	11	18
43	446	15	482	26	518	964	11	17
44	461	14	508	25	492	953	11	16
45	9.81 475	15	9.93 533	26	0.06 467	9.87 942	11	15
46	490	15	559	25	441	931	11	14
47	505	14	584	26	416	920	11	13
48	519	15	610	26	390	909	11	12
49	,534	15	636	25	364	898	11	11
50	9.81 549	14	9.93 661	26	0.06 339	9.87 887	10	10
51	563	15	687	25	313	877	11	9
52	578	14	712	26	288	866	11	8
53	592	15	738	25	262	855	11	7
54	607	15	763	26	237	844	11	6
55	9.81 622	14	9.93 789	25	0.06 211	9.87 833	11	5
56	636	15	814	26	186	822	11	4
57	651	14	840	25	160	811	11	3
58	665	15	865	26	135	800	11	2
59	680	14	891	25	109	789	11	1
60	9.81 694		9.93 916		0.06 084	9.87 778		0

| P. P. | L Cos | d | L Cot | c d | L Tan | L Sin | d | ′ |

49°

41°

′	L Sin	d	L Tan	c d	L Cot	L Cos	d		P. P.
0	9.81 694	15	9.93 916	26	0.06 084	9.87 778	11	60	
1	709	14	942	25	058	767	11	59	
2	723	15	967	26	033	756	11	58	
3	738	14	993	26	007	745	11	57	
4	752	15	.94 018	26	.05 982	734	11	56	
5	9.81 767	14	9.94 044	25	0.05 956	9.87 723	11	55	
6	781	15	069	26	931	712	11	54	
7	796	15	095	25	905	701	11	53	
8	810	15	120	26	880	690	11	52	
9	825	14	146	25	854	679	11	51	
10	9.81 839	15	9.94 171	26	0.05 829	9.87 668	11	50	
11	854	14	197	25	803	657	11	49	
12	868	14	222	26	778	646	11	48	
13	882	15	248	26	752	635	11	47	
14	897	14	273	25	727	624	11	46	
15	9.81 911	15	9.94 299	25	0.05 701	9.87 613	12	45	
16	926	14	324	26	676	601	11	44	
17	940	15	350	25	650	590	11	43	
18	955	14	375	26	625	579	11	42	
19	969	14	401	25	599	568	11	41	
20	9.81 983	15	9.94 426	26	0.05 574	9.87 557	11	40	
21	998	14	452	25	548	546	11	39	
22	.82 012	14	477	26	523	535	11	38	
23	026	15	503	26	497	524	11	37	
24	041	14	528	26	472	513	12	36	
25	9.82 055	14	9.94 554	25	0.05 446	9.87 501	11	35	
26	069	15	579	25	421	490	11	34	
27	084	14	604	26	396	479	11	33	
28	098	14	630	25	370	468	11	32	
29	112	14	655	26	345	457	11	31	
30	9.82 126	15	9.94 681	25	0.05 319	9.87 446	12	30	
31	141	14	706	26	294	434	11	29	
32	155	14	732	25	268	423	11	28	
33	169	15	757	25	243	412	11	27	
34	184	14	783	25	217	401	11	26	
35	9.82 198	14	9.94 808	26	0.05 192	9.87 390	12	25	
36	212	14	834	25	166	378	11	24	
37	226	14	859	25	141	367	11	23	
38	240	15	884	26	116	356	11	22	
39	255	14	910	25	090	345	11	21	
40	9.82 269	14	9.94 935	26	0.05 065	9.87 334	12	20	
41	283	14	961	25	039	322	11	19	
42	297	14	986	26	014	311	11	18	
43	311	15	.95 012	26	.04 988	300	12	17	
44	326	14	037	25	963	288	11	16	
45	9.82 340	14	9.95 062	26	0.04 938	9.87 277	11	15	
46	354	14	088	25	912	266	11	14	
47	368	14	113	26	887	255	12	13	
48	382	14	139	25	861	243	11	12	
49	396	14	164	26	836	232	11	11	
50	9.82 410	14	9.95 190	25	0.04 810	9.87 221	12	10	
51	424	15	215	25	785	209	11	9	
52	439	14	240	26	760	198	11	8	
53	453	14	266	25	734	187	12	7	
54	467	14	291	26	709	175	11	6	
55	9.82 481	14	9.95 317	25	0.04 683	9.87 164	11	5	
56	495	14	342	26	658	153	12	4	
57	509	14	368	25	632	141	11	3	
58	523	14	393	25	607	130	11	2	
59	537	14	418	26	582	119	12	1	
60	9.82 551		9.95 444		0.04 556	9.87 107		0	
	L Cos	d	L Cot	c d	L Tan	L Sin	d	′	P. P.

P. P.

	26	25
1	0.4	0.4
2	0.9	0.8
3	1.3	1.2
4	1.7	1.7
5	2.2	2.1
6	2.6	2.5
7	3.0	2.9
8	3.5	3.3
9	3.9	3.8
10	4.3	4.2
20	8.7	8.3
30	13.0	12.5
40	17.3	16.7
50	21.7	20.8

	15	14
1	0.2	0.2
2	0.5	0.5
3	0.8	0.7
4	1.0	0.9
5	1.2	1.2
6	1.5	1.4
7	1.8	1.6
8	2.0	1.9
9	2.2	2.1
10	2.5	2.3
20	5.0	4.7
30	7.5	7.0
40	10.0	9.3
50	12.5	11.7

48°

42°

P. P.	′	L Sin	d	L Tan	c d	L Cot	L Cos	d	
	0	9.82 551	14	9.95 444	25	0.04 556	9.87 107	11	60
	1	565	14	469	26	531	096	11	59
	2	579	14	495	25	505	085	11	58
	3	593	14	520	25	480	073	12	57
	4	607	14	545	26	455	062	11	56
								12	
	5	9.82 621	14	9.95 571	25	0.04 429	9.87 050	11	55
	6	635	14	596	26	404	039	11	54
	7	649	14	622	25	378	028	12	53
	8	663	14	647	25	353	016	11	52
	9	677	14	672	26	328	005	12	51
	10	9.82 691	14	9.95 698	25	0.04 302	9.86 993	11	50
	11	705	14	723	25	277	982	12	49
	12	719	14	748	26	252	970	11	48
	13	733	14	774	25	226	959	12	47
	14	747	14	799	26	201	947	11	46
	15	9.82 761	14	9.95 825	25	0.04 175	9.86 936	12	45
	16	775	13	850	25	150	924	11	44
	17	788	14	875	26	125	913	11	43
	18	802	14	901	25	099	902	12	42
	19	816	14	926	26	074	890	11	41
	20	9.82 830	14	9.95 952	25	0.04 048	9.86 879	12	40
	21	844	14	977	25	023	867	12	39
	22	858	14	.96 002	26	.03 998	855	11	38
	23	872	13	028	25	972	844	12	37
	24	885	14	053	25	947	832	11	36
13 12 11	25	9.82 899	14	9.96 078	26	0.03 922	9.86 821	12	35
1 0.2 0.2 0.2	26	913	14	104	25	896	809	11	34
2 0.4 0.4 0.4	27	927	14	129	26	871	798	12	33
3 0.6 0.6 0.6	28	941	14	155	25	845	786	11	32
4 0.9 0.8 0.7	29	955	13	180	25	820	775	12	31
5 1.1 1.0 0.9	30	9.82 968	14	9.96 205	26	0.03 795	9.86 763	11	30
6 1.3 1.2 1.1	31	982	14	231	25	769	752	12	29
7 1.5 1.4 1.3	32	996	14	256	25	744	740	12	28
8 1.7 1.6 1.5	33	.83 010	13	281	26	719	728	11	27
9 2.0 1.8 1.6	34	023	14	307	25	693	717	12	26
10 2.2 2.0 1.8	35	9.83 037	14	9.96 332	25	0.03 668	9.86 705	11	25
20 4.3 4.0 3.7	36	051	14	357	26	643	694	12	24
30 6.5 6.0 5.5	37	065	13	383	25	617	682	12	23
40 8.7 8.0 7.3	38	078	14	408	25	592	670	11	22
50 10.8 10.0 9.2	39	092	14	433	26	567	659	12	21
	40	9.83 106	14	9.96 459	25	0.03 541	9.86 647	12	20
	41	120	13	484	26	516	635	11	19
	42	133	14	510	25	490	624	12	18
	43	147	14	535	25	465	612	12	17
	44	161	13	560	26	440	600	11	16
	45	9.83 174	14	9.96 586	25	0.03 414	9.86 589	12	15
	46	188	14	611	25	389	577	12	14
	47	202	13	636	26	364	565	11	13
	48	215	14	662	25	338	554	12	12
	49	229	13	687	25	313	542	12	11
	50	9.83 242	14	9.96 712	26	0.03 288	9.86 530	12	10
	51	256	14	738	25	262	518	11	9
	52	270	13	763	25	237	507	12	8
	53	283	14	788	26	212	495	12	7
	54	297	13	814	25	186	483	11	6
	55	9.83 310	14	9.96 839	25	0.03 161	9.86 472	12	5
	56	324	14	864	26	136	460	12	4
	57	338	13	890	25	110	448	12	3
	58	351	14	915	25	085	436	11	2
	59	365	13	940	26	060	425	12	1
	60	9.83 378		9.96 966		0.03 034	9.86 413		0
P. P.		L Cos	d	L Cot	c d	L Tan	L Sin	d	′

47°

43°

'	L Sin	d	L Tan	c d	L Cot	L Cos	d	'	P. P.
0	9.83 378	14	9.96 966	25	0.03 034	9.86 413	12	60	
1	392	13	991	25	009	401	12	59	
2	405	14	.97 016	26	.02 984	389	12	58	
3	419	13	042	25	958	377	11	57	
4	432	14	067	25	933	366	12	56	
5	9.83 446	13	9.97 092	26	0.02 908	9.86 354	12	55	
6	459	14	118	25	882	342	12	54	
7	473	13	143	25	857	330	12	53	
8	486	14	168	25	832	318	12	52	
9	500	13	193	26	807	306	11	51	
10	9.83 513	14	9.97 219	25	0.02 781	9.86 295	12	50	
11	527	13	244	25	756	283	12	49	
12	540	14	269	26	731	271	12	48	
13	554	13	295	25	705	259	12	47	
14	567	14	320	25	680	247	12	46	
15	9.83 581	13	9.97 345	26	0.02 655	9.86 235	12	45	
16	594	14	371	25	629	223	12	44	
17	608	13	396	25	604	211	11	43	
18	621	13	421	26	579	200	12	42	
19	634	14	447	25	553	188	12	41	
20	9.83 648	13	9.97 472	25	0.02 528	9.86 176	12	40	
21	661	13	497	26	503	164	12	39	
22	674	14	523	25	477	152	12	38	
23	688	13	548	25	452	140	12	37	
24	701	14	573	25	427	128	12	36	
25	9.83 715	13	9.97 598	26	0.02 402	9.86 116	12	35	
26	728	13	624	25	376	104	12	34	
27	741	14	649	25	351	092	12	33	
28	755	13	674	26	326	080	12	32	
29	768	13	700	25	300	068	12	31	
30	9.83 781	14	9.97 725	25	0.02 275	9.86 056	12	30	
31	795	13	750	26	250	044	12	29	
32	808	13	776	25	224	032	12	28	
33	821	13	801	25	199	020	12	27	
34	834	14	826	25	174	008	12	26	
35	9.83 848	13	9.97 851	26	0.02 149	9.85 996	12	25	
36	861	13	877	25	123	984	12	24	
37	874	13	902	25	098	972	12	23	
38	887	14	927	26	073	960	12	22	
39	901	13	953	25	047	948	12	21	
40	9.83 914	13	9.97 978	25	0.02 022	9.85 936	12	20	
41	927	13	.98 003	26	.01 997	924	12	19	
42	940	14	029	25	971	912	12	18	
43	954	13	054	25	946	900	12	17	
44	967	13	079	25	921	888	12	16	
45	9.83 980	13	9.98 104	26	0.01 896	9.85 876	12	15	
46	993	13	130	25	870	864	12	14	
47	.84 006	14	155	25	845	851	13	13	
48	020	13	180	26	820	839	12	12	
49	033	13	206	25	794	827	12	11	
50	9.84 046	13	9.98 231	25	0.01 769	9.85 815	12	10	
51	059	13	256	25	744	803	12	9	
52	072	13	281	26	719	791	12	8	
53	085	13	307	25	693	779	13	7	
54	098	14	332	25	668	766	12	6	
55	9.84 112	13	9.98 357	26	0.01 643	9.85 754	12	5	
56	125	13	383	25	617	742	12	4	
57	138	13	408	25	592	730	12	3	
58	151	13	433	25	567	718	12	2	
59	164	13	458	26	542	706	13	1	
60	9.84 177		9.98 484		0.01 516	9.85 693		0	
	L Cos	d	L Cot	c d	L Tan	L Sin	d	'	P. P.

P. P.

	26	25
1	0.4	0.4
2	0.9	0.8
3	1.3	1.2
4	1.7	1.7
5	2.2	2.1
6	2.6	2.5
7	3.0	2.9
8	3.5	3.3
9	3.9	3.8
10	4.3	4.2
20	8.7	8.3
30	13.0	12.5
40	17.3	16.7
50	21.7	20.8

	14	13
1	0.2	0.2
2	0.5	0.4
3	0.7	0.6
4	0.9	0.9
5	1.2	1.1
6	1.4	1.3
7	1.6	1.5
8	1.9	1.7
9	2.1	2.0
10	2.3	2.2
20	4.7	4.3
30	7.0	6.5
40	9.3	8.7
50	11.7	10.8

46°

44°

P. P.	′	L Sin	d	L Tan	c d	L Cot	L Cos	d	
	0	9.84 177	13	9.98 484	25	0.01 516	9.85 693	12	**60**
	1	190	13	509	25	491	681	12	59
	2	203	13	534	26	466	669	12	58
	3	216	13	560	25	440	657	12	57
	4	229	13	585	25	415	645	13	56
	5	9.84 242	13	9.98 610	25	0.01 390	9.85 632	12	**55**
	6	255	14	635	26	365	620	12	54
	7	269	13	661	25	339	608	12	53
	8	282	13	686	25	314	596	13	52
	9	295	13	711	26	289	583	12	51
	10	9.84 308	13	9.98 737	25	0.01 263	9.85 571	12	**50**
	11	321	13	762	25	238	559	12	49
	12	334	13	787	25	213	547	13	48
	13	347	13	812	26	188	534	12	47
	14	360	13	838	25	162	522	12	46
	15	9.84 373	12	9.98 863	25	0.01 137	9.85 510	13	**45**
	16	385	13	888	25	112	497	12	44
	17	398	13	913	26	087	485	12	43
	18	411	13	939	25	061	473	13	42
	19	424	13	964	25	036	460	12	41
	20	9.84 437	13	9.98 989	26	0.01 011	9.85 448	12	**40**
	21	450	13	.99 015	25	.00 985	436	13	39
	22	463	13	040	25	960	423	12	38
	23	476	13	065	25	935	411	12	37
	24	489	13	090	26	910	399	13	36
	25	9.84 502	13	9.99 116	25	0.00 884	9.85 386	12	**35**
	26	515	13	141	25	859	374	13	34
	27	528	12	166	25	834	361	12	33
	28	540	13	191	26	809	349	12	32
	29	553	13	217	25	783	337	13	31
	30	9.84 566	13	9.99 242	25	0.00 758	9.85 324	12	**30**
	31	579	13	267	26	733	312	13	29
	32	592	13	293	25	707	299	12	28
	33	605	13	318	25	682	287	13	27
	34	618	12	343	25	657	274	12	26
	35	9.84 630	13	9.99 368	26	0.00 632	9.85 262	12	**25**
	36	643	13	394	25	606	250	13	24
	37	656	13	419	25	581	237	12	23
	38	669	13	444	25	556	225	13	22
	39	682	12	469	26	531	212	12	21
	40	9.84 694	13	9.99 495	25	0.00 505	9.85 200	13	**20**
	41	707	13	520	25	480	187	12	19
	42	720	13	545	25	455	175	13	18
	43	733	12	570	26	430	162	12	17
	44	745	13	596	25	404	150	13	16
	45	9.84 758	13	9.99 621	25	0.00 379	9.85 137	12	**15**
	46	771	13	646	26	354	125	13	14
	47	784	12	672	25	328	112	12	13
	48	796	13	697	25	303	100	13	12
	49	809	13	722	25	278	087	13	11
	50	9.84 822	13	9.99 747	26	0.00 253	9.85 074	12	**10**
	51	835	12	773	25	227	062	13	9
	52	847	13	798	25	202	049	12	8
	53	860	13	823	25	177	037	13	7
	54	873	12	848	26	152	024	12	6
	55	9.84 885	13	9.99 874	25	0.00 126	9.85 012	13	**5**
	56	898	13	899	25	101	.84 999	13	4
	57	911	12	924	25	076	986	12	3
	58	923	13	949	26	051	974	13	2
	59	936	13	975	25	025	961	12	1
	60	9.84 949		0.00 000		0.00 000	9.84 949		**0**
P. P.		L Cos	d	L Cot	c d	L Tan	L Sin	d	′

12 11

	12	11
1	0.2	0.2
2	0.4	0.4
3	0.6	0.6
4	0.8	0.7
5	1.0	0.9
6	1.2	1.1
7	1.4	1.3
8	1.6	1.5
9	1.8	1.6
10	2.0	1.8
20	4.0	3.7
30	6.0	5.5
40	8.0	7.3
50	10.0	9.2

45°

TABLE IV. LOG SIN AND LOG TAN OF SMALL ANGLES

Log Sin						0°					
′ ″	0″	1″	2″	3″	4″	5″	6″	7″	8″	9″	10″
0 0	4. —	68557	98660	*16270	*28763	*38454	*46373	*53067	*58866	*63982	*68557	50
10	5. 68557	72697	76476	79952	83170	86167	88969	91602	94085	96433	98660	40
20	98660	*00779	*02800	*04730	*06579	*08351	*10055	*11694	*13273	*14797	*16270	30
30	6. 16270	17694	19072	20409	21705	22964	24188	25378	26536	27664	28763	20
40	28763	29836	30882	31904	32903	33879	34833	35767	36682	37577	38454	10
50	38454	39315	40158	40985	41797	42594	43376	44145	44900	45643	46373	0 **59**
1 0	6.4 6373	7090	7797	8492	9175	9849	*0512	*1165	*1808	*2442	*3067	50
10	6.5 3067	3683	4291	4890	5481	6064	6639	7207	7767	8320	8866	40
20	8866	9406	9939	*0465	*0985	*1499	*2007	*2509	*3006	*3496	*3982	30
30	6.6 3982	4462	4936	5406	5870	6330	6785	7235	7680	8121	8557	20
40	8557	8990	9418	9841	*0261	*0676	*1088	*1496	*1900	*2300	*2697	10
50	6.7 2697	3090	3479	3865	4248	4627	5003	5376	5746	6112	6476	0 **58**
2 0	6476	6836	7193	7548	7900	8248	8595	8938	9278	9616	9952	50
10	9952	*0285	*0615	*0943	*1268	*1591	*1911	*2230	*2545	*2859	*3170	40
20	6.8 3170	3479	3786	4091	4394	4694	4993	5289	5584	5876	6167	30
30	6167	6455	6742	7027	7310	7591	7870	8147	8423	8697	8969	20
40	8969	9240	9509	9776	*0042	*0306	*0568	*0829	*1088	*1346	*1602	10
50	6.9 1602	1857	2110	2362	2612	2861	3109	3355	3599	3843	4085	0 **57**
3 0	4085	4325	4565	4803	5039	5275	5509	5742	5973	6204	6433	50
10	6433	6661	6888	7113	7338	7561	7783	8004	8224	8443	8660	40
20	8660	8877	9093	9307	9520	9733	9944	*0155	*0364	*0572	*0779	30
30	7.0 0779	0986	1191	1395	1599	1801	2003	2203	2403	2602	2800	20
40	2800	2997	3193	3388	3582	3776	3968	4160	4351	4541	4730	10
50	4730	4919	5106	5293	5479	5664	5849	6032	6215	6397	6579	0 **56**
4 0	6579	6759	6939	7118	7296	7474	7651	7827	8003	8177	8351	50
10	8351	8525	8698	8870	9041	9211	9381	9551	9719	9887	*0055	40
20	7.1 0055	0222	0388	0553	0718	0882	1046	1209	1371	1533	1694	30
30	1694	1854	2014	2174	2333	2491	2648	2805	2962	3118	3273	20
40	3273	3428	3582	3736	3889	4042	4194	4346	4497	4647	4797	10
50	4797	4947	5096	5244	5392	5540	5687	5833	5979	6125	6270	0 **55**
5 0	6270	6414	6558	6702	6845	6987	7130	7271	7413	7553	7694	50
10	7694	7834	7973	8112	8250	8389	8526	8663	8800	8937	9072	40
20	9072	9208	9343	9478	9612	9746	9879	*0012	*0145	*0277	*0409	30
30	7.2 0409	0540	0671	0802	0932	1062	1191	1320	1449	1577	1705	20
40	1705	1833	1960	2087	2213	2339	2465	2590	2715	2840	2964	10
50	2964	3088	3212	3335	3458	3580	3702	3824	3946	4067	4188	0 **54**
6 0	4188	4308	4428	4548	4668	4787	4906	5024	5142	5260	5378	50
10	5378	5495	5612	5728	5845	5961	6076	6192	6307	6421	6536	40
20	6536	6650	6764	6877	6991	7104	7216	7329	7441	7552	7664	30
30	7664	7775	7886	7997	8107	8217	8327	8437	8546	8655	8763	20
40	8763	8872	8980	9088	9196	9303	9410	9517	9623	9730	9836	10
50	9836	9942	*0047	*0152	*0257	*0362	*0467	*0571	*0675	*0779	*0882	0 **53**
7 0	7.3 0882	0986	1089	1191	1294	1396	1498	1600	1702	1803	1904	50
10	1904	2005	2106	2206	2306	2406	2506	2606	2705	2804	2903	40
20	2903	3001	3100	3198	3296	3393	3491	3588	3685	3782	3879	30
30	3879	3975	4071	4167	4263	4359	4454	4549	4644	4739	4833	20
40	4833	4928	5022	5116	5209	5303	5396	5489	5582	5675	5767	10
50	5767	5860	5952	6044	6135	6227	6318	6409	6500	6591	6682	0 **52**
8 0	6682	6772	6862	6952	7042	7132	7221	7310	7399	7488	7577	50
10	7577	7666	7754	7842	7930	8018	8106	8193	8280	8367	8454	40
20	8454	8541	8628	8714	8800	8887	8972	9058	9144	9229	9314	30
30	9314	9400	9484	9569	9654	9738	9822	9906	9990	*0074	*0158	20
40	7.4 0158	0241	0324	0408	0491	0573	0656	0739	0821	0903	0985	10
50	0985	1067	1149	1230	1312	1393	1474	1555	1636	1716	1797	0 **51**
9 0	1797	1877	1957	2037	2117	2197	2277	2356	2435	2515	2594	50
10	2594	2673	2751	2830	2908	2987	3065	3143	3221	3299	3376	40
20	3376	3454	3531	3608	3685	3762	3839	3916	3992	4069	4145	30
30	4145	4221	4297	4373	4449	4524	4600	4675	4750	4825	4900	20
40	4900	4975	5050	5124	5199	5273	5347	5421	5495	5569	5643	10
50	5643	5716	5790	5863	5936	6009	6082	6155	6228	6300	6373	0 **50**
	10″	9″	8″	7″	6″	5″	4″	3″	2″	1″	0″	″ ′
						89°					Log Cos

′ ″	0″	1″	2″	3″	4″	5″	6″	7″	8″	9″	10″	
	Log Tan						0°					
0 0	4. —	68557	98660	*16270	*28763	*38454	*46373	*53067	*58866	*63982	*68557	50
10	5. 68557	72697	76476	79952	83170	86167	88969	91602	94085	96433	98660	40
20	98660	*00779	*02800	*04730	*06579	*08351	*10055	*11694	*13273	*14797	*16270	30
30	6. 16270	17694	19072	20409	21705	22964	24188	25378	26536	27664	28763	20
40	28763	29836	30882	31904	32903	33879	34833	35767	36682	37577	38454	10
50	38454	39315	40158	40985	41797	42594	43376	44145	44900	45643	46373	0 **59**
1 0	6.4 6373	7090	7797	8492	9175	9849	*0512	*1165	*1808	*2442	*3067	50
10	6.5 3067	3683	4291	4890	5481	6064	6639	7207	7767	8320	8866	40
20	8866	9406	9939	*0465	*0985	*1499	*2007	*2509	*3006	*3496	*3982	30
30	6.6 3982	4462	4936	5406	5870	6330	6785	7235	7680	8121	8557	20
40	8557	8990	9418	9841	*0261	*0676	*1088	*1496	*1900	*2300	*2697	10
50	6.7 2697	3090	3479	3865	4248	4627	5003	5376	5746	6112	6476	0 **58**
2 0	6476	6836	7193	7548	7900	8248	8595	8938	9278	9616	9952	50
10	9952	*0285	*0615	*0943	*1268	*1591	*1911	*2230	*2545	*2859	*3170	40
20	6.8 3170	3479	3786	4091	4394	4694	4993	5289	5584	5876	6167	30
30	6167	6455	6742	7027	7310	7591	7870	8147	8423	8697	8969	20
40	8969	9240	9509	9776	*0042	*0306	*0568	*0829	*1088	*1346	*1602	10
50	6.9 1602	1857	2110	2362	2612	2861	3109	3355	3599	3843	4085	0 **57**
3 0	4085	4325	4565	4803	5039	5275	5509	5742	5973	6204	6433	50
10	6433	6661	6888	7113	7338	7561	7783	8004	8224	8443	8660	40
20	8660	8877	9093	9307	9521	9733	9944	*0155	*0364	*0572	*0779	30
30	7.0 0779	0986	1191	1395	1599	1801	2003	2203	2403	2602	2800	20
40	2800	2997	3193	3388	3582	3776	3968	4160	4351	4541	4730	10
50	4730	4919	5106	5293	5479	5664	5849	6032	6215	6397	6579	0 **56**
4 0	6579	6759	6939	7118	7296	7474	7651	7827	8003	8177	8352	50
10	8352	8525	8698	8870	9041	9211	9382	9551	9719	9887	*0055	40
20	7.1 0055	0222	0388	0553	0718	0882	1046	1209	1371	1533	1694	30
30	1694	1854	2014	2174	2333	2491	2648	2805	2962	3118	3273	20
40	3273	3428	3582	3736	3889	4042	4194	4346	4497	4647	4797	10
50	4797	4947	5096	5244	5392	5540	5687	5833	5979	6125	6270	0 **55**
5 0	6270	6414	6558	6702	6845	6988	7130	7271	7413	7553	7694	50
10	7694	7834	7973	8112	8250	8389	8526	8663	8800	8937	9073	40
20	9073	9208	9343	9478	9612	9746	9879	*0012	*0145	*0277	*0409	30
30	7.2 0409	0540	0671	0802	0932	1062	1191	1321	1449	1577	1705	20
40	1705	1833	1960	2087	2213	2339	2465	2590	2715	2840	2964	10
50	2964	3088	3212	3335	3458	3580	3703	3824	3946	4067	4188	0 **54**
6 0	4188	4308	4428	4548	4668	4787	4906	5024	5142	5260	5378	50
10	5378	5495	5612	5728	5845	5961	6076	6192	6307	6421	6536	40
20	6536	6650	6764	6877	6991	7104	7216	7329	7441	7552	7664	30
30	7664	7775	7886	7997	8107	8217	8327	8437	8546	8655	8764	20
40	8764	8872	8980	9088	9196	9303	9410	9517	9624	9730	9836	10
50	9836	9942	*0047	*0153	*0258	*0362	*0467	*0571	*0675	*0779	*0882	0 **53**
7 0	7.3 0882	0986	1089	1192	1294	1396	1499	1600	1702	1803	1904	50
10	1904	2005	2106	2206	2307	2406	2506	2606	2705	2804	2903	40
20	2903	3001	3100	3198	3296	3394	3491	3588	3685	3782	3879	30
30	3879	3975	4071	4167	4263	4359	4454	4549	4644	4739	4833	20
40	4833	4928	5022	5116	5209	5303	5396	5489	5582	5675	5767	10
50	5767	5860	5952	6044	6135	6227	6318	6409	6500	6591	6682	0 **52**
8 0	6682	6772	6862	6952	7042	7132	7221	7310	7400	7488	7577	50
10	7577	7666	7754	7842	7930	8018	8106	8193	8281	8368	8455	40
20	8455	8541	8628	8714	8801	8887	8973	9058	9144	9229	9315	30
30	9315	9400	9485	9569	9654	9738	9823	9907	9991	*0074	*0158	20
40	7.4 0158	0241	0325	0408	0491	0574	0656	0739	0821	0903	0985	10
50	0985	1067	1149	1230	1312	1393	1474	1555	1636	1716	1797	0 **51**
9 0	1797	1877	1958	2038	2117	2197	2277	2356	2436	2515	2594	50
10	2594	2673	2751	2830	2909	2987	3065	3143	3221	3299	3376	40
20	3376	3454	3531	3608	3686	3762	3839	3916	3992	4069	4145	30
30	4145	4221	4297	4373	4449	4524	4600	4675	4750	4825	4900	20
40	4900	4975	5050	5124	5199	5273	5347	5421	5495	5569	5643	10
50	5643	5716	5790	5863	5936	6009	6082	6155	6228	6300	6373	0 **50**
	10″	9″	8″	7″	6″	5″	4″	3″	2″	1″	0″	″ ′
			89°							Log Cot		

Log Sin						0°						
′ ″	0″	1″	2″	3″	4″	5″	6″	7″	8″	9″	10″	
10 0	7.46 373	445	517	589	661	733	805	876	948	*019	*090	50
10	7.47 090	162	233	303	374	445	515	586	656	726	797	40
20	797	867	936	*006	*076	*145	*215	*284	*353	*422	*491	30
30	7.48 491	560	629	698	766	835	903	971	*039	*108	*175	20
40	7.49 175	243	311	379	446	513	581	648	715	782	849	10
50	849	916	982	*049	*115	*182	*248	*314	*380	*446	*512	0 **49**
11 0	7.50 512	578	643	709	774	840	905	970	*035	*100	*165	50
10	7.51 165	230	294	359	423	488	552	616	680	744	808	40
20	808	872	936	999	*063	*126	*190	*253	*316	*379	*442	30
30	7.52 442	505	568	631	693	756	818	881	943	*005	*067	20
40	7.53 067	129	191	253	315	376	438	499	561	622	683	10
50	683	744	805	866	927	988	*049	*109	*170	*230	*291	0 **48**
12 0	7.54 291	351	411	471	531	591	651	711	771	830	890	50
10	890	949	*009	*068	*127	*186	*245	*304	*363	*422	*481	40
20	7.55 481	539	598	656	715	773	831	889	948	*006	*064	30
30	7.56 064	121	179	237	295	352	410	467	524	582	639	20
40	639	696	753	810	867	924	980	*037	*094	*150	*206	10
50	7.57 206	263	319	375	431	488	544	599	655	711	767	0 **47**
13 0	767	822	878	934	989	*044	*100	*155	*210	*265	*320	50
10	7.58 320	375	430	485	539	594	649	703	758	812	866	40
20	866	921	975	*029	*083	*137	*191	*245	*299	*352	*406	30
30	7.59 406	459	513	566	620	673	726	780	833	886	939	20
40	939	992	*045	*097	*150	*203	*255	*308	*360	*413	*465	10
50	7.60 465	517	570	622	674	726	778	830	882	934	985	0 **46**
14 0	985	*037	*089	*140	*192	*243	*294	*346	*397	*448	*499	50
10	7.61 499	550	601	652	703	754	805	855	906	957	*007	40
20	7.62 007	058	108	158	209	259	309	359	409	459	509	30
30	509	559	609	659	708	758	808	857	907	956	*006	20
40	7.63 006	055	104	153	203	252	301	350	399	448	496	10
50	496	545	594	642	691	740	788	837	885	933	982	0 **45**
15 0	982	*030	*078	*126	*174	*222	*270	*318	*366	*414	*461	50
10	7.64 461	509	557	604	652	699	747	794	842	889	936	40
20	936	983	*030	*078	*125	*172	*218	*265	*312	*359	*406	30
30	7.65 406	452	499	546	592	638	685	731	778	824	870	20
40	870	916	962	*009	*055	*101	*146	*192	*238	*284	*330	10
50	7.66 330	375	421	467	512	558	603	649	694	739	784	0 **44**
16 0	784	830	875	920	965	*010	*055	*100	*145	*190	*235	50
10	7.67 235	279	324	369	413	458	502	547	591	636	680	40
20	680	724	768	813	857	901	945	989	*033	*077	*121	30
30	7.68 121	165	208	252	296	340	383	427	470	514	557	20
40	557	601	644	687	731	774	817	860	903	946	989	10
50	989	*032	*075	*118	*161	*204	*247	*289	*332	*375	*417	0 **43**
17 0	7.69 417	460	502	545	587	630	672	714	757	799	841	50
10	841	883	925	967	*009	*051	*093	*135	*177	*219	*261	40
20	7.70 261	302	344	386	427	469	510	552	593	635	676	30
30	676	718	759	800	841	883	924	965	*006	*047	*088	20
40	7.71 088	129	170	211	251	292	333	374	414	455	496	10
50	496	536	577	617	658	698	739	779	819	859	900	0 **42**
18 0	900	940	980	*020	*060	*100	*140	*180	*220	*260	*300	50
10	7.72 300	340	380	419	459	499	538	578	618	657	697	40
20	697	736	775	815	854	894	933	972	*011	*050	*090	30
30	7.73 090	129	168	207	246	285	324	363	401	440	479	20
40	479	518	557	595	634	673	711	750	788	827	865	10
50	865	904	942	980	*019	*057	*095	*133	*171	*210	*248	0 **41**
19 0	7.74 248	286	324	362	400	438	476	514	551	589	627	50
10	627	665	703	740	778	815	853	891	928	966	*003	40
20	7.75 003	040	078	115	153	190	227	264	302	339	376	30
30	376	413	450	487	524	561	598	635	672	709	745	20
40	745	782	819	856	892	929	966	*002	*039	*075	*112	10
50	7.76 112	148	185	221	258	294	330	367	403	439	475	0 **40**
	10″	9″	8″	7″	6″	5″	4″	3″	2″	1″	0″	″ ′

Log Tan					0°								
′ ″	0″	1″	2″	3″	4″	5″	6″	7″	8″	9″	10″		
10 0	7.46 373	445	517	589	661	733	805	876	948	*019	*091	50	
10	7.47 091	162	233	304	374	445	516	586	656	727	797	40	
20	797	867	937	*006	*076	*146	*215	*284	*354	*423	*492	30	
30	7.48 492	561	629	698	767	835	903	972	*040	*108	*176	20	
40	7.49 176	243	311	379	446	514	581	648	715	782	849	10	
50	849	916	982	*049	*115	*182	*248	*314	*380	*446	*512	0	**49**
11 0	7.50 512	578	643	709	774	840	905	970	*035	*100	*165	50	
10	7.51 165	230	295	359	424	488	552	617	681	745	809	40	
20	809	872	936	*000	*063	*127	*190	*253	*316	*380	*443	30	
30	7.52 443	505	568	631	694	756	819	881	943	*005	*067	20	
40	7.53 067	129	191	253	315	377	438	500	561	622	683	10	
50	683	745	806	867	927	988	*049	*110	*170	*231	*291	0	**48**
12 0	7.54 291	351	411	471	532	591	651	711	771	830	890	50	
10	890	949	*009	*068	*127	*186	*245	*304	*363	*422	*481	40	
20	7.55 481	539	598	657	715	773	832	890	948	*006	*064	30	
30	7.56 064	122	179	237	295	352	410	467	525	582	639	20	
40	639	696	753	810	867	924	981	*037	*094	*150	*207	10	
50	7.57 207	263	319	376	432	488	544	600	656	711	767	0	**47**
13 0	767	823	878	934	989	*045	*100	*155	*210	*265	*320	50	
10	7.58 320	375	430	485	540	594	649	704	758	812	867	40	
20	867	921	975	*029	*083	*137	*191	*245	*299	*353	*406	30	
30	7.59 406	460	513	567	620	673	727	780	833	886	939	20	
40	939	992	*045	*098	*150	*203	*256	*308	*361	*413	*466	10	
50	7.60 466	518	570	622	674	726	778	830	882	934	986	0	**46**
14 0	986	*037	*089	*140	*192	*243	*295	*346	*397	*449	*500	50	
10	7.61 500	551	602	653	704	754	805	856	906	957	*008	40	
20	7.62 008	058	108	159	209	259	310	360	410	460	510	30	
30	510	560	609	659	709	759	808	858	907	957	*006	20	
40	7.63 006	055	105	154	203	252	301	350	399	448	497	10	
50	497	546	594	643	692	740	789	837	885	934	982	0	**45**
15 0	982	*030	*078	*127	*175	*223	*271	*318	*366	*414	*462	50	
10	7.64 462	510	557	605	652	700	747	795	842	889	937	40	
20	937	984	*031	*078	*125	*172	*219	*266	*313	*359	*406	30	
30	7.65 406	453	499	546	592	639	685	732	778	824	871	20	
40	871	917	963	*009	*055	*101	*147	*193	*239	*284	*330	10	
50	7.66 330	376	421	467	513	558	604	649	694	740	785	0	**44**
16 0	785	830	875	920	966	*011	*056	*100	*145	*190	*235	50	
10	7.67 235	280	324	369	414	458	503	547	592	636	680	40	
20	680	725	769	813	857	901	946	990	*034	*077	*121	30	
30	7.68 121	165	209	253	296	340	384	427	471	514	558	20	
40	558	601	645	688	731	774	818	861	904	947	990	10	
50	990	*033	*076	*119	*162	*204	*247	*290	*333	*375	*418	0	**43**
17 0	7.69 418	460	503	545	588	630	673	715	757	799	842	50	
10	842	884	926	968	*010	*052	*094	*136	*178	*219	*261	40	
20	7.70 261	303	345	386	428	469	511	553	594	635	677	30	
30	677	718	759	801	842	883	924	965	*006	*047	*088	20	
40	7.71 088	129	170	211	252	293	334	374	415	456	496	10	
50	496	537	577	618	658	699	739	779	820	860	900	0	**42**
18 0	900	940	981	*021	*061	*101	*141	*181	*221	*261	*301	50	
10	7.72 301	340	380	420	460	499	539	579	618	658	697	40	
20	697	737	776	815	855	894	933	973	*012	*051	*090	30	
30	7.73 090	129	168	207	246	285	324	363	402	441	480	20	
40	480	518	557	596	635	673	712	750	789	827	866	10	
50	866	904	943	981	*019	*058	*096	*134	*172	*210	*248	0	**41**
19 0	7.74 248	286	325	363	401	438	476	514	552	590	628	50	
10	628	665	703	741	779	816	854	891	929	966	*004	40	
20	7.75 004	041	079	116	153	191	228	265	302	339	377	30	
30	377	414	451	488	525	562	599	636	672	709	746	20	
40	746	783	820	856	893	930	966	*003	*040	*076	*113	10	
50	7.76 113	149	186	222	258	295	331	367	404	440	476	0	**40**
	10″	9″	8″	7″	6″	5″	4″	3″	2″	1″	0″	″	′

| 89° | Log Cot |

Log Sin   0°

′	″	0″	1″	2″	3″	4″	5″	6″	7″	8″	9″	10″		
20	0	7.76 475	512	548	584	620	656	692	728	764	800	836	50	
	10	836	872	907	943	979	*015	*051	*086	*122	*158	*193	40	
	20	7.77 193	229	264	300	335	371	406	442	477	512	548	30	
	30	548	583	618	654	689	724	759	794	829	864	899	20	
	40	899	934	969	*004	*039	*074	*109	*144	*179	*213	*248	10	
	50	7.78 248	283	318	352	387	422	456	491	525	560	594	0	39
21	0	594	629	663	698	732	766	801	835	869	903	938	50	
	10	938	972	*006	*040	*074	*1C3	*142	*176	*210	*244	*278	40	
	20	7.79 278	312	346	380	414	448	481	515	549	582	616	30	
	30	616	650	683	717	751	784	818	851	885	918	952	20	
	40	952	985	*018	*052	*085	*118	*152	*185	*218	*251	*284	10	
	50	7.80 284	317	351	384	417	450	483	516	549	582	615	0	38
22	0	615	647	680	713	746	779	812	844	877	910	942	50	
	10	942	975	*008	*040	*073	*105	*138	*170	*203	*235	*268	40	
	20	7.81 268	300	332	365	397	429	462	494	526	558	591	30	
	30	591	623	655	687	719	751	783	815	847	879	911	20	
	40	911	943	975	*007	*039	*070	*102	*134	*166	*198	*229	10	
	50	7.82 229	261	293	324	356	387	419	451	482	514	545	0	37
23	0	545	577	608	639	671	702	733	765	796	827	859	50	
	10	859	890	921	952	983	*015	*046	*077	*108	*139	*170	40	
	20	7.83 170	201	232	263	294	325	356	387	417	448	479	30	
	30	479	510	541	571	602	633	663	694	725	755	786	20	
	40	786	817	847	878	908	939	969	*000	*030	*060	*091	10	
	50	7.84 091	121	151	182	212	242	273	303	333	363	393	0	36
24	0	393	424	454	484	514	544	574	604	634	664	694	50	
	10	694	724	754	784	814	843	873	903	933	963	992	40	
	20	992	*022	*052	*082	*111	*141	*171	*200	*230	*259	*289	30	
	30	7.85 289	318	348	377	407	436	466	495	525	554	583	20	
	40	583	613	642	671	701	730	759	788	817	847	876	10	
	50	876	905	934	963	992	*021	*050	*079	*108	*137	*166	0	35
25	0	7.86 166	195	224	253	282	311	340	368	397	426	455	50	
	10	455	484	512	541	570	598	627	656	684	713	741	40	
	20	741	770	799	827	856	884	913	941	969	998	*026	30	
	30	7.87 026	055	083	111	140	168	196	224	253	281	309	20	
	40	309	337	366	394	422	450	478	506	534	562	590	10	
	50	590	618	646	674	702	730	758	786	814	842	870	0	34
26	0	870	897	925	953	981	*009	*036	*064	*092	*119	*147	50	
	10	7.88 147	175	202	230	258	285	313	340	368	395	423	40	
	20	423	450	478	505	533	560	587	615	642	669	697	30	
	30	697	724	751	779	806	833	860	888	915	942	969	20	
	40	969	996	*023	*050	*077	*105	*132	*159	*186	*213	*240	10	
	50	7.89 240	267	294	320	347	374	401	428	455	482	509	0	33
27	0	509	535	562	589	616	642	669	696	722	749	776	50	
	10	776	802	829	856	882	909	935	962	988	*015	*041	40	
	20	7.90 041	068	094	121	147	174	200	226	253	279	305	30	
	30	305	332	358	384	411	437	463	489	515	542	568	20	
	40	568	594	620	646	672	698	725	751	777	803	829	10	
	50	829	855	881	907	933	958	984	*010	*036	*062	*088	0	32
28	0	7.91 088	114	140	165	191	217	243	269	294	320	346	50	
	10	346	371	397	423	448	474	500	525	551	576	602	40	
	20	602	627	653	678	704	729	755	780	806	831	857	30	
	30	857	882	907	933	958	983	*009	*034	*059	*085	*110	20	
	40	7.92 110	135	160	186	211	236	261	286	311	336	362	10	
	50	362	387	412	437	462	487	512	537	562	587	612	0	31
29	0	612	637	662	687	712	737	761	786	811	836	861	50	
	10	861	886	910	935	960	985	*009	*034	*059	*084	*108	40	
	20	7.93 108	133	158	182	207	231	256	281	305	330	354	30	
	30	354	379	403	428	452	477	501	526	550	575	599	20	
	40	599	623	648	672	696	721	745	769	794	818	842	10	
	50	842	866	891	915	939	963	988	*012	*036	*060	*084	0	30
		10″	9″	8″	7″	6″	5″	4″	3″	2″	1″	0″	″	′

89°   Log Cos

Log Tan					0°						
′ ″	0″	1″	2″	3″	4″	5″	6″	7″	8″	9″	10″	
20 0	7.76 476	512	548	585	621	657	693	729	765	801	837	50
10	837	872	908	944	980	*016	*051	*087	*123	*158	*194	40
20	7.77 194	230	265	301	336	372	407	442	478	513	549	30
30	549	584	619	654	690	725	760	795	830	865	900	20
40	900	935	970	*005	*040	*075	*110	*145	*179	*214	*249	10
50	7.78 249	284	318	353	388	422	457	492	526	561	595	0 **39**
21 0	595	630	664	698	733	767	801	836	870	904	938	50
10	938	973	*007	*041	*075	*109	*143	*177	*211	*245	*279	40
20	7.79 279	313	347	381	415	448	482	516	550	583	617	30
30	617	651	684	718	751	785	819	852	886	919	952	20
40	952	986	*019	*053	*086	*119	*152	*186	*219	*252	*285	10
50	7.80 285	318	351	385	418	451	484	517	550	583	615	0 **38**
22 0	615	648	681	714	747	780	812	845	878	911	943	50
10	943	976	*009	*041	*074	*106	*139	*171	*204	*236	*269	40
20	7.81 269	301	333	366	398	430	463	495	527	559	591	30
30	591	624	656	688	720	752	784	816	848	880	912	20
40	912	944	976	*008	*040	*071	*103	*135	*167	*198	*230	10
50	7.82 230	262	294	325	357	388	420	452	483	515	546	0 **37**
23 0	546	578	609	640	672	703	734	766	797	828	860	50
10	860	891	922	953	984	*016	*047	*078	*109	*140	*171	40
20	7.83 171	202	233	264	295	326	357	388	418	449	480	30
30	480	511	542	572	603	634	664	695	726	756	787	20
40	787	818	848	879	909	940	970	*001	*031	*061	*092	10
50	7.84 092	122	152	183	213	243	274	304	334	364	394	0 **36**
24 0	394	425	455	485	515	545	575	605	635	665	695	50
10	695	725	755	785	815	845	874	904	934	964	993	40
20	993	*023	*053	*083	*112	*142	*172	*201	*231	*260	*290	30
30	7.85 290	319	349	378	408	437	467	496	526	555	584	20
40	584	614	643	672	702	731	760	789	819	848	877	10
50	877	906	935	964	993	*022	*051	*080	*109	*138	*167	0 **35**
25 0	7.86 167	196	225	254	283	312	341	370	398	427	456	50
10	456	485	513	542	571	600	628	657	685	714	743	40
20	743	771	800	828	857	885	914	942	971	999	*027	30
30	7.87 027	056	084	113	141	169	197	226	254	282	310	20
40	310	339	367	395	423	451	479	507	535	563	591	10
50	591	619	647	675	703	731	759	787	815	843	871	0 **34**
26 0	871	899	926	954	982	*010	*037	*065	*093	*121	*148	50
10	7.88 148	176	204	231	259	286	314	342	369	397	424	40
20	424	452	479	506	534	561	589	616	643	671	698	30
30	698	725	753	780	807	834	862	889	916	943	970	20
40	970	997	*025	*052	*079	*106	*133	*160	*187	*214	*241	10
50	7.89 241	268	295	322	349	376	403	429	456	483	510	0 **33**
27 0	510	537	563	590	617	644	670	697	724	750	777	50
10	777	804	830	857	884	910	937	963	990	*016	*043	40
20	7.90 043	069	096	122	149	175	201	228	254	280	307	30
30	307	333	359	386	412	438	464	491	517	543	569	20
40	569	595	622	648	674	700	726	752	778	804	830	10
50	830	856	882	908	934	960	986	*012	*038	*064	*089	0 **32**
28 0	7.91 089	115	141	167	193	218	244	270	296	321	347	50
10	347	373	398	424	450	475	501	527	552	578	603	40
20	603	629	654	680	705	731	756	782	807	833	858	30
30	858	883	909	934	960	985	*010	*036	*061	*086	*111	20
40	7.92 111	137	162	187	212	237	263	288	313	338	363	10
50	363	388	413	438	463	488	513	538	563	588	613	0 **31**
29 0	613	638	663	688	713	738	763	788	813	838	862	50
10	862	887	912	937	961	986	*011	*036	*060	*085	*110	40
20	7.93 110	134	159	184	208	233	258	282	307	331	356	30
30	356	380	405	429	454	478	503	527	552	576	601	20
40	601	625	649	674	698	722	747	771	795	820	844	10
50	844	868	892	917	941	965	989	*013	*038	*062	*086	0 **30**
	10″	9″	8″	7″	6″	5″	4″	3″	2″	1″	0″	″ ′

Log Sin					0°							
′ ″	0″	1″	2″	3″	4″	5″	6″	7″	8″	9″	10″	
30 0	7.94 084	108	132	157	181	205	229	253	277	301	325	50
10	325	349	373	397	421	445	469	492	516	540	564	40
20	564	588	612	636	659	683	707	731	755	778	802	30
30	802	826	849	873	897	921	944	968	991	*015	*039	20
40	7.95 039	062	086	109	133	157	180	204	227	251	274	10
50	274	298	321	344	368	391	415	438	461	485	508	0 **29**
31 0	508	532	555	578	601	625	648	671	695	718	741	50
10	741	764	787	811	834	857	880	903	926	950	973	40
20	973	996	*019	*042	*065	*088	*111	*134	*157	*180	*203	30
30	7.96 203	226	249	272	295	318	341	364	386	409	432	20
40	432	455	478	501	524	546	569	592	615	637	660	10
50	660	683	706	728	751	774	796	819	842	864	887	0 **28**
32 0	887	910	932	955	977	*000	*022	*045	*068	*090	*113	50
10	7.97 113	135	158	180	202	225	247	270	292	315	337	40
20	337	359	382	404	426	449	471	493	516	538	560	30
30	560	583	605	627	649	672	694	716	738	760	782	20
40	782	805	827	849	871	893	915	937	959	981	*003	10
50	7.98 003	025	048	070	092	114	136	157	179	201	223	0 **27**
33 0	223	245	267	289	311	333	355	377	398	420	442	50
10	442	464	486	508	529	551	573	595	616	638	660	40
20	660	682	703	725	747	768	790	812	833	855	876	30
30	876	898	920	941	963	984	*006	*027	*049	*070	*092	20
40	7.99 092	113	135	156	178	199	221	242	264	285	306	10
50	306	328	349	371	392	.413	435	456	477	499	520	0 **26**
34 0	520	541	562	584	605	626	647	669	690	711	732	50
10	732	753	775	796	817	838	859	880	901	922	943	40
20	943	965	986	*007	*028	*049	*070	*091	*112	*133	*154	30
30	8.00 154	175	196	217	238	259	279	300	321	342	363	20
40	363	384	405	426	447	467	488	509	530	551	571	10
50	571	592	613	634	654	675	696	717	737	758	779	0 **25**
35 0	779	799	820	841	861	882	903	923	944	964	985	50
10	985	*006	*026	*047	*067	*088	*108	*129	*149	*170	*190	40
20	8.01 190	211	231	252	272	293	313	333	354	374	395	30
30	395	415	435	456	476	496	517	537	557	578	598	20
40	598	618	639	659	679	699	720	740	760	780	801	10
50	801	821	841	861	881	901	922	942	962	982	*002	0 **24**
36 0	8.02 002	022	042	062	082	102	123	143	163	183	203	50
10	203	223	243	263	283	303	323	343	362	382	402	40
20	402	422	442	462	482	502	522	542	561	581	601	30
30	601	621	641	661	680	700	720	740	759	779	799	20
40	799	819	838	858	878	898	917	937	957	976	996	10
50	996	*016	*035	*055	*074	*094	*114	*133	*153	*172	*192	0 **23**
37 0	8.03 192	212	231	251	270	290	309	329	348	368	387	50
10	387	407	426	446	465	484	504	523	543	562	581	40
20	581	601	620	640	659	678	698	717	736	756	775	30
30	775	794	813	833	852	871	891	910	929	948	967	20
40	967	987	*006	*025	*044	*063	*083	*102	*121	*140	*159	10
50	8.04 159	178	197	217	236	255	274	293	312	331	350	0 **22**
38 0	350	369	388	407	426	445	464	483	502	521	540	50
10	540	559	578	597	616	635	654	673	692	710	729	40
20	729	748	767	786	805	824	843	861	880	899	918	30
30	918	937	955	974	993	*012	*030	*049	*068	*087	*105	20
40	8.05 105	124	143	161	180	199	218	236	255	274	292	10
50	292	311	329	348	367	385	404	422	441	460	478	0 **21**
39 0	478	497	515	534	552	571	589	608	626	645	663	50
10	663	682	700	719	737	756	774	792	811	829	848	40
20	848	866	885	903	921	940	958	976	995	*013	*031	30
30	8.06 031	050	068	086	105	123	141	159	178	196	214	20
40	214	232	251	269	287	305	324	342	360	378	396	10
50	396	414	433	451	469	487	505	523	541	560	578	0 **20**
	10″	9″	8″	7″	6″	5″	4″	3″	2″	1″	0″	″ ′

89°   Log Cos

Log Tan						0°						
′ ″	0″	1″	2″	3″	4″	5″	6″	7″	8″	9″	10″	
30 0	7.94 086	110	134	158	182	206	230	254	278	302	326	50
10	326	350	374	398	422	446	470	494	518	542	566	40
20	566	590	613	637	661	685	709	732	756	780	804	30
30	804	827	851	875	899	922	946	970	993	*017	*040	20
40	7.95 040	064	088	111	135	158	182	205	229	252	276	10
50	276	299	323	346	370	393	416	440	463	487	510	0 **29**
31 0	510	533	557	580	603	627	650	673	696	720	743	50
10	743	766	789	812	836	859	882	905	928	951	974	40
20	974	998	*021	*044	*067	*090	*113	*136	*159	*182	*205	30
30	7.96 205	228	251	274	297	320	343	365	388	411	434	20
40	434	457	480	503	525	548	571	594	617	639	662	10
50	662	685	708	730	753	776	798	821	844	866	889	0 **28**
32 0	889	911	934	957	979	*002	*024	*047	*069	*092	*114	50
10	7.97 114	137	159	182	204	227	249	272	294	317	339	40
20	339	361	384	406	428	451	473	495	518	540	562	30
30	562	585	607	629	651	673	696	718	740	762	784	20
40	784	807	829	851	873	895	917	939	961	983	*005	10
50	7.98 005	027	050	072	094	116	138	159	181	203	225	0 **27**
33 0	225	247	269	291	313	335	357	379	400	422	444	50
10	444	466	488	510	531	553	575	597	618	640	662	40
20	662	684	705	727	749	770	792	814	835	857	878	30
30	878	900	922	943	965	986	*008	*029	*051	*073	*094	20
40	7.99 094	116	137	158	180	201	223	244	266	287	308	10
50	308	330	351	373	394	415	437	458	479	501	522	0 **26**
34 0	522	543	564	586	607	628	649	671	692	713	734	50
10	734	755	777	798	819	840	861	882	903	925	946	40
20	946	967	988	*009	*030	*051	*072	*093	*114	*135	*156	30
30	8.00 156	177	198	219	240	261	282	303	324	344	365	20
40	365	386	407	428	449	470	490	511	532	553	574	10
50	574	594	615	636	657	677	698	719	740	760	781	0 **25**
35 0	781	802	822	843	864	884	905	925	946	967	987	50
10	987	*008	*028	*049	*070	*090	*111	*131	*152	*172	*193	40
20	8.01 193	213	234	254	274	295	315	336	356	377	397	30
30	397	417	438	458	478	499	519	539	560	580	600	20
40	600	621	641	661	682	702	722	742	762	783	803	10
50	803	823	843	863	884	904	924	944	964	984	*004	0 **24**
36 0	8.02 004	025	045	065	085	105	125	145	165	185	205	50
10	205	225	245	265	285	305	325	345	365	385	405	40
20	405	425	445	464	484	504	524	544	564	584	604	30
30	604	623	643	663	683	703	722	742	762	782	801	20
40	801	821	841	861	880	900	920	939	959	979	998	10
50	998	*018	*038	*057	*077	*097	*116	*136	*155	*175	*194	0 **23**
37 0	8.03 194	214	234	253	273	292	312	331	351	370	390	50
10	390	409	429	448	468	487	506	526	545	565	584	40
20	584	603	623	642	661	681	700	720	739	758	777	30
30	777	797	816	835	855	874	893	912	932	951	970	20
40	970	989	*008	*028	*047	*066	*085	*104	*124	*143	*162	10
50	8.04 162	181	200	219	238	257	276	296	315	334	353	0 **22**
38 0	353	372	391	410	429	448	467	486	505	524	543	50
10	543	562	581	600	619	638	656	675	694	713	732	40
20	732	751	770	789	808	826	845	864	883	902	921	30
30	921	939	958	977	996	*014	*033	*052	*071	*089	*108	20
40	8.05 108	127	146	164	183	202	220	239	258	276	295	10
50	295	314	332	351	369	388	407	425	444	462	481	0 **21**
39 0	481	499	518	537	555	574	592	611	629	648	666	50
10	666	685	703	722	740	758	777	795	814	832	851	40
20	851	869	887	906	924	943	961	979	998	*016	*034	30
30	8.06 034	053	071	089	107	126	144	162	181	199	217	20
40	217	235	254	272	290	308	326	345	363	381	399	10
50	399	417	436	454	472	490	508	526	544	562	581	0 **20**
	10″	9″	8″	7″	6″	5″	4″	3″	2″	1″	0″	″ ′

Log Sin   0°

′	″	0″	1″	2″	3″	4″	5″	6″	7″	8″	9″	10″		
40	0	8.06 578	596	614	632	650	668	686	704	722	740	758	50	
	10	758	776	794	812	830	848	866	884	902	920	938	40	
	20	938	956	974	992	*010	*028	*046	*063	*081	*099	*117	30	
	30	8.07 117	135	153	171	189	206	224	242	260	278	295	20	
	40	295	313	331	349	367	384	402	420	438	455	473	10	
	50	473	491	509	526	544	562	579	597	615	632	650	0	19
41	0	650	668	685	703	721	738	756	773	791	809	826	50	
	10	826	844	861	879	896	914	932	949	967	984	*002	40	
	20	8.08 002	019	037	054	072	089	107	124	141	159	176	30	
	30	176	194	211	229	246	263	281	298	316	333	350	20	
	40	350	368	385	403	420	437	455	472	489	506	524	10	
	50	524	541	558	576	593	610	627	645	662	679	696	0	18
42	0	696	714	731	748	765	783	800	817	834	851	868	50	
	10	868	886	903	920	937	954	971	988	*006	*023	*040	40	
	20	8.09 040	057	074	091	108	125	142	159	176	193	210	30	
	30	210	227	244	261	278	295	312	329	346	363	380	20	
	40	380	397	414	431	448	465	482	499	516	533	550	10	
	50	550	567	583	600	617	634	651	668	685	701	718	0	17
43	0	718	735	752	769	786	802	819	836	853	870	886	50	
	10	886	903	920	937	953	970	987	*004	*020	*037	*054	40	
	20	8.10 054	070	087	104	120	137	154	170	187	204	220	30	
	30	220	237	254	270	287	303	320	337	353	370	386	20	
	40	386	403	420	436	453	469	486	502	519	535	552	10	
	50	552	568	585	601	618	634	651	667	684	700	717	0	16
44	0	717	733	750	766	782	799	815	832	848	864	881	50	
	10	881	897	914	930	946	963	979	995	*012	*028	*044	40	
	20	8.11 044	061	077	093	110	126	142	159	175	191	207	30	
	30	207	224	240	256	272	289	305	321	337	354	370	20	
	40	370	386	402	418	435	451	467	483	499	515	531	10	
	50	531	548	564	580	596	612	628	644	660	677	693	0	15
45	0	693	709	725	741	757	773	789	805	821	837	853	50	
	10	853	869	885	901	917	933	949	965	981	997	*013	40	
	20	8.12 013	029	045	061	077	093	109	125	141	157	172	30	
	30	172	188	204	220	236	252	268	284	300	315	331	20	
	40	331	347	363	379	395	410	426	442	458	474	489	10	
	50	489	505	521	537	553	568	584	600	616	631	647	0	14
46	0	647	663	679	694	710	726	741	757	773	788	804	50	
	10	804	820	836	851	867	882	898	914	929	945	961	40	
	20	961	976	992	*007	*023	*039	*054	*070	*085	*101	*117	30	
	30	8.13 117	132	148	163	179	194	210	225	241	256	272	20	
	40	272	287	303	318	334	349	365	380	396	411	427	10	
	50	427	442	458	473	489	504	519	535	550	566	581	0	13
47	0	581	596	612	627	643	658	673	689	704	719	735	50	
	10	735	750	765	781	796	811	827	842	857	873	888	40	
	20	888	903	919	934	949	964	980	995	*010	*025	*041	30	
	30	8.14 041	056	071	086	101	117	132	147	162	178	193	20	
	40	193	208	223	238	253	269	284	299	314	329	344	10	
	50	344	359	375	390	405	420	435	450	465	480	495	0	12
48	0	495	510	525	541	556	571	586	601	616	631	646	50	
	10	646	661	676	691	706	721	736	751	766	781	796	40	
	20	796	811	826	841	856	871	886	901	915	930	945	30	
	30	945	960	975	990	*005	*020	*035	*050	*065	*079	*094	20	
	40	8.15 094	109	124	139	154	169	183	198	213	228	243	10	
	50	243	258	272	287	302	317	332	346	361	376	391	0	11
49	0	391	406	420	435	450	465	479	494	509	523	538	50	
	10	538	553	568	582	597	612	626	641	656	670	685	40	
	20	685	700	714	729	744	758	773	788	802	817	832	30	
	30	832	846	861	875	890	905	919	934	948	963	978	20	
	40	978	992	*007	*021	*036	*050	*065	*079	*094	*109	*123	10	
	50	8.16 123	138	152	167	181	196	210	225	239	254	268	0	10
		10″	9″	8″	7″	6″	5″	4″	3″	2″	1″	0″	″	′

89°   Log Cos

Log Tan						0°						
′ ″	0″	1″	2″	3″	4″	5″	6″	7″	8″	9″	10″	
40 0	8.06 581	599	617	635	653	671	689	707	725	743	761	50
10	761	779	797	815	833	851	869	887	905	923	941	40
20	941	959	977	995	*013	*031	*049	*066	*084	*102	*120	30
30	8.07 120	138	156	174	192	209	227	245	263	281	298	20
40	298	316	334	352	370	387	405	423	441	458	476	10
50	476	494	512	529	547	565	582	600	618	635	653	0 **19**
41 0	653	671	688	706	724	741	759	776	794	812	829	50
10	829	847	864	882	900	917	935	952	970	987	*005	40
20	8.08 005	022	040	057	075	092	110	127	145	162	180	30
30	180	197	214	232	249	267	284	301	319	336	354	20
40	354	371	388	406	423	440	458	475	492	510	527	10
50	527	544	562	579	596	613	631	648	665	682	700	0 **18**
42 0	700	717	734	751	769	786	803	820	837	855	872	50
10	872	889	906	923	940	957	975	992	*009	*026	*043	40
20	8.09 043	060	077	094	111	128	146	163	180	197	214	30
30	214	231	248	265	282	299	316	333	350	367	384	20
40	384	401	418	435	452	468	485	502	519	536	553	10
50	553	570	587	604	621	637	654	671	688	705	722	0 **17**
43 0	722	739	755	772	789	806	823	839	856	873	890	50
10	890	907	923	940	957	974	990	*007	*024	*040	*057	40
20	8.10 057	074	091	107	124	141	157	174	191	207	224	30
30	224	240	257	274	290	307	324	340	357	373	390	20
40	390	407	423	440	456	473	489	506	522	539	555	10
50	555	572	588	605	621	638	654	671	687	704	720	0 **16**
44 0	720	737	753	770	786	802	819	835	852	868	884	50
10	884	901	917	934	950	966	983	999	*015	*032	*048	40
20	8.11 048	064	081	097	113	130	146	162	178	195	211	30
30	211	227	244	260	276	292	309	325	341	357	373	20
40	373	390	406	422	438	454	471	487	503	519	535	10
50	535	551	567	584	600	616	632	648	664	680	696	0 **15**
45 0	696	712	729	745	761	777	793	809	825	841	857	50
10	857	873	889	905	921	937	953	969	985	*001	*017	40
20	8.12 017	033	049	065	081	097	113	129	144	160	176	30
30	176	192	208	224	240	256	272	288	303	319	335	20
40	335	351	367	383	398	414	430	446	462	478	493	10
50	493	509	525	541	556	572	588	604	620	635	651	0 **14**
46 0	651	667	682	698	714	730	745	761	777	792	808	50
10	808	824	839	855	871	886	902	918	933	949	965	40
20	965	980	996	*011	*027	*043	*058	*074	*089	*105	*121	30
30	8.13 121	136	152	167	183	198	214	229	245	260	276	20
40	276	291	307	322	338	353	369	384	400	415	431	10
50	431	446	462	477	493	508	523	539	554	570	585	0 **13**
47 0	585	601	616	631	647	662	677	693	708	724	739	50
10	739	754	770	785	800	816	831	846	861	877	892	40
20	892	907	923	938	953	968	984	999	*014	*029	*045	30
30	8.14 045	060	075	090	106	121	136	151	166	182	197	20
40	197	212	227	242	258	273	288	303	318	333	348	10
50	348	364	379	394	409	424	439	454	469	484	500	0 **12**
48 0	500	515	530	545	560	575	590	605	620	635	650	50
10	650	665	680	695	710	725	740	755	770	785	800	40
20	800	815	830	845	860	875	890	905	920	935	950	30
30	950	965	980	994	*009	*024	*039	*054	*069	*084	*099	20
40	8.15 099	114	128	143	158	173	188	203	218	232	247	10
50	247	262	277	292	306	321	336	351	366	380	395	0 **11**
49 0	395	410	425	439	454	469	484	498	513	528	543	50
10	543	557	572	587	602	616	631	646	660	675	690	40
20	690	704	719	734	748	763	778	792	807	822	836	30
30	836	851	865	880	895	909	924	938	953	968	982	20
40	982	997	*011	*026	*040	*055	*070	*084	*099	*113	*128	10
50	8.16 128	142	157	171	186	200	215	229	244	258	273	0 **10**
	10″	9″	8″	7″	6″	5″	4″	3″	2″	1″	0″	″ ′

89° Log Cot

Log Sin　　　　　　　　0°

′	″	0″	1″	2″	3″	4″	5″	6″	7″	8″	9″	10″		
50	0	8.16 268	283	297	311	326	340	355	369	384	398	413	50	
	10	413	427	441	456	470	485	499	513	528	542	557	40	
	20	557	571	585	600	614	628	643	657	672	686	700	30	
	30	700	715	729	743	757	772	786	800	815	829	843	20	
	40	843	858	872	886	900	915	929	943	957	972	986	10	
	50	986	*000	*014	*029	*043	*057	*071	*085	*100	*114	*128	0	9
51	0	8.17 128	142	156	171	185	199	213	227	241	256	270	50	
	10	270	284	298	312	326	340	355	369	383	397	411	40	
	20	411	425	439	453	467	481	495	510	524	538	552	30	
	30	552	566	580	594	608	622	636	650	664	678	692	20	
	40	692	706	720	734	748	762	776	790	804	818	832	10	
	50	832	846	860	874	888	902	916	930	943	957	971	0	8
52	0	971	985	999	*013	*027	*041	*055	*069	*082	*096	*110	50	
	10	8.18 110	124	138	152	166	180	193	207	221	235	249	40	
	20	249	263	276	290	304	318	332	345	359	373	387	30	
	30	387	401	414	428	442	456	469	483	497	511	524	20	
	40	524	538	552	566	579	593	607	621	634	648	662	10	
	50	662	675	689	703	716	730	744	757	771	785	798	0	7
53	0	798	812	826	839	853	867	880	894	908	921	935	50	
	10	935	948	962	976	989	*003	*016	*030	*044	*057	*071	40	
	20	8.19 071	084	098	111	125	139	152	166	179	193	206	30	
	30	206	220	233	247	260	274	287	301	314	328	341	20	
	40	341	355	368	382	395	409	422	436	449	463	476	10	
	50	476	489	503	516	530	543	557	570	583	597	610	0	6
54	0	610	624	637	650	664	677	691	704	717	731	744	50	
	10	744	757	771	784	797	811	824	837	851	864	877	40	
	20	877	891	904	917	931	944	957	971	984	997	*010	30	
	30	8.20 010	024	037	050	064	077	090	103	117	130	143	20	
	40	143	156	170	183	196	209	222	236	249	262	275	10	
	50	275	288	302	315	328	341	354	368	381	394	407	0	5
55	0	407	420	433	446	460	473	486	499	512	525	538	50	
	10	538	552	565	578	591	604	617	630	643	656	669	40	
	20	669	682	696	709	722	735	748	761	774	787	800	30	
	30	800	813	826	839	852	865	878	891	904	917	930	20	
	40	930	943	956	969	982	995	*008	*021	*034	*047	*060	10	
	50	8.21 060	073	086	099	112	125	138	151	164	177	189	0	4
56	0	189	202	215	228	241	254	267	280	293	306	319	50	
	10	319	331	344	357	370	383	396	409	422	434	447	40	
	20	447	460	473	486	499	511	524	537	550	563	576	30	
	30	576	588	601	614	627	640	652	665	678	691	703	20	
	40	703	716	729	742	754	767	780	793	805	818	831	10	
	50	831	844	856	869	882	895	907	920	933	945	958	0	3
57	0	958	971	983	996	*009	*022	*034	*047	*060	*072	*085	50	
	10	8.22 085	098	110	123	136	148	161	173	186	199	211	40	
	20	211	224	237	249	262	274	287	300	312	325	337	30	
	30	337	350	363	375	388	400	413	425	438	451	463	20	
	40	463	476	488	501	513	526	538	551	563	576	588	10	
	50	588	601	613	626	638	651	663	676	688	701	713	0	2
58	0	713	726	738	751	763	776	788	801	813	826	838	50	
	10	838	850	863	875	888	900	913	925	937	950	962	40	
	20	962	975	987	999	*012	*024	*037	*049	*061	*074	*086	30	
	30	8.23 086	098	111	123	136	148	160	173	185	197	210	20	
	40	210	222	234	247	259	271	284	296	308	321	333	10	
	50	333	345	357	370	382	394	407	419	431	443	456	0	1
59	0	456	468	480	492	505	517	529	541	554	566	578	50	
	10	578	590	603	615	627	639	652	664	676	688	700	40	
	20	700	713	725	737	749	761	773	786	798	810	822	30	
	30	822	834	846	859	871	883	895	907	919	931	944	20	
	40	944	956	968	980	992	*004	*016	*028	*041	*053	*065	10	
	50	8.24 065	077	089	101	113	125	137	149	161	173	186	0	0
		10″	9″	8″	7″	6″	5″	4″	3″	2″	1″	0″	″	′

89°　　　　　　　　　　　　　　Log Cos

Log Tan							0°						
′	″	0″	1″	2″	3″	4″	5″	6″	7″	8″	9″	10″	
50	0	8.16 273	287	302	316	331	345	359	374	388	403	417	50
	10	417	432	446	460	475	489	504	518	533	547	561	40
	20	561	576	590	604	619	633	647	662	676	691	705	30
	30	705	719	734	748	762	776	791	805	819	834	848	20
	40	848	862	877	891	905	919	934	948	962	976	991	10
	50	991	*005	*019	*033	*048	*062	*076	*090	*104	*119	*133	0 **9**
51	0	8.17 133	147	161	175	190	204	218	232	246	260	275	50
	10	275	289	303	317	331	345	359	373	388	402	416	40
	20	416	430	444	458	472	486	500	514	528	543	557	30
	30	557	571	585	599	613	627	641	655	669	683	697	20
	40	697	711	725	739	753	767	781	795	809	823	837	10
	50	837	851	865	879	893	907	921	934	948	962	976	0 **8**
52	0	976	990	*004	*018	*032	*046	*060	*074	*087	*101	*115	50
	10	8.18 115	129	143	157	171	185	198	212	226	240	254	40
	20	254	268	281	295	309	323	337	351	364	378	392	30
	30	392	406	419	433	447	461	475	488	502	516	530	20
	40	530	543	557	571	585	598	612	626	639	653	667	10
	50	667	681	694	708	722	735	749	763	776	790	804	0 **7**
53	0	804	817	831	845	858	872	886	899	913	926	940	50
	10	940	954	967	981	994	*008	*022	*035	*049	*062	*076	40
	20	8.19 076	090	103	117	130	144	157	171	184	198	211	30
	30	211	225	239	252	266	279	293	306	320	333	347	20
	40	347	360	374	387	401	414	427	441	454	468	481	10
	50	481	495	508	522	535	548	562	575	589	602	616	0 **6**
54	0	616	629	642	656	669	683	696	709	723	736	749	50
	10	749	763	776	789	803	816	830	843	856	870	883	40
	20	883	896	910	923	936	949	963	976	989	*003	*016	30
	30	8.20 016	029	042	056	069	082	096	109	122	135	149	20
	40	149	162	175	188	201	215	228	241	254	268	281	10
	50	281	294	307	320	334	347	360	373	386	399	413	0 **5**
55	0	413	426	439	452	465	478	491	505	518	531	544	50
	10	544	557	570	583	596	610	623	636	649	662	675	40
	20	675	688	701	714	727	740	753	767	780	793	806	30
	30	806	819	832	845	858	871	884	897	910	923	936	20
	40	936	949	962	975	988	*001	*014	*027	*040	*053	*066	10
	50	8.21 066	079	092	105	118	131	144	156	169	182	195	0 **4**
56	0	195	208	221	234	247	260	273	286	299	311	324	50
	10	324	337	350	363	376	389	402	414	427	440	453	40
	20	453	466	479	492	504	517	530	543	556	569	581	30
	30	581	594	607	620	633	645	658	671	684	697	709	20
	40	709	722	735	748	760	773	786	799	811	824	837	10
	50	837	850	862	875	888	901	913	926	939	951	964	0 **3**
57	0	964	977	989	*002	*015	*028	*040	*053	*066	*078	*091	50
	10	8.22 091	104	116	129	142	154	167	179	192	205	217	40
	20	217	230	243	255	268	280	293	306	318	331	343	30
	30	343	356	369	381	394	406	419	431	444	457	469	20
	40	469	482	494	507	519	532	544	557	569	582	595	10
	50	595	607	620	632	645	657	670	682	695	707	720	0 **2**
58	0	720	732	744	757	769	782	794	807	819	832	844	50
	10	844	857	869	881	894	906	919	931	944	956	968	40
	20	968	981	993	*006	*018	*030	*043	*055	*068	*080	*092	30
	30	8.23 092	105	117	130	142	154	167	179	191	204	216	20
	40	216	228	241	253	265	278	290	302	315	327	339	10
	50	339	352	364	376	388	401	413	425	438	450	462	0 **1**
59	0	462	474	487	499	511	523	536	548	560	572	585	50
	10	585	597	609	621	634	646	658	670	682	695	707	40
	20	707	719	731	743	756	768	780	792	804	816	829	30
	30	829	841	853	865	877	889	902	914	926	938	950	20
	40	950	962	974	987	999	*011	*023	*035	*047	*059	*071	10
	50	8.24 071	083	096	108	120	132	144	156	168	180	192	0 **0**
		10″	9″	8″	7″	6″	5″	4″	3″	2″	1″	0″	″ ′

89°   Log Cot

Log Sin   1°

′	0″	10″	20″	30″	40″	50″	60″		d.
0	8.24 186	306	426	546	665	785	903	59	120
1	903	*022	*140	*258	*375	*493	*609	58	118
2	8.25 609	726	842	958	*074	*189	*304	57	116
3	8.26 304	419	533	648	761	875	988	56	114
4	988	*101	*214	*326	*438	*550	*661	55	112
5	8.27 661	773	883	994	*104	*215	*324	54	110
6	8.28 324	434	543	652	761	869	977	53	109
7	977	*085	*193	*300	*407	*514	*621	52	107
8	8.29 621	727	833	939	*044	*150	*255	51	106
9	8.30 255	359	464	568	672	776	879	**50**	104
10	879	983	*086	*188	*291	*393	*495	49	103
11	8.31 495	597	699	800	901	*002	*103	48	101
12	8.32 103	203	303	403	503	602	702	47	100
13	702	801	899	998	*096	*195	*292	46	98
14	8.33 292	390	488	585	682	779	875	45	97
15	875	972	*068	*164	*260	*355	*450	44	96
16	8.34 450	546	640	735	830	924	*018	43	95
17	8.35 018	112	206	299	392	485	578	42	93
18	578	671	764	856	948	*040	*131	41	92
19	8.36 131	223	314	405	496	587	678	**40**	91
20	678	768	858	948	*038	*128	*217	39	90
21	8.37 217	306	395	484	573	662	750	38	89
22	750	838	926	*014	*101	*189	*276	37	88
23	8.38 276	363	450	537	624	710	796	36	87
24	796	882	968	*054	*139	*225	*310	35	86
25	8.39 310	395	480	565	649	734	818	34	85
26	818	902	986	*070	*153	*237	*320	33	84
27	8.40 320	403	486	569	651	734	816	32	83
28	816	898	980	*062	*144	*225	*307	31	82
29	8.41 307	388	469	550	631	711	792	**30**	81
30	792	872	952	*032	*112	*192	*272	29	80
31	8.42 272	351	430	510	589	667	746	28	79
32	746	825	903	982	*060	*138	*216	27	78
33	8.43 216	293	371	448	526	603	680	26	77
34	680	757	834	910	987	*063	*139	25	76
35	8.44 139	216	292	367	443	519	594	24	76
36	594	669	745	820	895	969	*044	23	75
37	8.45 044	119	193	267	341	415	489	22	74
38	489	563	637	710	784	857	930	21	74
39	930	*003	*076	*149	*222	*294	*366	**20**	73
40	8.46 366	439	511	583	655	727	799	19	72
41	799	870	942	*013	*084	*155	*226	18	71
42	8.47 226	297	368	439	509	580	650	17	71
43	650	720	790	860	930	*000	*069	16	70
44	8.48 069	139	208	278	347	416	485	15	69
45	485	554	622	691	760	828	896	14	68
46	896	965	*033	*101	*169	*236	*304	13	68
47	8.49 304	372	439	506	574	641	708	12	67
48	708	775	842	908	975	*042	*108	11	67
49	8.50 108	174	241	307	373	439	504	**10**	66
50	504	570	636	701	767	832	897	9	66
51	897	963	*028	*092	*157	*222	*287	8	65
52	8.51 287	351	416	480	544	609	673	7	64
53	673	737	801	864	928	992	*055	6	64
54	8.52 055	119	182	245	308	371	434	5	63
55	434	497	560	623	685	748	810	4	63
56	810	872	935	997	*059	*121	*183	3	62
57	8.53 183	245	306	368	429	491	552	2	62
58	552	614	675	736	797	858	919	1	61
59	919	979	*040	*101	*161	*222	*282	**0**	60
	60″	50″	40″	30″	20″	10″	0″	′	d.

88°   Log Cos

P. P.

	120	119	118
1	12.0	11.9	11.8
2	24.0	23.8	23.6
3	36.0	35.7	35.4
4	48.0	47.6	47.2
5	60.0	59.5	59.0
6	72.0	71.4	70.8
7	84.0	83.3	82.6
8	96.0	95.2	94.4
9	108.0	107.1	106.2

	117	116	115
1	11.7	11.6	11.5
2	23.4	23.2	23.0
3	35.1	34.8	34.5
4	46.8	46.4	46.0
5	58.5	58.0	57.5
6	70.2	69.6	69.0
7	81.9	81.2	80.5
8	93.6	92.8	92.0
9	105.3	104.4	103.5

	114	113	112	111
1	11.4	11.3	11.2	11.1
2	22.8	22.6	22.4	22.2
3	34.2	33.9	33.6	33.3
4	45.6	45.2	44.8	44.4
5	57.0	56.5	56.0	55.5
6	68.4	67.8	67.2	66.6
7	79.8	79.1	78.4	77.7
8	91.2	90.4	89.6	88.8
9	102.6	101.7	100.8	99.9

	110	109	108	107
1	11.0	10.9	10.8	10.7
2	22.0	21.8	21.6	21.4
3	33.0	32.7	32.4	32.1
4	44.0	43.6	43.2	42.8
5	55.0	54.5	54.0	53.5
6	66.0	65.4	64.8	64.2
7	77.0	76.3	75.6	74.9
8	88.0	87.2	86.4	85.6
9	99.0	98.1	97.2	96.3

	106	105	104	103
1	10.6	10.5	10.4	10.3
2	21.2	21.0	20.8	20.6
3	31.8	31.5	31.2	30.9
4	42.4	42.0	41.6	41.2
5	53.0	52.5	52.0	51.5
6	63.6	63.0	62.4	61.8
7	74.2	73.5	72.8	72.1
8	84.8	84.0	83.2	82.4
9	95.4	94.5	93.6	92.7

	102	101	100	99
1	10.2	10.1	10.0	9.9
2	20.4	20.2	20.0	19.8
3	30.6	30.3	30.0	29.7
4	40.8	40.4	40.0	39.6
5	51.0	50.5	50.0	49.5
6	61.2	60.6	60.0	59.4
7	71.4	70.7	70.0	69.3
8	81.6	80.8	80.0	79.2
9	91.8	90.9	90.0	89.1

	98	97	96	95
1	9.8	9.7	9.6	9.5
2	19.6	19.4	19.2	19.0
3	29.4	29.1	28.8	28.5
4	39.2	38.8	38.4	38.0
5	49.0	48.5	48.0	47.5
6	58.8	58.2	57.6	57.0
7	68.6	67.9	67.2	66.5
8	78.4	77.6	76.8	76.0
9	88.2	87.3	86.4	85.5

Log Tan					1°				
P. P.	'	0"	10"	20"	30"	40"	50"	60"		d.

Log Tan — 1°

'	0"	10"	20"	30"	40"	50"	60"		d.
0	8.24 192	313	433	553	672	791	910	59	120
1	910	*029	*147	*265	*382	*500	*616	58	118
2	8.25 616	733	849	965	*081	*196	*312	57	116
3	8.26 312	426	541	655	769	882	996	56	114
4	996	*109	*221	*334	*446	*558	*669	55	112
5	8.27 669	780	891	*002	*112	*223	*332	54	110
6	8.28 332	442	551	660	769	877	986	53	109
7	986	*094	*201	*309	*416	*523	*629	52	107
8	8.29 629	736	842	947	*053	*158	*263	51	106
9	8.30 263	368	473	577	681	785	888	**50**	104
10	888	992	*095	*198	*300	*403	*505	49	103
11	8.31 505	606	708	809	911	*012	*112	48	101
12	8.32 112	213	313	413	513	612	711	47	100
13	711	810	909	*008	*106	*205	*302	46	98
14	8.33 302	400	498	595	692	789	886	45	97
15	886	982	*078	*174	*270	*366	*461	44	96
16	8.34 461	556	651	746	840	935	*029	43	95
17	8.35 029	123	217	310	403	497	590	42	94
18	590	682	775	867	959	*051	*143	41	92
19	8.36 143	235	326	417	508	599	689	**40**	91
20	689	780	870	960	*050	*140	*229	39	90
21	8.37 229	318	408	497	585	674	762	38	89
22	762	850	938	*026	*114	*202	*289	37	88
23	8.38 289	376	463	550	636	723	809	36	87
24	809	895	981	*067	*153	*238	*323	35	86
25	8.39 323	408	493	578	663	747	832	34	85
26	832	916	*000	*083	*167	*250	*334	33	84
27	8.40 334	417	500	583	665	748	830	32	83
28	830	913	995	*077	*158	*240	*321	31	82
29	8.41 321	403	484	565	646	726	807	**30**	81
30	807	887	967	*048	*127	*207	*287	29	80
31	8.42 287	366	446	525	604	683	762	28	79
32	762	840	919	997	*075	*154	*232	27	78
33	8.43 232	309	387	464	542	619	696	26	77
34	696	773	850	927	*003	*080	*156	25	77
35	8.44 156	232	308	384	460	536	611	24	76
36	611	686	762	837	912	987	*061	23	75
37	8.45 061	136	210	285	359	433	507	22	74
38	507	581	655	728	802	875	948	21	74
39	948	*021	*094	*167	*240	*312	*385	**20**	73
40	8.46 385	457	529	602	674	745	817	19	72
41	817	889	960	*032	*103	*174	*245	18	71
42	8.47 245	316	387	458	528	599	669	17	71
43	669	740	810	880	950	*020	*089	16	70
44	8.48 089	159	228	298	367	436	505	15	69
45	505	574	643	711	780	849	917	14	69
46	917	985	*053	*121	*189	*257	*325	13	68
47	8.49 325	393	460	528	595	662	729	12	67
48	729	796	863	930	997	*063	*130	11	67
49	8.50 130	196	263	329	395	461	527	**10**	66
50	527	593	658	724	789	855	920	9	66
51	920	985	*050	*115	*180	*245	*310	8	65
52	8.51 310	374	439	503	568	632	696	7	64
53	696	760	824	888	952	*015	*079	6	64
54	8.52 079	143	206	269	332	396	459	5	63
55	459	522	584	647	710	772	835	4	63
56	835	897	960	*022	*084	*146	*208	3	62
57	8.53 208	270	332	393	455	516	578	2	62
58	578	639	700	762	823	884	945	1	61
59	945	*005	*066	*127	*187	*248	*308	**0**	60

| P. P. | | 60" | 50" | 40" | 30" | 20" | 10" | 0" | | ' | d. |

88°   Log Cot

P. P. (Proportional Parts)

	94	93	92	91	90
1	9.4	9.3	9.2	9.1	9.0
2	18.8	18.6	18.4	18.2	18.0
3	28.2	27.9	27.6	27.3	27.0
4	37.6	37.2	36.8	36.4	36.0
5	47.0	46.5	46.0	45.5	45.0
6	56.4	55.8	55.2	54.6	54.0
7	65.8	65.1	64.4	63.7	63.0
8	75.2	74.4	73.6	72.8	72.0
9	84.6	83.7	82.8	81.9	81.0

	89	88	87	86	85
1	8.9	8.8	8.7	8.6	8.5
2	17.8	17.6	17.4	17.2	17.0
3	26.7	26.4	26.1	25.8	25.5
4	35.6	35.2	34.8	34.4	34.0
5	44.5	44.0	43.5	43.0	42.5
6	53.4	52.8	52.2	51.6	51.0
7	62.3	61.6	60.9	60.2	59.5
8	71.2	70.4	69.6	68.8	68.0
9	80.1	79.2	78.3	77.4	76.5

	84	83	82	81	80
1	8.4	8.3	8.2	8.1	8.0
2	16.8	16.6	16.4	16.2	16.0
3	25.2	24.9	24.6	24.3	24.0
4	33.6	33.2	32.8	32.4	32.0
5	42.0	41.5	41.0	40.5	40.0
6	50.4	49.8	49.2	48.6	48.0
7	58.8	58.1	57.4	56.7	56.0
8	67.2	66.4	65.6	64.8	64.0
9	75.6	74.7	73.8	72.9	72.0

	79	78	77	76	75
1	7.9	7.8	7.7	7.6	7.5
2	15.8	15.6	15.4	15.2	15.0
3	23.7	23.4	23.1	22.8	22.5
4	31.6	31.2	30.8	30.4	30.0
5	39.5	39.0	38.5	38.0	37.5
6	47.4	46.8	46.2	45.6	45.0
7	55.3	54.6	53.9	53.2	52.5
8	63.2	62.4	61.6	60.8	60.0
9	71.1	70.2	69.3	68.4	67.5

	74	73	72	71	70
1	7.4	7.3	7.2	7.1	7.0
2	14.8	14.6	14.4	14.2	14.0
3	22.2	21.9	21.6	21.3	21.0
4	29.6	29.2	28.8	28.4	28.0
5	37.0	36.5	36.0	35.5	35.0
6	44.4	43.8	43.2	42.6	42.0
7	51.8	51.1	50.4	49.7	49.0
8	59.2	58.4	57.6	56.8	56.0
9	66.6	65.7	64.8	63.9	63.0

	69	68	67	66	65
1	6.9	6.8	6.7	6.6	6.5
2	13.8	13.6	13.4	13.2	13.0
3	20.7	20.4	20.1	19.8	19.5
4	27.6	27.2	26.8	26.4	26.0
5	34.5	34.0	33.5	33.0	32.5
6	41.4	40.8	40.2	39.6	39.0
7	48.3	47.6	46.9	46.2	45.5
8	55.2	54.4	53.6	52.8	52.0
9	62.1	61.2	60.3	59.4	58.5

	64	63	62	61	60
1	6.4	6.3	6.2	6.1	6.0
2	12.8	12.6	12.4	12.2	12.0
3	19.2	18.9	18.6	18.3	18.0
4	25.6	25.2	24.8	24.4	24.0
5	32.0	31.5	31.0	30.5	30.0
6	38.4	37.8	37.2	36.6	36.0
7	44.8	44.1	43.4	42.7	42.0
8	51.2	50.4	49.6	48.8	48.0
9	57.6	56.7	55.8	54.9	54.0

Log Sin				2°						P. P.
′	0″	10″	20″	30″	40″	50″	60″		d.	
0	8.54 282	342	402	462	522	582	642	59	60	
1	642	702	762	821	881	940	999	58	60	
2	999	*059	*118	*177	*236	*295	*354	57	59	
3	8.55 354	413	471	530	589	647	705	56	58	
4	705	764	822	880	938	996	*054	55	58	
5	8.56 054	112	170	227	285	342	400	54	58	
6	400	457	515	572	629	686	743	53	57	
7	743	800	857	914	970	*027	*084	52	57	
8	8.57 084	140	196	253	309	365	421	51	56	
9	421	477	533	589	645	701	757	**50**	56	
10	757	812	868	923	979	*034	*089	49	55	**61 60**
11	8.58 089	144	200	255	310	364	419	48	55	1 6.1 6.0
12	419	474	529	583	638	693	747	47	55	2 12.2 12.0
13	747	801	856	910	964	*018	*072	46	54	3 18.3 18.0
14	8.59 072	126	180	234	288	341	395	45	54	4 24.4 24.0
15	395	448	502	555	609	662	715	44	53	5 30.5 30.0
16	715	768	821	874	927	980	*033	43	53	6 36.6 36.0
17	8.60 033	086	139	191	244	296	349	42	53	7 42.7 42.0
18	349	401	454	506	558	610	662	41	52	8 48.8 48.0
19	662	714	766	818	870	922	973	**40**	52	9 54.9 54.0
20	973	*025	*077	*128	*180	*231	*282	39	52	**59 58**
21	8.61 282	334	385	436	487	538	589	38	51	1 5.9 5.8
22	589	640	691	742	792	843	894	37	51	2 11.8 11.6
23	894	944	995	*045	*096	*146	*196	36	50	3 17.7 17.4
24	8.62 196	246	297	347	397	447	497	35	50	4 23.6 23.2
25	497	546	596	646	696	745	795	34	50	5 29.5 29.0
26	795	844	894	943	993	*042	*091	33	49	6 35.4 34.8
27	8.63 091	140	189	238	288	336	385	32	49	7 41.3 40.6
28	385	434	483	532	580	629	678	31	49	8 47.2 46.4
29	678	726	775	823	871	920	968	**30**	48	9 53.1 52.2
30	968	*016	*064	*112	*160	*208	*256	29	48	**57 56**
31	8.64 256	304	352	400	448	495	543	28	48	1 5.7 5.6
32	543	590	638	685	733	780	827	27	47	2 11.4 11.2
33	827	875	922	969	*016	*063	*110	26	47	3 17.1 16.8
34	8.65 110	157	204	251	298	344	391	25	47	4 22.8 22.4
35	391	438	484	531	577	624	670	24	46	5 28.5 28.0
34	670	717	763	809	855	901	947	23	46	6 34.2 33.6
37	947	994	*040	*085	*131	*177	*223	22	46	7 39.9 39.2
38	8.66 223	269	314	360	406	451	497	21	46	8 45.6 44.8
39	497	542	588	633	678	724	769	**20**	45	9 51.3 50.4
40	769	814	859	904	949	994	*039	19	45	**55 54**
41	8.67 039	084	129	174	219	263	308	18	45	1 5.5 5.4
42	308	353	397	442	486	531	575	17	44	2 11.0 10.8
43	575	619	664	708	752	796	841	16	44	3 16.5 16.2
44	841	885	929	973	*017	*060	*104	15	44	4 22.0 21.6
45	8.68 104	148	192	236	279	323	367	14	44	5 27.5 27.0
46	367	410	454	497	540	584	627	13	43	6 33.0 32.4
47	627	670	714	757	800	843	886	12	43	7 38.5 37.8
48	886	929	972	*015	*058	*101	*144	11	43	8 44.0 43.2
49	8.69 144	187	229	272	315	357	400	**10**	43	9 49.5 48.6
50	400	442	485	527	570	612	654	9	42	**53 52**
51	654	697	739	781	823	865	907	8	42	1 5.3 5.2
52	907	949	991	*033	*075	*117	*159	7	42	2 10.6 10.4
53	8.70 159	201	242	284	326	367	409	6	42	3 15.9 15.6
54	409	451	492	534	575	616	658	5	42	4 21.2 20.8
55	658	699	740	781	823	864	905	4	41	5 26.5 26.0
56	905	946	987	*028	*069	*110	*151	3	41	6 31.8 31.2
57	8.71 151	192	232	273	314	355	395	2	41	7 37.1 36.4
58	395	436	476	517	557	598	638	1	40	8 42.4 41.6
59	638	679	719	759	800	840	880	**0**	40	9 47.7 46.8
	60″	50″	40″	30″	20″	10″	0″	′	d.	P. P.

87°   Log Cos

Log Tan		2°								
P. P.	'	0''	10''	20''	30''	40''	50''	60''		d.
	0	8.54 308	369	429	489	549	609	669	59	60
	1	669	729	789	848	908	967	*027	58	60
	2	8.55 027	086	145	205	264	323	382	57	59
	3	382	441	499	558	617	675	734	56	59
	4	734	792	850	909	967	*025	*083	55	58
	5	8.56 083	141	199	256	314	372	429	54	58
	6	429	487	544	601	659	716	773	53	57
	7	773	830	887	944	*000	*057	*114	52	57
	8	8.57 114	170	227	283	340	396	452	51	56
	9	452	508	564	620	676	732	788	**50**	56
	10	788	843	899	955	*010	*065	*121	49	56
	11	8.58 121	176	231	286	341	396	451	48	55
	12	451	506	561	616	670	725	779	47	55
	13	779	834	888	943	997	*051	*105	46	54
	14	8.59 105	159	213	267	321	375	428	45	54
	15	428	482	536	589	642	696	749	44	54
	16	749	802	856	909	962	*015	*068	43	53
	17	8.60 068	121	173	226	279	331	384	42	53
	18	384	436	489	541	593	646	698	41	52
	19	698	750	802	854	906	958	*009	**40**	52
	20	8.61 009	061	113	164	216	267	319	39	52
	21	319	370	422	473	524	575	626	38	51
	22	626	677	728	779	830	881	931	37	51
	23	931	982	*033	*083	*134	*184	*234	36	50
	24	8.62 234	285	335	385	435	485	535	35	50
	25	535	585	635	685	735	784	834	34	50
	26	834	884	933	983	*032	*081	*131	33	50
	27	8.63 131	180	229	278	328	377	426	32	49
	28	426	475	523	572	621	670	718	31	49
	29	718	767	816	864	913	961	*009	**30**	48
	30	8.64 009	058	106	154	202	250	298	29	48
	31	298	346	394	442	490	538	585	28	48
	32	585	633	681	728	776	823	870	27	48
	33	870	918	965	*012	*060	*107	*154	26	47
	34	8.65 154	201	248	295	342	388	435	25	47
	35	435	482	529	575	622	668	715	24	47
	36	715	761	808	854	900	947	993	23	46
	37	993	*039	*085	*131	*177	*223	*269	22	46
	38	8.66 269	315	361	406	452	498	543	21	46
	39	543	589	634	680	725	771	816	**20**	46
	40	816	861	906	952	997	*042	*087	19	45
	41	8.67 087	132	177	222	267	312	356	18	45
	42	356	401	446	490	535	579	624	17	45
	43	624	668	713	757	801	846	890	16	44
	44	890	934	978	*022	*066	*110	*154	15	44
	45	8.68 154	198	242	286	330	373	417	14	44
	46	417	461	504	548	592	635	678	13	44
	47	678	722	765	808	852	895	938	12	43
	48	938	981	*024	*067	*110	*153	*196	11	43
	49	8.69 196	239	282	325	368	410	453	**10**	43
	50	453	496	538	581	623	666	708	9	42
	51	708	750	793	835	877	920	962	8	42
	52	962	*004	*046	*088	*130	*172	*214	7	42
	53	8.70 214	256	298	339	381	423	465	6	42
	54	465	506	548	589	631	673	714	5	42
	55	714	755	797	838	879	921	962	4	41
	56	962	*003	*044	*085	*126	*167	*208	3	41
	57	8.71 208	249	290	331	372	413	453	2	41
	58	453	494	535	575	616	657	697	1	41
	59	697	738	778	819	859	899	940	**0**	40
P. P.		60''	50''	40''	30''	20''	10''	0''	'	d.
				87°			Log Cot			

P. P. columns:

	51	50
1	5.1	5.0
2	10.2	10.0
3	15.3	15.0
4	20.4	20.0
5	25.5	25.0
6	30.6	30.0
7	35.7	35.0
8	40.8	40.0
9	45.9	45.0

	49	48
1	4.9	4.8
2	9.8	9.6
3	14.7	14.4
4	19.6	19.2
5	24.5	24.0
6	29.4	28.8
7	34.3	33.6
8	39.2	38.4
9	44.1	43.2

	47	46	45
1	4.7	4.6	4.5
2	9.4	9.2	9.0
3	14.1	13.8	13.5
4	18.8	18.4	18.0
5	23.5	23.0	22.5
6	28.2	27.6	27.0
7	32.9	32.2	31.5
8	37.6	36.8	36.0
9	42.3	41.4	40.5

	44	43
1	4.4	4.3
2	8.8	8.6
3	13.2	12.9
4	17.6	17.2
5	22.0	21.5
6	26.4	25.8
7	30.8	30.1
8	35.2	34.4
9	39.6	38.7

	42	41	40
1	4.2	4.1	4.0
2	8.4	8.2	8.0
3	12.6	12.3	12.0
4	16.8	16.4	16.0
5	21.0	20.5	20.0
6	25.2	24.6	24.0
7	29.4	28.7	28.0
8	33.6	32.8	32.0
9	37.8	36.9	36.0

	Log Sin				3°					P. P.
'	0"	10"	20"	30"	40"	50"	60"		d.	P. P.
0	8.71 880	920	960	*000	*040	*080	*120	59	40	
1	8.72 120	160	200	240	280	320	359	58	40	
2	359	399	439	478	518	558	597	57	40	
3	597	637	676	716	755	794	834	56	40	
4	834	873	912	951	991	*030	*069	55	39	
5	8.73 069	108	147	186	225	264	303	54	39	
6	303	342	380	419	458	497	535	53	39	
7	535	574	613	651	690	728	767	52	39	
8	767	805	844	882	920	959	997	51	38	
9	997	*035	*073	*112	*150	*188	*226	**50**	38	
10	8.74 226	264	302	340	378	416	454	49	38	
11	454	491	529	567	605	642	680	48	38	
12	680	718	755	793	831	868	906	47	38	**41 40**
13	906	943	980	*018	*055	*092	*130	46	37	1 4.1 4.0
14	8.75 130	167	204	241	279	316	353	45	37	2 8.2 8.0
										3 12.3 12.0
15	353	390	427	464	501	538	575	44	37	4 16.4 16.0
16	575	612	648	685	722	759	795	43	37	5 20.5 20.0
17	795	832	869	905	942	979	*015	42	37	6 24.6 24.0
18	8.76 015	052	088	125	161	197	234	41	36	7 28.7 28.0
19	234	270	306	343	379	415	451	**40**	36	8 32.8 32.0
										9 36.9 36.0
20	451	487	523	559	595	631	667	39	36	
21	667	703	739	775	811	847	883	38	36	
22	883	919	954	990	*026	*061	*097	37	36	
23	8.77 097	133	168	204	239	275	310	36	36	
24	310	346	381	416	452	487	522	35	35	
25	522	558	593	628	663	698	733	34	35	
26	733	768	803	838	873	908	943	33	35	**39 38**
27	943	978	*013	*048	*083	*118	*152	32	35	1 3.9 3.8
28	8.78 152	187	222	257	291	326	360	31	35	2 7.8 7.6
29	360	395	430	464	499	533	568	**30**	35	3 11.7 11.4
										4 15.6 15.2
30	568	602	636	671	705	739	774	29	34	5 19.5 19.0
31	774	808	842	876	910	945	979	28	34	6 23.4 22.8
32	979	*013	*047	*081	*115	*149	*183	27	34	7 27.3 26.6
33	8.79 183	217	251	284	318	352	386	26	34	8 31.2 30.4
34	386	420	453	487	521	555	588	25	34	9 35.1 34.2
35	588	622	655	689	722	756	789	24	34	
36	789	823	856	890	923	956	990	23	34	
37	990	*023	*056	*090	*123	*156	*189	22	33	
38	8.80 189	222	255	289	322	355	388	21	33	
39	388	421	454	487	519	552	585	**20**	33	**37 36**
40	585	618	651	684	716	749	782	19	33	1 3.7 3.6
41	782	815	847	880	913	945	978	18	33	2 7.4 7.2
42	978	*010	*043	*075	*108	*140	*173	17	32	3 11.1 10.8
43	8.81 173	205	237	270	302	334	367	16	32	4 14.8 14.4
44	367	399	431	463	496	528	560	15	32	5 18.5 18.0
										6 22.2 21.6
45	560	592	624	656	688	720	752	14	32	7 25.9 25.2
46	752	784	816	848	880	912	944	13	32	8 29.6 28.8
47	944	975	*007	*039	*071	*103	*134	12	32	9 33.3 32.4
48	8.82 134	166	198	229	261	292	324	11	32	
49	324	356	387	419	450	482	513	**10**	32	
50	513	544	576	607	639	670	701	9	31	
51	701	732	764	795	826	857	888	8	31	
52	888	920	951	982	*013	*044	*075	7	31	
53	8.83 075	106	137	168	199	230	261	6	31	
54	261	292	322	353	384	415	446	5	31	
55	446	476	507	538	568	599	630	4	31	
56	630	660	691	721	752	783	813	3	30	
57	813	844	874	904	935	965	996	2	30	
58	996	*026	*056	*087	*117	*147	*177	1	30	
59	8.84 177	208	238	268	298	328	358	**0**	30	
	60"	50"	40"	30"	20"	10"	0"	**'**	d.	P. P.
				86°						Log Cos

Log Tan		3°								
P. P.	′	0″	10″	20″	30″	40″	50″	60″		d.
	0	8.71 940	980	*020	*060	*100	*141	*181	59	40
	1	8.72 181	221	261	301	341	380	420	58	40
	2	420	460	500	540	579	619	659	57	40
	3	659	698	738	777	817	856	896	56	40
	4	896	935	975	*014	*053	*093	*132	55	39
	5	8.73 132	171	210	249	288	327	366	54	39
	6	366	405	444	483	522	561	600	53	39
	7	600	638	677	716	754	793	832	52	39
	8	832	870	909	947	986	*024	*063	51	38
	9	8.74 063	101	139	178	216	254	292	**50**	38
	10	292	330	369	407	445	483	521	49	38
	11	521	559	597	634	672	710	748	48	38
	12	748	786	823	861	899	936	974	47	38
	13	974	*012	*049	*087	*124	*162	*199	46	38
	14	8.75 199	236	274	311	348	385	423	45	37
	15	423	460	497	534	571	608	645	44	37
	16	645	682	719	756	793	830	867	43	37
	17	867	904	940	977	*014	*051	*087	42	37
	18	8.76 087	124	160	197	233	270	306	41	36
	19	306	343	379	416	452	488	525	**40**	36
	20	525	561	597	633	669	706	742	39	36
	21	742	778	814	850	886	922	958	38	36
	22	958	994	*030	*065	*101	*137	*173	37	36
	23	8.77 173	208	244	280	315	351	387	36	36
	24	387	422	458	493	529	564	600	35	36
	25	600	635	670	706	741	776	811	34	35
	26	811	847	882	917	952	987	*022	33	35
	27	8.78 022	057	092	127	162	197	232	32	35
	28	232	267	302	337	371	406	441	31	35
	29	441	475	510	545	579	614	649	**30**	35
	30	649	683	718	752	787	821	855	29	34
	31	855	890	924	958	993	*027	*061	28	34
	32	8.79 061	096	130	164	198	232	266	27	34
	33	266	300	334	368	402	436	470	26	34
	34	470	504	538	572	606	639	673	25	34
	35	673	707	741	774	808	842	875	24	34
	36	875	909	942	976	*009	*043	*076	23	34
	37	8.80 076	110	143	177	210	243	277	22	34
	38	277	310	343	376	409	443	476	21	33
	39	476	509	542	575	608	641	674	**20**	33
	40	674	707	740	773	806	839	872	19	33
	41	872	905	937	970	*003	*036	*068	18	33
	42	8.81 068	101	134	166	199	232	264	17	33
	43	264	297	329	362	394	427	459	16	32
	44	459	491	524	556	588	621	653	15	32
	45	653	685	717	750	782	814	846	14	32
	46	846	878	910	942	974	*006	*038	13	32
	47	8.82 038	070	102	134	166	198	230	12	32
	48	230	262	293	325	357	389	420	11	32
	49	420	452	484	515	547	579	610	**10**	32
	50	610	642	673	705	736	768	799	9	32
	51	799	831	862	893	925	956	987	8	31
	52	987	*019	*050	*081	*112	*144	*175	7	31
	53	8.83 175	206	237	268	299	330	361	6	31
	54	361	392	423	454	485	516	547	5	31
	55	547	578	609	640	671	701	732	4	31
	56	732	763	794	824	855	886	916	3	31
	57	916	947	978	*008	*039	*069	*100	2	31
	58	8.84 100	130	161	191	222	252	282	1	30
	59	282	313	343	374	404	434	464	**0**	30
P. P.		60″	50″	40″	30″	20″	10″	0″	′	d.

P.P. side blocks:

	35	34
1	3.5	3.4
2	7.0	6.8
3	10.5	10.2
4	14.0	13.6
5	17.5	17.0
6	21.0	20.4
7	24.5	23.8
8	28.0	27.2
9	31.5	30.6

	33	32
1	3.3	3.2
2	6.6	6.4
3	9.9	9.6
4	13.2	12.8
5	16.5	16.0
6	19.8	19.2
7	23.1	22.4
8	26.4	25.6
9	29.7	28.8

	31	30
1	3.1	3.0
2	6.2	6.0
3	9.3	9.0
4	12.4	12.0
5	15.5	15.0
6	18.6	18.0
7	21.7	21.0
8	24.8	24.0
9	27.9	27.0

86°   Log Cot

TABLE V. ANGLES WHICH ARE MULTIPLES OF 15°

°	Rad	Sin	Cos	Tan	Cot	Sec	Csc
0	0	0	1	0	$\pm\infty$	1	$\pm\infty$
15	$\frac{1}{12}\pi$	$\frac{1}{4}(-\sqrt{2}+\sqrt{6})$	$\frac{1}{4}(\sqrt{2}+\sqrt{6})$	$2-\sqrt{3}$	$2+\sqrt{3}$	$-\sqrt{2}+\sqrt{6}$	$\sqrt{2}+\sqrt{6}$
30	$\frac{1}{6}\pi$	$\frac{1}{2}$	$\frac{1}{2}\sqrt{3}$	$\frac{1}{3}\sqrt{3}$	$\sqrt{3}$	$\frac{2}{3}\sqrt{3}$	2
45	$\frac{1}{4}\pi$	$\frac{1}{2}\sqrt{2}$	$\frac{1}{2}\sqrt{2}$	1	1	$\sqrt{2}$	$\sqrt{2}$
60	$\frac{1}{3}\pi$	$\frac{1}{2}\sqrt{3}$	$\frac{1}{2}$	$\sqrt{3}$	$\frac{1}{3}\sqrt{3}$	2	$\frac{2}{3}\sqrt{3}$
75	$\frac{5}{12}\pi$	$\frac{1}{4}(\sqrt{2}+\sqrt{6})$	$\frac{1}{4}(-\sqrt{2}+\sqrt{6})$	$2+\sqrt{3}$	$2-\sqrt{3}$	$\sqrt{2}+\sqrt{6}$	$-\sqrt{2}+\sqrt{6}$
90	$\frac{1}{2}\pi$	1	0	$\pm\infty$	0	$\pm\infty$	1
105	$\frac{7}{12}\pi$	$\frac{1}{4}(\sqrt{2}+\sqrt{6})$	$\frac{1}{4}(\sqrt{2}-\sqrt{6})$	$-2-\sqrt{3}$	$-2+\sqrt{3}$	$-\sqrt{2}-\sqrt{6}$	$-\sqrt{2}+\sqrt{6}$
120	$\frac{2}{3}\pi$	$\frac{1}{2}\sqrt{3}$	$-\frac{1}{2}$	$-\sqrt{3}$	$-\frac{1}{3}\sqrt{3}$	-2	$\frac{2}{3}\sqrt{3}$
135	$\frac{3}{4}\pi$	$\frac{1}{2}\sqrt{2}$	$-\frac{1}{2}\sqrt{2}$	-1	-1	$-\sqrt{2}$	$\sqrt{2}$
150	$\frac{5}{6}\pi$	$\frac{1}{2}$	$-\frac{1}{2}\sqrt{3}$	$-\frac{1}{3}\sqrt{3}$	$-\sqrt{3}$	$-\frac{2}{3}\sqrt{3}$	2
165	$\frac{11}{12}\pi$	$\frac{1}{4}(-\sqrt{2}+\sqrt{6})$	$\frac{1}{4}(-\sqrt{2}-\sqrt{6})$	$-2+\sqrt{3}$	$-2-\sqrt{3}$	$\sqrt{2}-\sqrt{6}$	$\sqrt{2}+\sqrt{6}$
180	π	0	-1	0	$\pm\infty$	-1	$\pm\infty$
195	$\frac{13}{12}\pi$	$\frac{1}{4}(\sqrt{2}-\sqrt{6})$	$\frac{1}{4}(-\sqrt{2}-\sqrt{6})$	$2-\sqrt{3}$	$2+\sqrt{3}$	$\sqrt{2}-\sqrt{6}$	$-\sqrt{2}-\sqrt{6}$
210	$\frac{7}{6}\pi$	$-\frac{1}{2}$	$-\frac{1}{2}\sqrt{3}$	$\frac{1}{3}\sqrt{3}$	$\sqrt{3}$	$-\frac{2}{3}\sqrt{3}$	-2
225	$\frac{5}{4}\pi$	$-\frac{1}{2}\sqrt{2}$	$-\frac{1}{2}\sqrt{2}$	1	1	$-\sqrt{2}$	$-\sqrt{2}$
240	$\frac{4}{3}\pi$	$-\frac{1}{2}\sqrt{3}$	$-\frac{1}{2}$	$\sqrt{3}$	$\frac{1}{3}\sqrt{3}$	-2	$-\frac{2}{3}\sqrt{3}$
255	$\frac{17}{12}\pi$	$\frac{1}{4}(-\sqrt{2}-\sqrt{6})$	$\frac{1}{4}(\sqrt{2}-\sqrt{6})$	$2+\sqrt{3}$	$2-\sqrt{3}$	$-\sqrt{2}-\sqrt{6}$	$\sqrt{2}-\sqrt{6}$
270	$\frac{3}{2}\pi$	-1	0	$\pm\infty$	0	$\pm\infty$	-1
285	$\frac{19}{12}\pi$	$\frac{1}{4}(-\sqrt{2}-\sqrt{6})$	$\frac{1}{4}(-\sqrt{2}+\sqrt{6})$	$-2-\sqrt{3}$	$-2+\sqrt{3}$	$\sqrt{2}+\sqrt{6}$	$\sqrt{2}-\sqrt{6}$
300	$\frac{5}{3}\pi$	$-\frac{1}{2}\sqrt{3}$	$\frac{1}{2}$	$-\sqrt{3}$	$-\frac{1}{3}\sqrt{3}$	2	$-\frac{2}{3}\sqrt{3}$
315	$\frac{7}{4}\pi$	$-\frac{1}{2}\sqrt{2}$	$\frac{1}{2}\sqrt{2}$	-1	-1	$\sqrt{2}$	$-\sqrt{2}$
330	$\frac{11}{6}\pi$	$-\frac{1}{2}$	$\frac{1}{2}\sqrt{3}$	$-\frac{1}{3}\sqrt{3}$	$-\sqrt{3}$	$\frac{2}{3}\sqrt{3}$	-2
345	$\frac{23}{12}\pi$	$\frac{1}{4}(\sqrt{2}-\sqrt{6})$	$\frac{1}{4}(\sqrt{2}+\sqrt{6})$	$-2+\sqrt{3}$	$-2-\sqrt{3}$	$-\sqrt{2}+\sqrt{6}$	$-\sqrt{2}-\sqrt{6}$
360	2π	0	1	0	$\pm\infty$	1	$\pm\infty$

TABLE VI. NATURAL TRIGONOMETRIC FUNCTIONS

0°

Proportional Parts:

30

1	0.5
2	1.0
3	1.5
4	2.0
5	2.5
6	3.0
7	3.5
8	4.0
9	4.5
10	5.0
20	10.0
30	15.0
40	20.0
50	25.0

29

1	0.5
2	1.0
3	1.4
4	1.9
5	2.4
6	2.9
7	3.4
8	3.9
9	4.4
10	4.8
20	9.7
30	14.5
40	19.3
50	24.2

'	Sin.	d.	Cos.	d.	Tan.	d.	Cot.	d.	'
0	.00000	29	1.0000	0	.00000	29	∞		60
1	029	29	000	0	029	29	3437.7	1718.8	59
2	058	29	000	0	058	29	1718.9	573.0	58
3	087	29	000	0	087	29	1145.9	286.46	57
4	116	29	000	0	116	29	859.44	171.89	56
5	.00145	30	1.0000	0	.00145	30	687.55	114.59	55
6	175	29	000	0	175	29	572.96	81.85	54
7	204	29	000	0	204	29	491.11	61.39	53
8	233	29	000	0	233	29	429.72	47.75	52
9	262	29	000	0	262	29	381.97	38.20	51
10	.00291	29	1.0000	1	.00291	29	343.77	31.25	50
11	320	29	.99999	0	320	29	312.52	26.04	49
12	349	29	999	0	349	29	286.48	22.04	48
13	378	29	999	0	378	29	264.44	18.89	47
14	407	29	999	0	407	29	245.55	16.37	46
15	.00436	29	.99999	0	.00436	29	229.18	14.32	45
16	465	30	999	0	466	30	214.86	12.64	44
17	495	29	999	0	495	29	202.22	11.24	43
18	524	29	999	1	524	29	190.98	10.05	42
19	553	29	998	0	553	29	180.93	9.04	41
20	.00582	29	.99998	0	.00582	29	171.89	8.19	40
21	611	29	998	0	611	29	163.70	7.44	39
22	640	29	998	0	640	29	156.26	6.79	38
23	669	29	998	0	669	29	149.47	6.23	37
24	698	29	998	1	698	29	143.24	5.73	36
25	.00727	29	.99997	0	.00727	29	137.51	5.29	35
26	756	29	997	0	756	29	132.22	4.90	34
27	785	29	997	0	785	30	127.32	4.55	33
28	814	30	997	1	815	29	122.77	4.23	32
29	844	29	996	0	844	29	118.54	3.95	31
30	.00873	29	.99996	0	.00873	29	114.59	3.70	30
31	902	29	996	0	902	29	110.89	3.46	29
32	931	29	996	1	931	29	107.43	3.26	28
33	960	29	995	0	960	29	104.17	3.06	27
34	989	29	995	0	989	29	101.11	2.892	26
35	.01018	29	.99995	0	.01018	29	98.218	2.729	25
36	047	29	995	1	047	29	95.489	2.581	24
37	076	29	994	0	076	29	92.908	2.445	23
38	105	29	994	0	105	30	90.463	2.319	22
39	134	30	994	1	135	29	88.144	2.204	21
40	.01164	29	.99993	0	.01164	29	85.940	2.096	20
41	193	29	993	0	193	29	83.844	1.997	19
42	222	29	993	1	222	29	81.847	1.904	18
43	251	29	992	0	251	29	79.943	1.817	17
44	280	29	992	1	280	29	78.126	1.736	16
45	.01309	29	.99991	0	.01309	29	76.390	1.661	15
46	338	29	991	0	338	29	74.729	1.590	14
47	367	29	991	1	367	29	73.139	1.524	13
48	396	29	990	0	396	29	71.615	1.462	12
49	425	29	990	1	425	30	70.153	1.403	11
50	.01454	29	.99989	0	.01455	29	68.750	1.348	10
51	483	30	989	0	484	29	67.402	1.297	9
52	513	29	989	1	513	29	66.105	1.247	8
53	542	29	988	0	542	29	64.858	1.201	7
54	571	29	988	1	571	29	63.657	1.158	6
55	.01600	29	.99987	0	.01600	29	62.499	1.116	5
56	629	29	987	1	629	29	61.383	1.077	4
57	658	29	986	0	658	29	60.306	1.040	3
58	687	29	986	1	687	29	59.266	1.005	2
59	716	29	985	0	716	30	58.261	.971	1
60	.01745		.99985		.01746		57.290		0

P. P.		Cos.	d.	Sin.	d.	Cot.	d.	Tan.	d	'

89°

115

1°

′	Sin.	d.	Cos.	d.	Tan.	d.	Cot.	d.		P. P.
0	.01745	29	.99985	1	.01746	29	57.290	939	60	
1	774	29	984	0	775	29	56.351	909	59	
2	803	29	984	1	804	29	55.442	881	58	
3	832	30	983	0	833	29	54.561	852	57	
4	862	29	983	1	862	29	53.709	827	56	
5	.01891	29	.99982	0	.01891	29	52.882	801	55	
6	920	29	982	1	920	29	.081	778	54	
7	949	29	981	1	949	29	51.303	754	53	
8	978	29	980	0	978	29	50.549	733	52	
9	.02007	29	980	1	.02007	29	49.816	712	51	
10	.02036	29	.99979	0	.02036	30	49.104	692	50	
11	065	29	979	1	066	29	48.412	672	49	
12	094	29	978	1	095	29	47.740	655	48	
13	123	29	977	0	124	29	.085	636	47	
14	152	29	977	1	153	29	46.449	620	46	
15	.02181	30	.99976	0	.02182	29	45.829	603	45	
16	211	29	976	1	211	29	.226	587	44	
17	240	29	975	1	240	29	44.639	573	43	
18	269	29	974	0	269	29	.066	558	42	
19	298	29	974	1	298	30	43.508	544	41	
20	.02327	29	.99973	1	.02328	29	42.964	531	40	
21	356	29	972	0	357	29	.433	517	39	
22	385	29	972	1	386	29	41.916	505	38	30 · 29
23	414	29	971	1	415	29	.411	494	37	1 · 0.5 · 0.5
24	443	29	970	1	444	29	40.917	481	36	2 · 1.0 · 1.0
25	.02472	29	.99969	0	.02473	29	40.436	471	35	3 · 1.5 · 1.4
26	501	29	969	1	502	29	39.965	459	34	4 · 2.0 · 1.9
27	530	30	968	1	531	29	.506	449	33	5 · 2.5 · 2.4
28	560	29	967	1	560	29	.057	439	32	6 · 3.0 · 2.9
29	589	29	966	0	589	30	38.618	430	31	7 · 3.5 · 3.4
30	.02618	29	.99966	1	.02619	29	38.188	419	30	8 · 4.0 · 3.9
31	647	29	965	1	648	29	37.769	411	29	9 · 4.5 · 4.4
32	676	29	964	1	677	29	.358	402	28	10 · 5.0 · 4.8
33	705	29	963	0	706	29	36.956	393	27	20 · 10.0 · 9.7
34	734	29	963	1	735	29	.563	385	26	30 · 15.0 · 14.5
35	.02763	29	.99962	1	.02764	29	36.178	377	25	40 · 20.0 · 19.3
36	792	29	961	1	793	29	35.801	370	24	50 · 25.0 · 24.2
37	821	29	960	1	822	29	.431	361	23	
38	850	29	959	0	851	30	.070	355	22	
39	879	29	959	1	881	29	34.715	347	21	
40	.02908	30	.99958	1	.02910	29	34.368	341	20	
41	938	29	957	1	939	29	.027	333	19	
42	967	29	956	1	968	29	33.694	328	18	
43	996	29	955	1	997	29	.366	321	17	
44	.03025	29	954	1	.03026	29	.045	315	16	
45	.03054	29	.99953	1	.03055	29	32.730	309	15	
46	083	29	952	0	084	30	.421	303	14	
47	112	29	952	1	114	29	.118	297	13	
48	141	29	951	1	143	29	31.821	293	12	
49	170	29	950	1	172	29	.528	286	11	
50	.03199	29	.99949	1	.03201	29	31.242	282	10	
51	228	29	948	1	230	29	30.960	277	9	
52	257	29	947	1	259	29	.683	271	8	
53	286	30	946	1	288	29	.412	267	7	
54	316	29	945	1	317	29	.145	263	6	
55	.03345	29	.99944	1	.03346	30	29.882	258	5	
56	374	29	943	1	376	29	.624	253	4	
57	403	29	942	1	405	29	.371	249	3	
58	432	29	941	1	434	29	.122	245	2	
59	461	29	940	1	463	29	28.877	241	1	
60	.03490		.99939		.03492		28.636		0	
	Cos.	d.	Sin.	d.	Cot.	d.	Tan.	d.	′	P. P.

88°

2°

P. P.	′	Sin.	d.	Cos.	d.	Tan.	d.	Cot.	d.	
	0	.03490	29	.99939	1	.03492	29	28.636	237	60
	1	519	29	938	1	521	29	.399	233	59
	2	548	29	937	1	550	29	.166	229	58
	3	577	29	936	1	579	30	27.937	225	57
	4	606	29	935	1	609	29	.712	222	56
	5	.03635	29	.99934	1	.03638	29	27.490	219	55
	6	664	29	933	1	667	29	.271	214	54
	7	693	30	932	1	696	29	.057	212	53
	8	723	29	931	1	725	29	26.845	208	52
	9	752	29	930	1	754	29	.637	205	51
	10	.03781	29	.99929	2	.03783	29	26.432	202	50
	11	810	29	927	1	812	30	.230	199	49
	12	839	29	926	1	842	29	.031	196	48
	13	868	29	925	1	871	29	25.835	193	47
	14	897	29	924	1	900	29	.642	190	46
	15	.03926	29	.99923	1	.03929	29	25.452	188	45
	16	955	29	922	1	958	29	.264	184	44
	17	984	29	921	2	987	29	.080	182	43
	18	.04013	29	919	1	.04016	30	24.898	179	42
	19	042	29	918	1	046	29	.719	177	41
	20	.04071	29	.99917	1	.04075	29	24.542	174	40
	21	100	29	916	1	104	29	.368	172	39
	22	129	30	915	2	133	29	.196	170	38
	23	159	29	913	1	162	29	.026	167	37
	24	188	29	912	1	191	29	23.859	164	36
	25	.04217	29	.99911	1	.04220	30	23.695	163	35
	26	246	29	910	1	250	29	.532	160	34
	27	275	29	909	2	279	29	.372	158	33
	28	304	29	907	1	308	29	.214	156	32
	29	333	29	906	1	337	29	.058	154	31
	30	.04362	29	.99905	1	.04366	29	22.904	152	30
	31	391	29	904	2	395	29	.752	150	29
	32	420	29	902	1	424	30	.602	148	28
	33	449	29	901	1	454	29	.454	146	27
	34	478	29	900	2	483	29	.308	144	26
	35	.04507	29	.99898	1	.04512	29	22.164	142	25
	36	536	29	897	1	541	29	.022	141	24
	37	565	29	896	2	570	29	21.881	138	23
	38	594	29	894	1	599	30	.743	137	22
	39	623	30	893	1	628	29	.606	136	21
	40	.04653	29	.99892	2	.04658	29	21.470	133	20
	41	682	29	890	1	687	29	.337	132	19
	42	711	29	889	1	716	29	.205	130	18
	43	740	29	888	2	745	29	.075	129	17
	44	769	29	886	1	774	29	20.946	127	16
	45	.04798	29	.99885	2	.04803	30	20.819	126	15
	46	827	29	883	1	833	29	.693	124	14
	47	856	29	882	1	862	29	.569	123	13
	48	885	29	881	2	891	29	.446	121	12
	49	914	29	879	1	920	29	.325	119	11
	50	.04943	29	.99878	2	.04949	29	20.206	119	10
	51	972	29	876	1	978	29	.087	117	9
	52	.05001	29	875	2	.05007	30	19.970	115	8
	53	030	29	873	1	037	29	.855	115	7
	54	059	29	872	2	066	29	.740	113	6
	55	.05088	29	.99870	1	.05095	29	19.627	111	5
	56	117	29	869	2	124	29	.516	111	4
	57	146	29	867	1	153	29	.405	109	3
	58	175	30	866	2	182	30	.296	108	2
	59	205	29	864	1	212	29	.188	107	1
	60	.05234		.99863		.05241		19.081		0
P. P.		Cos.	d.	Sin.	d.	Cot.	d.	Tan.	d.	′

87°

P. P.

2	
1	0.0
2	0.1
3	0.1
4	0.1
5	0.2
6	0.2
7	0.2
8	0.3
9	0.3
10	0.3
20	0.7
30	1.0
40	1.3
50	1.7

3°

′	Sin.	d.	Cos.	d.	Tan.	d.	Cot.	d.		P. P.
0	.05234	29	.99863	2	.05241	29	19.081	105	60	
1	263	29	861	1	270	29	18.976	105	59	
2	292	29	860	2	299	29	.871	103	58	
3	321	29	858	1	328	29	.768	102	57	
4	350	29	857	2	357	30	.666	102	56	
5	.05379	29	.99855	1	.05387	29	18.564	100	55	
6	408	29	854	2	416	29	.464	98	54	
7	437	29	852	1	445	29	.366	98	53	
8	466	29	851	2	474	29	.268	97	52	
9	495	29	849	2	503	30	.171	96	51	
10	.05524	29	.99847	1	.05533	29	18.075	95	50	
11	553	29	846	2	562	29	17.980	94	49	
12	582	29	844	2	591	29	.886	93	48	
13	611	29	842	1	620	29	.793	91	47	
14	640	29	841	2	649	29	.702	91	46	
15	.05669	29	.99839	1	.05678	30	17.611	90	45	
16	698	29	838	2	708	29	.521	90	44	
17	727	29	836	2	737	29	.431	88	43	
18	756	29	834	1	766	29	.343	87	42	
19	785	29	833	2	795	29	.256	87	41	
20	.05814	30	.99831	2	.05824	30	17.169	85	40	
21	844	29	829	2	854	29	.084	85	39	
22	873	29	827	1	883	29	16.999	84	38	
23	902	29	826	2	912	29	.915	83	37	
24	931	29	824	2	941	29	.832	82	36	30 29
25	.05960	29	.99822	1	.05970	29	16.750	82	35	1 0.5 0.5
26	989	29	821	2	999	30	.668	81	34	2 1.0 1.0
27	.06018	29	819	2	.06029	29	.587	80	33	3 1.5 1.4
28	047	29	817	2	058	29	.507	79	32	4 2.0 1.9
29	076	29	815	2	087	29	.428	78	31	5 2.5 2.4
										6 3.0 2.9
30	.06105	29	.99813	1	.06116	29	16.350	78	30	7 3.5 3.4
31	134	29	812	2	145	30	.272	77	29	8 4.0 3.9
32	163	29	810	2	175	29	.195	76	28	9 4.5 4.4
33	192	29	808	2	204	29	.119	76	27	10 5.0 4.8
34	221	29	806	2	233	29	.043	74	26	20 10.0 9.7
35	.06250	29	.99804	1	.06262	29	15.969	74	25	30 15.0 14.5
36	279	29	803	2	291	30	.895	74	24	40 20.0 19.3
37	308	29	801	2	321	29	.821	73	23	50 25.0 24.2
38	337	29	799	2	350	29	.748	72	22	
39	366	29	797	2	379	29	.676	71	21	
40	.06395	29	.99795	2	.06408	30	15.605	71	20	
41	424	29	793	1	438	29	.534	70	19	
42	453	29	792	2	467	29	.464	70	18	
43	482	29	790	2	496	29	.394	69	17	
44	511	29	788	2	525	29	.325	68	16	
45	.06540	29	.99786	2	.06554	30	15.257	68	15	
46	569	29	784	2	584	29	.189	67	14	
47	598	29	782	2	613	29	.122	66	13	
48	627	29	780	2	642	29	.056	66	12	
49	656	29	778	2	671	29	14.990	66	11	
50	.06685	29	.99776	2	.06700	30	14.924	64	10	
51	714	29	774	2	730	29	.860	65	9	
52	743	30	772	2	759	29	.795	63	8	
53	773	29	770	2	788	29	.732	63	7	
54	802	29	768	2	817	30	.669	63	6	
55	.06831	29	.99766	2	.06847	29	14.606	62	5	
56	860	29	764	2	876	29	.544	62	4	
57	889	29	762	2	905	29	.482	61	3	
58	918	29	760	2	934	29	.421	60	2	
59	947	29	758	2	963	30	.361	60	1	
60	.06976		.99756		.06993		14.301		0	

	Cos.	d.	Sin.	d.	Cot.	d.	Tan.	d.	′	P. P.

86°

					4°					
P. P.	′	Sin.	d.	Cos.	d.	Tan.	d.	Cot.	d.	
	0	.06976		.99756		.06993		14.301		**60**
	1	.07005	29	754	2	.07022	29	.241	60	59
	2	034	29	752	2	051	29	.182	59	58
	3	063	29	750	2	080	29	.124	58	57
	4	092	29	748	2	110	30	.065	59	56
			29		2		29		57	
	5	.07121		.99746		.07139		14.008		**55**
	6	150	29	744	2	168	29	13.951	57	54
	7	179	29	742	2	197	29	.894	57	53
	8	208	29	740	2	227	30	.838	56	52
	9	237	29	738	2	256	29	.782	56	51
			29		2		29		55	
	10	.07266		.99736		.07285		13.727		**50**
	11	295	29	734	2	314	29	.672	55	49
	12	324	29	731	3	344	30	.617	55	48
	13	353	29	729	2	373	29	.563	54	47
	14	382	29	727	2	402	29	.510	53	46
			29		2		29		53	
	15	.07411		.99725		.07431		13.457		**45**
	16	440	29	723	2	461	30	.404	53	44
	17	469	29	721	2	490	29	.352	52	43
	18	498	29	719	2	519	29	.300	52	42
	19	527	29	716	3	548	29	.248	52	41
			29		2		30		51	
	20	.07556		.99714		.07578		13.197		**40**
	21	585	29	712	2	607	29	.146	51	39
	22	614	29	710	2	636	29	.096	50	38
	23	643	29	708	2	665	29	.046	50	37
	24	672	29	705	3	695	30	12.996	50	36
			29		2		29		49	
3 2	**25**	.07701		.99703		.07724		12.947		**35**
1 0.0 0.0	26	730	29	701	2	753	29	.898	49	34
2 0.1 0.1	27	759	29	699	2	782	29	.850	48	33
3 0.2 0.1	28	788	29	696	3	812	30	.801	49	32
4 0.2 0.1	29	817	29	694	2	841	29	.754	47	31
5 0.2 0.2			29		2		29		48	
6 0.3 0.2	**30**	.07846		.99692		.07870		12.706		**30**
7 0.4 0.2	31	875	29	689	3	899	29	.659	47	29
8 0.4 0.3	32	904	29	687	2	929	30	.612	47	28
9 0.4 0.3	33	933	29	685	2	958	29	.566	46	27
10 0.5 0.3	34	962	29	683	2	987	29	.520	46	26
20 1.0 0.7			29		3		30		46	
30 1.5 1.0	**35**	.07991		.99680		.08017		12.474		**25**
40 2.0 1.3	36	.08020	29	678	2	046	29	.429	45	24
50 2.5 1.7	37	049	29	676	2	075	29	.384	45	23
	38	078	29	673	3	104	29	.339	45	22
	39	107	29	671	2	134	30	.295	44	21
			29		3		29		44	
	40	.08136		.99668		.08163		12.251		**20**
	41	165	29	666	2	192	29	.207	44	19
	42	194	29	664	2	221	29	.163	44	18
	43	223	29	661	3	251	30	.120	43	17
	44	252	29	659	2	280	29	.077	43	16
			29		2		29		42	
	45	.08281		.99657		.08309		12.035		**15**
	46	310	29	654	3	339	30	11.992	43	14
	47	339	29	652	2	368	29	.950	42	13
	48	368	29	649	3	397	29	.909	41	12
	49	397	29	647	2	427	30	.867	42	11
			29		3		29		41	
	50	.08426		.99644		.08456		11.826		**10**
	51	455	29	642	2	485	29	.785	41	9
	52	484	29	639	3	514	29	.745	40	8
	53	513	29	637	2	544	30	.705	40	7
	54	542	29	635	2	573	29	.664	41	6
			29		3		29		39	
	55	.08571		.99632		.08602		11.625		**5**
	56	600	29	630	2	632	30	.585	40	4
	57	629	29	627	3	661	29	.546	39	3
	58	658	29	625	2	690	29	.507	39	2
	59	687	29	622	3	720	30	.468	39	1
			29		3		29		38	
	60	.08716		.99619		.08749		11.430		**0**
P. P.		Cos.	d.	Sin.	d.	Cot.	d.	Tan.	d.	′
						85°				

5°

′	Sin.	d.	Cos.	d.	Tan.	d.	Cot.	d.		P. P.
0	.08716		.99619		.08749		11.430		60	
1	745	29	617	2	778	29	.392	38	59	
2	774	29	614	3	807	29	.354	38	58	
3	803	29	612	2	837	30	.316	38	57	
4	831	28	609	3	866	28	.279	37	56	
		29		2		29		37		
5	.08860		.99607		.08895		11.242		55	
6	889	29	604	3	925	30	.205	37	54	
7	918	29	602	2	954	29	.168	37	53	
8	947	29	599	3	983	29	.132	36	52	
9	976	29	596	3	.09013	30	.095	37	51	
		29		2		29		36		**30 29**
10	.09005		.99594		.09042		11.059		50	1 0.5 0.5
11	034	29	591	3	071	29	.024	35	49	2 1.0 1.0
12	063	29	588	3	101	30	10.988	36	48	3 1.5 1.4
13	092	29	586	2	130	29	.953	35	47	4 2.0 1.9
14	121	29	583	3	159	29	.918	35	46	5 2.5 2.4
		29				30		35		6 3.0 2.9
										7 3.5 3.4
15	.09150		.99580		.09189		10.883		45	8 4.0 3.9
16	179	29	578	2	218	29	.848	35	44	9 4.5 4.4
17	208	29	575	3	247	29	.814	34	43	10 5.0 4.8
18	237	29	572	3	277	30	.780	34	42	20 10.0 9.7
19	266	29	570	2	306	29	.746	34	41	30 15.0 14.5
		29		3		29		34		40 20.0 19.3
20	.09295		.99567		.09335		10.712		40	50 25.0 24.2
21	324	29	564	3	365	30	.678	34	39	
22	353	29	562	2	394	29	.645	33	38	
23	382	29	559	3	423	29	.612	33	37	
24	411	29	556	3	453	30	.579	33	36	
		29		3		29		33		
25	.09440		.99553		.09482		10.546		35	
26	469	29	551	2	511	29	.514	32	34	
27	498	29	548	3	541	30	.481	33	33	
28	527	29	545	3	570	29	.449	32	32	
29	556	29	542	3	600	30	.417	32	31	
		29		2		29		32		
30	.09585		.99540		.09629		10.385		30	
31	614	29	537	3	658	29	.354	31	29	
32	642	28	534	3	688	30	.322	32	28	
33	671	29	531	3	717	29	.291	31	27	
34	700	29	528	3	746	29	.260	31	26	
		29		2		30		31		
35	.09729		.99526		.09776		10.229		25	
36	758	29	523	3	805	29	.199	30	24	
37	787	29	520	3	834	29	.168	31	23	
38	816	29	517	3	864	30	.138	30	22	
39	845	29	514	3	893	29	.108	30	21	
		29		3		30		30		**28 4**
40	.09874		.99511		.09923		10.078		20	1 0.5 0.1
41	903	29	508	3	952	29	.048	30	19	2 0.9 0.1
42	932	29	506	2	981	30	.019	29	18	3 1.4 0.2
43	961	29	503	3	.10011	29	9.9893	297	17	4 1.9 0.3
44	990	29	500	3	040	30	.9601	292	16	5 2.3 0.3
		29		3		29		291		6 2.8 0.4
										7 3.3 0.5
45	.10019		.99497		.10069		9.9310		15	8 3.7 0.5
46	048	29	494	3	099	30	.9021	289	14	9 4.2 0.6
47	077	29	491	3	128	29	.8734	287	13	10 4.7 0.7
48	106	29	488	3	158	30	.8448	286	12	20 9.3 1.3
49	135	29	485	3	187	29	.8164	284	11	30 14.0 2.0
		29		3		29		282		40 18.7 2.7
50	.10164		.99482		.10216		9.7882		10	50 23.3 3.3
51	192	28	479	3	246	30	.7601	281	9	
52	221	29	476	3	275	29	.7322	279	8	
53	250	29	473	3	305	30	.7044	278	7	
54	279	29	470	3	334	29	.6768	276	6	
		29		3		29		275		
55	.10308		.99467		.10363		9.6493		5	
56	337	29	464	3	393	30	.6220	273	4	
57	366	29	461	3	422	29	.5949	271	3	
58	395	29	458	3	452	30	.5679	270	2	
59	424	29	455	3	481	29	.5411	268	1	
		29		3		29		267		
60	.10453		.99452		.10510		9.5144		0	
	Cos.	d.	Sin.	d.	Cot.	d.	Tan.	d.	′	P. P.

84°

6°										
P. P.	′	Sin.	d.	Cos.	d.	Tan.	d.	Cot.	d.	
	0	.10453	29	.99452	3	.10510	30	9.5144	266	**60**
	1	482	29	449	3	540	29	.4878	264	59
	2	511	29	446	3	569	30	.4614	262	58
	3	540	29	443	3	599	29	.4352	262	57
	4	569	28	440	3	628	29	.4090	259	56
	5	.10597	29	.99437	3	.10657	30	9.3831	259	**55**
	6	626	29	434	3	687	29	.3572	257	54
	7	655	29	431	3	716	30	.3315	255	53
	8	684	29	428	4	746	29	.3060	254	52
	9	713	29	424	3	775	30	.2806	253	51
	10	.10742	29	.99421	3	.10805	29	9.2553	251	**50**
	11	771	29	418	3	834	29	.2302	250	49
	12	800	29	415	3	863	30	.2052	249	48
	13	829	29	412	3	893	29	.1803	248	47
	14	858	29	409	3	922	30	.1555	246	46
	15	.10887	29	.99406	4	.10952	29	9.1309	244	**45**
	16	916	29	402	3	981	30	.1065	244	44
	17	945	28	399	3	.11011	29	.0821	242	43
	18	973	29	396	3	040	30	.0579	241	42
	19	.11002	29	393	3	070	29	.0338	240	41
	20	.11031	29	.99390	4	.11099	29	9.0098	238	**40**
	21	060	29	386	3	128	30	8.9860	237	39
	22	089	29	383	3	158	29	.9623	236	38
	23	118	29	380	3	187	30	.9387	235	37
	24	147	29	377	3	217	29	.9152	233	36
	25	.11176	29	.99374	4	.11246	30	8.8919	233	**35**
	26	205	29	370	3	276	29	.8686	231	34
	27	234	29	367	3	305	30	.8455	230	33
	28	263	28	364	4	335	29	.8225	229	32
	29	291	29	360	3	364	30	.7996	227	31
	30	.11320	29	.99357	3	.11394	29	8.7769	227	**30**
	31	349	29	354	3	423	29	.7542	225	29
	32	378	29	351	4	452	30	.7317	224	28
	33	407	29	347	3	482	29	.7093	223	27
	34	436	29	344	3	511	30	.6870	222	26
	35	.11465	29	.99341	4	.11541	29	8.6648	221	**25**
	36	494	29	337	3	570	30	.6427	219	24
	37	523	29	334	3	600	29	.6208	219	23
	38	552	28	331	4	629	30	.5989	217	22
	39	580	29	327	3	659	29	.5772	217	21
	40	.11609	29	.99324	4	.11688	30	8.5555	215	**20**
	41	638	29	320	3	718	29	.5340	214	19
	42	667	29	317	3	747	30	.5126	213	18
	43	696	29	314	4	777	29	.4913	212	17
	44	725	29	310	3	806	30	.4701	211	16
	45	.11754	29	.99307	4	.11836	29	8.4490	210	**15**
	46	783	29	303	3	865	30	.4280	209	14
	47	812	28	300	3	895	29	.4071	208	13
	48	840	29	297	4	924	30	.3863	207	12
	49	869	29	293	3	954	29	.3656	206	11
	50	.11898	29	.99290	4	.11983	30	8.3450	205	**10**
	51	927	29	286	3	.12013	30	.3245	204	9
	52	956	29	283	4	042	30	.3041	203	8
	53	985	29	279	3	072	30	.2838	202	7
	54	.12014	29	276	4	101	29	.2636	202	6
	55	.12043	28	.99272	3	.12131	29	8.2434	200	**5**
	56	071	29	269	4	160	30	.2234	199	4
	57	100	29	265	3	190	29	.2035	198	3
	58	129	29	262	4	219	30	.1837	197	2
	59	158	29	258	3	249	29	.1640	197	1
	60	.12187		.99255		.12278		8.1443		**0**
P. P.		Cos.	d.	Sin.	d.	Cot.	d.	Tan.	d.	′
83°										

P. P.

	3	2
1	0.0	0.0
2	0.1	0.1
3	0.2	0.1
4	0.2	0.1
5	0.2	0.2
6	0.3	0.2
7	0.4	0.2
8	0.4	0.3
9	0.4	0.3
10	0.5	0.3
20	1.0	0.7
30	1.5	1.0
40	2.0	1.3
50	2.5	1.7

										P. P.	
						7°					

′	Sin.	d.	Cos.	d.	Tan.	d.	Cot.	d.		P. P.	
0	.12187		.99255		.12278		8.1443		60		
1	216	29	251	4	308	30	.1248	195	59		
2	245	29	248	3	338	30	.1054	194	58		
3	274	29	244	4	367	29	.0860	194	57		
4	302	28	240	4	397	30	.0667	193	56		
		29		3		29		191			
5	.12331	29	.99237	4	.12426	30	8.0476	191	55		
6	360	29	233	3	456	29	.0285	190	54		
7	389	29	230	4	485	30	.0095	189	53		
8	418	29	226	4	515	29	7.9906	188	52		
9	447	29	222	3	544	30	.9718	188	51		
10	.12476	28	.99219	4	.12574	29	7.9530		50		
11	504	29	215	4	603	30	.9344	186	49	30	29
12	533	29	211	3	633	29	.9158	186	48	1 0.5	0.5
13	562	29	208	4	662	30	.8973	185	47	2 1.0	1.0
14	591	29	204	4	692	30	.8789	184	46	3 1.5	1.4
								183		4 2.0	1.9
15	.12620	29	.99200	3	.12722	29	7.8606	182	45	5 2.5	2.4
16	649	29	197	4	751	30	.8424	181	44	6 3.0	2.9
17	678	28	193	4	781	30	.8243	181	43	7 3.5	3.4
18	706	29	189	3	810	30	.8062	180	42	8 4.0	3.9
19	735	29	186	4	840	29	.7882	178	41	9 4.5	4.4
										10 5.0	4.8
										20 10.0	9.7
20	.12764	29	.99182	4	.12869	30	7.7704	179	40	30 15.0	14.5
21	793	29	178	3	899	30	.7525	177	39	40 20.0	19.3
22	822	29	175	4	929	29	.7348	177	38	50 25.0	24.2
23	851	29	171	4	958	30	.7171	175	37		
24	880	28	167	4	988	29	.6996	175	36		
25	.12908	29	.99163	3	.13017	30	7.6821	174	35		
26	937	29	160	4	047	29	.6647	174	34		
27	966	29	156	4	076	30	.6473	172	33		
28	995	29	152	4	106	30	.6301	172	32		
29	.13024	29	148	4	136	29	.6129	171	31		
30	.13053	28	.99144	3	.13165	30	7.5958	171	30		
31	081	29	141	4	195	29	.5787	169	29		
32	110	29	137	4	224	30	.5618	169	28		
33	139	29	133	4	254	30	.5449	168	27		
34	168	29	129	4	284	29	.5281	168	26		
35	.13197	29	.99125	3	.13313	30	7.5113	166	25		
36	226	28	122	4	343	29	.4947	166	24	28	5
37	254	29	118	4	372	30	.4781	166	23	1 0.5	0.1
38	283	29	114	4	402	30	.4615	164	22	2 0.9	0.2
39	312	29	110	4	432	29	.4451	164	21	3 1.4	0.2
										4 1.9	0.3
40	.13341	29	.99106	4	.13461	30	7.4287	163	20	5 2.3	0.4
41	370	29	102	4	491	30	.4124	162	19	6 2.8	0.5
42	399	28	098	4	521	29	.3962	162	18	7 3.3	0.6
43	427	29	094	3	550	30	.3800	161	17	8 3.7	0.7
44	456	29	091	4	580	29	.3639	160	16	9 4.2	0.8
										10 4.7	0.8
										20 9.3	1.7
45	.13485	29	.99087	4	.13609	30	7.3479	160	15	30 14.0	2.5
46	514	29	083	4	639	30	.3319	159	14	40 18.7	3.3
47	543	29	079	4	669	29	.3160	158	13	50 23.3	4.2
48	572	28	075	4	698	30	.3002	158	12		
49	600	29	071	4	728	30	.2844	157	11		
50	.13629	29	.99067	4	.13758	29	7.2687	156	10		
51	658	29	063	4	787	30	.2531	156	9		
52	687	29	059	4	817	29	.2375	155	8		
53	716	28	055	4	846	30	.2220	154	7		
54	744	29	051	4	876	30	.2066	154	6		
55	.13773	29	.99047	4	.13906	29	7.1912	153	5		
56	802	29	043	4	935	30	.1759	152	4		
57	831	29	039	4	965	30	.1607	152	3		
58	860	29	035	4	995	29	.1455	151	2		
59	889	28	031	4	.14024	30	.1304	150	1		
60	.13917		.99027		.14054		7.1154		0		

	Cos.	d.	Sin.	d.	Cot.	d.	Tan.	d.	′	P. P.	

| | | | | | | 82° | | | | | |

8°

P. P.	′	Sin.	d.	Cos.	d.	Tan.	d.	Cot.	d.	
	0	.13917		.99027		.14054		7.1154		60
	1	946	29	023	4	084	30	.1004	150	59
	2	975	29	019	4	113	29	.0855	149	58
	3	.14004	29	015	4	143	30	.0706	149	57
	4	033	29	011	4	173	30	.0558	148	56
	5	.14061	28	.99006	5	.14202	29	7.0410	148	55
	6	090	29	002	4	232	30	.0264	146	54
	7	119	29	.98998	4	262	30	.0117	147	53
	8	148	29	994	4	291	29	6.9972	145	52
	9	177	29	990	4	321	30	.9827	145	51
	10	.14205	28	.98986	4	.14351	30	6.9682	145	50
	11	234	29	982	4	381	30	.9538	144	49
	12	263	29	978	4	410	29	.9395	143	48
	13	292	29	973	5	440	30	.9252	143	47
	14	320	28	969	4	470	30	.9110	142	46
	15	.14349	29	.98965	4	.14499	29	6.8969	141	45
	16	378	29	961	4	529	30	.8828	141	44
	17	407	29	957	4	559	30	.8687	141	43
	18	436	29	953	4	588	29	.8548	139	42
	19	464	28	948	5	618	30	.8408	140	41
	20	.14493	29	.98944	4	.14648	30	6.8269	139	40
	21	522	29	940	4	678	30	.8131	138	39
	22	551	29	936	4	707	29	.7994	137	38
	23	580	29	931	5	737	30	.7856	138	37
	24	608	28	927	4	767	30	.7720	136	36
	25	.14637	29	.98923	4	.14796	29	6.7584	136	35
	26	666	29	919	4	826	30	.7448	136	34
	27	695	29	914	5	856	30	.7313	135	33
	28	723	28	910	4	886	30	.7179	134	32
	29	752	29	906	4	915	29	.7045	134	31
	30	.14781	29	.98902	4	.14945	30	6.6912	133	30
	31	810	29	897	5	975	30	.6779	133	29
	32	838	28	893	4	.15005	30	.6646	133	28
	33	867	29	889	4	034	29	.6514	132	27
	34	896	29	884	5	064	30	.6383	131	26
	35	.14925	29	.98880	4	.15094	30	6.6252	131	25
	36	954	29	876	4	124	30	.6122	130	24
	37	982	28	871	5	153	29	.5992	130	23
	38	.15011	29	867	4	183	30	.5863	129	22
	39	040	29	863	4	213	30	.5734	129	21
	40	.15069	29	.98858	5	.15243	30	6.5606	128	20
	41	097	28	854	4	272	29	.5478	128	19
	42	126	29	849	5	302	30	.5350	128	18
	43	155	29	845	4	332	30	.5223	127	17
	44	184	29	841	4	362	30	.5097	126	16
	45	.15212	28	.98836	5	.15391	29	6.4971	126	15
	46	241	29	832	4	421	30	.4846	125	14
	47	270	29	827	5	451	30	.4721	125	13
	48	299	29	823	4	481	30	.4596	125	12
	49	327	29	818	5	511	30	.4472	124	11
	50	.15356	29	.98814	4	.15540	29	6.4348	124	10
	51	385	29	809	5	570	30	.4225	123	9
	52	414	29	805	4	600	30	.4103	122	8
	53	442	28	800	5	630	30	.3980	123	7
	54	471	29	796	4	660	30	.3859	121	6
	55	.15500	29	.98791	5	.15689	29	6.3737	122	5
	56	529	29	787	4	719	30	.3617	120	4
	57	557	28	782	5	749	30	.3496	121	3
	58	586	29	778	4	779	30	.3376	120	2
	59	615	29	773	5	809	30	.3257	119	1
	60	.15643	28	.98769	4	.15838	29	6.3138	119	0
P. P.		Cos.	d.	Sin.	d.	Cot.	d.	Tan.	d.	′

P. P.

	4	3
1	0.1	0.0
2	0.1	0.1
3	0.2	0.2
4	0.3	0.2
5	0.3	0.2
6	0.4	0.3
7	0.5	0.4
8	0.5	0.4
9	0.6	0.4
10	0.7	0.5
20	1.3	1.0
30	2.0	1.5
40	2.7	2.0
50	3.3	2.5

81°

9°

′	Sin.	d.	Cos.	d.	Tan.	d.	Cot.	d.	
0	.15643	29	.98769	5	.15838	30	6.3138	119	60
1	672	29	764	4	868	30	.3019	118	59
2	701	29	760	5	898	30	.2901	118	58
3	730	28	755	4	928	30	.2783	117	57
4	758	29	751	5	958	30	.2666	117	56
5	.15787	29	.98746	5	.15988	29	6.2549	117	55
6	816	29	741	4	.16017	30	.2432	116	54
7	845	28	737	5	047	30	.2316	116	53
8	873	29	732	4	077	30	.2200	115	52
9	902	29	728	5	107	30	.2085	115	51
10	.15931	28	.98723	5	.16137	30	6.1970	114	50
11	959	29	718	4	167	29	.1856	114	49
12	988	29	714	5	196	30	.1742	114	48
13	.16017	29	709	5	226	30	.1628	113	47
14	046	28	704	4	256	30	.1515	113	46
15	.16074	29	.98700	5	.16286	30	6.1402	112	45
16	103	29	695	5	316	30	.1290	112	44
17	132	28	690	4	346	30	.1178	112	43
18	160	29	686	5	376	29	.1066	111	42
19	189	29	681	5	405	30	.0955	111	41
20	.16218	28	.98676	5	.16435	30	6.0844	110	40
21	246	29	671	4	465	30	.0734	110	39
22	275	29	667	5	495	30	.0624	110	38
23	304	29	662	5	525	30	.0514	109	37
24	333	28	657	5	555	30	.0405	109	36
25	.16361	29	.98652	4	.16585	30	6.0296	108	35
26	390	29	648	5	615	30	.0188	108	34
27	419	28	643	5	645	29	.0080	108	33
28	447	29	638	5	674	30	5.9972	107	32
29	476	29	633	4	704	30	.9865	107	31
30	.16505	28	.98629	5	.16734	30	5.9758	107	30
31	533	29	624	5	764	30	.9651	106	29
32	562	29	619	5	794	30	.9545	106	28
33	591	29	614	5	824	30	.9439	106	27
34	620	28	609	5	854	30	.9333	105	26
35	.16648	29	.98604	4	.16884	30	5.9228	104	25
36	677	29	600	5	914	30	.9124	105	24
37	706	28	595	5	944	30	.9019	104	23
38	734	29	590	5	974	30	.8915	104	22
39	763	29	585	5	.17004	29	.8811	103	21
40	.16792	28	.98580	5	.17033	30	5.8708	103	20
41	820	29	575	5	063	30	.8605	103	19
42	849	29	570	5	093	30	.8502	102	18
43	878	28	565	4	123	30	.8400	102	17
44	906	29	561	5	153	30	.8298	101	16
45	.16935	29	.98556	5	.17183	30	5.8197	102	15
46	964	28	551	5	213	30	.8095	101	14
47	992	29	546	5	243	30	.7994	100	13
48	.17021	29	541	5	273	30	.7894	100	12
49	050	28	536	5	303	30	.7794	100	11
50	.17078	29	.98531	5	.17333	30	5.7694	100	10
51	107	29	526	5	363	30	.7594	99	9
52	136	28	521	5	393	30	.7495	99	8
53	164	29	516	5	423	30	.7396	99	7
54	193	29	511	5	453	30	.7297	98	6
55	.17222	28	.98506	5	.17483	30	5.7199	98	5
56	250	29	501	5	513	30	.7101	97	4
57	279	29	496	5	543	30	.7004	98	3
58	308	28	491	5	573	30	.6906	97	2
59	336	29	486	5	603	30	.6809	96	1
60	.17365		.98481		.17633		5.6713		0
	Cos.	d.	Sin.	d.	Cot.	d.	Tan.	d.	′

P. P.

	31	30
1	0.5	0.5
2	1.0	1.0
3	1.6	1.5
4	2.1	2.0
5	2.6	2.5
6	3.1	3.0
7	3.6	3.5
8	4.1	4.0
9	4.6	4.5
10	5.2	5.0
20	10.3	10.0
30	15.5	15.0
40	20.7	20.0
50	25.8	25.0

	29	28
1	0.5	0.5
2	1.0	0.9
3	1.4	1.4
4	1.9	1.9
5	2.4	2.3
6	2.9	2.8
7	3.4	3.3
8	3.9	3.7
9	4.4	4.2
10	4.8	4.7
20	9.7	9.3
30	14.5	14.0
40	19.3	18.7
50	24.2	23.3

80°

10°										
P. P.	′	Sin.	d.	Cos.	d.	Tan.	d.	Cot.	d.	
	0	.17365	28	.98481	5	.17633	30	5.6713	96	**60**
	1	393	29	476	5	663	30	.6617	96	59
	2	422	29	471	5	693	30	.6521	96	58
	3	451	28	466	5	723	30	.6425	96	57
	4	479	29	461	6	753	30	.6329	95	56
	5	.17508	29	.98455	5	.17783	30	5.6234	94	**55**
	6	537	28	450	5	813	30	.6140	95	54
	7	565	29	445	5	843	30	.6045	94	53
	8	594	29	440	5	873	30	.5951	94	52
	9	623	28	435	5	903	30	.5857	93	51
	10	.17651	29	.98430	5	.17933	30	5.5764	93	**50**
	11	680	28	425	5	963	30	.5671	93	49
	12	708	29	420	6	993	30	.5578	93	48
6 5	13	737	29	414	5	.18023	30	.5485	92	47
1 0.1 0.1	14	766	28	409	5	053	30	.5393	92	46
2 0.2 0.2										
3 0.3 0.2	**15**	.17794	29	.98404	5	.18083	30	5.5301	92	**45**
4 0.4 0.3	16	823	29	399	5	113	30	.5209	91	44
5 0.5 0.4	17	852	28	394	5	143	30	.5118	92	43
6 0.6 0.5	18	880	29	389	6	173	30	.5026	90	42
7 0.7 0.6	19	909	28	383	5	203	30	.4936	91	41
8 0.8 0.7										
9 0.9 0.8	**20**	.17937	29	.98378	5	.18233	30	5.4845	90	**40**
10 1.0 0.8	21	966	29	373	5	263	30	.4755	90	39
20 2.0 1.7	22	995	28	368	6	293	30	.4665	90	38
30 3.0 2.5	23	.18023	29	362	5	323	30	.4575	89	37
40 4.0 3.3	24	052	29	357	5	353	31	.4486	89	36
50 5.0 4.2										
	25	.18081	28	.98352	5	.18384	30	5.4397	89	**35**
	26	109	29	347	6	414	30	.4308	89	34
	27	138	28	341	5	444	30	.4219	88	33
	28	166	29	336	5	474	30	.4131	88	32
	29	195	29	331	6	504	30	.4043	88	31
	30	.18224	28	.98325	5	.18534	30	5.3955	87	**30**
	31	252	29	320	5	564	30	.3868	87	29
	32	281	28	315	5	594	30	.3781	87	28
	33	309	29	310	6	624	30	.3694	87	27
	34	338	29	304	5	654	30	.3607	86	26
4	**35**	.18367	28	.98299	5	.18684	30	5.3521	86	**25**
1 0.1	36	395	29	294	6	714	31	.3435	86	24
2 0.1	37	424	28	288	5	745	30	.3349	86	23
4 0.3	38	452	29	283	6	775	30	.3263	85	22
5 0.3	39	481	28	277	5	805	30	.3178	85	21
6 0.4										
7 0.5	**40**	.18509	29	.98272	5	.18835	30	5.3093	85	**20**
8 0.5	41	538	29	267	6	865	30	.3008	84	19
9 0.6	42	567	28	261	5	895	30	.2924	85	18
10 0.7	43	595	29	256	6	925	30	.2839	84	17
20 1.3	44	624	28	250	5	955	31	.2755	83	16
30 2.0										
40 2.7	**45**	.18652	29	.98245	5	.18986	30	5.2672	84	**15**
50 3.3	46	681	29	240	6	.19016	30	.2588	83	14
	47	710	28	234	5	046	30	.2505	83	13
	48	738	29	229	6	076	30	.2422	83	12
	49	767	28	223	5	106	30	.2339	82	11
	50	.18795	29	.98218	6	.19136	30	5.2257	83	**10**
	51	824	28	212	5	166	31	.2174	82	9
	52	852	29	207	6	197	30	.2092	81	8
	53	881	29	201	5	227	30	.2011	82	7
	54	910	28	196	6	257	30	.1929	81	6
	55	.18938	29	.98190	5	.19287	30	5.1848	81	**5**
	56	967	28	185	6	317	30	.1767	81	4
	57	995	29	179	5	347	31	.1686	80	3
	58	.19024	28	174	6	378	30	.1606	80	2
	59	052	29	168	5	408	30	.1526	80	1
	60	.19081		.98163		.19438		5.1446		**0**
P. P.		Cos.	d.	Sin.	d.	Cot.	d.	Tan.	d.	′

11°

'	Sin.	d.	Cos.	d.	Tan.	d.	Cot.	d.	
0	.19081	28	.98163	6	.19438	30	5.1446	80	60
1	109	29	157	5	468	30	.1366	80	59
2	138	29	152	6	498	31	.1286	79	58
3	167	28	146	6	529	30	.1207	79	57
4	195	29	140	5	559	30	.1128	79	56
5	.19224	28	.98135	6	.19589	30	5.1049	79	55
6	252	29	129	5	619	30	.0970	78	54
7	281	28	124	6	649	31	.0892	78	53
8	309	29	118	6	680	30	.0814	78	52
9	338	28	112	5	710	30	.0736	78	51
10	.19366	29	.98107	6	.19740	30	5.0658	77	50
11	395	28	101	5	770	31	.0581	77	49
12	423	29	096	6	801	30	.0504	77	48
13	452	29	090	6	831	30	.0427	77	47
14	481	28	084	5	861	30	.0350	77	46
15	.19509	29	.98079	6	.19891	30	5.0273	76	45
16	538	28	073	6	921	31	.0197	76	44
17	566	29	067	6	952	30	.0121	76	43
18	595	28	061	5	982	30	.0045	76	42
19	623	29	056	6	.20012	30	4.9969	75	41
20	.19652	28	.98050	6	.20042	31	4.9894	75	40
21	680	29	044	5	073	30	.9819	75	39
22	709	28	039	6	103	30	.9744	75	38
23	737	29	033	6	133	31	.9669	75	37
24	766	28	027	6	164	30	.9594	74	36
25	.19794	29	.98021	5	.20194	30	4.9520	74	35
26	823	28	016	6	224	30	.9446	74	34
27	851	29	010	6	254	31	.9372	74	33
28	880	28	004	6	285	30	.9298	73	32
29	908	29	.97998	6	315	30	.9225	73	31
30	.19937	28	.97992	5	.20345	31	4.9152	74	30
31	965	29	987	6	376	30	.9078	72	29
32	994	28	981	6	406	30	.9006	73	28
33	.20022	29	975	6	436	30	.8933	73	27
34	051	28	969	6	466	31	.8860	72	26
35	.20079	29	.97963	5	.20497	30	4.8788	72	25
36	108	28	958	6	527	30	.8716	72	24
37	136	29	952	6	557	31	.8644	71	23
38	165	28	946	6	588	30	.8573	72	22
39	193	29	940	6	618	30	.8501	71	21
40	.20222	28	.97934	6	.20648	31	4.8430	71	20
41	250	29	928	6	679	30	.8359	71	19
42	279	28	922	6	709	30	.8288	70	18
43	307	29	916	6	739	31	.8218	71	17
44	336	28	910	5	770	30	.8147	70	16
45	.20364	29	.97905	6	.20800	30	4.8077	70	15
46	393	28	899	6	830	31	.8007	70	14
47	421	29	893	6	861	30	.7937	70	13
48	450	28	887	6	891	30	.7867	69	12
49	478	29	881	6	921	31	.7798	69	11
50	.20507	28	.97875	6	.20952	30	4.7729	70	10
51	535	28	869	6	982	31	.7659	68	9
52	563	28	863	6	.21013	30	.7591	69	8
53	592	28	857	6	043	30	.7522	69	7
54	620	29	851	6	073	31	.7453	68	6
55	.20649	28	.97845	6	.21104	30	4.7385	68	5
56	677	29	839	6	134	30	.7317	68	4
57	706	28	833	6	164	31	.7249	68	3
58	734	29	827	6	195	30	.7181	67	2
59	763	28	821	6	225	31	.7114	68	1
60	.20791		.97815		.21256		4.7046		0

| | Cos. | d. | Sin. | d. | Cot. | d. | Tan. | d. | ' |

78°

P. P.

	31	30
1	0.5	0.5
2	1.0	1.0
3	1.6	1.5
4	2.1	2.0
5	2.6	2.5
6	3.1	3.0
7	3.6	3.5
8	4.1	4.0
9	4.6	4.5
10	5.2	5.0
20	10.3	10.0
30	15.5	15.0
40	20.7	20.0
50	25.8	25.0

	29	28
1	0.5	0.5
2	1.0	0.9
3	1.4	1.4
4	1.9	1.9
5	2.4	2.3
6	2.9	2.8
7	3.4	3.3
8	3.9	3.7
9	4.4	4.2
10	4.8	4.7
20	9.7	9.3
30	14.5	14.0
40	19.3	18.7
50	24.2	23.3

12°

P. P.	′	Sin.	d.	Cos.	d.	Tan.	d.	Cot.	d.	
	0	.20791	29	.97815	6	.21256	30	4.7046	67	**60**
	1	820	28	809	6	286	30	.6979	67	59
	2	848	29	803	6	316	31	.6912	67	58
	3	877	28	797	6	347	30	.6845	66	57
	4	905	28	791	7	377	31	.6779	67	56
	5	.20933	29	.97784	6	.21408	30	4.6712	66	**55**
	6	962	28	778	6	438	31	.6646	66	54
	7	990	29	772	6	469	30	.6580	66	53
	8	.21019	28	766	6	499	30	.6514	66	52
	9	047	29	760	6	529	31	.6448	66	51
	10	.21076	28	.97754	6	.21560	30	4.6382	65	**50**
	11	104	28	748	6	590	31	.6317	65	49
7 **6**	12	132	29	742	7	621	30	.6252	65	48
1 0.1 0.1	13	161	28	735	6	651	31	.6187	65	47
2 0.2 0.2	14	189	29	729	6	682	30	.6122	65	46
3 0.4 0.3	**15**	.21218	28	.97723	6	.21712	31	4.6057	64	**45**
4 0.5 0.4	16	246	29	717	6	743	30	.5993	65	44
5 0.6 0.5	17	275	28	711	6	773	31	.5928	64	43
6 0.7 0.6	18	303	28	705	7	804	30	.5864	64	42
7 0.8 0.7	19	331	29	698	6	834	30	.5800	64	41
8 0.9 0.8	**20**	.21360	28	.97692	6	.21864	31	4.5736	63	**40**
9 1.0 0.9	21	388	29	686	6	895	30	.5673	64	39
10 1.2 1.0	22	417	28	680	7	925	31	.5609	63	38
20 2.3 2.0	23	445	29	673	6	956	30	.5546	63	37
30 3.5 3.0	24	474	28	667	6	986	31	.5483	63	36
40 4.7 4.0	**25**	.21502	28	.97661	6	.22017	30	4.5420	63	**35**
50 5.8 5.0	26	530	29	655	7	047	31	.5357	63	34
	27	559	28	648	6	078	30	.5294	62	33
	28	587	29	642	6	108	31	.5232	63	32
	29	616	28	636	6	139	30	.5169	62	31
	30	.21644	28	.97630	7	.22169	31	4.5107	62	**30**
	31	672	29	623	6	200	31	.5045	62	29
	32	701	28	617	6	231	30	.4983	61	28
	33	729	29	611	7	261	31	.4922	62	27
	34	758	28	604	6	292	30	.4860	61	26
5	**35**	.21786	28	.97598	6	.22322	31	4.4799	62	**25**
1 0.1	36	814	29	592	7	353	30	.4737	61	24
2 0.2	37	843	28	585	6	383	31	.4676	61	23
3 0.2	38	871	28	579	6	414	30	.4615	60	22
4 0.3	39	899	29	573	7	444	31	.4555	61	21
5 0.4	**40**	.21928	28	.97566	6	.22475	30	4.4494	60	**20**
6 0.5	41	956	29	560	7	505	31	.4434	61	19
7 0.6	42	985	28	553	6	536	31	.4373	60	18
8 0.7	43	.22013	28	547	6	567	30	.4313	60	17
9 0.8	44	041	29	541	7	597	31	.4253	59	16
10 0.8	**45**	.22070	28	.97534	6	.22628	30	4.4194	60	**15**
20 1.7	46	098	28	528	7	658	31	.4134	59	14
30 2.5	47	126	29	521	6.	689	30	.4075	60	13
40 3.3	48	155	28	515	7	719	31	.4015	59	12
50 4.2	49	183	29	508	6	750	31	.3956	59	11
	50	.22212	28	.97502	6	.22781	30	4.3897	59	**10**
	51	240	28	496	7	811	31	.3838	59	9
	52	268	29	489	6	842	30	.3779	58	8
	53	297	28	483	7	872	31	.3721	59	7
	54	325	28	476	6	903	31	.3662	58	6
	55	.22353	29	.97470	7	.22934	30	4.3604	58	**5**
	56	382	28	463	6	964	31	.3546	58	4
	57	410	28	457	7	995	31	.3488	58	3
	58	438	29	450	6	.23026	30	.3430	58	2
	59	467	28	444	7	056	31	.3372	57	1
	60	.22495		.97437		.23087		4.3315		**0**
P. P.		Cos.	d.	Sin.	d.	Cot.	d.	Tan.	d.	′

77°

13°

'	Sin.	d.	Cos.	d.	Tan.	d.	Cot.	d.		P. P.
0	.22495	28	.97437	7	.23087	30	4.3315	58	60	
1	523	29	430	6	117	31	.3257	57	59	
2	552	28	424	7	148	31	.3200	57	58	
3	580	28	417	6	179	31	.3143	57	57	
4	608	29	411	7	209	30	.3086	57	56	
5	.22637	28	.97404	6	.23240	31	4.3029	57	55	
6	665	28	398	7	271	30	.2972	56	54	
7	693	29	391	7	301	31	.2916	57	53	
8	722	28	384	6	332	31	.2859	56	52	
9	750	28	378	7	363	30	.2803	56	51	
10	.22778	29	.97371	6	.23393	31	4.2747	56	50	
11	807	28	365	7	424	31	.2691	56	49	
12	835	28	358	7	455	30	.2635	55	48	**32** **31**
13	863	29	351	6	485	31	.2580	56	47	1 0.5 0.5
14	892	28	345	7	516	31	.2524	56	46	2 1.1 1.0
15	.22920	28	.97338	7	.23547	31	4.2468	55	45	3 1.6 1.6
16	948	29	331	6	578	30	.2413	55	44	4 2.1 2.1
17	977	28	325	7	608	31	.2358	55	43	5 2.7 2.6
18	.23005	28	318	7	639	31	.2303	55	42	6 3.2 3.1
19	033	29	311	7	670	30	.2248	55	41	7 3.7 3.6
20	.23062	28	.97304	6	.23700	31	4.2193	54	40	8 4.3 4.1
21	090	28	298	7	731	31	.2139	55	39	9 4.8 4.6
22	118	28	291	7	762	31	.2084	54	38	10 5.3 5.2
23	146	29	284	6	793	30	.2030	54	37	20 10.7 10.3
24	175	28	278	7	823	31	.1976	54	36	30 16.0 15.5
25	.23203	28	.97271	7	.23854	31	4.1922	54	35	40 21.3 20.7
26	231	29	264	7	885	31	.1868	54	34	50 26.7 25.8
27	260	28	257	6	916	30	.1814	54	33	
28	288	28	251	7	946	31	.1760	54	32	
29	316	29	244	7	977	31	.1706	53	31	
30	.23345	28	.97237	7	.24008	31	4.1653	53	30	
31	373	28	230	7	039	30	.1600	53	29	
32	401	28	223	6	069	31	.1547	54	28	
33	429	29	217	7	100	31	.1493	52	27	
34	458	28	210	7	131	31	.1441	53	26	
35	.23486	28	.97203	7	.24162	31	4.1388	53	25	**30** **29**
36	514	28	196	7	193	30	.1335	53	24	1 0.5 0.5
37	542	29	189	7	223	31	.1282	52	23	2 1.0 1.0
38	571	28	182	6	254	31	.1230	52	22	3 1.5 1.4
39	599	28	176	7	285	31	.1178	52	21	4 2.0 1.9
40	.23627	29	.97169	7	.24316	31	4.1126	52	20	5 2.5 2.4
41	656	28	162	7	347	30	.1074	52	19	6 3.0 2.9
42	684	28	155	7	377	31	.1022	52	18	7 3.5 3.4
43	712	28	148	7	408	31	.0970	52	17	8 4.0 3.9
44	740	29	141	7	439	31	.0918	51	16	9 4.5 4.4
45	.23769	28	.97134	7	.24470	31	4.0867	52	15	10 5.0 4.8
46	797	28	127	7	501	31	.0815	51	14	20 10.0 9.7
47	825	28	120	7	532	30	.0764	51	13	30 15.0 14.5
48	853	29	113	7	562	31	.0713	51	12	40 20.0 19.3
49	882	28	106	6	593	31	.0662	51	11	50 25.0 24.2
50	.23910	28	.97100	7	.24624	31	4.0611	51	10	
51	938	28	093	7	655	31	.0560	51	9	
52	966	29	086	7	686	31	.0509	50	8	
53	995	28	079	7	717	30	.0459	51	7	
54	.24023	28	072	7	747	31	.0408	50	6	
55	.24051	28	.97065	7	.24778	31	4.0358	50	5	
56	079	29	058	7	809	31	.0308	51	4	
57	108	28	051	7	840	31	.0257	50	3	
58	136	28	044	7	871	31	.0207	49	2	
59	164	28	037	7	902	31	.0158	50	1	
60	.24192		.97030		.24933		4.0108		0	
	Cos.	d.	Sin.	d.	Cot.	d.	Tan.	d.	'	P. P.

76°

14°

'	Sin.	d.	Cos.	d.	Tan.	d.	Cot.	d.	
0	.24192		.97030		.24933		4.0108		60
1	220	28	023	7	964	31	.0058	50	59
2	249	29	015	8	995	31	.0009	49	58
3	277	28	008	7	.25026	31	3.9959	50	57
4	305	28	001	7	056	30	.9910	49	56
		28		7		31		49	
5	.24333	29	.96994	7	.25087	31	3.9861	49	55
6	362	28	987	7	118	31	.9812	49	54
7	390	28	980	7	149	31	.9763	49	53
8	418	28	973	7	180	31	.9714	49	52
9	446	28	966	7	211	31	.9665	48	51
10	.24474	29	.96959	7	.25242	31	3.9617	49	50
11	503	28	952	7	273	31	.9568	48	49
12	531	28	945	8	304	31	.9520	49	48
13	559	28	937	7	335	31	.9471	48	47
14	587	28	930	7	366	31	.9423	48	46
15	.24615	29	.96923	7	.25397	31	3.9375	48	45
16	644	28	916	7	428	31	.9327	48	44
17	672	28	909	7	459	31	.9279	47	43
18	700	28	902	8	490	31	.9232	48	42
19	728	28	894	7	521	31	.9184	48	41
20	.24756	28	.96887	7	.25552	31	3.9136	47	40
21	784	29	880	7	583	31	.9089	47	39
22	813	28	873	7	614	31	.9042	47	38
23	841	28	866	8	645	31	.8995	48	37
24	869	28	858	7	676	31	.8947	47	36
25	.24897	28	.96851	7	.25707	31	3.8900	46	35
26	925	29	844	7	738	31	.8854	47	34
27	954	28	837	8	769	31	.8807	47	33
28	982	28	829	7	800	31	.8760	46	32
29	.25010	28	822	7	831	31	.8714	47	31
30	.25038	28	.96815	8	.25862	31	3.8667	46	30
31	066	28	807	7	893	31	.8621	46	29
32	094	28	800	7	924	31	.8575	47	28
33	122	29	793	7	955	31	.8528	46	27
34	151	28	786	8	986	31	.8482	46	26
35	.25179	28	.96778	7	.26017	31	3.8436	45	25
36	207	28	771	7	048	31	.8391	46	24
37	235	28	764	8	079	31	.8345	46	23
38	263	28	756	7	110	31	.8299	45	22
39	291	29	749	7	141	31	.8254	46	21
40	.25320	28	.96742	8	.26172	31	3.8208	45	20
41	348	28	734	7	203	32	.8163	45	19
42	376	28	727	8	235	31	.8118	45	18
43	404	28	719	7	266	31	.8073	45	17
44	432	28	712	7	297	31	.8028	45	16
45	.25460	28	.96705	8	.26328	31	3.7983	45	15
46	488	28	697	7	359	31	.7938	45	14
47	516	29	690	8	390	31	.7893	45	13
48	545	28	682	7	421	31	.7848	44	12
49	573	28	675	8	452	31	.7804	44	11
50	.25601	28	.96667	7	.26483	32	3.7760	45	10
51	629	28	660	7	515	31	.7715	44	9
52	657	28	653	8	546	31	.7671	44	8
53	685	28	645	7	577	31	.7627	44	7
54	713	28	638	8	608	31	.7583	44	6
55	.25741	28	.96630	7	.26639	31	3.7539	44	5
56	769	29	623	8	670	31	.7495	44	4
57	798	28	615	7	701	32	.7451	43	3
58	826	28	608	8	733	31	.7408	44	2
59	854	28	600	7	764	31	.7364	43	1
60	.25882		.96593		.26795		3.7321		0

| P. P. | | Cos. | d. | Sin. | d. | Cot. | d. | Tan. | d. | ' |

75°

P. P.

28 8
1	0.5	0.1
2	0.9	0.3
3	1.4	0.4
4	1.9	0.5
5	2.3	0.7
6	2.8	0.8
7	3.3	0.9
8	3.7	1.1
9	4.2	1.2
10	4.7	1.3
20	9.3	2.7
30	14.0	4.0
40	18.7	5.3
50	23.3	6.7

7 6
1	0.1	0.1
2	0.2	0.2
3	0.4	0.3
4	0.5	0.4
5	0.6	0.5
6	0.7	0.6
7	0.8	0.7
8	0.9	0.8
9	1.0	0.9
10	1.2	1.0
20	2.3	2.0
30	3.5	3.0
40	4.7	4.0
50	5.8	5.0

15°

′	Sin.	d.	Cos.	d.	Tan.	d.	Cot.	d.		P. P.
0	.25882		.96593		.26795		3.7321		**60**	
1	910	28	585	8	826	31	.7277	44	59	
2	938	28	578	7	857	31	.7234	43	58	
3	966	28	570	8	888	31	.7191	43	57	
4	994	28	562	8	920	32	.7148	43	56	
		28		7		31		43		
5	.26022		.96555		.26951		3.7105		**55**	
6	050	28	547	8	982	31	.7062	43	54	
7	079	29	540	7	.27013	31	.7019	43	53	
8	107	28	532	8	044	31	.6976	43	52	
9	135	28	524	8	076	32	.6933	43	51	
		28		7		31		42		
10	.26163		.96517		.27107		3.6891		**50**	
11	191	28	509	8	138	31	.6848	43	49	
12	219	28	502	7	169	31	.6806	42	48	
13	247	28	494	8	201	32	.6764	42	47	**32** **31**
14	275	28	486	8	232	31	.6722	42	46	1 0.5 0.5
		28		7		31		42		2 1.1 1.0
15	.26303		.96479		.27263		3.6680		**45**	3 1.6 1.6
16	331	28	471	8	294	31	.6638	42	44	4 2.1 2.1
17	359	28	463	8	326	32	.6596	42	43	5 2.7 2.6
18	387	28	456	7	357	31	.6554	42	42	6 3.2 3.1
19	415	28	448	8	388	31	.6512	42	41	7 3.7 3.6
		28		8		31		42		8 4.3 4.1
20	.26443		.96440		.27419		3.6470		**40**	9 4.8 4.6
21	471	28	433	7	451	32	.6429	41	39	10 5.3 5.2
22	500	29	425	8	482	31	.6387	42	38	20 10.7 10.3
23	528	28	417	8	513	31	.6346	41	37	30 16.0 15.5
24	556	28	410	7	545	32	.6305	41	36	40 21.3 20.7
		28		8		31		41		50 26.7 25.8
25	.26584		.96402		.27576		3.6264		**35**	
26	612	28	394	8	607	31	.6222	42	34	
27	640	28	386	8	638	31	.6181	41	33	
28	668	28	379	7	670	32	.6140	41	32	
29	696	28	371	8	701	31	.6100	40	31	
		28		8		31		41		
30	.26724		.96363		.27732		3.6059		**30**	
31	752	28	355	8	764	32	.6018	41	29	
32	780	28	347	8	795	31	.5978	40	28	
33	808	28	340	7	826	31	.5937	41	27	
34	836	28	332	8	858	32	.5897	40	26	
		28		8		31		41		
35	.26864		.96324		.27889		3.5856		**25**	
36	892	28	316	8	921	32	.5816	40	24	**29** **28**
37	920	28	308	8	952	31	.5776	40	23	1 0.5 0.5
38	948	28	301	7	983	31	.5736	40	22	2 1.0 0.9
39	976	28	293	8	.28015	32	.5696	40	21	3 1.4 1.4
		28		8		31		40		4 1.9 1.9
40	.27004		.96285		.28046		3.5656		**20**	5 2.4 2.3
41	032	28	277	8	077	31	.5616	40	19	6 2.9 2.8
42	060	28	269	8	109	32	.5576	40	18	7 3.4 3.3
43	088	28	261	8	140	31	.5536	40	17	8 3.9 3.7
44	116	28	253	7	172	32	.5497	39	16	9 4.4 4.2
		28		8		31		40		10 4.8 4.7
45	.27144		.96246		.28203		3.5457		**15**	20 9.7 9.3
46	172	28	238	8	234	31	.5418	39	14	30 14.5 14.0
47	200	28	230	8	266	32	.5379	39	13	40 19.3 18.7
48	228	28	222	8	297	31	.5339	40	12	50 24.2 23.3
49	256	28	214	8	329	32	.5300	39	11	
		28		8		31		39		
50	.27284		.96206		.28360		3.5261		**10**	
51	312	28	198	8	391	31	.5222	39	9	
52	340	28	190	8	423	32	.5183	39	8	
53	368	28	182	8	454	31	.5144	39	7	
54	396	28	174	8	486	32	.5105	39	6	
		28		8		31		38		
55	.27424		.96166		.28517		3.5067		**5**	
56	452	28	158	8	549	32	.5028	39	4	
57	480	28	150	8	580	31	.4989	39	3	
58	508	28	142	8	612	32	.4951	38	2	
59	536	28	134	8	643	31	.4912	39	1	
		28		8		32		38		
60	.27564		.96126		.28675		3.4874		**0**	
	Cos.	d.	Sin.	d.	Cot.	d.	Tan.	d.	′	P. P.

74°

16°

P. P.	′	Sin.	d.	Cos.	d.	Tan.	d.	Cot.	d.	
	0	.27564		.96126		.28675		3.4874		60
	1	592	28	118	8	706	31	.4836	38	59
	2	620	28	110	8	738	32	.4798	38	58
	3	648	28	102	8	769	31	.4760	38	57
	4	676	28	094	8	801	32	.4722	38	56
	5	.27704	27	.96086	8	.28832	32	3.4684	38	55
	6	731	28	078	8	864	31	.4646	38	54
	7	759	28	070	8	895	32	.4608	38	53
	8	787	28	062	8	927	31	.4570	37	52
	9	815	28	054	8	958	32	.4533	38	51
	10	.27843	28	.96046	9	.28990	31	3.4495	37	50
	11	871	28	037	8	.29021	32	.4458	38	49
27 9	12	899	28	029	8	053	31	.4420	37	48
1 0.4 0.2	13	927	28	021	8	084	32	.4383	37	47
2 0.9 0.3	14	955	28	013	8	116	31	.4346	38	46
3 1.4 0.4	15	.27983	28	.96005	8	.29147	32	3.4308	37	45
4 1.8 0.6	16	.28011	28	.95997	8	179	31	.4271	37	44
5 2.2 0.8	17	039	28	989	8	210	32	.4234	37	43
6 2.7 0.9	18	067	28	981	9	242	32	.4197	37	42
7 3.2 1.0	19	095	28	972	8	274	31	.4160	36	41
8 3.6 1.2	20	.28123	27	.95964	8	.29305	32	3.4124	37	40
9 4.0 1.4	21	150	28	956	8	337	31	.4087	37	39
10 4.5 1.5	22	178	28	948	8	368	32	.4050	36	38
20 9.0 3.0	23	206	28	940	9	400	32	.4014	37	37
30 13.5 4.5	24	234	28	931	8	432	31	.3977	36	36
40 18.0 6.0	25	.28262	28	.95923	8	.29463	32	3.3941	37	35
50 22.5 7.5	26	290	28	915	8	495	31	.3904	36	34
	27	318	28	907	9	526	32	.3868	36	33
	28	346	28	898	8	558	32	.3832	36	32
	29	374	28	890	8	590	31	.3796	37	31
	30	.28402	27	.95882	8	.29621	32	3.3759	36	30
	31	429	28	874	9	653	32	.3723	36	29
	32	457	28	865	8	685	31	.3687	35	28
	33	485	28	857	8	716	32	.3652	36	27
	34	513	28	849	8	748	32	.3616	36	26
8 7	35	.28541	28	.95841	9	.29780	31	3.3580	36	25
1 0.1 0.1	36	569	28	832	8	811	32	.3544	35	24
2 0.3 0.2	37	597	28	824	8	843	32	.3509	36	23
3 0.4 0.4	38	625	27	816	9	875	31	.3473	35	22
4 0.5 0.5	39	652	28	807	8	906	32	.3438	36	21
5 0.7 0.6	40	.28680	28	.95799	8	.29938	32	3.3402	35	20
6 0.8 0.7	41	708	28	791	9	970	31	.3367	35	19
7 0.9 0.8	42	736	28	782	8	.30001	32	.3332	35	18
8 1.1 0.9	43	764	28	774	8	033	32	.3297	36	17
9 1.2 1.0	44	792	28	766	9	065	32	.3261	35	16
10 1.3 1.2	45	.28820	27	.95757	8	.30097	31	3.3226	35	15
20 2.7 2.3	46	847	28	749	9	128	32	.3191	35	14
30 4.0 3.5	47	875	28	740	8	160	32	.3156	34	13
40 5.3 4.7	48	903	28	732	8	192	32	.3122	35	12
50 6.7 5.8	49	931	28	724	9	224	31	.3087	35	11
	50	.28959	28	.95715	8	.30255	32	3.3052	35	10
	51	987	28	707	9	287	32	.3017	34	9
	52	.29015	27	698	8	319	32	.2983	35	8
	53	042	28	690	9	351	31	.2948	34	7
	54	070	28	681	8	382	32	.2914	35	6
	55	.29098	28	.95673	9	.30414	32	3.2879	34	5
	56	126	28	664	8	446	32	.2845	34	4
	57	154	28	656	9	478	31	.2811	34	3
	58	182	27	647	8	509	32	.2777	34	2
	59	209	28	639	9	541	32	.2743	34	1
	60	.29237		.95630		.30573		3.2709		0
P. P.		Cos.	d.	Sin.	d.	Cot.	d.	Tan.	d.	′

73°

17°

′	Sin.	d.	Cos.	d.	Tan.	d.	Cot.	d.	
0	.29237	28	.95630	8	.30573	32	3.2709	34	**60**
1	265	28	622	9	605	32	.2675	34	59
2	293	28	613	8	637	32	.2641	34	58
3	321	27	605	9	669	31	.2607	34	57
4	348	28	596	8	700	32	.2573	34	56
5	.29376	28	.95588	9	.30732	32	3.2539	33	**55**
6	404	28	579	8	764	32	.2506	34	54
7	432	28	571	9	796	32	.2472	34	53
8	460	27	562	8	828	32	.2438	34	52
9	487	28	554	9	860	31	.2405	34	51
10	.29515	28	.95545	9	.30891	32	3.2371	33	**50**
11	543	28	536	8	923	32	.2338	33	49
12	571	28	528	9	955	32	.2305	33	48
13	599	27	519	8	987	32	.2272	34	47
14	626	28	511	9	.31019	32	.2238	33	46
15	.29654	28	.95502	9	.31051	32	3.2205	33	**45**
16	682	28	493	8	083	32	.2172	33	44
17	710	27	485	9	115	32	.2139	33	43
18	737	28	476	9	147	31	.2106	33	42
19	765	28	467	8	178	32	.2073	32	41
20	.29793	28	.95459	9	.31210	32	3.2041	33	**40**
21	821	28	450	9	242	32	.2008	33	39
22	849	27	441	8	274	32	.1975	32	38
23	876	28	433	9	306	32	.1943	33	37
24	904	28	424	9	338	32	.1910	33	36
25	.29932	28	.95415	8	.31370	32	3.1878	33	**35**
26	960	27	407	9	402	32	.1845	32	34
27	987	28	398	9	434	32	.1813	33	33
28	.30015	28	389	9	466	32	.1780	32	32
29	043	28	380	8	498	32	.1748	32	31
30	.30071	27	.95372	9	.31530	32	3.1716	32	**30**
31	098	28	363	9	562	32	.1684	32	29
32	126	28	354	9	594	32	.1652	32	28
33	154	28	345	8	626	32	.1620	32	27
34	182	27	337	9	658	32	.1588	32	26
35	.30209	28	.95328	9	.31690	32	3.1556	32	**25**
36	237	28	319	9	722	32	.1524	32	24
37	265	27	310	9	754	32	.1492	32	23
38	292	28	301	8	786	32	.1460	31	22
39	320	28	293	9	818	32	.1429	32	21
40	.30348	28	.95284	9	.31850	32	3.1397	31	**20**
41	376	27	275	9	882	32	.1366	32	19
42	403	28	266	9	914	32	.1334	31	18
43	431	28	257	9	946	32	.1303	32	17
44	459	27	248	8	978	32	.1271	31	16
45	.30486	28	.95240	9	.32010	32	3.1240	31	**15**
46	514	28	231	9	042	32	.1209	31	14
47	542	28	222	9	074	32	.1178	32	13
48	570	27	213	9	106	33	.1146	31	12
49	597	28	204	9	139	32	.1115	31	11
50	.30625	28	.95195	9	.32171	32	3.1084	31	**10**
51	653	27	186	9	203	32	.1053	31	9
52	680	28	177	9	235	32	.1022	31	8
53	708	28	168	9	267	32	.0991	30	7
54	736	27	159	9	299	32	.0961	31	6
55	.30763	28	.95150	8	.32331	32	3.0930	31	**5**
56	791	28	142	9	363	33	.0899	31	4
57	819	27	133	9	396	32	.0868	30	3
58	846	28	124	9	428	32	.0838	31	2
59	874	28	115	9	460	32	.0807	30	1
60	.30902		.95106		.32492		3.0777		**0**
	Cos.	d.	Sin.	d.	Cot.	d.	Tan.	d.	′

P. P.

	34	33
1	0.6	0.6
2	1.1	1.1
3	1.7	1.6
4	2.3	2.2
5	2.8	2.8
6	3.4	3.3
7	4.0	3.8
8	4.5	4.4
9	5.1	5.0
10	5.7	5.5
20	11.3	11.0
30	17.0	16.5
40	22.7	22.0
50	28.3	27.5

	32	31
1	0.5	0.5
2	1.1	1.0
3	1.6	1.6
4	2.1	2.1
5	2.7	2.6
6	3.2	3.1
7	3.7	3.6
8	4.3	4.1
9	4.8	4.6
10	5.3	5.2
20	10.7	10.3
30	16.0	15.5
40	21.3	20.7
50	26.7	25.8

	30	29
1	0.5	0.5
2	1.0	1.0
3	1.5	1.4
4	2.0	1.9
5	2.5	2.4
6	3.0	2.9
7	3.5	3.4
8	4.0	3.9
9	4.5	4.4
10	5.0	4.8
20	10.0	9.7
30	15.0	14.5
40	20.0	19.3
50	25.0	24.2

72°

18°

P. P.	'	Sin.	d.	Cos.	d.	Tan.	d.	Cot.	d.	
	0	.30902	27	.95106	9	.32492	32	3.0777	31	60
	1	929	28	097	9	524	32	.0746	30	59
	2	957	28	088	9	556	32	.0716	30	58
	3	985	27	079	9	588	33	.0686	31	57
	4	.31012	28	070	9	621	32	.0655	30	56
	5	.31040	28	.95061	9	.32653	32	3.0625	30	55
	6	068	27	052	9	685	32	.0595	30	54
	7	095	28	043	10	717	32	.0565	30	53
	8	123	28	033	9	749	33	.0535	30	52
	9	151	27	024	9	782	32	.0505	30	51
	10	.31178	28	.95015	9	.32814	32	3.0475	30	50
	11	206	27	006	9	846	32	.0445	30	49
	12	233	28	.94997	9	878	33	.0415	30	48
	13	261	28	988	9	911	32	.0385	29	47
	14	289	27	979	9	943	32	.0356	30	46
	15	.31316	28	.94970	9	.32975	32	3.0326	30	45
	16	344	28	961	9	.33007	33	.0296	29	44
	17	372	27	952	9	040	32	.0267	30	43
	18	399	28	943	10	072	32	.0237	29	42
	19	427	27	933	9	104	32	.0208	30	41
	20	.31454	28	.94924	9	.33136	33	3.0178	29	40
	21	482	28	915	9	169	32	.0149	29	39
	22	510	27	906	9	201	32	.0120	30	38
	23	537	28	897	9	233	33	.0090	29	37
	24	565	28	888	10	266	32	.0061	29	36
	25	.31593	27	.94878	9	.33298	32	3.0032	29	35
	26	620	28	869	9	330	33	.0003	29	34
	27	648	27	860	9	363	32	2.9974	29	33
	28	675	28	851	9	395	32	.9945	29	32
	29	703	27	842	10	427	33	.9916	29	31
	30	.31730	28	.94832	9	.33460	32	2.9887	29	30
	31	758	28	823	9	492	32	.9858	29	29
	32	786	27	814	9	524	33	.9829	29	28
	33	813	28	805	10	557	32	.9800	28	27
	34	841	27	795	9	589	32	.9772	29	26
	35	.31868	28	.94786	9	.33621	33	2.9743	29	25
	36	896	27	777	9	654	32	.9714	28	24
	37	923	28	768	10	686	32	.9686	29	23
	38	951	28	758	9	718	33	.9657	28	22
	39	979	27	749	9	751	32	.9629	29	21
	40	.32006	28	.94740	10	.33783	33	2.9600	28	20
	41	034	27	730	9	816	32	.9572	28	19
	42	061	28	721	9	848	33	.9544	29	18
	43	089	27	712	10	881	32	.9515	28	17
	44	116	28	702	9	913	32	.9487	28	16
	45	.32144	27	.94693	9	.33945	33	2.9459	28	15
	46	171	28	684	10	978	32	.9431	28	14
	47	199	28	674	9	.34010	33	.9403	28	13
	48	227	27	665	9	043	32	.9375	28	12
	49	254	28	656	10	075	33	.9347	28	11
	50	.32282	27	.94646	9	.34108	32	2.9319	28	10
	51	309	28	637	10	140	33	.9291	28	9
	52	337	27	627	9	173	32	.9263	28	8
	53	364	28	618	9	205	33	.9235	27	7
	54	392	27	609	10	238	32	.9208	28	6
	55	.32419	28	.94599	9	.34270	33	2.9180	28	5
	56	447	27	590	10	303	32	.9152	27	4
	57	474	28	580	9	335	33	.9125	28	3
	58	502	27	571	10	368	32	.9097	27	2
	59	529	28	561	9	400	33	.9070	28	1
	60	.32557		.94552		.34433		2.9042		0

P. P.

	28	27
1	0.5	0.4
2	0.9	0.9
3	1.4	1.4
4	1.9	1.8
5	2.3	2.2
6	2.8	2.7
7	3.3	3.2
8	3.7	3.6
9	4.2	4.0
10	4.7	4.5
20	9.3	9.0
30	14.0	13.5
40	18.7	18.0
50	23.3	22.5

	10	9
1	0.2	0.2
2	0.3	0.3
3	0.5	0.4
4	0.7	0.6
5	0.8	0.8
6	1.0	0.9
7	1.2	1.0
8	1.3	1.2
9	1.5	1.4
10	1.7	1.5
20	3.3	3.0
30	5.0	4.5
40	6.7	6.0
50	8.3	7.5

	8
1	0.1
2	0.3
3	0.4
4	0.5
5	0.7
6	0.8
7	0.9
8	1.1
9	1.2
10	1.3
20	2.7
30	4.0
40	5.3
50	6.7

P. P.		Cos.	d.	Sin.	d.	Cot.	d.	Tan.	d.	'

71°

19°

′	Sin.	d.	Cos.	d.	Tan.	d.	Cot.	d.		P. P.
0	.32557	27	.94552	10	.34433	32	2.9042	27	**60**	
1	584	28	542	9	465	33	.9015	28	59	
2	612	27	533	10	498	32	.8987	27	58	
3	639	28	523	9	530	33	.8960	27	57	
4	667	27	514	10	563	33	.8933	28	56	**34**　**33**
5	.32694	28	.94504	9	.34596	32	2.8905	27	**55**	1　0.6　0.6
6	722	27	495	10	628	33	.8878	27	54	2　1.1　1.1
7	749	28	485	9	661	32	.8851	27	53	3　1.7　1.6
8	777	27	476	10	693	33	.8824	27	52	4　2.3　2.2
9	804	28	466	9	726	32	.8797	27	51	5　2.8　2.8
10	.32832	27	.94457	10	.34758	33	2.8770	27	**50**	6　3.4　3.3
11	859	28	447	9	791	33	.8743	27	49	7　4.0　3.8
12	887	27	438	10	824	32	.8716	27	48	8　4.5　4.4
13	914	28	428	10	856	33	.8689	27	47	9　5.1　5.0
14	942	27	418	9	889	33	.8662	26	46	10　5.7　5.5
15	.32969	28	.94409	10	.34922	32	2.8636	27	**45**	20　11.3　11.0
16	997	27	399	9	954	33	.8609	27	44	30　17.0　16.5
17	.33024	27	390	10	987	33	.8582	26	43	40　22.7　22.0
18	051	28	380	10	.35020	32	.8556	27	42	50　28.3　27.5
19	079	27	370	9	052	33	.8529	27	41	
20	.33106	28	.94361	10	.35085	33	2.8502	26	**40**	
21	134	27	351	9	118	32	.8476	27	39	**32**　**28**
22	161	28	342	10	150	33	.8449	26	38	1　0.5　0.5
23	189	27	332	10	183	32	.8423	26	37	2　1.1　0.9
24	216	28	322	9	216	33	.8397	27	36	3　1.6　1.4
25	.33244	27	.94313	10	.35248	33	2.8370	26	**35**	4　2.1　1.9
26	271	27	303	10	281	33	.8344	26	34	5　2.7　2.3
27	298	28	293	9	314	32	.8318	27	33	6　3.2　2.8
28	326	27	284	10	346	33	.8291	26	32	7　3.7　3.3
29	353	28	274	10	379	33	.8265	26	31	8　4.3　3.7
30	.33381	27	.94264	10	.35412	33	2.8239	26	**30**	9　4.8　4.2
31	408	28	254	9	445	32	.8213	26	29	10　5.3　4.7
32	436	27	245	10	477	33	.8187	26	28	20　10.7　9.3
33	463	27	235	10	510	33	.8161	26	27	30　16.0　14.0
34	490	28	225	10	543	33	.8135	26	26	40　21.3　18.7
35	.33518	27	.94215	9	.35576	32	2.8109	26	**25**	50　26.7　23.3
36	545	28	206	10	608	33	.8083	26	24	
37	573	27	196	10	641	33	.8057	25	23	
38	600	27	186	10	674	33	.8032	26	22	**27**　**26**
39	627	28	176	9	707	33	.8006	26	21	1　0.4　0.4
40	.33655	27	.94167	10	.35740	32	2.7980	25	**20**	2　0.9　0.9
41	682	28	157	10	772	33	.7955	26	19	3　1.4　1.3
42	710	27	147	10	805	33	.7929	26	18	4　1.8　1.7
43	737	27	137	10	838	33	.7903	25	17	5　2.2　2.2
44	764	28	127	9	871	33	.7878	26	16	6　2.7　2.6
45	.33792	27	.94118	10	.35904	33	2.7852	25	**15**	7　3.2　3.0
46	819	27	108	10	937	32	.7827	26	14	8　3.6　3.5
47	846	28	098	10	969	33	.7801	25	13	9　4.0　3.9
48	874	27	088	10	.36002	33	.7776	25	12	10　4.5　4.3
49	901	28	078	10	035	33	.7751	26	11	20　9.0　8.7
50	.33929	27	.94068	10	.36068	33	2.7725	25	**10**	30　13.5　13.0
51	956	27	058	9	101	33	.7700	25	9	40　18.0　17.3
52	983	28	049	10	134	33	.7675	25	8	50　22.5　21.7
53	.34011	27	039	10	167	32	.7650	25	7	
54	038	27	029	10	199	33	.7625	25	6	
55	.34065	28	.94019	10	.36232	33	2.7600	25	**5**	**25**　**24**
56	093	27	009	10	265	33	.7575	25	4	1　0.4　0.4
57	120	27	.93999	10	298	33	.7550	25	3	2　0.8　0.8
58	147	28	989	10	331	33	.7525	25	2	3　1.2　1.2
59	175	27	979	10	364	33	.7500	25	1	4　1.7　1.6
60	.34202		.93969		.36397		2.7475		**0**	5　2.1　2.0
										6　2.5　2.4
										7　2.9　2.8
										8　3.3　3.2
										9　3.8　3.6
										10　4.2　4.0
										20　8.3　8.0
										30　12.5　12.0
										40　16.7　16.0
										50　20.8　20.0
	Cos.	d.	Sin.	d.	Cot.	d.	Tan.	d.	′	P. P.

70°

20°

'	Sin.	d.	Cos.	d.	Tan.	d.	Cot.	d.	
0	.34202	27	.93969	10	.36397	33	2.7475	25	60
1	229	28	959	10	430	33	.7450	25	59
2	257	27	949	10	463	33	.7425	25	58
3	284	27	939	10	496	33	.7400	24	57
4	311	28	929	10	529	33	.7376	25	56
5	.34339	27	.93919	10	.36562	33	2.7351	25	55
6	366	27	909	10	595	33	.7326	25	54
7	393	28	899	10	628	33	.7302	24	53
8	421	27	889	10	661	33	.7277	25	52
9	448	27	879	10	694	33	.7253	24	51
10	.34475	28	.93869	10	.36727	33	2.7228	24	50
11	503	27	859	10	760	33	.7204	25	49
12	530	27	849	10	793	33	.7179	24	48
13	557	27	839	10	826	33	.7155	25	47
14	584	28	829	10	859	33	.7130	24	46
15	.34612	27	.93819	10	.36892	33	2.7106	24	45
16	639	27	809	10	925	33	.7082	24	44
17	666	28	799	10	958	33	.7058	24	43
18	694	27	789	10	991	33	.7034	25	42
19	721	27	779	10	.37024	33	.7009	24	41
20	.34748	27	.93769	10	.37057	33	2.6985	24	40
21	775	28	759	11	090	33	.6961	24	39
22	803	27	748	10	123	34	.6937	24	38
23	830	27	738	10	157	33	.6913	24	37
24	857	27	728	10	190	33	.6889	24	36
25	.34884	28	.93718	10	.37223	33	2.6865	24	35
26	912	27	708	10	256	33	.6841	23	34
27	939	27	698	10	289	33	.6818	23	33
28	966	27	688	10	322	33	.6794	24	32
29	993	28	677	11	355	33	.6770	24	31
30	.35021	27	.93667	10	.37388	34	2.6746	23	30
31	048	27	657	10	422	33	.6723	24	29
32	075	27	647	10	455	33	.6699	24	28
33	102	28	637	11	488	33	.6675	23	27
34	130	27	626	10	521	33	.6652	24	26
35	.35157	27	.93616	10	.37554	34	2.6628	23	25
36	184	27	606	10	588	33	.6605	24	24
37	211	28	596	11	621	33	.6581	23	23
38	239	27	585	10	654	33	.6558	24	22
39	266	27	575	10	687	33	.6534	23	21
40	.35293	27	.93565	10	.37720	34	2.6511	23	20
41	320	27	555	11	754	33	.6488	24	19
42	347	28	544	10	787	33	.6464	23	18
43	375	27	534	10	820	33	.6441	23	17
44	402	27	524	10	853	34	.6418	23	16
45	.35429	27	.93514	11	.37887	33	2.6395	24	15
46	456	28	503	10	920	33	.6371	23	14
47	484	27	493	10	953	33	.6348	23	13
48	511	27	483	11	986	34	.6325	23	12
49	538	27	472	10	.38020	33	.6302	23	11
50	.35565	27	.93462	10	.38053	33	2.6279	23	10
51	592	27	452	11	086	34	.6256	23	9
52	619	28	441	10	120	33	.6233	23	8
53	647	27	431	11	153	33	.6210	23	7
54	674	27	420	10	186	34	.6187	22	6
55	.35701	27	.93410	10	.38220	33	2.6165	23	5
56	728	27	400	11	253	33	.6142	23	4
57	755	27	389	10	286	34	.6119	23	3
58	782	28	379	11	320	33	.6096	22	2
59	810	27	368	10	353	33	.6074	23	1
60	.35837		.93358		.38386		2.6051		0

| P. P. | Cos. | d. | Sin. | d. | Cot. | d. | Tan. | d. | ' |

P. P.

	23	22
1	0.4	0.4
2	0.8	0.7
3	1.2	1.1
4	1.5	1.5
5	1.9	1.8
6	2.3	2.2
7	2.7	2.6
8	3.1	2.9
9	3.4	3.3
10	3.8	3.7
20	7.7	7.3
30	11.5	11.0
40	15.3	14.7
50	19.2	18.3

	10	11
1	0.2	0.2
2	0.3	0.4
3	0.5	0.6
4	0.7	0.7
5	0.8	0.9
6	1.0	1.1
7	1.2	1.3
8	1.3	1.5
9	1.5	1.6
10	1.7	1.8
20	3.3	3.7
30	5.0	5.5
40	6.7	7.3
50	8.3	9.2

	9
1	0.2
2	0.3
3	0.4
4	0.6
5	0.8
6	0.9
7	1.0
8	1.2
9	1.4
10	1.5
20	3.0
30	4.5
40	6.0
50	7.5

21°

′	Sin.	d.	Cos.	d.	Tan.	d.	Cot.	d.		P. P.
0	.35837	27	.93358	10	.38386	34	2.6051	23	60	
1	864	27	348	11	420	33	.6028	22	59	
2	891	27	337	10	453	34	.6006	23	58	
3	918	27	327	11	487	33	.5983	22	57	**35** **34**
4	945	28	316	10	520	33	.5961	23	56	1　0.6　0.6
										2　1.2　1.1
5	.35973	27	.93306	11	.38553	34	2.5938	22	55	3　1.8　1.7
6	.36000	27	295	10	587	33	.5916	23	54	4　2.3　2.3
7	027	27	285	11	620	34	.5893	23	53	5　2.9　2.8
8	054	27	274	10	654	33	.5871	22	52	6　3.5　3.4
9	081	27	264	11	687	34	.5848	23	51	7　4.1　4.0
										8　4.7　4.6
10	.36108	27	.93253	10	.38721	33	2.5826	22	50	9　5.2　5.1
11	135	27	243	11	754	33	.5804	22	49	10　5.8　5.7
12	162	28	232	10	787	34	.5782	22	48	20　11.7　11.3
13	190	27	222	10	821	33	.5759	23	47	30　17.5　17.0
14	217	27	211	10	854	34	.5737	22	46	40　23.3　22.7
										50　29.2　28.3
15	.36244	27	.93201	11	.38888	33	2.5715	22	45	
16	271	27	190	10	921	34	.5693	22	44	
17	298	27	180	11	955	33	.5671	22	43	**33** **28**
18	325	27	169	10	988	34	.5649	22	42	
19	352	27	159	11	.39022	33	.5627	22	41	1　0.6　0.5
										2　1.1　0.9
20	.36379	27	.93148	11	.39055	34	2.5605	22	40	3　1.6　1.4
21	406	28	137	10	089	33	.5583	22	39	4　2.2　1.9
22	434	27	127	11	122	34	.5561	22	38	5　2.8　2.3
23	461	27	116	10	156	34	.5539	22	37	6　3.3　2.8
24	488	27	106	11	190	33	.5517	22	36	7　3.8　3.3
										8　4.4　3.7
25	.36515	27	.93095	11	.39223	34	2.5495	22	35	9　5.0　4.2
26	542	27	084	10	257	33	.5473	21	34	10　5.5　4.7
27	569	27	074	11	290	34	.5452	22	33	20　11.0　9.3
28	596	27	063	11	324	33	.5430	22	32	30　16.5　14.0
29	623	27	052	10	357	34	.5408	22	31	40　22.0　18.7
										50　27.5　23.3
30	.36650	27	.93042	11	.39391	34	2.5386	21	30	
31	677	27	031	11	425	33	.5365	22	29	
32	704	27	020	10	458	34	.5343	21	28	**27** **26**
33	731	27	010	11	492	34	.5322	22	27	
34	758	27	.92999	11	526	33	.5300	21	26	1　0.4　0.4
										2　0.9　0.9
35	.36785	27	.92988	10	.39559	34	2.5279	22	25	3　1.4　1.3
36	812	27	978	11	593	33	.5257	21	24	4　1.8　1.7
37	839	28	967	11	626	34	.5236	22	23	5　2.2　2.2
38	867	27	956	11	660	34	.5214	21	22	6　2.7　2.6
39	894	27	945	10	694	33	.5193	21	21	7　3.2　3.0
										8　3.6　3.5
40	.36921	27	.92935	11	.39727	34	2.5172	22	20	9　4.0　3.9
41	948	27	924	11	761	34	.5150	21	19	10　4.5　4.3
42	975	27	913	11	795	34	.5129	21	18	20　9.0　8.7
43	.37002	27	902	10	829	33	.5108	22	17	30　13.5　13.0
44	029	27	892	11	862	34	.5086	21	16	40　18.0　17.3
										50　22.5　21.7
45	.37056	27	.92881	11	.39896	34	2.5065	21	15	
46	083	27	870	11	930	33	.5044	21	14	
47	110	27	859	10	963	34	.5023	21	13	**23** **22**
48	137	27	849	11	997	34	.5002	21	12	
49	164	27	838	11	.40031	34	.4981	21	11	1　0.4　0.4
										2　0.8　0.7
50	.37191	27	.92827	11	.40065	33	2.4960	21	10	3　1.2　1.1
51	218	27	816	11	098	34	.4939	21	9	4　1.5　1.5
52	245	27	805	11	132	34	.4918	21	8	5　1.9　1.8
53	272	27	794	10	166	34	.4897	21	7	6　2.3　2.2
54	299	27	784	11	200	34	.4876	21	6	7　2.7　2.6
										8　3.1　2.9
55	.37326	27	.92773	11	.40234	33	2.4855	21	5	9　3.4　3.3
56	353	27	762	11	267	34	.4834	21	4	10　3.8　3.7
57	380	27	751	11	301	34	.4813	21	3	20　7.7　7.3
58	407	27	740	11	335	34	.4792	20	2	30　11.5　11.0
59	434	27	729	11	369	34	.4772	21	1	40　15.3　14.7
60	.37461		.92718		.40403		2.4751		0	50　19.2　18.3

	Cos.	d.	Sin.	d.	Cot.	d.	Tan.	d.	′	P. P.

68°

22°

′	Sin.	d.	Cos.	d.	Tan.	d.	Cot.	d.	′
0	.37461	27	.92718	11	.40403	33	2.4751	21	60
1	488	27	707	10	436	34	.4730	21	59
2	515	27	697	11	470	34	.4709	20	58
3	542	27	686	11	504	34	.4689	21	57
4	569	26	675	11	538	34	.4668	20	56
5	.37595	27	.92664	11	.40572	34	2.4648	21	55
6	622	27	653	11	606	34	.4627	21	54
7	649	27	642	11	640	34	.4606	21	53
8	676	27	631	11	674	33	.4586	20	52
9	703	27	620	11	707		.4566	21	51
10	.37730	27	.92609	11	.40741	34	2.4545	20	50
11	757	27	598	11	775	34	.4525	21	49
12	784	27	587	11	809	34	.4504	20	48
13	811	27	576	11	843	34	.4484	20	47
14	838	27	565	11	877	34	.4464	21	46
15	.37865	27	.92554	11	.40911	34	2.4443	20	45
16	892	27	543	11	945	34	.4423	20	44
17	919	27	532	11	979	34	.4403	20	43
18	946	27	521	11	.41013	34	.4383	21	42
19	973	26	510	11	047	34	.4362	20	41
20	.37999	27	.92499	11	.41081	34	2.4342	20	40
21	.38026	27	488	11	115	34	.4322	20	39
22	053	27	477	11	149	34	.4302	20	38
23	080	27	466	11	183	34	.4282	20	37
24	107	27	455	11	217	34	.4262	20	36
25	.38134	27	.92444	12	.41251	34	2.4242	20	35
26	161	27	432	11	285	34	.4222	20	34
27	188	27	421	11	319	34	.4202	20	33
28	215	26	410	11	353	34	.4182	20	32
29	241	27	399	11	387	34	.4162	20	31
30	.38268	27	.92388	11	.41421	34	2.4142	20	30
31	295	27	377	11	455	35	.4122	20	29
32	322	27	366	11	490	34	.4102	19	28
33	349	27	355	12	524	34	.4083	20	27
34	376	27	343	11	558	34	.4063	20	26
35	.38403	27	.92332	11	.41592	34	2.4043	20	25
36	430	26	321	11	626	34	.4023	19	24
37	456	27	310	11	660	34	.4004	20	23
38	483	27	299	12	694	34	.3984	20	22
39	510	27	287	11	728	35	.3964	19	21
40	.38537	27	.92276	11	.41763	34	2.3945	20	20
41	564	27	265	11	797	34	.3925	19	19
42	591	26	254	11	831	34	.3906	20	18
43	617	27	243	12	865	34	.3886	19	17
44	644	27	231	11	899	34	.3867	20	16
45	.38671	27	.92220	11	.41933	35	2.3847	19	15
46	698	27	209	11	968	34	.3828	20	14
47	725	27	198	12	.42002	34	.3808	19	13
48	752	26	186	11	036	34	.3789	19	12
49	778	27	175	11	070	35	.3770	20	11
50	.38805	27	.92164	12	.42105	34	2.3750	19	10
51	832	27	152	11	139	34	.3731	19	9
52	859	27	141	11	173	34	.3712	19	8
53	886	26	130	11	207	35	.3693	20	7
54	912	27	119	12	242	34	.3673	19	6
55	.38939	27	.92107	11	.42276	34	2.3654	19	5
56	966	27	096	11	310	35	.3635	19	4
57	993	27	085	12	345	34	.3616	19	3
58	.39020	26	073	11	379	34	.3597	19	2
59	046	27	062	12	413	34	.3578	19	1
60	.39073		.92050		.42447		2.3559		0

Bottom: P. P. | Cos. | d. | Sin. | d. | Cot. | d. | Tan. | d. | ′

67°

P. P. tables:

	21	20
1	0.4	0.3
2	0.7	0.7
3	1.0	1.0
4	1.4	1.3
5	1.8	1.7
6	2.1	2.0
7	2.4	2.3
8	2.8	2.7
9	3.2	3.0
10	3.5	3.3
20	7.0	6.7
30	10.5	10.0
40	14.0	13.3
50	17.5	16.7

	19	12
1	0.3	0.2
2	0.6	0.4
3	1.0	0.6
4	1.3	0.8
5	1.6	1.0
6	1.9	1.2
7	2.2	1.4
8	2.5	1.6
9	2.8	1.8
10	3.2	2.0
20	6.3	4.0
30	9.5	6.0
40	12.7	8.0
50	15.8	10.0

	11	10
1	0.2	0.2
2	0.4	0.3
3	0.6	0.5
4	0.7	0.7
5	0.9	0.8
6	1.1	1.0
7	1.3	1.2
8	1.5	1.3
9	1.6	1.5
10	1.8	1.7
20	3.7	3.3
30	5.5	5.0
40	7.3	6.7
50	9.2	8.3

23°

′	Sin.	d.	Cos.	d.	Tan.	d.	Cot.	d.	
0	.39073	27	.92050	11	.42447	35	2.3559	20	60
1	100	27	039	11	482	35	.3539	20	59
2	127	26	028	11	516	34	.3520	19	58
3	153	27	016	12	551	35	.3501	19	57
4	180	27	005	11	585	34	.3483	18	56
5	.39207	27	.91994	11	.42619	34	2.3464	19	55
6	234	27	982	12	654	35	.3445	19	54
7	260	26	971	11	688	34	.3426	19	53
8	287	27	959	12	722	34	.3407	19	52
9	314	27	948	11	757	35	.3388	19	51
10	.39341	27	.91936	12	.42791	34	2.3369	19	50
11	367	26	925	11	826	35	.3351	18	49
12	394	27	914	11	860	34	.3332	19	48
13	421	27	902	12	894	34	.3313	19	47
14	448	27	891	11	929	35	.3294	19	46
15	.39474	26	.91879	12	.42963	34	2.3276	18	45
16	501	27	868	11	998	35	.3257	19	44
17	528	27	856	12	.43032	34	.3238	19	43
18	555	27	845	11	067	35	.3220	18	42
19	581	26	833	12	101	34	.3201	19	41
20	.39608	27	.91822	11	.43136	35	2.3183	18	40
21	635	27	810	12	170	34	.3164	19	39
22	661	26	799	11	205	35	.3146	18	38
23	688	27	787	12	239	34	.3127	19	37
24	715	27	775	12	274	35	.3109	18	36
25	.39741	26	.91764	11	.43308	34	2.3090	19	35
26	768	27	752	12	343	35	.3072	18	34
27	795	27	741	11	378	35	.3053	19	33
28	822	27	729	12	412	34	.3035	18	32
29	848	26	718	11	447	35	.3017	18	31
30	.39875	27	.91706	12	.43481	34	2.2998	19	30
31	902	27	694	12	516	35	.2980	18	29
32	928	26	683	11	550	34	.2962	18	28
33	955	27	671	12	585	35	.2944	18	27
34	982	27	660	11	620	35	.2925	19	26
35	.40008	26	.91648	12	.43654	34	2.2907	18	25
36	035	27	636	12	689	35	.2889	18	24
37	062	27	625	11	724	35	.2871	18	23
38	088	26	613	12	758	34	.2853	18	22
39	115	27	601	12	793	35	.2835	18	21
40	.40141	26	.91590	11	.43828	35	2.2817	18	20
41	168	27	578	12	862	34	.2799	18	19
42	195	27	566	12	897	35	.2781	18	18
43	221	26	555	11	932	35	.2763	18	17
44	248	27	543	12	966	34	.2745	18	16
45	.40275	27	.91531	12	.44001	35	2.2727	18	15
46	301	26	519	12	036	35	.2709	18	14
47	328	27	508	11	071	35	.2691	18	13
48	355	27	496	12	105	34	.2673	18	12
49	381	26	484	12	140	35	.2655	18	11
50	.40408	27	.91472	12	.44175	35	2.2637	18	10
51	434	26	461	11	210	35	.2620	17	9
52	461	27	449	12	244	34	.2602	18	8
53	488	27	437	12	279	35	.2584	18	7
54	514	26	425	12	314	35	.2566	18	6
55	.40541	27	.91414	11	.44349	35	2.2549	17	5
56	567	26	402	12	384	35	.2531	18	4
57	594	27	390	12	418	34	.2513	18	3
58	621	27	378	12	453	35	.2496	17	2
59	647	26	366	12	488	35	.2478	18	1
60	.40674	27	.91355	11	.44523	35	2.2460	18	0

| | Cos. | d. | Sin. | d. | Cot. | d. | Tan. | d. | ′ |

P. P.

	36	35
1	0.6	0.6
2	1.2	1.2
3	1.8	1.8
4	2.4	2.3
5	3.0	2.9
6	3.6	3.5
7	4.2	4.1
8	4.8	4.7
9	5.4	5.2
10	6.0	5.8
20	12.0	11.7
30	18.0	17.5
40	24.0	23.3
50	30.0	29.2

	34	27
1	0.6	0.4
2	1.1	0.9
3	1.7	1.4
4	2.3	1.8
5	2.8	2.2
6	3.4	2.7
7	4.0	3.2
8	4.5	3.6
9	5.1	4.0
10	5.7	4.5
20	11.3	9.0
30	17.0	13.5
40	22.7	18.0
50	28.3	22.5

	26	20
1	0.4	0.3
2	0.9	0.7
3	1.3	1.0
4	1.7	1.3
5	2.2	1.7
6	2.6	2.0
7	3.0	2.3
8	3.5	2.7
9	3.9	3.0
10	4.3	3.3
20	8.7	6.7
30	13.0	10.0
40	17.3	13.3
50	21.7	16.7

	19	18
1	0.3	0.3
2	0.6	0.6
3	1.0	0.9
4	1.3	1.2
5	1.6	1.5
6	1.9	1.8
7	2.2	2.1
8	2.5	2.4
9	2.8	2.7
10	3.2	3.0
20	6.3	6.0
30	9.5	9.0
40	12.7	12.0
50	15.8	15.0

66°

24°

P. P. tables:

	17	16
1	0.3	0.3
2	0.6	0.5
3	0.8	0.8
4	1.1	1.1
5	1.4	1.3
6	1.7	1.6
7	2.0	1.9
8	2.3	2.1
9	2.6	2.4
10	2.8	2.7
20	5.7	5.3
30	8.5	8.0
40	11.3	10.7
50	14.2	13.3

	13	12
1	0.2	0.2
2	0.4	0.4
3	0.6	0.6
4	0.9	0.8
5	1.1	1.0
6	1.3	1.2
7	1.5	1.4
8	1.7	1.6
9	2.0	1.8
10	2.2	2.0
20	4.3	4.0
30	6.5	6.0
40	8.7	8.0
50	10.8	10.0

	11
1	0.2
2	0.4
3	0.6
4	0.7
5	0.9
6	1.1
7	1.3
8	1.5
9	1.6
10	1.8
20	3.7
30	5.5
40	7.3
50	9.2

′	Sin.	d.	Cos.	d.	Tan.	d.	Cot.	d.	
0	.40674	26	.91355	12	.44523	35	2.2460	17	60
1	700	27	343	12	558	35	.2443	18	59
2	727	26	331	12	593	34	.2425	17	58
3	753	27	319	12	627	35	.2408	18	57
4	780	26	307	12	662	35	.2390	17	56
5	.40806	27	.91295	12	.44697	35	2.2373	18	55
6	833	27	283	11	732	35	.2355	17	54
7	860	26	272	12	767	35	.2338	18	53
8	886	27	260	12	802	35	.2320	17	52
9	913	26	248	12	837	35	.2303	17	51
10	.40939	27	.91236	12	.44872	35	2.2286	18	50
11	966	26	224	12	907	35	.2268	17	49
12	992	27	212	12	942	35	.2251	17	48
13	.41019	26	200	12	977	35	.2234	18	47
14	045	27	188	12	.45012	35	.2216	17	46
15	.41072	26	.91176	12	.45047	35	2.2199	17	45
16	098	27	164	12	082	35	.2182	17	44
17	125	26	152	12	117	35	.2165	17	43
18	151	27	140	12	152	35	.2148	18	42
19	178	26	128	12	187	35	.2130	17	41
20	.41204	27	.91116	12	.45222	35	2.2113	17	40
21	231	26	104	12	257	35	.2096	17	39
22	257	27	092	12	292	35	.2079	17	38
23	284	26	080	12	327	35	.2062	17	37
24	310	27	068	12	362	35	.2045	17	36
25	.41337	26	.91056	12	.45397	35	2.2028	17	35
26	363	27	044	12	432	35	.2011	17	34
27	390	26	032	12	467	35	.1994	17	33
28	416	27	020	12	502	36	.1977	17	32
29	443	26	008	12	538	35	.1960	17	31
30	.41469	27	.90996	12	.45573	35	2.1943	17	30
31	496	26	984	12	608	35	.1926	17	29
32	522	27	972	12	643	35	.1909	17	28
33	549	26	960	12	678	35	.1892	16	27
34	575	27	948	12	713	35	.1876	17	26
35	.41602	26	.90936	12	.45748	36	2.1859	17	25
36	628	27	924	13	784	35	.1842	17	24
37	655	26	911	12	819	35	.1825	17	23
38	681	26	899	12	854	35	.1808	16	22
39	707	27	887	12	889	35	.1792	17	21
40	.41734	26	.90875	12	.45924	36	2.1775	17	20
41	760	27	863	12	960	35	.1758	16	19
42	787	26	851	12	995	35	.1742	17	18
43	813	27	839	13	.46030	35	.1725	17	17
44	840	26	826	12	065	36	.1708	16	16
45	.41866	26	.90814	12	.46101	35	2.1692	17	15
46	892	27	802	12	136	35	.1675	16	14
47	919	26	790	12	171	35	.1659	17	13
48	945	27	778	12	206	36	.1642	17	12
49	972	26	766	13	242	35	.1625	16	11
50	.41998	26	.90753	12	.46277	35	2.1609	17	10
51	.42024	27	741	12	312	36	.1592	16	9
52	051	26	729	12	348	35	.1576	16	8
53	077	27	717	13	383	35	.1560	17	7
54	104	26	704	12	418	36	.1543	16	6
55	.42130	26	.90692	12	.46454	35	2.1527	17	5
56	156	27	680	12	489	36	.1510	16	4
57	183	26	668	13	525	35	.1494	16	3
58	209	26	655	12	560	35	.1478	17	2
59	235	27	643	12	595	36	.1461	16	1
60	.42262		.90631		.46631		2.1445		0

P. P.			Cos.	d.	Sin.	d.	Cot.	d.	Tan.	d.	′

65°

25°

′	Sin.	d.	Cos.	d.	Tan.	d.	Cot.	d.	
0	.42262	26	.90631	13	.46631	35	2.1445	16	60
1	288	27	618	12	666	36	.1429	16	59
2	315	26	606	12	702	35	.1413	17	58
3	341	26	594	12	737	35	.1396	16	57
4	367	27	582	13	772	36	.1380	16	56
5	.42394	26	.90569	12	.46808	35	2.1364	16	55
6	420	26	557	12	843	36	.1348	16	54
7	446	27	545	13	879	35	.1332	17	53
8	473	26	532	12	914	36	.1315	16	52
9	499	26	520	13	950	35	.1299	16	51
10	.42525	27	.90507	12	.46985	36	2.1283	16	50
11	552	26	495	12	.47021	35	.1267	16	49
12	578	26	483	13	056	36	.1251	16	48
13	604	27	470	12	092	36	.1235	16	47
14	631	26	458	12	128	35	.1219	16	46
15	.42657	26	.90446	13	.47163	36	2.1203	16	45
16	683	26	433	12	199	35	.1187	16	44
17	709	27	421	13	234	36	.1171	16	43
18	736	26	408	12	270	35	.1155	16	42
19	762	26	396	13	305	36	.1139	16	41
20	.42788	27	.90383	12	.47341	36	2.1123	16	40
21	815	26	371	13	377	35	.1107	15	39
22	841	26	358	12	412	36	.1092	16	38
23	867	27	346	12	448	35	.1076	16	37
24	894	26	334	13	483	36	.1060	16	36
25	.42920	26	.90321	12	.47519	36	2.1044	16	35
26	946	26	309	13	555	35	.1028	15	34
27	972	27	296	12	590	36	.1013	16	33
28	999	26	284	13	626	36	.0997	16	32
29	.43025	26	271	12	662	36	.0981	16	31
30	.43051	26	.90259	13	.47698	35	2.0965	15	30
31	077	27	246	13	733	36	.0950	16	29
32	104	26	233	12	769	36	.0934	16	28
33	130	26	221	13	805	35	.0918	15	27
34	156	26	208	12	840	36	.0903	16	26
35	.43182	27	.90196	13	.47876	36	2.0887	15	25
36	209	26	183	12	912	36	.0872	16	24
37	235	26	171	13	948	36	.0856	16	23
38	261	26	158	12	984	35	.0840	15	22
39	287	26	146	13	.48019	36	.0825	16	21
40	.43313	27	.90133	13	.48055	36	2.0809	15	20
41	340	26	120	12	091	36	.0794	16	19
42	366	26	108	13	127	36	.0778	15	18
43	392	26	095	13	163	35	.0763	15	17
44	418	27	082	12	198	36	.0748	16	16
45	.43445	26	.90070	13	.48234	36	2.0732	15	15
46	471	26	057	12	270	36	.0717	16	14
47	497	26	045	13	306	36	.0701	15	13
48	523	26	032	13	342	36	.0686	15	12
49	549	26	019	12	378	36	.0671	16	11
50	.43575	27	.90007	13	.48414	36	2.0655	15	10
51	602	26	.89994	13	450	36	.0640	15	9
52	628	26	981	13	486	35	.0625	16	8
53	654	26	968	12	521	36	.0609	15	7
54	680	26	956	13	557	36	.0594	15	6
55	.43706	27	.89943	13	.48593	36	2.0579	15	5
56	733	26	930	12	629	36	.0564	15	4
57	759	26	918	13	665	36	.0549	16	3
58	785	26	905	13	701	36	.0533	15	2
59	811	26	892	13	737	36	.0518	15	1
60	.43837		.89879		.48773		2.0503		0
	Cos.	d.	Sin.	d.	Cot.	d.	Tan.	d.	′

P. P.

	37	36
1	0.6	0.6
2	1.2	1.2
3	1.8	1.8
4	2.5	2.4
5	3.1	3.0
6	3.7	3.6
7	4.3	4.2
8	4.9	4.8
9	5.6	5.4
10	6.2	6.0
20	12.3	12.0
30	18.5	18.0
40	24.7	24.0
50	30.8	30.0

	35	27
1	0.6	0.4
2	1.2	0.9
3	1.8	1.4
4	2.3	1.8
5	2.9	2.2
6	3.5	2.7
7	4.1	3.2
8	4.7	3.6
9	5.2	4.0
10	5.8	4.5
20	11.7	9.0
30	17.5	13.5
40	23.3	18.0
50	29.2	22.5

	26	25
1	0.4	0.4
2	0.9	0.8
3	1.3	1.2
4	1.7	1.7
5	2.2	2.1
6	2.6	2.5
7	3.0	2.9
8	3.5	3.3
9	3.9	3.8
10	4.3	4.2
20	8.7	8.3
30	13.0	12.5
40	17.3	16.7
50	21.7	20.8

64°

26°

'	Sin.	d.	Cos.	d.	Tan.	d.	Cot.	d.	
0	.43837	26	.89879	12	.48773	36	2.0503	15	60
1	863	26	867	13	809	36	.0488	15	59
2	889	26	854	13	845	36	.0473	15	58
3	916	27	841	13	881	36	.0458	15	57
4	942	26	828	13	917	36	.0443	15	56
		26		12		36		15	
5	.43968	26	.89816	13	.48953	36	2.0428	15	55
6	994	26	803	13	989	37	.0413	15	54
7	.44020	26	790	13	.49026	36	.0398	15	53
8	046	26	777	13	062	36	.0383	15	52
9	072	26	764	12	098	36	.0368	15	51
10	.44098	26	.89752	13	.49134	36	2.0353	15	50
11	124	26	739	13	170	36	.0338	15	49
12	151	27	726	13	206	36	.0323	15	48
13	177	26	713	13	242	36	.0308	15	47
14	203	26	700	13	278	37	.0293	15	46
15	.44229	26	.89687	13	.49315	36	2.0278	15	45
16	255	26	674	12	351	36	.0263	15	44
17	281	26	662	13	387	36	.0248	15	43
18	307	26	649	13	423	36	.0233	14	42
19	333	26	636	13	459	36	.0219	15	41
20	.44359	26	.89623	13	.49495	37	2.0204	15	40
21	385	26	610	13	532	36	.0189	15	39
22	411	26	597	13	568	36	.0174	14	38
23	437	27	584	13	604	36	.0160	15	37
24	464	26	571	13	640	37	.0145	15	36
25	.44490	26	.89558	13	.49677	36	2.0130	15	35
26	516	26	545	13	713	36	.0115	14	34
27	542	26	532	13	749	37	.0101	15	33
28	568	26	519	13	786	36	.0086	14	32
29	594	26	506	13	822	36	.0072	15	31
30	.44620	26	.89493	13	.49858	36	2.0057	15	30
31	646	26	480	13	894	37	.0042	14	29
32	672	26	467	13	931	36	.0028	15	28
33	698	26	454	13	967	37	.0013	14	27
34	724	26	441	13	.50004	36	1.9999	15	26
35	.44750	26	.89428	13	.50040	36	1.9984	14	25
36	776	26	415	13	076	37	.9970	15	24
37	802	26	402	13	113	36	.9955	14	23
38	828	26	389	13	149	36	.9941	15	22
39	854	26	376	13	185	37	.9926	14	21
40	.44880	26	.89363	13	.50222	36	1.9912	15	20
41	906	26	350	13	258	37	.9897	14	19
42	932	26	337	13	295	36	.9883	15	18
43	958	26	324	13	331	37	.9868	14	17
44	984	26	311	13	368	36	.9854	14	16
45	.45010	26	.89298	13	.50404	37	1.9840	15	15
46	036	26	285	13	441	36	.9825	14	14
47	062	26	272	13	477	37	.9811	14	13
48	088	26	259	14	514	36	.9797	15	12
49	114	26	245	13	550	37	.9782	14	11
50	.45140	26	.89232	13	.50587	36	1.9768	14	10
51	166	26	219	13	623	37	.9754	14	9
52	192	26	206	13	660	36	.9740	15	8
53	218	25	193	13	696	37	.9725	14	7
54	243	26	180	13	733	36	.9711	14	6
55	.45269	26	.89167	14	.50769	37	1.9697	14	5
56	295	26	153	13	806	37	.9683	14	4
57	321	26	140	13	843	36	.9669	15	3
58	347	26	127	13	879	37	.9654	14	2
59	373	26	114	13	916	37	.9640	14	1
60	.45399		.89101		.50953		1.9626		0

Bottom headings: P. P. | | Cos. | d. | Sin. | d. | Cot. | d. | Tan. | d. | '

63°

P. P.

	17	16
1	0.3	0.3
2	0.6	0.5
3	0.8	0.8
4	1.1	1.1
5	1.4	1.3
6	1.7	1.6
7	2.0	1.9
8	2.3	2.1
9	2.6	2.4
10	2.8	2.7
20	5.7	5.3
30	8.5	8.0
40	11.3	10.7
50	14.2	13.3

	15	14
1	0.2	0.2
2	0.5	0.5
3	0.8	0.7
4	1.0	0.9
5	1.2	1.2
6	1.5	1.4
7	1.8	1.6
8	2.0	1.9
9	2.2	2.1
10	2.5	2.3
20	5.0	4.7
30	7.5	7.0
40	10.0	9.3
50	12.5	11.7

	13	12
1	0.2	0.2
2	0.4	0.4
3	0.6	0.6
4	0.9	0.8
5	1.1	1.0
6	1.3	1.2
7	1.5	1.4
8	1.7	1.6
9	2.0	1.8
10	2.2	2.0
20	4.3	4.0
30	6.5	6.0
40	8.7	8.0
50	10.8	10.0

27°

′	Sin.	d.	Cos.	d.	Tan.	d.	Cot.	d.		P. P.
0	.45399		.89101		.50953		1.9626		60	
1	425	26	087	14	989	36	.9612	14	59	
2	451	26	074	13	.51026	37	.9598	14	58	
3	477	26	061	13	063	37	.9584	14	57	
4	503	26	048	13	099	36	.9570	14	56	
		26		13		37		14		
5	.45529	25	.89035	14	.51136	37	1.9556	14	55	
6	554	26	021	13	173	37	.9542	14	54	**38** **37**
7	580	26	008	13	209	36	.9528	14	53	1 0.6 0.6
8	606	26	.88995	13	246	37	.9514	14	52	2 1.3 1.2
9	632	26	981	14	283	37	.9500	14	51	3 1.9 1.8
		26		13		36		14		4 2.5 2.5
10	.45658	26	.88968	13	.51319	37	1.9486	14	50	5 3.2 3.1
11	684	26	955	13	356	37	.9472	14	49	6 3.8 3.7
12	710	26	942	14	393	37	.9458	14	48	7 4.4 4.3
13	736	26	928	13	430	37	.9444	14	47	8 5.1 4.9
14	762	25	915	13	467	36	.9430	14	46	9 5.7 5.6
										10 6.3 6.2
15	.45787	26	.88902	14	.51503	37	1.9416	14	45	20 12.7 12.3
16	813	26	888	13	540	37	.9402	14	44	30 19.0 18.5
17	839	26	875	13	577	37	.9388	14	43	40 25.3 24.7
18	865	26	862	13	614	37	.9375	13	42	50 31.7 30.8
19	891	26	848	14	651	37	.9361	14	41	
		26		13		37		14		
20	.45917	25	.88835	13	.51688	36	1.9347	14	40	
21	942	26	822	14	724	37	.9333	14	39	
22	968	26	808	13	761	37	.9319	13	38	
23	994	26	795	13	798	37	.9306	14	37	
24	.46020	26	782	14	835	37	.9292	14	36	**36** **26**
25	.46046	26	.88768	13	.51872	37	1.9278	14	35	1 0.6 0.4
26	072	25	755	14	909	37	.9265	13	34	2 1.2 0.9
27	097	26	741	13	946	37	.9251	14	33	3 1.8 1.3
28	123	26	728	13	983	37	.9237	14	32	4 2.4 1.7
29	149	26	715	14	.52020	37	.9223	14	31	5 3.0 2.2
		26				37		13		6 3.6 2.6
30	.46175	26	.88701	13	.52057	37	1.9210	14	30	7 4.2 3.0
31	201	25	688	14	094	37	.9196	13	29	8 4.8 3.5
32	226	26	674	13	131	37	.9183	14	28	9 5.4 3.9
33	252	26	661	14	168	37	.9169	14	27	10 6.0 4.3
34	278	26	647	13	205	37	.9155	13	26	20 12.0 8.7
		26				37				30 18.0 13.0
35	.46304	26	.88634	14	.52242	37	1.9142	14	25	40 24.0 17.3
36	330	25	620	13	279	37	.9128	13	24	50 30.0 21.7
37	355	26	607	14	316	37	.9115	14	23	
38	381	26	593	13	353	37	.9101	13	22	
39	407	26	580	14	390	37	.9088	13	21	
		26				37		14		
40	.46433	25	.88566	13	.52427	37	1.9074	13	20	
41	458	26	553	14	464	37	.9061	14	19	
42	484	26	539	13	501	37	.9047	13	18	**25** **15**
43	510	26	526	14	538	37	.9034	14	17	1 0.4 0.2
44	536	25	512	13	575	38	.9020	13	16	2 0.8 0.5
45	.46561	26	.88499	14	.52613	37	1.9007	14	15	3 1.2 0.8
46	587	26	485	13	650	37	.8993	13	14	4 1.7 1.0
47	613	26	472	14	687	37	.8980	13	13	5 2.1 1.2
48	639	25	458	13	724	37	.8967	14	12	6 2.5 1.5
49	664	26	445	14	761	37	.8953	13	11	7 2.9 1.8
										8 3.3 2.0
50	.46690	26	.88431	14	.52798	38	1.8940	13	10	9 3.8 2.2
51	716	26	417	13	836	37	.8927	14	9	10 4.2 2.5
52	742	25	404	14	873	37	.8913	13	8	20 8.3 5.0
53	767	26	390	13	910	37	.8900	13	7	30 12.5 7.5
54	793	26	377	14	947	38	.8887	13	6	40 16.7 10.0
		26						14		50 20.8 12.5
55	.46819	25	.88363	14	.52985	37	1.8873	13	5	
56	844	26	349	13	.53022	37	.8860	13	4	
57	870	26	336	14	059	37	.8847	13	3	
58	896	25	322	14	096	38	.8834	14	2	
59	921	26	308	13	134	37	.8820	13	1	
60	.46947		.88295		.53171		1.8807		0	
	Cos.	d.	Sin.	d.	Cot.	d.	Tan.	d.	′	P. P.

62°

28°

P. P.	′	Sin.	d.	Cos.	d.	Tan.	d.	Cot.	d.	
	0	.46947		.88295		.53171		1.8807		**60**
	1	973	26	281	14	208	37	.8794	13	59
	2	999	26	267	14	246	38	.8781	13	58
	3	.47024	25	254	13	283	37	.8768	13	57
	4	050	26	240	14	320	37	.8755	13	56
			26		14		38		14	
	5	.47076	25	.88226	13	.53358	37	1.8741	13	**55**
	6	101	26	213	14	395	37	.8728	13	54
	7	127	26	199	14	432	38	.8715	13	53
	8	153	25	185	13	470	37	.8702	13	52
	9	178	26	172	14	507	38	.8689	13	51
	10	.47204	25	.88158	14	.53545	37	1.8676	13	**50**
	11	229	26	144	14	582	38	.8663	13	49
	12	255	26	130	13	620	37	.8650	13	48
14 13	13	281	25	117	14	657	37	.8637	13	47
1 | 0.2 | 0.2	14	306	26	103	14	694	38	.8624	13	46
2 | 0.5 | 0.4	**15**	.47332	26	.88089	14	.53732	37	1.8611	13	**45**
3 | 0.7 | 0.6	16	358	25	075	13	769	38	.8598	13	44
4 | 0.9 | 0.9	17	383	26	062	14	807	37	.8585	13	43
5 | 1.2 | 1.1	18	409	25	048	14	844	38	.8572	13	42
6 | 1.4 | 1.3	19	434	26	034	14	882	38	.8559	13	41
7 | 1.6 | 1.5	**20**	.47460	26	.88020	14	.53920	37	1.8546	13	**40**
8 | 1.9 | 1.7	21	486	25	006	13	957	38	.8533	13	39
9 | 2.1 | 2.0	22	511	26	.87993	14	995	37	.8520	13	38
10 | 2.3 | 2.2	23	537	25	979	14	.54032	38	.8507	12	37
20 | 4.7 | 4.3	24	562	26	965	14	070	37	.8495	13	36
30 | 7.0 | 6.5	**25**	.47588	26	.87951	14	.54107	38	1.8482	13	**35**
40 | 9.3 | 8.7	26	614	25	937	14	145	38	.8469	13	34
50 | 11.7 | 10.8	27	639	26	923	14	183	37	.8456	13	33
	28	665	25	909	13	220	38	.8443	13	32
	29	690	26	896	14	258	38	.8430	12	31
	30	.47716	25	.87882	14	.54296	37	1.8418	13	**30**
	31	741	26	868	14	333	38	.8405	13	29
	32	767	26	854	14	371	38	.8392	13	28
	33	793	25	840	14	409	37	.8379	12	27
	34	818	26	826	14	446	38	.8367	13	26
	35	.47844	25	.87812	14	.54484	38	1.8354	13	**25**
12	36	869	26	798	14	522	38	.8341	12	24
1 | 0.2	37	895	25	784	14	560	38	.8329	13	23
2 | 0.4	38	920	26	770	14	597	38	.8316	13	22
3 | 0.6	39	946	25	756	13	635	38	.8303	12	21
4 | 0.8	**40**	.47971	26	.87743	14	.54673	38	1.8291	13	**20**
5 | 1.0	41	997	25	729	14	711	38	.8278	13	19
6 | 1.2	42	.48022	26	715	14	748	38	.8265	12	18
7 | 1.4	43	048	25	701	14	786	38	.8253	13	17
8 | 1.6	44	073	26	687	14	824	38	.8240	12	16
9 | 1.8	**45**	.48099	25	.87673	14	.54862	38	1.8228	13	**15**
10 | 2.0	46	124	26	659	14	900	38	.8215	13	14
20 | 4.0	47	150	25	645	14	938	37	.8202	12	13
30 | 6.0	48	175	26	631	14	975	38	.8190	13	12
40 | 8.0	49	201	25	617	14	.55013	38	.8177	12	11
50 | 10.0	**50**	.48226	26	.87603	14	.55051	38	1.8165	13	**10**
	51	252	25	589	14	089	38	.8152	12	9
	52	277	26	575	14	127	38	.8140	13	8
	53	303	25	561	15	165	38	.8127	12	7
	54	328	26	546	14	203	38	.8115	12	6
	55	.48354	25	.87532	14	.55241	38	1.8103	13	**5**
	56	379	26	518	14	279	38	.8090	12	4
	57	405	25	504	14	317	38	.8078	13	3
	58	430	26	490	14	355	38	.8065	12	2
	59	456	25	476	14	393	38	.8053	13	1
	60	.48481		.87462		.55431		1.8040		**0**
P. P.		**Cos.**	**d.**	**Sin.**	**d.**	**Cot.**	**d.**	**Tan.**	**d.**	**′**

61°

29°

′	Sin.	d.	Cos.	d.	Tan.	d.	Cot.	d.	′
0	.48481	25	.87462	14	.55431	38	1.8040	12	60
1	506	26	448	14	469	38	.8028	12	59
2	532	25	434	14	507	38	.8016	13	58
3	557	26	420	14	545	38	.8003	12	57
4	583	25	406	15	583	38	.7991	12	56
5	.48608	26	.87391	14	.55621	38	1.7979	13	55
6	634	25	377	14	659	38	.7966	12	54
7	659	25	363	14	697	39	.7954	12	53
8	684	26	349	14	736	38	.7942	12	52
9	710	25	335	14	774	38	.7930	13	51
10	.48735	26	.87321	15	.55812	38	1.7917	12	50
11	761	25	306	14	850	38	.7905	12	49
12	786	25	292	14	888	38	.7893	12	48
13	811	26	278	14	926	38	.7881	13	47
14	837	25	264	14	964	39	.7868	12	46
15	.48862	26	.87250	15	.56003	38	1.7856	12	45
16	888	25	235	14	041	38	.7844	12	44
17	913	25	221	14	079	38	.7832	12	43
18	938	26	207	14	117	39	.7820	12	42
19	964	25	193	15	156	38	.7808	12	41
20	.48989	25	.87178	14	.56194	38	1.7796	13	40
21	.49014	26	164	14	232	38	.7783	12	39
22	040	25	150	14	270	39	.7771	12	38
23	065	25	136	15	309	38	.7759	12	37
24	090	26	121	14	347	38	.7747	12	36
25	.49116	25	.87107	14	.56385	39	1.7735	12	35
26	141	25	093	14	424	38	.7723	12	34
27	166	26	079	15	462	38	.7711	12	33
28	192	25	064	14	501	38	.7699	12	32
29	217	25	050	14	539	38	.7687	12	31
30	.49242	26	.87036	15	.56577	39	1.7675	12	30
31	268	25	021	14	616	38	.7663	12	29
32	293	25	007	14	654	39	.7651	12	28
33	318	26	.86993	15	693	38	.7639	12	27
34	344	25	978	14	731	38	.7627	12	26
35	.49369	25	.86964	15	.56769	39	1.7615	12	25
36	394	25	949	14	808	38	.7603	12	24
37	419	26	935	14	846	39	.7591	12	23
38	445	25	921	15	885	38	.7579	12	22
39	470	25	906	14	923	39	.7567	11	21
40	.49495	26	.86892	14	.56962	38	1.7556	12	20
41	521	25	878	15	.57000	39	.7544	12	19
42	546	25	863	14	039	39	.7532	12	18
43	571	25	849	15	078	38	.7520	12	17
44	596	26	834	14	116	39	.7508	12	16
45	.49622	25	.86820	15	.57155	38	1.7496	11	15
46	647	25	805	14	193	39	.7485	12	14
47	672	26	791	14	232	39	.7473	12	13
48	697	25	777	15	271	38	.7461	12	12
49	723	25	762	14	309	39	.7449	12	11
50	.49748	25	.86748	15	.57348	38	1.7437	11	10
51	773	25	733	14	386	39	.7426	12	9
52	798	26	719	15	425	39	.7414	12	8
53	824	25	704	14	464	38	.7402	11	7
54	849	25	690	15	503	39	.7391	12	6
55	.49874	25	.86675	14	.57541	39	1.7379	12	5
56	899	25	661	15	580	39	.7367	12	4
57	924	26	646	14	619	38	.7355	11	3
58	950	25	632	15	657	39	.7344	12	2
59	975	25	617	14	696	39	.7332	11	1
60	.50000		.86603		.57735		1.7321		0
	Cos.	d.	Sin.	d.	Cot.	d.	Tan.	d.	′

60°

P. P.

	40	39
1	0.7	0.6
2	1.3	1.3
3	2.0	2.0
4	2.7	2.6
5	3.3	3.2
6	4.0	3.9
7	4.7	4.6
8	5.3	5.2
9	6.0	5.8
10	6.7	6.5
20	13.3	13.0
30	20.0	19.5
40	26.7	26.0
50	33.3	32.5

	38	26
1	0.6	0.4
2	1.3	0.9
3	1.9	1.3
4	2.5	1.7
5	3.2	2.2
6	3.8	2.6
7	4.4	3.0
8	5.1	3.5
9	5.7	3.9
10	6.3	4.3
20	12.7	8.7
30	19.0	13.0
40	25.3	17.3
50	31.7	21.7

	25	15
1	0.4	0.2
2	0.8	0.5
3	1.2	0.8
4	1.7	1.0
5	2.1	1.2
6	2.5	1.5
7	2.9	1.8
8	3.3	2.0
9	3.8	2.2
10	4.2	2.5
20	8.3	5.0
30	12.5	7.5
40	16.7	10.0
50	20.8	12.5

30°

'	Sin.	d.	Cos.	d.	Tan.	d.	Cot.	d.	
0	.50000	25	.86603	15	.57735	39	1.7321	12	60
1	025	25	588	15	774	39	.7309	12	59
2	050	26	573	14	813	38	.7297	11	58
3	076	25	559	15	851	39	.7286	12	57
4	101	25	544	14	890	39	.7274	12	56
5	.50126	25	.86530	15	.57929	39	1.7262	11	55
6	151	25	515	14	968	39	.7251	12	54
7	176	25	501	15	.58007	39	.7239	11	53
8	201	26	486	15	046	39	.7228	12	52
9	227	25	471	14	085	39	.7216	11	51
10	.50252	25	.86457	15	.58124	38	1.7205	12	50
11	277	25	442	15	162	39	.7193	11	49
12	302	25	427	14	201	39	.7182	12	48
13	327	25	413	15	240	39	.7170	11	47
14	352	25	398	14	279	39	.7159	12	46
15	.50377	26	.86384	15	.58318	39	1.7147	11	45
16	403	25	369	15	357	39	.7136	12	44
17	428	25	354	14	396	39	.7124	11	43
18	453	25	340	15	435	39	.7113	11	42
19	478	25	325	15	474	39	.7102	12	41
20	.50503	25	.86310	15	.58513	39	1.7090	11	40
21	528	25	295	14	552	39	.7079	12	39
22	553	25	281	15	591	40	.7067	11	38
23	578	25	266	15	631	39	.7056	11	37
24	603	25	251	14	670	39	.7045	12	36
25	.50628	26	.86237	15	.58709	39	1.7033	11	35
26	654	25	222	15	748	39	.7022	11	34
27	679	25	207	15	787	39	.7011	12	33
28	704	25	192	14	826	39	.6999	11	32
29	729	25	178	15	865	40	.6988	11	31
30	.50754	25	.86163	15	.58905	39	1.6977	12	30
31	779	25	148	15	944	39	.6965	11	29
32	804	25	133	14	983	39	.6954	11	28
33	829	25	119	15	.59022	39	.6943	11	27
34	854	25	104	15	061	40	.6932	12	26
35	.50879	25	.86089	15	.59101	39	1.6920	11	25
36	904	25	074	15	140	39	.6909	11	24
37	929	25	059	14	179	39	.6898	11	23
38	954	25	045	15	218	40	.6887	12	22
39	979	25	030	15	258	39	.6875	11	21
40	.51004	25	.86015	15	.59297	39	1.6864	11	20
41	029	25	000	15	336	40	.6853	11	19
42	054	25	.85985	15	376	39	.6842	11	18
43	079	25	970	14	415	39	.6831	11	17
44	104	25	956	15	454	40	.6820	12	16
45	.51129	25	.85941	15	.59494	39	1.6808	11	15
46	154	25	926	15	533	40	.6797	11	14
47	179	25	911	15	573	39	.6786	11	13
48	204	25	896	15	612	39	.6775	11	12
49	229	25	881	15	651	40	.6764	11	11
50	.51254	25	.85866	15	.59691	39	1.6753	11	10
51	279	25	851	15	730	40	.6742	11	9
52	304	25	836	15	770	39	.6731	11	8
53	329	25	821	15	809	40	.6720	11	7
54	354	25	806	14	849	39	.6709	11	6
55	.51379	25	.85792	15	.59888	40	1.6698	11	5
56	404	25	777	15	928	39	.6687	11	4
57	429	25	762	15	967	40	.6676	11	3
58	454	25	747	15	.60007	39	.6665	11	2
59	479	25	732	15	046	40	.6654	11	1
60	.51504		.85717		.60086		1.6643		0

P. P.

	14	13
1	0.2	0.2
2	0.5	0.4
3	0.7	0.6
4	0.9	0.9
5	1.2	1.1
6	1.4	1.3
7	1.6	1.5
8	1.9	1.7
9	2.1	2.0
10	2.3	2.2
20	4.7	4.3
30	7.0	6.5
40	9.3	8.7
50	11.7	10.8

	12	11
1	0.2	0.2
2	0.4	0.4
3	0.6	0.6
4	0.8	0.7
5	1.0	0.9
6	1.2	1.1
7	1.4	1.3
8	1.6	1.5
9	1.8	1.6
10	2.0	1.8
20	4.0	3.7
30	6.0	5.5
40	8.0	7.3
50	10.0	9.2

P. P.		Cos.	d.	Sin.	d.	Cot.	d.	Tan.	d.	'

59°

31°

′	Sin.	d.	Cos.	d.	Tan.	d.	Cot.	d.	′
0	.51504	25	.85717	15	.60086	40	1.6643	11	60
1	529	25	702	15	126	39	.6632	11	59
2	554	25	687	15	165	40	.6621	11	58
3	579	25	672	15	205	40	.6610	11	57
4	604	24	657	15	245	39	.6599	11	56
5	.51628	25	.85642	15	.60284	40	1.6588	11	55
6	653	25	627	15	324	40	.6577	11	54
7	678	25	612	15	364	39	.6566	11	53
8	703	25	597	15	403	40	.6555	10	52
9	728	25	582	15	443	40	.6545	11	51
10	.51753	25	.85567	16	.60483	39	1.6534	11	50
11	778	25	551	15	522	40	.6523	11	49
12	803	25	536	15	562	40	.6512	11	48
13	828	24	521	15	602	40	.6501	11	47
14	852	25	506	15	642	39	.6490	11	46
15	.51877	25	.85491	15	.60681	40	1.6479	10	45
16	902	25	476	15	721	40	.6469	11	44
17	927	25	461	15	761	40	.6458	11	43
18	952	25	446	15	801	40	.6447	11	42
19	977	25	431	15	841	40	.6436	10	41
20	.52002	24	.85416	15	.60881	40	1.6426	11	40
21	026	25	401	16	921	39	.6415	11	39
22	051	25	385	15	960	40	.6404	11	38
23	076	25	370	15	.61000	40	.6393	10	37
24	101	25	355	15	040	40	.6383	11	36
25	.52126	25	.85340	15	.61080	40	1.6372	11	35
26	151	24	325	15	120	40	.6361	10	34
27	175	25	310	16	160	40	.6351	11	33
28	200	25	294	15	200	40	.6340	11	32
29	225	25	279	15	240	40	.6329	10	31
30	.52250	25	.85264	15	.61280	40	1.6319	11	30
31	275	24	249	15	320	40	.6308	11	29
32	299	25	234	16	360	40	.6297	10	28
33	324	25	218	15	400	40	.6287	11	27
34	349	25	203	15	440	40	.6276	11	26
35	.52374	25	.85188	15	.61480	40	1.6265	10	25
36	399	24	173	16	520	41	.6255	11	24
37	423	25	157	15	561	40	.6244	10	23
38	448	25	142	15	601	40	.6234	11	22
39	473	25	127	15	641	40	.6223	11	21
40	.52498	24	.85112	16	.61681	40	1.6212	10	20
41	522	25	096	15	721	40	.6202	11	19
42	547	25	081	15	761	40	.6191	10	18
43	572	25	066	15	801	41	.6181	11	17
44	597	24	051	16	842	40	.6170	10	16
45	.52621	25	.85035	15	.61882	40	1.6160	11	15
46	646	25	020	15	922	40	.6149	10	14
47	671	25	005	16	962	41	.6139	11	13
48	696	24	.84989	15	.62003	40	.6128	10	12
49	720	25	974	15	043	40	.6118	11	11
50	.52745	25	.84959	16	.62083	41	1.6107	10	10
51	770	24	943	15	124	40	.6097	10	9
52	794	25	928	15	164	40	.6087	11	8
53	819	25	913	16	204	41	.6076	10	7
54	844	25	897	15	245	40	.6066	11	6
55	.52869	24	.84882	16	.62285	40	1.6055	10	5
56	893	25	866	15	325	41	.6045	11	4
57	918	25	851	15	366	40	.6034	10	3
58	943	24	836	16	406	40	.6024	10	2
59	967	25	820	15	446	41	.6014	11	1
60	.52992		.84805		.62487		1.6003		0

| | Cos. | d. | Sin. | d. | Cot. | d. | Tan. | d. | ′ |

58°

P. P.

	42	41
1	0.7	0.7
2	1.4	1.4
3	2.1	2.0
4	2.8	2.7
5	3.5	3.4
6	4.2	4.1
7	4.9	4.8
8	5.6	5.5
9	6.3	6.2
10	7.0	6.8
20	14.0	13.7
30	21.0	20.5
40	28.0	27.3
50	35.0	34.2

	40	39
1	0.7	0.6
2	1.3	1.3
3	2.0	2.0
4	2.7	2.6
5	3.3	3.2
6	4.0	3.9
7	4.7	4.6
8	5.3	5.2
9	6.0	5.8
10	6.7	6.5
20	13.3	13.0
30	20.0	19.5
40	26.7	26.0
50	33.3	32.5

	25	24
1	0.4	0.4
2	0.8	0.8
3	1.2	1.2
4	1.7	1.6
5	2.1	2.0
6	2.5	2.4
7	2.9	2.8
8	3.3	3.2
9	3.8	3.6
10	4.2	4.0
20	8.3	8.0
30	12.5	12.0
40	16.7	16.0
50	20.8	20.0

32°

P. P.	′	Sin.	d.	Cos.	d.	Tan.	d.	Cot.	d.	
	0	.52992	25	.84805	16	.62487	40	1.6003	10	60
	1	.53017	24	789	15	527	41	.5993	10	59
	2	041	25	774	15	568	40	.5983	11	58
	3	066	25	759	16	608	41	.5972	10	57
	4	091	24	743	15	649	40	.5962	10	56
	5	.53115	25	.84728	16	.62689	41	1.5952	11	55
	6	140	24	712	15	730	40	.5941	10	54
	7	164	25	697	16	770	41	.5931	10	53
	8	189	25	681	15	811	41	.5921	10	52
	9	214	24	666	16	852	40	.5911	11	51
	10	.53238	25	.84650	15	.62892	41	1.5900	10	50
	11	263	25	635	16	933	40	.5890	10	49
	12	288	24	619	15	973	41	.5880	11	48
	13	312	25	604	16	.63014	41	.5869	10	47
	14	337	24	588	15	055	40	.5859	10	46
	15	.53361	25	.84573	16	.63095	41	1.5849	10	45
	16	386	25	557	15	136	41	.5839	10	44
	17	411	24	542	16	177	40	.5829	11	43
	18	435	25	526	15	217	41	.5818	10	42
	19	460	24	511	16	258	41	.5808	10	41
	20	.53484	25	.84495	15	.63299	41	1.5798	10	40
	21	509	25	480	16	340	40	.5788	10	39
	22	534	24	464	16	380	41	.5778	10	38
	23	558	25	448	15	421	41	.5768	11	37
	24	583	24	433	16	462	41	.5757	10	36
	25	.53607	25	.84417	15	.63503	41	1.5747	10	35
	26	632	24	402	16	544	40	.5737	10	34
	27	656	25	386	16	584	41	.5727	10	33
	28	681	24	370	15	625	41	.5717	10	32
	29	705	25	355	16	666	41	.5707	10	31
	30	.53730	24	.84339	15	.63707	41	1.5697	10	30
	31	754	25	324	16	748	41	.5687	10	29
	32	779	25	308	16	789	41	.5677	10	28
	33	804	24	292	15	830	41	.5667	10	27
	34	828	25	277	16	871	41	.5657	10	26
	35	.53853	24	.84261	16	.63912	41	1.5647	10	25
	36	877	25	245	15	953	41	.5637	10	24
	37	902	24	230	16	994	41	.5627	10	23
	38	926	25	214	16	.64035	41	.5617	10	22
	39	951	24	198	16	076	41	.5607	10	21
	40	.53975	25	.84182	15	.64117	41	1.5597	10	20
	41	.54000	24	167	16	158	41	.5587	10	19
	42	024	25	151	16	199	41	.5577	10	18
	43	049	24	135	15	240	41	.5567	10	17
	44	073	24	120	16	281	41	.5557	10	16
	45	.54097	25	.84104	16	.64322	41	1.5547	10	15
	46	122	24	088	16	363	41	.5537	10	14
	47	146	25	072	15	404	42	.5527	10	13
	48	171	24	057	16	446	41	.5517	10	12
	49	195	25	041	16	487	41	.5507	10	11
	50	.54220	24	.84025	16	.64528	41	1.5497	10	10
	51	244	25	009	15	569	41	.5487	10	9
	52	269	24	.83994	16	610	42	.5477	9	8
	53	293	24	978	16	652	41	.5468	10	7
	54	317	25	962	16	693	41	.5458	10	6
	55	.54342	24	.83946	16	.64734	41	1.5448	10	5
	56	366	25	930	15	775	42	.5438	10	4
	57	391	24	915	16	817	41	.5428	10	3
	58	415	25	899	16	858	41	.5418	10	2
	59	440	24	883	16	899	42	.5408	9	1
	60	.54464		.83867		.64941		1.5399		0

| P. P. | | Cos. | d. | Sin. | d. | Cot. | d. | Tan. | d. | ′ |

57°

P. P. tables:

	16	15
1	0.3	0.2
2	0.5	0.5
3	0.8	0.8
4	1.1	1.0
5	1.3	1.2
6	1.6	1.5
7	1.9	1.8
8	2.1	2.0
9	2.4	2.2
10	2.7	2.5
20	5.3	5.0
30	8.0	7.5
40	10.7	10.0
50	13.3	12.5

	11	10
1	0.2	0.2
2	0.4	0.3
3	0.6	0.5
4	0.7	0.7
5	0.9	0.8
6	1.1	1.0
7	1.3	1.2
8	1.5	1.3
9	1.6	1.5
10	1.8	1.7
20	3.7	3.3
30	5.5	5.0
40	7.3	6.7
50	9.2	8.3

	9
1	0.2
2	0.3
3	0.4
4	0.6
5	0.8
6	0.9
7	1.0
8	1.2
9	1.4
10	1.5
20	3.0
30	4.5
40	6.0
50	7.5

33°

′	Sin.	d.	Cos.	d.	Tan.	d.	Cot.	d.	
0	.54464		.83867		.64941		1.5399		60
1	488	24	851	16	982	41	.5389	10	59
2	513	25	835	16	.65024	42	.5379	10	58
3	537	24	819	16	065	41	.5369	10	57
4	561	24	804	15	106	41	.5359	10	56
		25		16		42		9	
5	.54586		.83788		.65148		1.5350		55
6	610	24	772	16	189	41	.5340	10	54
7	635	25	756	16	231	42	.5330	10	53
8	659	24	740	16	272	41	.5320	10	52
9	683	24	724	16	314	42	.5311	9	51
		25		16		41		10	
10	.54708		.83708		.65355		1.5301		50
11	732	24	692	16	397	42	.5291	10	49
12	756	24	676	16	438	41	.5282	9	48
13	781	25	660	16	480	42	.5272	10	47
14	805	24	645	15	521	41	.5262	10	46
		24		16		42		9	
15	.54829		.83629		.65563		1.5253		45
16	854	25	613	16	604	41	.5243	10	44
17	878	24	597	16	646	42	.5233	10	43
18	902	24	581	16	688	42	.5224	9	42
19	927	25	565	16	729	41	.5214	10	41
		24		16		42		10	
20	.54951		.83549		.65771		1.5204		40
21	975	24	533	16	813	42	.5195	9	39
22	999	25	517	16	854	41	.5185	10	38
23	.55024	24	501	16	896	42	.5175	9	37
24	048	24	485	16	938	42	.5166	10	36
		25		16		42		10	
25	.55072		.83469		.65980		1.5156		35
26	097	24	453	16	.66021	41	.5147	10	34
27	121	24	437	16	063	42	.5137	10	33
28	145	24	421	16	105	42	.5127	9	32
29	169	25	405	16	147	42	.5118	10	31
		24		16		41			
30	.55194		.83389		.66189		1.5108		30
31	218	24	373	16	230	41	.5099	10	29
32	242	24	356	17	272	42	.5089	9	28
33	266	25	340	16	314	42	.5080	10	27
34	291	24	324	16	356	42	.5070	9	26
35	.55315		.83308		.66398		1.5061		25
36	339	24	292	16	440	42	.5051	10	24
37	363	24	276	16	482	42	.5042	9	23
38	388	25	260	16	524	42	.5032	10	22
39	412	24	244	16	566	42	.5023	9	21
40	.55436		.83228		.66608		1.5013		20
41	460	24	212	16	650	42	.5004	9	19
42	484	24	195	17	692	42	.4994	10	18
43	509	25	179	16	734	42	.4985	9	17
44	533	24	163	16	776	42	.4975	10	16
		24		16		42		9	
45	.55557		.83147		.66818		1.4966		15
46	581	24	131	16	860	42	.4957	9	14
47	605	24	115	16	902	42	.4947	10	13
48	630	25	098	17	944	42	.4938	9	12
49	654	24	082	16	986	42	.4928	10	11
		24		16		42		9	
50	.55678		.83066		.67028		1.4919		10
51	702	24	050	16	071	43	.4910	9	9
52	726	24	034	16	113	42	.4900	10	8
53	750	24	017	17	155	42	.4891	9	7
54	775	25	001	16	197	42	.4882	9	6
		24		16		42		10	
55	.55799		.82985		.67239		1.4872		5
56	823	24	969	16	282	43	.4863	9	4
57	847	24	953	16	324	42	.4854	9	3
58	871	24	936	17	366	42	.4844	10	2
59	895	24	920	16	409	43	.4835	9	1
		24		16		42		9	
60	.55919		.82904		.67451		1.4826		0
	Cos.	d.	Sin.	d.	Cot.	d.	Tan.	d.	′

P. P.

	44	43
1	0.7	0.7
2	1.5	1.4
3	2.2	2.2
4	2.9	2.9
5	3.7	3.6
6	4.4	4.3
7	5.1	5.0
8	5.9	5.7
9	6.6	6.4
10	7.3	7.2
20	14.7	14.3
30	22.0	21.5
40	29.3	28.7
50	36.7	35.8

	42	41
1	0.7	0.7
2	1.4	1.4
3	2.1	2.0
4	2.8	2.7
5	3.5	3.4
6	4.2	4.1
7	4.9	4.8
8	5.6	5.5
9	6.3	6.2
10	7.0	6.8
20	14.0	13.7
30	21.0	20.5
40	28.0	27.3
50	35.0	34.2

	25	24
1	0.4	0.4
2	0.8	0.8
3	1.2	1.2
4	1.7	1.6
5	2.1	2.0
6	2.5	2.4
7	2.9	2.8
8	3.3	3.2
9	3.8	3.6
10	4.2	4.0
20	8.3	8.0
30	12.5	12.0
40	16.7	16.0
50	20.8	20.0

	23	17
1	0.4	0.3
2	0.8	0.6
3	1.2	0.8
4	1.5	1.1
5	1.9	1.4
6	2.3	1.7
7	2.7	2.0
8	3.1	2.3
9	3.4	2.6
10	3.8	2.8
20	7.7	5.7
30	11.5	8.5
40	15.3	11.3
50	19.2	14.2

56°

34°

'	Sin.	d.	Cos.	d.	Tan.	d.	Cot.	d.	
0	.55919	24	.82904	17	.67451	42	1.4826	10	60
1	943	25	887	16	493	43	.4816	9	59
2	968	24	871	16	536	42	.4807	9	58
3	992	24	855	16	578	42	.4798	10	57
4	.56016	24	839	17	620	43	.4788	9	56
5	.56040	24	.82822	16	.67663	42	1.4779	9	55
6	064	24	806	16	705	43	.4770	9	54
7	088	24	790	17	748	42	.4761	10	53
8	112	24	773	16	790	42	.4751	9	52
9	136	24	757	16	832	43	.4742	9	51
10	.56160	24	.82741	17	.67875	42	1.4733	9	50
11	184	24	724	16	917	43	.4724	9	49
12	208	24	708	16	960	42	.4715	10	48
13	232	24	692	17	.68002	43	.4705	9	47
14	256	24	675	16	045	43	.4696	9	46
15	.56280	25	.82659	16	.68088	42	1.4687	9	45
16	305	24	643	17	130	43	.4678	9	44
17	329	24	626	16	173	43	.4669	10	43
18	353	24	610	17	215	43	.4659	9	42
19	377	24	593	16	258	43	.4650	9	41
20	.56401	24	.82577	16	.68301	42	1.4641	9	40
21	425	24	561	17	343	43	.4632	9	39
22	449	24	544	16	386	43	.4623	9	38
23	473	24	528	17	429	42	.4614	9	37
24	497	24	511	16	471	43	.4605	9	36
25	.56521	24	.82495	17	.68514	43	1.4596	10	35
26	545	24	478	16	557	43	.4586	9	34
27	569	24	462	16	600	42	.4577	9	33
28	593	24	446	17	642	43	.4568	9	32
29	617	24	429	16	685	43	.4559	9	31
30	.56641	24	.82413	17	.68728	43	1.4550	9	30
31	665	24	396	16	771	43	.4541	9	29
32	689	24	380	17	814	43	.4532	9	28
33	713	23	363	16	857	43	.4523	9	27
34	736	24	347	17	900	42	.4514	9	26
35	.56760	24	.82330	16	.68942	43	1.4505	9	25
36	784	24	314	17	985	43	.4496	9	24
37	808	24	297	16	.69028	43	.4487	9	23
38	832	24	281	17	071	43	.4478	9	22
39	856	24	264	16	114	43	.4469	9	21
40	.56880	24	.82248	17	.69157	43	1.4460	9	20
41	904	24	231	17	200	43	.4451	9	19
42	928	24	214	16	243	43	.4442	9	18
43	952	24	198	17	286	43	.4433	9	17
44	976	24	181	16	329	43	.4424	9	16
45	.57000	24	.82165	17	.69372	44	1.4415	9	15
46	024	23	148	16	416	43	.4406	9	14
47	047	24	132	17	459	43	.4397	9	13
48	071	24	115	17	502	43	.4388	9	12
49	095	24	098	16	545	43	.4379	9	11
50	.57119	24	.82082	17	.69588	43	1.4370	9	10
51	143	24	065	17	631	44	.4361	9	9
52	167	24	048	16	675	44	.4352	8	8
53	191	24	032	17	718	43	.4344	9	7
54	215	23	015	16	761	43	.4335	9	6
55	.57238	24	.81999	17	.69804	43	1.4326	9	5
56	262	24	982	17	847	44	.4317	9	4
57	286	24	965	16	891	43	.4308	9	3
58	310	24	949	17	934	43	.4299	9	2
59	334	24	932	17	977	44	.4290	9	1
60	.57358		.81915		.70021		1.4281		0

| P. P. | | Cos. | d. | Sin. | d. | Cot. | d. | Tan. | d. | ' |

55°

P. P.

	16	15
1	0.3	0.2
2	0.5	0.5
3	0.8	0.8
4	1.1	1.0
5	1.3	1.2
6	1.6	1.5
7	1.9	1.8
8	2.1	2.0
9	2.4	2.2
10	2.7	2.5
20	5.3	5.0
30	8.0	7.5
40	10.7	10.0
50	13.3	12.5

	10	9
1	0.2	0.2
2	0.3	0.3
3	0.5	0.4
4	0.7	0.6
5	0.8	0.8
6	1.0	0.9
7	1.2	1.0
8	1.3	1.2
9	1.5	1.4
10	1.7	1.5
20	3.3	3.0
30	5.0	4.5
40	6.7	6.0
50	8.3	7.5

	8
1	0.1
2	0.3
3	0.4
4	0.5
5	0.7
6	0.8
7	0.9
8	1.1
9	1.2
10	1.3
20	2.7
30	4.0
40	5.3
50	6.7

35°

′	Sin.	d.	Cos.	d.	Tan.	d.	Cot.	d.	
0	.57358		.81915		.70021		1.4281		**60**
1	381	23	899	16	064	43	.4273	8	59
2	405	24	882	17	107	43	.4264	9	58
3	429	24	865	17	151	44	.4255	9	57
4	453	24	848	17	194	43	.4246	9	56
		24		16		44		9	
5	.57477		.81832		.70238		1.4237		**55**
6	501	24	815	17	281	43	.4229	8	54
7	524	23	798	17	325	44	.4220	9	53
8	548	24	782	16	368	43	.4211	9	52
9	572	24	765	17	412	44	.4202	9	51
		24		17		43		9	
10	.57596		.81748		.70455		1.4193		**50**
11	619	23	731	17	499	44	.4185	8	49
12	643	24	714	17	542	43	.4176	9	48
13	667	24	698	16	586	44	.4167	9	47
14	691	24	681	17	629	43	.4158	9	46
		24		17		44		8	
15	.57715		.81664		.70673		1.4150		**45**
16	738	23	647	17	717	44	.4141	9	44
17	762	24	631	16	760	43	.4132	9	43
18	786	24	614	17	804	44	.4124	8	42
19	810	24	597	17	848	44	.4115	9	41
		23		17		43		9	
20	.57833		.81580		.70891		1.4106		**40**
21	857	24	563	17	935	44	.4097	9	39
22	881	24	546	17	979	44	.4089	8	38
23	904	23	530	16	.71023	44	.4080	9	37
24	928	24	513	17	066	43	.4071	9	36
		24		17		44		8	
25	.57952		.81496		.71110		1.4063		**35**
26	976	24	479	17	154	44	.4054	9	34
27	999	23	462	17	198	44	.4045	9	33
28	.58023	24	445	17	242	44	.4037	8	32
29	047	24	428	17	285	43	.4028	9	31
		23		16		44			
30	.58070		.81412		.71329		1.4019		**30**
31	094	24	395	17	373	44	.4011	8	29
32	118	24	378	17	417	44	.4002	9	28
33	141	23	361	17	461	44	.3994	8	27
34	165	24	344	17	505	44	.3985	9	26
		24		17		44		9	
35	.58189		.81327		.71549		1.3976		**25**
36	212	23	310	17	593	44	.3968	8	24
37	236	24	293	17	637	44	.3959	9	23
38	260	24	276	17	681	44	.3951	8	22
39	283	23	259	17	725	44	.3942	9	21
		24		17		44		8	
40	.58307		.81242		.71769		1.3934		**20**
41	330	23	225	17	813	44	.3925	9	19
42	354	24	208	17	857	44	.3916	9	18
43	378	24	191	17	901	44	.3908	8	17
44	401	23	174	17	946	45	.3899	9	16
		24		17		44		8	
45	.58425		.81157		.71990		1.3891		**15**
46	449	24	140	17	.72034	44	.3882	9	14
47	472	23	123	17	078	44	.3874	8	13
48	496	24	106	17	122	44	.3865	9	12
49	519	23	089	17	167	45	.3857	8	11
		24		17		44		9	
50	.58543		.81072		.72211		1.3848		**10**
51	567	24	055	17	255	44	.3840	8	9
52	590	23	038	17	299	44	.3831	9	8
53	614	24	021	17	344	45	.3823	8	7
54	637	23	004	17	388	44	.3814	9	6
		24		17		44		8	
55	.58661		.80987		.72432		1.3806		**5**
56	684	23	970	17	477	45	.3798	8	4
57	708	24	953	17	521	44	.3789	9	3
58	731	23	936	17	565	44	.3781	8	2
59	755	24	919	17	610	45	.3772	9	1
		24		17		44		8	
60	.58779		.80902		.72654		1.3764		**0**
	Cos.	d.	Sin.	d.	Cot.	d.	Tan.	d.	′

P. P.

	46	45
1	0.8	0.8
2	1.5	1.5
3	2.3	2.2
4	3.1	3.0
5	3.8	3.8
6	4.6	4.5
7	5.4	5.2
8	6.1	6.0
9	6.9	6.8
10	7.7	7.5
20	15.3	15.0
30	23.0	22.5
40	30.7	30.0
50	38.3	37.5

	44	43
1	0.7	0.7
2	1.5	1.4
3	2.2	2.2
4	2.9	2.9
5	3.7	3.6
6	4.4	4.3
7	5.1	5.0
8	5.9	5.7
9	6.6	6.4
10	7.3	7.2
20	14.7	14.3
30	22.0	21.5
40	29.3	28.7
50	36.7	35.8

	24	23
1	0.4	0.4
2	0.8	0.8
3	1.2	1.2
4	1.6	1.5
5	2.0	1.9
6	2.4	2.3
7	2.8	2.7
8	3.2	3.1
9	3.6	3.4
10	4.0	3.8
20	8.0	7.7
30	12.0	11.5
40	16.0	15.3
50	20.0	19.2

36°

P. P.	'	Sin.	d.	Cos.	d.	Tan.	d.	Cot.	d.	
	0	.58779		.80902		.72654		1.3764		60
	1	802	23	885	17	699	45	.3755	9	59
	2	826	24	867	18	743	44	.3747	8	58
	3	849	23	850	17	788	45	.3739	8	57
	4	873	24	833	17	832	44	.3730	9	56
			23		17		45		8	
	5	.58896		.80816		.72877		1.3722		55
	6	920	24	799	17	921	44	.3713	9	54
	7	943	23	782	17	966	45	.3705	8	53
	8	967	24	765	17	.73010	44	.3697	8	52
	9	990	23	748	17	055	45	.3688	9	51
			24		18		45		8	
	10	.59014		.80730		.73100		1.3680		50
	11	037	23	713	17	144	44	.3672	8	49
	12	061	24	696	17	189	45	.3663	9	48
	13	084	23	679	17	234	45	.3655	8	47
	14	108	24	662	17	278	44	.3647	8	46
			23		18		45		9	
	15	.59131		.80644		.73323		1.3638		45
	16	154	23	627	17	368	45	.3630	8	44
	17	178	24	610	17	413	45	.3622	8	43
	18	201	23	593	17	457	44	.3613	9	42
	19	225	24	576	17	502	45	.3605	8	41
			23		18		45		8	
	20	.59248		.80558		.73547		1.3597		40
	21	272	24	541	17	592	45	.3588	9	39
	22	295	23	524	17	637	45	.3580	8	38
	23	318	23	507	17	681	44	.3572	8	37
	24	342	24	489	18	726	45	.3564	8	36
			23		17		45		9	
	25	.59365		.80472		.73771		1.3555		35
	26	389	24	455	17	816	45	.3547	8	34
	27	412	23	438	17	861	45	.3539	8	33
	28	436	24	420	18	906	45	.3531	8	32
	29	459	23	403	17	951	45	.3522	9	31
			23		17		45		8	
	30	.59482		.80386		.73996		1.3514		30
	31	506	24	368	18	.74041	45	.3506	8	29
	32	529	23	351	17	086	45	.3498	8	28
	33	552	23	334	17	131	45	.3490	8	27
	34	576	24	316	18	176	45	.3481	9	26
			23		17		45		8	
	35	.59599		.80299		.74221		1.3473		25
	36	622	23	282	17	267	46	.3465	8	24
	37	646	24	264	18	312	45	.3457	8	23
	38	669	23	247	17	357	45	.3449	8	22
	39	693	24	230	17	402	45	.3440	9	21
			23		18		45		8	
	40	.59716		.80212		.74447		1.3432		20
	41	739	23	195	17	492	45	.3424	8	19
	42	763	24	178	17	538	46	.3416	8	18
	43	786	23	160	18	583	45	.3408	8	17
	44	809	23	143	17	628	45	.3400	8	16
			23		18		46			
	45	.59832		.80125		.74674		1.3392		15
	46	856	24	108	17	719	45	.3384	8	14
	47	879	23	091	17	764	45	.3375	9	13
	48	902	23	073	18	810	46	.3367	8	12
	49	926	24	056	17	855	45	.3359	8	11
			23		18		45		8	
	50	.59949		.80038		.74900		1.3351		10
	51	972	23	021	17	946	46	.3343	8	9
	52	995	23	003	18	991	45	.3335	8	8
	53	.60019	24	.79986	17	.75037	46	.3327	8	7
	54	042	23	968	18	082	45	.3319	8	6
			23		17		46			
	55	.60065		.79951		.75128		1.3311		5
	56	089	24	934	17	173	45	.3303	8	4
	57	112	23	916	18	219	46	.3295	8	3
	58	135	23	899	17	264	45	.3287	8	2
	59	158	23	881	18	310	46	.3278	9	1
			24		17		45			
	60	.60182		.79864		.75355		1.3270		0
P. P.		Cos.	d.	Sin.	d.	Cot.	d.	Tan.	d.	'

53°

P. P.

	18	17
1	0.3	0.3
2	0.6	0.6
3	0.9	0.8
4	1.2	1.1
5	1.5	1.4
6	1.8	1.7
7	2.1	2.0
8	2.4	2.3
9	2.7	2.6
10	3.0	2.8
20	6.0	5.7
30	9.0	8.5
40	12.0	11.3
50	15.0	14.2

	16	9
1	0.3	0.2
2	0.5	0.3
3	0.8	0.4
4	1.1	0.6
5	1.3	0.8
6	1.6	0.9
7	1.9	1.0
8	2.1	1.2
9	2.4	1.4
10	2.7	1.5
20	5.3	3.0
30	8.0	4.5
40	10.7	6.0
50	13.3	7.5

	8
1	0.1
2	0.3
3	0.4
4	0.5
5	0.7
6	0.8
7	0.9
8	1.1
9	1.2
10	1.3
20	2.7
30	4.0
40	5.3
50	6.7

	Sin.	d.	Cos.	d.	Tan.	d.	Cot.	d.		P. P.	
0	.60182	23	.79864	18	.75355	46	1.3270	8	**60**		
1	205	23	846	17	401	46	.3262	8	59		
2	228	23	829	18	447	45	.3254	8	58		
3	251	23	811	18	492	46	.3246	8	57		
4	274	24	793	17	538	46	.3238	8	56	**48**	**47**
5	.60298	23	.79776	18	.75584	45	1.3230	8	**55**	1 0.8	0.8
6	321	23	758	17	629	46	.3222	8	54	2 1.6	1.6
7	344	23	741	18	675	46	.3214	8	53	3 2.4	2.4
8	367	23	723	17	721	46	.3206	8	52	4 3.2	3.1
9	390	24	706	18	767	45	.3198	8	51	5 4.0	3.9
										6 4.8	4.7
10	.60414	23	.79688	17	.75812	46	1.3190	8	**50**	7 5.6	5.5
11	437	23	671	18	858	46	.3182	7	49	8 6.4	6.3
12	460	23	653	18	904	46	.3175	8	48	9 7.2	7.0
13	483	23	635	17	950	46	.3167	8	47	10 8.0	7.8
14	506	23	618	18	996	46	.3159	8	46	20 16.0	15.7
										30 24.0	23.5
15	.60529	24	.79600	17	.76042	46	1.3151	8	**45**	40 32.0	31.3
16	553	23	583	18	088	46	.3143	8	44	50 40.0	39.2
17	576	23	565	18	134	46	.3135	8	43		
18	599	23	547	17	180	46	.3127	8	42		
19	622	23	530	18	226	46	.3119	8	41		
20	.60645	23	.79512	18	.76272	46	1.3111	8	**40**		
21	668	23	494	17	318	46	.3103	8	39		
22	691	23	477	18	364	46	.3095	8	38		
23	714	24	459	18	410	46	.3087	8	37		
24	738	23	441	17	456	46	.3079	7	36	**46**	**45**
25	.60761	23	.79424	18	.76502	46	1.3072	8	**35**	1 0.8	0.8
26	784	23	406	18	548	46	.3064	8	34	2 1.5	1.5
27	807	23	388	17	594	46	.3056	8	33	3 2.3	2.2
28	830	23	371	18	640	46	.3048	8	32	4 3.1	3.0
29	853	23	353	18	686	47	·.3040	8	31	5 3.8	3.8
										6 4.6	4.5
30	.60876	23	.79335	17	.76733	46	1.3032	8	**30**	7 5.4	5.2
31	899	23	318	18	779	46	.3024	7	29	8 6.1	6.0
32	922	23	300	18	825	46	.3017	8	28	9 6.9	6.8
33	945	23	282	18	871	47	.3009	8	27	10 7.7	7.5
34	968	23	264	17	918	46	.3001	8	26	20 15.3	15.0
										30 23.0	22.5
35	.60991	24	.79247	18	.76964	46	1.2993	8	**25**	40 30.7	30.0
36	.61015	23	229	18	.77010	47	.2985	8	24	50 38.3	37.5
37	038	23	211	18	057	46	.2977	8	23		
38	061	23	193	17	103	46	.2970	7	22		
39	084	23	176	18	149	47	.2962	8	21		
40	.61107	23	.79158	18	.77196	46	1.2954	8	**20**		
41	130	23	140	18	242	47	.2946	8	19		
42	153	23	122	17	289	46	.2938	7	18		
43	176	23	105	18	335	47	.2931	8	17		
44	199	23	087	18	382	46	.2923	8	16	**24**	**23**
45	.61222	23	.79069	18	.77428	47	1.2915	8	**15**	1 0.4	0.4
46	245	23	051	18	475	46	.2907	7	14	2 0.8	0.8
47	268	23	033	17	521	47	.2900	8	13	3 1.2	1.2
48	291	23	016	18	568	47	.2892	8	12	4 1.6	1.5
49	314	23	.78998	18	615	46	.2884	8	11	5 2.0	1.9
										6 2.4	2.3
50	.61337	23	.78980	18	.77661	47	1.2876	7	**10**	7 2.8	2.7
51	360	23	962	18	708	46	.2869	8	9	8 3.2	3.1
52	383	23	944	18	754	47	.2861	8	8	9 3.6	3.4
53	406	23	926	18	801	47	.2853	7	7	10 4.0	3.8
54	429	22	908	17	848	47	.2846	8	6	20 8.0	7.7
										30 12.0	11.5
55	.61451	23	.78891	18	.77895	46	1.2838	8	**5**	40 16.0	15.3
56	474	23	873	18	941	47	.2830	8	4	50 20.0	19.2
57	497	23	855	18	988	47	.2822	7	3		
58	520	23	837	18	.78035	47	.2815	8	2		
59	543	23	819	18	082	47	.2807	8	1		
60	.61566		.78801		.78129		1.2799		**0**		
	Cos.	d.	Sin.	d.	Cot.	d.	Tan.	d.	'	P. P.	

52°

38°

P. P.	'	Sin.	d.	Cos.	d.	Tan.	d.	Cot.	d.	
	0	.61566		.78801		.78129		1.2799		**60**
	1	589	23	783	18	175	46	.2792	7	59
	2	612	23	765	18	222	47	.2784	8	58
	3	635	23	747	18	269	47	.2776	8	57
	4	658	23	729	18	316	47	.2769	7	56
	5	.61681	23	.78711	17	.78363	47	1.2761	8	**55**
	6	704	23	694	22	410	47	.2753	8	54
	7	726	22	676	18	457	47	.2746	7	53
	8	749	23	658	18	504	47	.2738	8	52
	9	772	23	640	18	551	47	.2731	7	51
	10	.61795	23	.78622	18	.78598	47	1.2723	8	**50**
	11	818	23	604	18	645	47	.2715	8	49
	12	841	23	586	18	692	47	.2708	7	48
	13	864	23	568	18	739	47	.2700	8	47
	14	887	23	550	22	786	47	.2693	7	46
	15	.61909	22	.78532	18	.78834	48	1.2685	8	**45**
	16	932	23	514	18	881	47	.2677	8	44
	17	955	23	496	18	928	47	.2670	7	43
	18	978	23	478	18	975	47	.2662	8	42
	19	.62001	23	460	18	.79022	47	.2655	7	41
	20	.62024	23	.78442	18	.79070	48	1.2647	8	**40**
	21	046	22	424	18	117	47	.2640	7	39
	22	069	23	405	19	164	47	.2632	8	38
	23	092	23	387	18	212	48	.2624	8	37
	24	115	23	369	18	259	47	.2617	7	36
	25	.62138	23	.78351	18	.79306	47	1.2609	8	**35**
	26	160	22	333	18	354	48	.2602	7	34
	27	183	23	315	18	401	47	.2594	8	33
	28	206	23	297	18	449	48	.2587	7	32
	29	229	23	279	18	496	47	.2579	8	31
	30	.62251	22	.78261	18	.79544	48	1.2572	7	**30**
	31	274	23	243	18	591	47	.2564	8	29
	32	297	23	225	18	639	48	.2557	7	28
	33	320	23	206	19	686	47	.2549	8	27
	34	342	22	188	18	734	48	.2542	7	26
	35	.62365	23	.78170	18	.79781	47	1.2534	8	**25**
	36	388	23	152	18	829	48	.2527	7	24
	37	411	23	134	18	877	48	.2519	8	23
	38	433	22	116	18	924	47	.2512	7	22
	39	456	23	098	18	972	48	.2504	8	21
	40	.62479	23	.78079	19	.80020	48	1.2497	7	**20**
	41	502	23	061	18	067	47	.2489	8	19
	42	524	22	043	18	115	48	.2482	7	18
	43	547	23	025	18	163	48	.2475	7	17
	44	570	23	007	18	211	48	.2467	8	16
	45	.62592	22	.77988	19	.80258	47	1.2460	7	**15**
	46	615	23	970	18	306	48	.2452	8	14
	47	638	23	952	18	354	48	.2445	7	13
	48	660	22	934	18	402	48	.2437	8	12
	49	683	23	916	18	450	48	.2430	7	11
	50	.62706	23	.77897	19	.80498	48	1.2423	7	**10**
	51	728	22	879	18	546	48	.2415	8	9
	52	751	23	861	18	594	48	.2408	7	8
	53	774	23	843	18	642	48	.2401	7	7
	54	796	22	824	19	690	48	.2393	8	6
	55	.62819	23	.77806	18	.80738	48	1.2386	7	**5**
	56	842	23	788	18	786	48	.2378	8	4
	57	864	22	769	19	834	48	.2371	7	3
	58	887	23	751	18	882	48	.2364	7	2
	59	909	22	733	18	930	48	.2356	8	1
	60	.62932	23	.77715	18	.80978	48	1.2349	7	**0**

| P. P. | | Cos. | d. | Sin. | d. | Cot. | d. | Tan. | d. | ' |

51°

P. P. columns (left margin):

	22	19
1	0.4	0.3
2	0.7	0.6
3	1.1	1.0
4	1.5	1.3
5	1.8	1.6
6	2.2	1.9
7	2.6	2.2
8	2.9	2.5
9	3.3	2.8
10	3.7	3.2
20	7.3	6.3
30	11.0	9.5
40	14.7	12.7
50	18.3	15.8

	18	17
1	0.3	0.3
2	0.6	0.6
3	0.9	0.8
4	1.2	1.1
5	1.5	1.4
6	1.8	1.7
7	2.1	2.0
8	2.4	2.3
9	2.7	2.6
10	3.0	2.8
20	6.0	5.7
30	9.0	8.5
40	12.0	11.3
50	15.0	14.2

	8	7
1	0.1	0.1
2	0.3	0.2
3	0.4	0.4
4	0.5	0.5
5	0.7	0.6
6	0.8	0.7
7	0.9	0.8
8	1.1	0.9
9	1.2	1.0
10	1.3	1.2
20	2.7	2.3
30	4.0	3.5
40	5.3	4.7
50	6.7	5.8

39°

'	Sin.	d.	Cos.	d.	Tan.	d.	Cot.	d.		P. P.
0	.62932	23	.77715	19	.80978	49	1.2349	7	60	
1	955	22	696	18	.81027	48	.2342	8	59	
2	977	23	678	18	075	48	.2334	7	58	
3	.63000	22	660	19	123	48	.2327	7	57	
4	022	23	641	18	171	49	.2320	8	56	
5	.63045	23	.77623	18	.81220	48	1.2312	7	55	
6	068	22	605	19	268	48	.2305	7	54	51 50
7	090	23	586	18	316	48	.2298	8	53	1 0.8 0.8
8	113	22	568	18	364	49	.2290	7	52	2 1.7 1.7
9	135	23	550	19	413	48	.2283	7	51	3 2.6 2.5
10	.63158	22	.77531	18	.81461	49	1.2276	8	50	4 3.4 3.3
11	180	23	513	19	510	48	.2268	7	49	5 4.2 4.2
12	203	22	494	18	558	48	.2261	7	48	6 5.1 5.0
13	225	23	476	18	606	49	.2254	7	47	7 6.0 5.8
14	248	23	458	19	655	48	.2247	8	46	8 6.8 6.7
15	.63271	22	.77439	18	.81703	49	1.2239	7	45	9 7.6 7.5
16	293	23	421	19	752	48	.2232	7	44	10 8.5 8.3
17	316	22	402	18	800	49	.2225	7	43	20 17.0 16.7
18	338	23	384	18	849	49	.2218	8	42	30 25.5 25.0
19	361	22	366	19	898	48	.2210	7	41	40 34.0 33.3
20	.63383	23	.77347	18	.81946	49	1.2203	7	40	50 42.5 41.7
21	406	22	329	19	995	49	.2196	7	39	
22	428	23	310	18	.82044	48	.2189	8	38	
23	451	22	292	19	092	49	.2181	7	37	
24	473	23	273	18	141	49	.2174	7	36	49 48
25	.63496	22	.77255	19	.82190	48	1.2167	7	35	1 0.8 0.8
26	518	22	236	18	238	49	.2160	7	34	2 1.6 1.6
27	540	23	218	19	287	48	.2153	8	33	3 2.4 2.4
28	563	22	199	18	336	49	.2145	7	32	4 3.3 3.2
29	585	23	181	19	385	49	.2138	7	31	5 4.1 4.0
30	.63608	22	.77162	18	.82434	49	1.2131	7	30	6 4.9 4.8
31	630	23	144	19	483	48	.2124	7	29	7 5.7 5.6
32	653	22	125	18	531	49	.2117	8	28	8 6.5 6.4
33	675	23	107	19	580	49	.2109	7	27	9 7.4 7.2
34	698	22	088	18	629	49	.2102	7	26	10 8.2 8.0
35	.63720	22	.77070	19	.82678	49	1.2095	7	25	20 16.3 16.0
36	742	23	051	18	727	49	.2088	7	24	30 24.5 24.0
37	765	22	033	19	776	49	.2081	7	23	40 32.7 32.0
38	787	23	014	18	825	49	.2074	8	22	50 40.8 40.0
39	810	22	.76996	19	874	49	.2066	7	21	
40	.63832	22	.76977	18	.82923	49	1.2059	7	20	
41	854	23	959	19	972	50	.2052	7	19	
42	877	22	940	19	.83022	49	.2045	7	18	23 22
43	899	23	921	18	071	49	.2038	7	17	1 0.4 0.4
44	922	22	903	19	120	49	.2031	7	16	2 0.8 0.7
45	.63944	22	.76884	18	.83169	49	1.2024	7	15	3 1.2 1.1
46	966	23	866	19	218	50	.2017	8	14	4 1.5 1.5
47	989	22	847	19	268	49	.2009	7	13	5 1.9 1.8
48	.64011	22	828	18	317	49	.2002	7	12	6 2.3 2.2
49	033	23	810	19	366	49	.1995	7	11	7 2.7 2.6
50	.64056	22	.76791	19	.83415	50	1.1988	7	10	8 3.1 2.9
51	078	22	772	18	465	49	.1981	7	9	9 3.4 3.3
52	100	23	754	19	514	50	.1974	7	8	10 3.8 3.7
53	123	22	735	18	564	49	.1967	7	7	20 7.7 7.3
54	145	22	717	19	613	49	.1960	7	6	30 11.5 11.0
55	.64167	23	.76698	19	.83662	50	1.1953	7	5	40 15.3 14.7
56	190	22	679	18	712	49	.1946	7	4	50 19.2 18.3
57	212	22	661	19	761	50	.1939	7	3	
58	234	22	642	19	811	49	.1932	7	2	
59	256	23	623	19	860	50	.1925	7	1	
60	.64279		.76604		.83910		1.1918		0	
	Cos.	d.	Sin.	d.	Cot.	d.	Tan.	d.	'	P. P.

50°

40°

P. P.

	20	19
1	0.3	0.3
2	0.7	0.6
3	1.0	1.0
4	1.3	1.3
5	1.7	1.6
6	2.0	1.9
7	2.3	2.2
8	2.7	2.5
9	3.0	2.8
10	3.3	3.2
20	6.7	6.3
30	10.0	9.5
40	13.3	12.7
50	16.7	15.8

	18	8
1	0.3	0.1
2	0.6	0.3
3	0.9	0.4
4	1.2	0.5
5	1.5	0.7
6	1.8	0.8
7	2.1	0.9
8	2.4	1.1
9	2.7	1.2
10	3.0	1.3
20	6.0	2.7
30	9.0	4.0
40	12.0	5.3
50	15.0	6.7

	7	6
1	0.1	0.1
2	0.2	0.2
3	0.4	0.3
4	0.5	0.4
5	0.6	0.5
6	0.7	0.6
7	0.8	0.7
8	0.9	0.8
9	1.0	0.9
10	1.2	1.0
20	2.3	2.0
30	3.5	3.0
40	4.7	4.0
50	5.8	5.0

′	Sin.	d.	Cos.	d.	Tan.	d.	Cot.	d.	
0	.64279	22	.76604	18	.83910	50	1.1918	8	60
1	301	22	586	19	960	49	.1910	7	59
2	323	23	567	19	.84009	50	.1903	7	58
3	346	22	548	18	059	49	.1896	7	57
4	368	22	530	19	108	50	.1889	7	56
5	.64390	22	.76511	19	.84158	50	1.1882	7	55
6	412	23	492	19	208	50	.1875	7	54
7	435	22	473	18	258	49	.1868	7	53
8	457	22	455	19	307	50	.1861	7	52
9	479	22	436	19	357	50	.1854	7	51
10	.64501	23	.76417	19	.84407	50	1.1847	7	50
11	524	22	398	18	457	50	.1840	7	49
12	546	22	380	19	507	49	.1833	7	48
13	568	22	361	19	556	50	.1826	7	47
14	590	22	342	19	606	50	.1819	7	46
15	.64612	23	.76323	19	.84656	50	1.1812	6	45
16	635	22	304	18	706	50	.1806	7	44
17	657	22	286	19	756	50	.1799	7	43
18	679	22	267	19	806	50	.1792	7	42
19	701	22	248	19	856	50	.1785	7	41
20	.64723	23	.76229	19	.84906	50	1.1778	7	40
21	746	22	210	18	956	50	.1771	7	39
22	768	22	192	19	.85006	51	.1764	7	38
23	790	22	173	19	057	50	.1757	7	37
24	812	22	154	19	107	50	.1750	7	36
25	.64834	22	.76135	19	.85157	50	1.1743	7	35
26	856	22	116	19	207	50	.1736	7	34
27	878	23	097	19	257	51	.1729	7	33
28	901	22	078	19	308	50	.1722	7	32
29	923	22	059	18	358	50	.1715	7	31
30	.64945	22	.76041	19	.85408	50	1.1708	6	30
31	967	22	022	19	458	51	.1702	7	29
32	989	22	003	19	509	50	.1695	7	28
33	.65011	22	.75984	19	559	50	.1688	7	27
34	033	22	965	19	609	51	.1681	7	26
35	.65055	22	.75946	19	.85660	50	1.1674	7	25
36	077	23	927	19	710	51	.1667	7	24
37	100	22	908	19	761	50	.1660	7	23
38	122	22	889	19	811	51	.1653	6	22
39	144	22	870	19	862	50	.1647	7	21
40	.65166	22	.75851	19	.85912	51	1.1640	7	20
41	188	22	832	19	963	51	.1633	7	19
42	210	22	813	19	.86014	50	.1626	7	18
43	232	22	794	19	064	51	.1619	7	17
44	254	22	775	19	115	51	.1612	6	16
45	.65276	22	.75756	18	.86166	50	1.1606	7	15
46	298	22	738	19	216	51	.1599	7	14
47	320	22	719	19	267	51	.1592	7	13
48	342	22	700	20	318	50	.1585	7	12
49	364	22	680	19	368	51	.1578	7	11
50	.65386	22	.75661	19	.86419	51	1.1571	6	10
51	408	22	642	19	470	51	.1565	7	9
52	430	22	623	19	521	51	.1558	7	8
53	452	22	604	19	572	51	.1551	7	7
54	474	22	585	19	623	51	.1544	6	6
55	.65496	22	.75566	19	.86674	51	1.1538	7	5
56	518	22	547	19	725	51	.1531	7	4
57	540	22	528	19	776	51	.1524	7	3
58	562	22	509	19	827	51	.1517	7	2
59	584	22	490	19	878	51	.1510	6	1
60	.65606		.75471		.86929		1.1504		0

P. P.		Cos.	d.	Sin.	d.	Cot.	d.	Tan.	d.	′

49°

41°

′	Sin.	d.	Cos.	d.	Tan.	d.	Cot.	d.	
0	.65606	22	.75471	19	.86929	51	1.1504	7	60
1	628	22	452	19	980	51	.1497	7	59
2	650	22	433	19	.87031	51	.1490	7	58
3	672	22	414	19	082	51	.1483	6	57
4	694	22	395	20	133	51	.1477	7	56
5	.65716	22	.75375	19	.87184	52	1.1470	7	55
6	738	21	356	19	236	51	.1463	7	54
7	759	22	337	19	287	51	.1456	6	53
8	781	22	318	19	338	51	.1450	7	52
9	803	22	299	19	389	52	.1443	7	51
10	.65825	22	.75280	19	.87441	51	1.1436	6	50
11	847	22	261	20	492	51	.1430	7	49
12	869	22	241	19	543	52	.1423	7	48
13	891	22	222	19	595	51	.1416	6	47
14	913	22	203	19	646	52	.1410	7	46
15	.65935	21	.75184	19	.87698	51	1.1403	7	45
16	956	22	165	19	749	52	.1396	7	44
17	978	22	146	20	801	51	.1389	6	43
18	.66000	22	126	19	852	52	.1383	7	42
19	022	22	107	19	904	51	.1376	7	41
20	.66044	22	.75088	19	.87955	52	1.1369	6	40
21	066	22	069	19	.88007	52	.1363	7	39
22	088	21	050	20	059	51	.1356	7	38
23	109	22	030	19	110	52	.1349	6	37
24	131	22	011	19	162	52	.1343	7	36
25	.66153	22	.74992	19	.88214	51	1.1336	7	35
26	175	22	973	20	265	52	.1329	6	34
27	197	21	953	19	317	52	.1323	7	33
28	218	22	934	19	369	52	.1316	6	32
29	240	22	915	19	421	52	.1310	7	31
30	.66262	22	.74896	20	.88473	51	1.1303	7	30
31	284	22	876	19	524	52	.1296	6	29
32	306	21	857	19	576	52	.1290	7	28
33	327	22	838	20	628	52	.1283	7	27
34	349	22	818	19	680	52	.1276	6	26
35	.66371	22	.74799	19	.88732	52	1.1270	7	25
36	393	21	780	20	784	52	.1263	6	24
37	414	22	760	19	836	52	.1257	7	23
38	436	22	741	19	888	52	.1250	7	22
39	458	22	722	19	940	52	.1243	6	21
40	.66480	21	.74703	20	.88992	53	1.1237	7	20
41	501	22	683	19	.89045	52	.1230	6	19
42	523	22	664	20	097	52	.1224	7	18
43	545	21	644	19	149	52	.1217	6	17
44	566	22	625	19	201	52	.1211	7	16
45	.66588	22	.74606	20	.89253	53	1.1204	7	15
46	610	22	586	19	306	52	.1197	6	14
47	632	21	567	19	358	52	.1191	7	13
48	653	22	548	20	410	53	.1184	6	12
49	675	22	528	19	463	52	.1178	7	11
50	.66697	21	.74509	20	.89515	52	1.1171	6	10
51	718	22	489	19	567	53	.1165	7	9
52	740	22	470	19	620	52	.1158	6	8
53	762	21	451	20	672	53	.1152	7	7
54	783	22	431	19	725	52	.1145	6	6
55	.66805	22	.74412	20	.89777	53	1.1139	7	5
56	827	21	392	19	830	53	.1132	6	4
57	848	22	373	20	883	52	.1126	7	3
58	870	21	353	19	935	53	.1119	6	2
59	891	22	334	20	988	52	.1113	7	1
60	.66913		.74314		.90040		1.1106		0
	Cos.	d.	Sin.	d.	Cot.	d.	Tan.	d.	′

P. P.

	55	54
1	0.9	0.9
2	1.8	1.8
3	2.8	2.7
4	3.7	3.6
5	4.6	4.5
6	5.5	5.4
7	6.4	6.3
8	7.3	7.2
9	8.2	8.1
10	9.2	9.0
20	18.3	18.0
30	27.5	27.0
40	36.7	36.0
50	45.8	45.0

	53	52
1	0.9	0.9
2	1.8	1.7
3	2.6	2.6
4	3.5	3.5
5	4.4	4.3
6	5.3	5.2
7	6.2	6.1
8	7.1	6.9
9	8.0	7.8
10	8.8	8.7
20	17.7	17.3
30	26.5	26.0
40	35.3	34.7
50	44.2	43.3

	51
1	0.8
2	1.7
3	2.6
4	3.4
5	4.2
6	5.1
7	6.0
8	6.8
9	7.6
10	8.5
20	17.0
30	25.5
40	34.0
50	42.5

48°

42°

P. P.	′	Sin.	d.	Cos.	d.	Tan.	d.	Cot.	d.	
	0	.66913	22	.74314	19	.90040	53	1.1106	6	**60**
	1	935	21	295	19	093	53	.1100	7	59
	2	956	22	276	20	146	53	.1093	6	58
	3	978	21	256	19	199	52	.1087	7	57
	4	999	22	237	20	251	53	.1080	6	56
	5	.67021	22	.74217	19	.90304	53	1.1074	7	**55**
	6	043	21	198	20	357	53	.1067	6	54
22 21	7	064	22	178	19	410	53	.1061	7	53
1 0.4 0.4	8	086	21	159	20	463	53	.1054	6	52
2 0.7 0.7	9	107	22	139	19	516	53	.1048	7	51
3 1.1 1.0	**10**	.67129	22	.74120	20	.90569	52	1.1041	6	**50**
4 1.5 1.4	11	151	21	100	20	621	53	.1035	7	49
5 1.8 1.8	12	172	22	080	19	674	53	.1028	6	48
6 2.2 2.1	13	194	21	061	20	727	53	.1022	6	47
7 2.6 2.4	14	215	22	041	19	781	54	.1016	7	46
8 2.9 2.8	**15**	.67237	21	.74022	20	.90834	53	1.1009	6	**45**
9 3.3 3.2	16	258	22	002	19	887	53	.1003	7	44
10 3.7 3.5	17	280	21	.73983	20	940	53	.0996	6	43
20 7.3 7.0	18	301	22	963	19	993	53	.0990	6	42
30 11.0 10.5	19	323	21	944	20	.91046	53	.0983	7	41
40 14.7 14.0	**20**	.67344	22	.73924	20	.91099	54	1.0977	6	**40**
50 18.3 17.5	21	366	21	904	19	153	53	.0971	7	39
	22	387	22	885	20	206	53	.0964	6	38
	23	409	21	865	19	259	54	.0958	7	37
	24	430	22	846	20	313	53	.0951	6	36
	25	.67452	21	.73826	20	.91366	53	1.0945	6	**35**
20 19	26	473	22	806	19	419	54	.0939	7	34
1 0.3 0.3	27	495	21	787	20	473	53	.0932	6	33
2 0.7 0.6	28	516	22	767	20	526	54	.0926	7	32
3 1.0 1.0	29	538	21	747	19	580	53	.0919	6	31
4 1.3 1.3	**30**	.67559	21	.73728	20	.91633	54	1.0913	6	**30**
5 1.7 1.6	31	580	22	708	20	687	53	.0907	7	29
6 2.0 1.9	32	602	21	688	19	740	54	.0900	6	28
7 2.3 2.2	33	623	22	669	20	794	53	.0894	6	27
8 2.7 2.5	34	645	22	649	20	847	54	.0888	7	26
9 3.0 2.8	**35**	.67666	22	.73629	19	.91901	54	1.0881	6	**25**
10 3.3 3.2	36	688	21	610	20	955	53	.0875	6	24
20 6.7 6.3	37	709	21	590	20	.92008	54	.0869	7	23
30 10.0 9.5	38	730	21	570	19	062	54	.0862	6	22
40 13.3 12.7	39	752	21	551	20	116	54	.0856	6	21
50 16.7 15.8	**40**	.67773	22	.73531	20	.92170	54	1.0850	7	**20**
	41	795	21	511	20	224	53	.0843	6	19
	42	816	21	491	19	277	54	.0837	6	18
	43	837	22	472	20	331	54	.0831	7	17
	44	859	21	452	20	385	54	.0824	6	16
7 6	**45**	.67880	21	.73432	19	.92439	54	1.0818	6	**15**
1 0.1 0.1	46	901	22	413	20	493	54	.0812	7	14
2 0.2 0.2	47	923	21	393	20	547	54	.0805	6	13
3 0.4 0.3	48	944	21	373	20	601	54	.0799	6	12
4 0.5 0.4	49	965	22	353	20	655	54	.0793	7	11
5 0.6 0.5	**50**	.67987	21	.73333	19	.92709	54	1.0786	6	**10**
6 0.7 0.6	51	.68008	21	314	20	763	54	.0780	6	9
7 0.8 0.7	52	029	22	294	20	817	55	.0774	6	8
8 0.9 0.8	53	051	21	274	20	872	54	.0768	7	7
9 1.0 0.9	54	072	21	254	20	926	54	.0761	6	6
10 1.2 1.0	**55**	.68093	22	.73234	19	.92980	54	1.0755	6	**5**
20 2.3 2.0	56	115	21	215	20	.93034	54	.0749	7	4
30 3.5 3.0	57	136	21	195	20	088	55	.0742	6	3
40 4.7 4.0	58	157	22	175	20	143	54	.0736	6	2
50 5.8 5.0	59	179	21	155	20	197	55	.0730	6	1
	60	.68200		.73135		.93252		1.0724		**0**
P. P.		Cos.	d.	Sin.	d.	Cot.	d.	Tan.	d.	′

47°

43°

'	Sin.	d.	Cos.	d.	Tan.	d.	Cot.	d.		P. P.
0	.68200	21	.73135	19	.93252	54	1.0724	7	60	
1	221	21	116	20	306	54	.0717	6	59	
·2	242	22	096	20	360	55	.0711	6	58	
3	264	21	076	20	415	54	.0705	6	57	
4	285	21	056	20	469	55	.0699	7	56	
5	.68306	21	.73036	20	.93524	54	1.0692	6	55	
6	327	22	016	20	578	55	.0686	6	54	58 · 57
7	349	21	.72996	20	633	55	.0680	6	53	1 1.0 1.0
8	370	21	976	19	688	54	.0674	6	52	2 1.9 1.9
9	391	21	957	20	742	55	.0668	7	51	3 2.9 2.9
										4 3.9 3.8
10	.68412	22	.72937	20	.93797	55	1.0661	6	50	5 4.8 4.8
11	434	21	917	20	852	54	.0655	6	49	6 5.8 5.7
12	455	21	897	20	906	55	.0649	6	48	7 6.8 6.7
13	476	21	877	20	961	55	.0643	6	47	8 7.7 7.6
14	497	21	857	20	.94016	55	.0637	7	46	9 8.7 8.6
										10 9.7 9.5
15	.68518	21	.72837	20	.94071	54	1.0630	6	45	20 19.3 19.0
16	539	22	817	20	125	55	.0624	6	44	30 29.0 28.5
17	561	21	797	20	180	55	.0618	6	43	40 38.7 38.0
18	582	21	777	20	235	55	.0612	6	42	50 48.3 47.5
19	603	21	757	20	290	55	.0606	7	41	
20	.68624	21	.72737	20	.94345	55	1.0599	6	40	
21	645	21	717	20	400	55	.0593	6	39	
22	666	22	697	20	455	55	.0587	6	38	
23	688	21	677	20	510	55	.0581	6	37	
24	709	21	657	20	565	55	.0575	6	36	
										56 · 55
25	.68730	21	.72637	20	.94620	56	1.0569	7	35	1 0.9 0.9
26	751	21	617	20	676	55	.0562	6	34	2 1.9 1.8
27	772	21	597	20	731	55	.0556	6	33	3 2.8 2.8
28	793	21	577	20	786	55	.0550	6	32	4 3.7 3.7
29	814	21	557	20	841	55	.0544	6	31	5 4.7 4.6
										6 5.6 5.5
30	.68835	22	.72537	20	.94896	56	1.0538	6	30	7 6.5 6.4
31	857	21	517	20	952	55	.0532	6	29	8 7.5 7.3
32	878	21	497	20	.95007	55	.0526	7	28	9 8.4 8.2
33	899	21	477	20	062	56	.0519	6	27	10 9.3 9.2
34	920	21	457	20	118	55	.0513	6	26	20 18.7 18.3
										30 28.0 27.5
35	.68941	21	.72437	20	.95173	56	1.0507	6	25	40 37.3 36.7
36	962	21	417	20	229	55	.0501	6	24	50 46.7 45.8
37	983	21	397	20	284	56	.0495	6	23	
38	.69004	21	377	20	340	55	.0489	6	22	
39	025	21	357	20	395	56	.0483	6	21	
40	.69046	21	.72337	20	.95451	55	1.0477	7	20	
41	067	21	317	20	506	56	.0470	6	19	
42	088	21	297	20	562	56	.0464	6	18	54
43	109	21	277	20	618	55	.0458	6	17	1 0.9
44	130	21	257	21	673	56	.0452	6	16	2 1.8
										3 2.7
45	.69151	21	.72236	20	.95729	56	1.0446	6	15	4 3.6
46	172	21	216	20	785	56	.0440	6	14	5 4.5
47	193	21	196	20	841	56	.0434	6	13	6 5.4
48	214	21	176	20	897	55	.0428	6	12	7 6.3
49	235	21	156	20	952	56	.0422	6	11	8 7.2
										9 8.1
50	.69256	21	.72136	20	.96008	56	1.0416	6	10	10 9.0
51	277	21	116	21	064	56	.0410	6	9	20 18.0
52	298	21	095	20	120	56	.0404	6	8	30 27.0
53	319	21	075	20	176	56	.0398	7	7	40 36.0
54	340	21	055	20	232	56	.0392	6	6	50 45.0
55	.69361	21	.72035	20	.96288	56	1.0385	6	5	
56	382	21	015	20	344	56	.0379	6	4	
57	403	21	.71995	21	400	57	.0373	6	3	
58	424	21	974	20	457	56	.0367	6	2	
59	445	21	954	20	513	56	.0361	6	1	
60	.69466		.71934		.96569		1.0355		0	
	Cos.	d.	Sin.	d.	Cot.	d.	Tan.	d.	'	P. P.

46°

44°

P. P.

22	21
1	0.4	0.4
2	0.7	0.7
3	1.1	1.0
4	1.5	1.4
5	1.8	1.8
6	2.2	2.1
7	2.6	2.4
8	2.9	2.8
9	3.3	3.2
10	3.7	3.5
20	7.3	7.0
30	11.0	10.5
40	14.7	14.0
50	18.3	17.5

20	7
1	0.3	0.1
2	0.7	0.2
3	1.0	0.4
4	1.3	0.5
5	1.7	0.6
6	2.0	0.7
7	2.3	0.8
8	2.7	0.9
9	3.0	1.0
10	3.3	1.2
20	6.7	2.3
30	10.0	3.5
40	13.3	4.7
50	16.7	5.8

6	5
1	0.1	0.1
2	0.2	0.2
3	0.3	0.2
4	0.4	0.3
5	0.5	0.4
6	0.6	0.5
7	0.7	0.6
8	0.8	0.7
9	0.9	0.8
10	1.0	0.8
20	2.0	1.7
30	3.0	2.5
40	4.0	3.3
50	5.0	4.2

'	Sin.	d.	Cos.	d.	Tan.	d.	Cot.	d.	'
0	.69466	21	.71934	20	.96569	56	1.0355	6	60
1	487	21	914	20	625	56	.0349	6	59
2	508	21	894	21	681	57	.0343	6	58
3	529	20	873	20	738	56	.0337	6	57
4	549	21	853	20	794	56	.0331	6	56
5	.69570	21	.71833	20	.96850	57	1.0325	6	55
6	591	21	813	20	907	56	.0319	6	54
7	612	21	792	21	963	56	.0313	6	53
8	633	21	772	20	.97020	56	.0307	6	52
9	654	21	752	20	076	57	.0301	6	51
10	.69675	21	.71732	21	.97133	56	1.0295	6	50
11	696	21	711	20	189	57	.0289	6	49
12	717	20	691	20	246	56	.0283	6	48
13	737	21	671	21	302	57	.0277	6	47
14	758	21	650	20	359	57	.0271	6	46
15	.69779	21	.71630	20	.97416	56	1.0265	6	45
16	800	21	610	20	472	57	.0259	6	44
17	821	21	590	21	529	57	.0253	6	43
18	842	20	569	20	586	57	.0247	6	42
19	862	21	549	20	643	57	.0241	6	41
20	.69883	21	.71529	21	.97700	56	1.0235	5	40
21	904	21	508	20	756	57	.0230	6	39
22	925	21	488	20	813	57	.0224	6	38
23	946	20	468	21	870	57	.0218	6	37
24	966	21	447	20	927	57	.0212	6	36
25	.69987	21	.71427	20	.97984	57	1.0206	6	35
26	.70008	21	407	21	.98041	57	.0200	6	34
27	029	20	386	20	098	57	.0194	6	33
28	049	21	366	21	155	58	.0188	6	32
29	070	21	345	20	213	57	.0182	6	31
30	.70091	21	.71325	20	.98270	57	1.0176	6	30
31	112	20	305	21	327	57	.0170	6	29
32	132	21	284	20	384	57	.0164	6	28
33	153	21	264	21	441	58	.0158	6	27
34	174	21	243	20	499	57	.0152	5	26
35	.70195	20	.71223	20	.98556	57	1.0147	6	25
36	215	21	203	21	613	58	.0141	6	24
37	236	21	182	20	671	57	.0135	6	23
38	257	20	162	21	728	58	.0129	6	22
39	277	21	141	20	786	57	.0123	6	21
40	.70298	21	.71121	21	.98843	58	1.0117	6	20
41	319	20	100	20	901	57	.0111	6	19
42	339	21	080	21	958	58	.0105	6	18
43	360	21	059	20	.99016	57	.0099	5	17
44	381	20	039	20	073	58	.0094	6	16
45	.70401	21	.71019	21	.99131	58	1.0088	6	15
46	422	21	.70998	20	189	58	.0082	6	14
47	443	20	978	21	247	57	.0076	6	13
48	463	21	957	20	304	58	.0070	6	12
49	484	21	937	21	362	58	.0064	6	11
50	.70505	20	.70916	20	.99420	58	1.0058	6	10
51	525	21	896	21	478	58	.0052	5	9
52	546	21	875	20	536	58	.0047	6	8
53	567	20	855	21	594	58	.0041	6	7
54	587	21	834	21	652	58	.0035	6	6
55	.70608	20	.70813	20	.99710	58	1.0029	6	5
56	628	21	793	21	768	58	.0023	6	4
57	649	21	772	20	826	58	.0017	5	3
58	670	20	752	21	884	58	.0012	6	2
59	690	21	731	20	942	58	.0006	6	1
60	.70711		.70711		1.0000		1.0000		0

| P. P. | | | Cos. | d. | Sin. | d. | Cot. | d. | Tan. | d. | ' |

45°

TABLE VII. FOUR–PLACE LOGARITHMS

No.	0	1	2	3	4	5	6	7	8	9
10	0000	0043	0086	0128	0170	0212	0253	0294	0334	0374
11	0414	0453	0492	0531	0569	0607	0645	0682	0719	0755
12	0792	0828	0864	0899	0934	0969	1004	1038	1072	1106
13	1139	1173	1206	1239	1271	1303	1335	1367	1399	1430
14	1461	1492	1523	1553	1584	1614	1644	1673	1703	1732
15	1761	1790	1818	1847	1875	1903	1931	1959	1987	2014
16	2041	2068	2095	2122	2148	2175	2201	2227	2253	2279
17	2304	2330	2355	2380	2405	2430	2455	2480	2504	2529
18	2553	2577	2601	2625	2648	2672	2695	2718	2742	2765
19	2788	2810	2833	2856	2878	2900	2923	2945	2967	2989
20	3010	3032	3054	3075	3096	3118	3139	3160	3181	3201
21	3222	3243	3263	3284	3304	3324	3345	3365	3385	3404
22	3424	3444	3464	3483	3502	3522	3541	3560	3579	3598
23	3617	3636	3655	3674	3692	3711	3729	3747	3766	3784
24	3802	3820	3838	3856	3874	3892	3909	3927	3945	3962
25	3979	3997	4014	4031	4048	4065	4082	4099	4116	4133
26	4150	4166	4183	4200	4216	4232	4249	4265	4281	4298
27	4314	4330	4346	4362	4378	4393	4409	4425	4440	4456
28	4472	4487	4502	4518	4533	4548	4564	4579	4594	4609
29	4624	4639	4654	4669	4683	4698	4713	4728	4742	4757
30	4771	4786	4800	4814	4829	4843	4857	4871	4886	4900
31	4914	4928	4942	4955	4969	4983	4997	5011	5024	5038
32	5051	5065	5079	5092	5105	5119	5132	5145	5159	5172
33	5185	5198	5211	5224	5237	5250	5263	5276	5289	5302
34	5315	5328	5340	5353	5366	5378	5391	5403	5416	5428
35	5441	5453	5465	5478	5490	5502	5514	5527	5539	5551
36	5563	5575	5587	5599	5611	5623	5635	5647	5658	5670
37	5682	5694	5705	5717	5729	5740	5752	5763	5775	5786
38	5798	5809	5821	5832	5843	5855	5866	5877	5888	5899
39	5911	5922	5933	5944	5955	5966	5977	5988	5999	6010
40	6021	6031	6042	6053	6064	6075	6085	6096	6107	6117
41	6128	6138	6149	6160	6170	6180	6191	6201	6212	6222
42	6232	6243	6253	6263	6274	6284	6294	6304	6314	6325
43	6335	6345	6355	6365	6375	6385	6395	6405	6415	6425
44	6435	6444	6454	6464	6474	6484	6493	6503	6513	6522
45	6532	6542	6551	6561	6571	6580	6590	6599	6609	6618
46	6628	6637	6646	6656	6665	6675	6684	6693	6702	6712
47	6721	6730	6739	6749	6758	6767	6776	6785	6794	6803
48	6812	6821	6830	6839	6848	6857	6866	6875	6884	6893
49	6902	6911	6920	6928	6937	6946	6955	6964	6972	6981
50	6990	6998	7007	7016	7024	7033	7042	7050	7059	7067
51	7076	7084	7093	7101	7110	7118	7126	7135	7143	7152
52	7160	7168	7177	7185	7193	7202	7210	7218	7226	7235
53	7243	7251	7259	7267	7275	7284	7292	7300	7308	7316
54	7324	7332	7340	7348	7356	7364	7372	7380	7388	7396
No.	0	1	2	3	4	5	6	7	8	9

No.	0	1	2	3	4	5	6	7	8	9
55	7404	7412	7419	7427	7435	7443	7451	7459	7466	7474
56	7482	7490	7497	7505	7513	7520	7528	7536	7543	7551
57	7559	7566	7574	7582	7589	7597	7604	7612	7619	7627
58	7634	7642	7649	7657	7664	7672	7679	7686	7694	7701
59	7709	7716	7723	7731	7738	7745	7752	7760	7767	7774
60	7782	7789	7796	7803	7810	7818	7825	7832	7839	7846
61	7853	7860	7868	7875	7882	7889	7896	7903	7910	7917
62	7924	7931	7938	7945	7952	7959	7966	7973	7980	7987
63	7993	8000	8007	8014	8021	8028	8035	8041	8048	8055
64	8062	8069	8075	8082	8089	8096	8102	8109	8116	8122
65	8129	8136	8142	8149	8156	8162	8169	8176	8182	8189
66	8195	8202	8209	8215	8222	8228	8235	8241	8248	8254
67	8261	8267	8274	8280	8287	8293	8299	8306	8312	8319
68	8325	8331	8338	8344	8351	8357	8363	8370	8376	8382
69	8388	8395	8401	8407	8414	8420	8426	8432	8439	8445
70	8451	8457	8463	8470	8476	8482	8488	8494	8500	8506
71	8513	8519	8525	8531	8537	8543	8549	8555	8561	8567
72	8573	8579	8585	8591	8597	8603	8609	8615	8621	8627
73	8633	8639	8645	8651	8657	8663	8669	8675	8681	8686
74	8692	8698	8704	8710	8716	8722	8727	8733	8739	8745
75	8751	8756	8762	8768	8774	8779	8785	8791	8797	8802
76	8808	8814	8820	8825	8831	8837	8842	8848	8854	8859
77	8865	8871	8876	8882	8887	8893	8899	8904	8910	8915
78	8921	8927	8932	8938	8943	8949	8954	8960	8965	8971
79	8976	8982	8987	8993	8998	9004	9009	9015	9020	9025
80	9031	9036	9042	9047	9053	9058	9063	9069	9074	9079
81	9085	9090	9096	9101	9106	9112	9117	9122	9128	9133
82	9138	9143	9149	9154	9159	9165	9170	9175	9180	9186
83	9191	9196	9201	9206	9212	9217	9222	9227	9232	9238
84	9243	9248	9253	9258	9263	9269	9274	9279	9284	9289
85	9294	9299	9304	9309	9315	9320	9325	9330	9335	9340
86	9345	9350	9355	9360	9365	9370	9375	9380	9385	9390
87	9395	9400	9405	9410	9415	9420	9425	9430	9435	9440
88	9445	9450	9455	9460	9465	9469	9474	9479	9484	9489
89	9494	9499	9504	9509	9513	9518	9523	9528	9533	9538
90	9542	9547	9552	9557	9562	9566	9571	9576	9581	9586
91	9590	9595	9600	9605	9609	9614	9619	9624	9628	9633
92	9638	9643	9647	9652	9657	9661	9666	9671	9675	9680
93	9685	9689	9694	9699	9703	9708	9713	9717	9722	9727
94	9731	9736	9741	9745	9750	9754	9759	9763	9768	9773
95	9777	9782	9786	9791	9795	9800	9805	9809	9814	9818
96	9823	9827	9832	9836	9841	9845	9850	9854	9859	9863
97	9868	9872	9877	9881	9886	9890	9894	9899	9903	9908
98	9912	9917	9921	9926	9930	9934	9939	9943	9948	9952
99	9956	9961	9965	9969	9974	9978	9983	9987	9991	9996
No.	0	1	2	3	4	5	6	7	8	9

TABLE VIII. FOUR-PLACE ANTILOGARITHMS

Log	0	1	2	3	4	5	6	7	8	9
.00	1000	1002	1005	1007	1009	1012	1014	1016	1019	1021
.01	1023	1026	1028	1030	1033	1035	1038	1040	1042	1045
.02	1047	1050	1052	1054	1057	1059	1062	1064	1067	1069
.03	1072	1074	1076	1079	1081	1084	1086	1089	1091	1094
.04	1096	1099	1102	1104	1107	1109	1112	1114	1117	1119
.05	1122	1125	1127	1130	1132	1135	1138	1140	1143	1146
.06	1148	1151	1153	1156	1159	1161	1164	1167	1169	1172
.07	1175	1178	1180	1183	1186	1189	1191	1194	1197	1199
.08	1202	1205	1208	1211	1213	1216	1219	1222	1225	1227
.09	1230	1233	1236	1239	1242	1245	1247	1250	1253	1256
.10	1259	1262	1265	1268	1271	1274	1276	1279	1282	1285
.11	1288	1291	1294	1297	1300	1303	1306	1309	1312	1315
.12	1318	1321	1324	1327	1330	1334	1337	1340	1343	1346
.13	1349	1352	1355	1358	1361	1365	1368	1371	1374	1377
.14	1380	1384	1387	1390	1393	1396	1400	1403	1406	1409
.15	1413	1416	1419	1422	1426	1429	1432	1435	1439	1442
.16	1445	1449	1452	1455	1459	1462	1466	1469	1472	1476
.17	1479	1483	1486	1489	1493	1496	1500	1503	1507	1510
.18	1514	1517	1521	1524	1528	1531	1535	1538	1542	1545
.19	1549	1552	1556	1560	1563	1567	1570	1574	1578	1581
.20	1585	1589	1592	1596	1600	1603	1607	1611	1614	1618
.21	1622	1626	1629	1633	1637	1641	1644	1648	1652	1656
.22	1660	1663	1667	1671	1675	1679	1683	1687	1690	1694
.23	1698	1702	1706	1710	1714	1718	1722	1726	1730	1734
.24	1738	1742	1746	1750	1754	1758	1762	1766	1770	1774
.25	1778	1782	1786	1791	1795	1799	1803	1807	1811	1816
.26	1820	1824	1828	1832	1837	1841	1845	1849	1854	1858
.27	1862	1866	1871	1875	1879	1884	1888	1892	1897	1901
.28	1905	1910	1914	1919	1923	1928	1932	1936	1941	1945
.29	1950	1954	1959	1963	1968	1972	1977	1982	1986	1991
.30	1995	2000	2004	2009	2014	2018	2023	2028	2032	2037
.31	2042	2046	2051	2056	2061	2065	2070	2075	2080	2084
.32	2089	2094	2099	2104	2109	2113	2118	2123	2128	2133
.33	2138	2143	2148	2153	2158	2163	2168	2173	2178	2183
.34	2188	2193	2198	2203	2208	2213	2218	2223	2228	2234
.35	2239	2244	2249	2254	2259	2265	2270	2275	2280	2286
.36	2291	2296	2301	2307	2312	2317	2323	2328	2333	2339
.37	2344	2350	2355	2360	2366	2371	2377	2382	2388	2393
.38	2399	2404	2410	2415	2421	2427	2432	2438	2443	2449
.39	2455	2460	2466	2472	2477	2483	2489	2495	2500	2506
.40	2512	2518	2523	2529	2535	2541	2547	2553	2559	2564
.41	2570	2576	2582	2588	2594	2600	2606	2612	2618	2624
.42	2630	2636	2642	2649	2655	2661	2667	2673	2679	2685
.43	2692	2698	2704	2710	2716	2723	2729	2735	2742	2748
.44	2754	2761	2767	2773	2780	2786	2793	2799	2805	2812
.45	2818	2825	2831	2838	2844	2851	2858	2864	2871	2877
.46	2884	2891	2897	2904	2911	2917	2924	2931	2938	2944
.47	2951	2958	2965	2972	2979	2985	2992	2999	3006	3013
.48	3020	3027	3034	3041	3048	3055	3062	3069	3076	3083
.49	3090	3097	3105	3112	3119	3126	3133	3141	3148	3155
Log	0	1	2	3	4	5	6	7	8	9

Log	0	1	2	3	4	5	6	7	8	9
.50	3162	3170	3177	3184	3192	3199	3206	3214	3221	3228
.51	3236	3243	3251	3258	3266	3273	3281	3289	3296	3304
.52	3311	3319	3327	3334	3342	3350	3357	3365	3373	3381
.53	3388	3396	3404	3412	3420	3428	3436	3443	3451	3459
.54	3467	3475	3483	3491	3499	3508	3516	3524	3532	3540
.55	3548	3556	3565	3573	3581	3589	3597	3606	3614	3622
.56	3631	3639	3648	3656	3664	3673	3681	3690	3698	3707
.57	3715	3724	3733	3741	3750	3758	3767	3776	3784	3793
.58	3802	3811	3819	3828	3837	3846	3855	3864	3873	3882
.59	3890	3899	3908	3917	3926	3936	3945	3954	3963	3972
.60	3981	3990	3999	4009	4018	4027	4036	4046	4055	4064
.61	4074	4083	4093	4102	4111	4121	4130	4140	4150	4159
.62	4169	4178	4188	4198	4207	4217	4227	4236	4246	4256
.63	4266	4276	4285	4295	4305	4315	4325	4335	4345	4355
.64	4365	4375	4385	4395	4406	4416	4426	4436	4446	4457
.65	4467	4477	4487	4498	4508	4519	4529	4539	4550	4560
.66	4571	4581	4592	4603	4613	4624	4634	4645	4656	4667
.67	4677	4688	4699	4710	4721	4732	4742	4753	4764	4775
.68	4786	4797	4808	4819	4831	4842	4853	4864	4875	4887
.69	4898	4909	4920	4932	4943	4955	4966	4977	4989	5000
.70	5012	5023	5035	5047	5058	5070	5082	5093	5105	5117
.71	5129	5140	5152	5164	5176	5188	5200	5212	5224	5236
.72	5248	5260	5272	5284	5297	5309	5321	5333	5346	5358
.73	5370	5383	5395	5408	5420	5433	5445	5458	5470	5483
.74	5495	5508	5521	5534	5546	5559	5572	5585	5598	5610
.75	5623	5636	5649	5662	5675	5689	5702	5715	5728	5741
.76	5754	5768	5781	5794	5808	5821	5834	5848	5861	5875
.77	5888	5902	5916	5929	5943	5957	5970	5984	5998	6012
.78	6026	6039	6053	6067	6081	6095	6109	6124	6138	6152
.79	6166	6180	6194	6209	6223	6237	6252	6266	6281	6295
.80	6310	6324	6339	6353	6368	6383	6397	6412	6427	6442
.81	6457	6471	6486	6501	6516	6531	6546	6561	6577	6592
.82	6607	6622	6637	6653	6668	6683	6699	6714	6730	6745
.83	6761	6776	6792	6808	6823	6839	6855	6871	6887	6902
.84	6918	6934	6950	6966	6982	6998	7015	7031	7047	7063
.85	7079	7096	7112	7129	7145	7161	7178	7194	7211	7228
.86	7244	7261	7278	7295	7311	7328	7345	7362	7379	7396
.87	7413	7430	7447	7464	7482	7499	7516	7534	7551	7568
.88	7586	7603	7621	7638	7656	7674	7691	7709	7727	7745
.89	7762	7780	7798	7816	7834	7852	7870	7889	7907	7925
.90	7943	7962	7980	7998	8017	8035	8054	8072	8091	8110
.91	8128	8147	8166	8185	8204	8222	8241	8260	8279	8299
.92	8318	8337	8356	8375	8395	8414	8433	8453	8472	8492
.93	8511	8531	8551	8570	8590	8610	8630	8650	8670	8690
.94	8710	8730	8750	8770	8790	8810	8831	8851	8872	8892
.95	8913	8933	8954	8974	8995	9016	9036	9057	9078	9099
.96	9120	9141	9162	9183	9204	9226	9247	9268	9290	9311
.97	9333	9354	9376	9397	9419	9441	9462	9484	9506	9528
.98	9550	9572	9594	9616	9638	9661	9683	9705	9727	9750
.99	9772	9795	9817	9840	9863	9886	9908	9931	9954	9977
Log	0	1	2	3	4	5	6	7	8	9

TABLE IX. FOUR-PLACE LOGARITHMIC FUNCTIONS

Rad	°	L Sin	L Cos	L Tan	L Cot	°	Rad
.0000	**0**	− ∞	0.0000	− ∞	∞	**90**	1.5708
.0175	1	8.2419	9.9999	8.2419	1.7581	89	1.5533
.0349	2	.5428	.9997	.5431	.4569	88	1.5359
.0524	3	.7188	.9994	.7194	.2806	87	1.5184
.0698	4	.8436	.9989	.8446	.1554	86	1.5010
.0873	**5**	8.9403	9.9983	8.9420	1.0580	**85**	1.4835
.1047	6	9.0192	.9976	9.0216	0.9784	84	1.4661
.1222	7	.0859	.9968	.0891	.9109	83	1.4486
.1396	8	.1436	.9958	.1478	.8522	82	1.4312
.1571	9	.1943	.9946	.1997	.8003	81	1.4137
.1745	**10**	9.2397	9.9934	9.2463	0.7537	**80**	1.3963
.1920	11	.2806	.9919	.2887	.7113	79	1.3788
.2094	12	.3179	.9904	.3275	.6725	78	1.3614
.2269	13	.3521	.9887	.3634	.6366	77	1.3439
.2443	14	.3837	.9869	.3968	.6032	76	1.3265
.2618	**15**	9.4130	9.9849	9.4281	0.5719	**75**	1.3090
.2793	16	.4403	.9828	.4575	.5425	74	1.2915
.2967	17	.4659	.9806	.4853	.5147	73	1.2741
.3142	18	.4900	.9782	.5118	.4882	72	1.2566
.3316	19	.5126	.9757	.5370	.4630	71	1.2392
.3491	**20**	9.5341	9.9730	9.5611	0.4389	**70**	1.2217
.3665	21	.5543	.9702	.5842	.4158	69	1.2043
.3840	22	.5736	.9672	.6064	.3936	68	1.1868
.4014	23	.5919	.9640	.6279	.3721	67	1.1694
.4189	24	.6093	.9607	.6486	.3514	66	1.1519
.4363	**25**	9.6259	9.9573	9.6687	0.3313	**65**	1.1345
.4538	26	.6418	.9537	.6882	.3118	64	1.1170
.4712	27	.6570	.9499	.7072	.2928	63	1.0996
.4887	28	.6716	.9459	.7257	.2743	62	1.0821
.5061	29	.6856	.9418	.7438	.2562	61	1.0647
.5236	**30**	9.6990	9.9375	9.7614	0.2386	**60**	1.0472
.5411	31	.7118	.9331	.7788	.2212	59	1.0297
.5585	32	.7242	.9284	.7958	.2042	58	1.0123
.5760	33	.7361	.9236	.8125	.1875	57	.9948
.5934	34	.7476	.9186	.8290	.1710	56	.9774
.6109	**35**	9.7586	9.9134	9.8452	0.1548	**55**	.9599
.6283	36	.7692	.9080	.8613	.1387	54	.9425
.6458	37	.7795	.9023	.8771	.1229	53	.9250
.6632	38	.7893	.8965	.8928	.1072	52	.9076
.6807	39	.7989	.8905	.9084	.0916	51	.8901
.6981	**40**	9.8081	9.8843	9.9238	0.0762	**50**	.8727
.7156	41	.8169	.8778	.9392	.0608	49	.8552
.7330	42	.8255	.8711	.9544	.0456	48	.8378
.7505	43	.8338	.8641	.9697	.0303	47	.8203
.7679	44	.8418	.8569	.9848	.0152	46	.8029
.7854	**45**	9.8495	9.8495	0.0000	0.0000	**45**	.7854
Rad	°	L Cos	L Sin	L Cot	L Tan	°	Rad

TABLE X.   FOUR–PLACE FUNCTIONS

°	Sin	Cos	Tan	Cot	Sec	Csc	
0	.0000	1.0000	.0000	∞	1.0000	∞	**90**
1	.0175	.9998	.0175	57.2900	1.0002	57.2987	89
2	.0349	.9994	.0349	28.6363	1.0006	28.6537	88
3	.0523	.9986	.0524	19.0811	1.0014	19.1073	87
4	.0698	.9976	.0699	14.3007	1.0024	14.3356	86
5	.0872	.9962	.0875	11.4301	1.0038	11.4737	**85**
6	.1045	.9945	.1051	9.5144	1.0055	9.5668	84
7	.1219	.9925	.1228	8.1443	1.0075	8.2055	83
8	.1392	.9903	.1405	7.1154	1.0098	7.1853	82
9	.1564	.9877	.1584	6.3138	1.0125	6.3925	81
10	.1736	.9848	.1763	5.6713	1.0154	5.7588	**80**
11	.1908	.9816	.1944	5.1446	1.0187	5.2408	79
12	.2079	.9781	.2126	4.7046	1.0223	4.8097	78
13	.2250	.9744	.2309	4.3315	1.0263	4.4454	77
14	.2419	.9703	.2493	4.0108	1.0306	4.1336	76
15	.2588	.9659	.2679	3.7321	1.0353	3.8637	**75**
16	.2756	.9613	.2867	3.4874	1.0403	3.6280	74
17	.2924	.9563	.3057	3.2709	1.0457	3.4203	73
18	.3090	.9511	.3249	3.0777	1.0515	3.2361	72
19	.3256	.9455	.3443	2.9042	1.0576	3.0716	71
20	.3420	.9397	.3640	2.7475	1.0642	2.9238	**70**
21	.3584	.9336	.3839	2.6051	1.0711	2.7904	69
22	.3746	.9272	.4040	2.4751	1.0785	2.6695	68
23	.3907	.9205	.4245	2.3559	1.0864	2.5593	67
24	.4067	.9135	.4452	2.2460	1.0946	2.4586	66
25	.4226	.9063	.4663	2.1445	1.1034	2.3662	**65**
26	.4384	.8988	.4877	2.0503	1.1126	2.2812	64
27	.4540	.8910	.5095	1.9626	1.1223	2.2027	63
28	.4695	.8829	.5317	1.8807	1.1326	2.1301	62
29	.4848	.8746	.5543	1.8040	1.1434	2.0627	61
30	.5000	.8660	.5774	1.7321	1.1547	2.0000	**60**
31	.5150	.8572	.6009	1.6643	1.1666	1.9416	59
32	.5299	.8480	.6249	1.6003	1.1792	1.8871	58
33	.5446	.8387	.6494	1.5399	1.1924	1.8361	57
34	.5592	.8290	.6745	1.4826	1.2062	1.7883	56
35	.5736	.8192	.7002	1.4281	1.2208	1.7434	**55**
36	.5878	.8090	.7265	1.3764	1.2361	1.7013	54
37	.6018	.7986	.7536	1.3270	1.2521	1.6616	53
38	.6157	.7880	.7813	1.2799	1.2690	1.6243	52
39	.6293	.7771	.8098	1.2349	1.2868	1.5890	51
40	.6428	.7660	.8391	1.1918	1.3054	1.5557	**50**
41	.6561	.7547	.8693	1.1504	1.3250	1.5243	49
42	.6691	.7431	.9004	1.1106	1.3456	1.4945	48
43	.6820	.7314	.9325	1.0724	1.3673	1.4663	47
44	.6947	.7193	.9657	1.0355	1.3902	1.4396	46
45	.7071	.7071	1.0000	1.0000	1.4142	1.4142	**45**
	Cos	Sin	Cot ·	Tan	Csc	Sec	°

TABLE XI. FUNCTIONS IN RADIAN MEASURE

Rad	L Sin	L Cos	L Tan	L Cot		Rad	L Sin	L Cos	L Tan	L Cot
.00	. . . . . .	0.0000	. . . . . .	. . . . . .		**.50**	9.6807	9.9433	9.7374	0.2626
.01	8.0000	.0000	8.0000	2.0000		.51	.6886	.9409	.7477	.2523
.02	.3010	9.9999	.3011	1.6989		.52	.6963	.9384	.7578	.2422
.03	.4771	.9998	.4773	.5227		.53	.7037	.9359	.7678	.2322
.04	.6019	.9997	.6023	.3977		.54	.7111	.9333	.7777	.2223
.05	8.6988	9.9995	8.6993	1.3007		**.55**	9.7182	9.9307	9.7875	0.2125
.06	.7779	.9992	.7787	.2213		.56	.7252	.9280	.7972	.2028
.07	.8447	.9989	.8458	.1542		.57	.7321	.9253	.8068	.1932
.08	.9026	.9986	.9040	.0960		.58	.7388	.9224	.8164	.1836
.09	.9537	.9982	.9554	.0446		.59	.7454	.9196	.8258	.1742
.10	8.9993	9.9978	9.0015	0.9985		**.60**	9.7518	9.9166	9.8351	0.1649
.11	9.0405	.9974	.0431	.9569		.61	.7581	.9136	.8444	.1556
.12	.0781	.9969	.0813	.9187		.62	.7642	.9106	.8536	.1464
.13	.1127	.9963	.1164	.8836		.63	.7702	.9074	.8628	.1372
.14	.1447	.9957	.1490	.8510		.64	.7761	.9042	.8719	.1281
.15	9.1745	9.9951	9.1794	0.8206		**.65**	9.7819	9.9010	9.8809	0.1191
.16	.2023	.9944	.2078	.7922		.66	.7875	.8976	.8899	.1101
.17	.2284	.9937	.2347	.7653		.67	.7931	.8942	.8989	.1011
.18	.2529	.9929	.2600	.7400		.68	.7985	.8907	.9078	.0922
.19	.2761	.9921	.2840	.7160		.69	.8038	.8872	.9166	.0834
.20	9.2981	9.9913	9.3069	0.6931		**.70**	9.8090	9.8836	9.9255	0.0745
.21	.3190	.9904	.3287	.6713		.71	.8141	.8799	.9343	.0657
.22	.3389	.9894	.3495	.6505		.72	.8191	.8761	.9430	.0570
.23	.3579	.9884	.3695	.6305		.73	.8240	.8723	.9518	.0482
.24	.3760	.9874	.3887	.6113		.74	.8288	.8683	.9605	.0395
.25	9.3934	9.9863	9.4071	0.5929		**.75**	9.8336	9.8643	9.9692	0.0308
.26	.4101	.9852	.4249	.5751		.76	.8382	.8602	.9779	.0221
.27	.4261	.9840	.4421	.5579		.77	.8427	.8561	.9866	.0134
.28	.4415	.9827	.4587	.5413		.78	.8471	.8518	.9953	.0047
.29	.4563	.9815	.4748	.5252		.79	.8515	.8475	0.0040	9.9960
.30	9.4706	9.9802	9.4904	0.5096		**.80**	9.8557	9.8431	0.0127	9.9873
.31	.4844	.9788	.5056	.4944		.81	.8599	.8385	.0214	.9786
.32	.4977	.9774	.5203	.4797		.82	.8640	.8339	.0301	.9699
.33	.5106	.9759	.5347	.4653		.83	.8680	.8292	.0388	.9612
.34	.5231	.9744	.5487	.4513		.84	.8719	.8244	.0475	.9525
.35	9.5352	9.9728	9.5623	0.4377		**.85**	9.8758	9.8195	0.0563	9.9437
.36	.5469	.9712	.5757	.4243		.86	.8796	.8145	.0650	.9350
.37	.5582	.9696	.5887	.4113		.87	.8833	.8094	.0738	.9262
.38	.5693	.9679	.6014	.3986		.88	.8869	.8042	.0827	.9173
.39	.5800	.9661	.6139	.3861		.89	.8905	.7989	.0915	.9085
.40	9.5904	9.9643	9.6261	0.3739		**.90**	9.8939	9.7935	0.1004	9.8996
.41	.6005	.9624	.6381	.3619		.91	.8974	.7880	.1094	.8906
.42	.6104	.9605	.6499	.3501		.92	.9007	.7823	.1184	.8816
.43	.6200	.9585	.6615	.3385		.93	.9040	.7766	.1274	.8726
.44	.6293	.9565	.6728	.3272		.94	.9072	.7707	.1365	.8635
.45	9.6385	9.9545	9.6840	0.3160		**.95**	9.9103	9.7647	0.1456	9.8544
.46	.6473	.9523	.6950	.3050		.96	.9134	.7585	.1548	.8452
.47	.6560	.9502	.7058	.2942		.97	.9164	.7523	.1641	.8359
.48	.6644	.9479	.7165	.2835		.98	.9193	.7459	.1735	.8265
.49	.6727	.9456	.7270	.2730		.99	.9222	.7393	.1829	.8171
.50	9.6807	9.9433	9.7374	0.2626		**1.00**	9.9250	9.7326	0.1924	9.8076

Rad	L Sin	L Cos	L Tan	L Cot
1.00	9.9250	9.7326	0.1924	9.8076
1.01	.9278	.7258	.2020	.7980
1.02	.9305	.7188	.2117	.7883
1.03	.9331	.7117	.2215	.7785
1.04	.9357	.7043	.2314	.7686
1.05	9.9382	9.6969	0.2414	9.7586
1.06	.9407	.6892	.2515	.7485
1.07	.9431	.6814	.2617	.7383
1.08	.9454	.6733	.2721	.7279
1.09	.9477	.6651	.2826	.7174
1.10	9.9500	9.6567	0.2933	9.7067
1.11	.9522	.6480	.3041	.6959
1.12	.9543	.6392	.3151	.6849
1.13	.9564	.6301	.3263	.6737
1.14	.9584	.6208	.3376	.6624
1.15	9.9604	9.6112	0.3492	9.6508
1.16	.9623	.6013	.3609	.6391
1.17	.9641	.5912	.3729	.6271
1.18	.9660	.5808	.3851	.6149
1.19	.9677	.5701	.3976	.6024
1.20	9.9694	9.5591	0.4103	9.5897
1.21	.9711	.5478	.4233	.5767
1.22	.9727	.5361	.4366	.5634
1.23	.9743	.5241	.4502	.5498
1.24	.9758	.5116	.4642	.5358
1.25	9.9773	9.4988	0.4785	9.5215
1.26	.9787	.4855	.4932	.5068
1.27	.9800	.4717	.5083	.4917
1.28	.9814	.4575	.5239	.4761
1.29	.9826	.4427	.5400	.4600
1.30	9.9839	9.4273	0.5566	9.4434

Rad	L Sin	L Cos	L Tan	L Cot
1.30	9.9839	9.4273	0.5566	9.4434
1.31	.9851	.4114	.5737	.4263
1.32	.9862	.3948	.5914	.4086
1.33	.9873	.3774	.6098	.3902
1.34	.9883	.3594	.6290	.3710
1.35	9.9893	9.3405	0.6489	9.3511
1.36	.9903	.3206	.6696	.3304
1.37	.9912	.2998	.6914	.3086
1.38	.9920	.2779	.7141	.2859
1.39	.9929	.2548	.7380	.2620
1.40	9.9936	9.2304	0.7633	9.2367
1.41	.9944	.2044	.7900	.2100
1.42	.9950	.1767	.8183	.1817
1.43	.9957	.1472	.8485	.1515
1.44	.9963	.1154	.8809	.1191
1.45	9.9968	9.0810	0.9158	9.0842
1.46	.9973	.0436	.9537	.0463
1.47	.9978	.0027	.9951	.0049
1.48	.9982	8.9575	1.0407	8.9593
1.49	.9986	.9069	.0917	.9083
1.50	9.9989	8.8496	1.1493	8.8507
1.51	.9992	.7836	.2156	.7844
1.52	.9994	.7056	.2938	.7062
1.53	.9996	.6105	.3891	.6109
1.54	.9998	.4884	.5114	.4886
1.55	9.9999	8.3180	1.6820	8.3180
1.56	0.0000	8.0333	1.9667	8.0333
1.57	0.0000	6.9011	3.0989	6.9011
1.58	0.0000	7.9640n	3.0360n	7.9640n
1.59	9.9999	8.2834n	2.7166n	8.2834n
1.60	9.9998	8.4654n	1.5344n	9.4656n

TABLE XII. ANGLES IN RADIANS

°	Rad	′	Rad	″	Rad
1	0.01745 33	1	0.00029 09	1	0.00000 48
2	0.03490 66	2	0.00058 18	2	0.00000 97
3	0.05235 99	3	0.00087 27	3	0.00001 45
4	0.06981 32	4	0.00116 36	4	0.00001 94
5	0.08726 65	5	0.00145 44	5	0.00002 42
6	0.10471 98	6	0.00174 53	6	0.00002 91
7	0.12217 30	7	0.00203 62	7	0.00003 39
8	0.13962 63	8	0.00232 71	8	0.00003 88
9	0.15707 96	9	0.00261 80	9	0.00004 36
10	0.17453 29	10	0.00290 89	10	0.00004 85
20	0.34906 59	20	0.00581 78	20	0.00009 70
30	0.52359 88	30	0.00872 66	30	0.00014 54
40	0.69813 17	40	0.01163 55	40	0.00019 39
50	0.87266 46	50	0.01454 44	50	0.00024 24
60	1.04719 76	60	0.01745 33	60	0.00029 09
70	1.22173 05				
80	1.39626 34				
90	1.57079 63				

TABLE XIII. FUNCTIONS IN RADIAN MEASURE

Rad	Sin	Cos	Tan	Cot	Rad	Sin	Cos	Tan	Cot
.00	.0000	1.0000	.0000	......	**.50**	.4794	.8776	.5463	1.830
.01	.0100	1.0000	.0100	99.997	.51	.4882	.8727	.5594	1.788
.02	.0200	.9998	.0200	49.993	.52	.4969	.8678	.5726	1.747
.03	.0300	.9996	.0300	33.323	.53	.5055	.8628	.5859	1.707
.04	.0400	.9992	.0400	24.987	.54	.5141	.8577	.5994	1.668
.05	.0500	.9988	.0500	19.983	**.55**	.5227	.8525	.6131	1.631
.06	.0600	.9982	.0601	16.647	.56	.5312	.8473	.6269	1.595
.07	.0699	.9976	.0701	14.262	.57	.5396	.8419	.6410	1.560
.08	.0799	.9968	.0802	12.473	.58	.5480	.8365	.6552	1.526
.09	.0899	.9960	.0902	11.081	.59	.5564	.8309	.6696	1.494
.10	.0998	.9950	.1003	9.967	**.60**	.5646	.8253	.6841	1.462
.11	.1098	.9940	.1104	9.054	.61	.5729	.8196	.6989	1.431
.12	.1197	.9928	.1206	8.293	.62	.5810	.8139	.7139	1.401
.13	.1296	.9916	.1307	7.649	.63	.5891	.8080	.7291	1.372
.14	.1395	.9902	.1409	7.096	.64	.5972	.8021	.7445	1.343
.15	.1494	.9888	.1511	6.617	**.65**	.6052	.7961	.7602	1.315
.16	.1593	.9872	.1614	6.197	.66	.6131	.7900	.7761	1.288
.17	.1692	.9856	.1717	5.826	.67	.6210	.7838	.7923	1.262
.18	.1790	.9838	.1820	5.495	.68	.6288	.7776	.8087	1.237
.19	.1889	.9820	.1923	5.200	.69	.6365	.7712	.8253	1.212
.20	.1987	.9801	.2027	4.933	**.70**	.6442	.7648	.8423	1.187
.21	.2085	.9780	.2131	4.692	.71	.6518	.7584	.8595	1.163
.22	.2182	.9759	.2236	4.472	.72	.6594	.7518	.8771	1.140
.23	.2280	.9737	.2341	4.271	.73	.6669	.7452	.8949	1.117
.24	.2377	.9713	.2447	4.086	.74	.6743	.7385	.9131	1.095
.25	.2474	.9689	.2553	3.916	**.75**	.6816	.7317	.9316	1.073
.26	.2571	.9664	.2660	3.759	.76	.6889	.7248	.9505	1.052
.27	.2667	.9638	.2768	3.613	.77	.6961	.7179	.9697	1.031
.28	.2764	.9611	.2876	3.478	.78	.7033	.7109	.9893	1.011
.29	.2860	.9582	.2984	3.351	.79	.7104	.7038	1.009	.9908
.30	.2955	.9553	.3093	3.233	**.80**	.7174	.6967	1.030	.9712
.31	.3051	.9523	.3203	3.122	.81	.7243	.6895	1.050	.9520
.32	.3146	.9492	.3314	3.018	.82	.7311	.6822	1.072	.9331
.33	.3240	.9460	.3425	2.920	.83	.7379	.6749	1.093	.9146
.34	.3335	.9428	.3537	2.827	.84	.7446	.6675	1.116	.8964
.35	.3429	.9394	.3650	2.740	**.85**	.7513	.6600	1.138	.8785
.36	.3523	.9359	.3764	2.657	.86	.7578	.6524	1.162	.8609
.37	.3616	.9323	.3879	2.578	.87	.7643	.6448	1.185	.8437
.38	.3709	.9287	.3994	2.504	.88	.7707	.6372	1.210	.8267
.39	.3802	.9249	.4111	2.433	.89	.7771	.6294	1.235	.8100
.40	.3894	.9211	.4228	2.365	**.90**	.7833	.6216	1.260	.7936
.41	.3986	.9171	.4346	2.301	.91	.7895	.6137	1.286	.7774
.42	.4078	.9131	.4466	2.239	.92	.7956	.6058	1.313	.7615
.43	.4169	.9090	.4586	2.180	.93	.8016	.5978	1.341	.7458
.44	.4259	.9048	.4708	2.124	.94	.8076	.5898	1.369	.7303
.45	.4350	.9004	.4831	2.070	**.95**	.8134	.5817	1.398	.7151
.46	.4439	.8961	.4954	2.018	.96	.8192	.5735	1.428	.7001
.47	.4529	.8916	.5080	1.969	.97	.8249	.5653	1.459	.6853
.48	.4618	.8870	.5206	1.921	.98	.8305	.5570	1.491	.6707
.49	.4706	.8823	.5334	1.875	.99	.8360	.5487	1.524	.6563
.50	.4794	.8776	.5463	1.830	**1.00**	.8415	.5403	1.557	.6421

Rad	Sin	Cos	Tan	Cot
1.00	.8415	.5403	1.557	.6421
1.01	.8468	.5319	1.592	.6281
1.02	.8521	.5234	1.628	.6142
1.03	.8573	.5148	1.665	.6005
1.04	.8624	.5062	1.704	.5870
1.05	.8674	.4976	1.743	.5736
1.06	.8724	.4889	1.784	.5604
1.07	.8772	.4801	1.827	.5473
1.08	.8820	.4713	1.871	.5344
1.09	.8866	.4625	1.917	.5216
1.10	.8912	.4536	1.965	.5090
1.11	.8957	.4447	2.014	.4964
1.12	.9001	.4357	2.066	.4840
1.13	.9044	.4267	2.120	.4718
1.14	.9086	.4176	2.176	.4596
1.15	.9128	.4085	2.234	.4475
1.16	.9168	.3993	2.296	.4356
1.17	.9208	.3902	2.360	.4237
1.18	.9246	.3809	2.427	.4120
1.19	.9284	.3717	2.498	.4003
1.20	.9320	.3624	2.572	.3888
1.21	.9356	.3530	2.650	.3773
1.22	.9391	.3436	2.733	.3659
1.23	.9425	.3342	2.820	.3546
1.24	.9458	.3248	2.912	.3434
1.25	.9490	.3153	3.010	.3323
1.26	.9521	.3058	3.113	.3212
1.27	.9551	.2963	3.224	.3102
1.28	.9580	.2867	3.341	.2993
1.29	.9608	.2771	3.467	.2884
1.30	.9636	.2675	3.602	.2776

Rad	Sin	Cos	Tan	Cot
1.30	.9636	.2675	3.602	.2776
1.31	.9662	.2579	3.747	.2669
1.32	.9687	.2482	3.903	.2562
1.33	.9711	.2385	4.072	.2456
1.34	.9735	.2288	4.256	.2350
1.35	.9757	.2190	4.455	.2245
1.36	.9779	.2092	4.673	.2140
1.37	.9799	.1994	4.913	.2035
1.38	.9819	.1896	5.177	.1931
1.39	.9837	.1798	5.471	.1828
1.40	.9854	.1700	5.798	.1725
1.41	.9871	.1601	6.165	.1622
1.42	.9887	.1502	6.581	.1519
1.43	.9901	.1403	7.055	.1417
1.44	.9915	.1304	7.602	.1315
1.45	.9927	.1205	8.238	.1214
1.46	.9939	.1106	8.989	.1113
1.47	.9949	.1006	9.887	.1011
1.48	.9959	.0907	10.983	.0910
1.49	.9967	.0807	12.350	.0810
1.50	.9975	.0707	14.101	.0709
1.51	.9982	.0608	16.428	.0609
1.52	.9987	.0508	19.670	.0508
1.53	.9992	.0408	24.498	.0408
1.54	.9995	.0308	32.461	.0308
1.55	.9998	.0208	48.078	.0208
1.56	.9999	.0108	92.620	.0108
1.57	1.0000	.0008	1255.8	.0008
1.58	1.0000	−.0092	−108.65	−.0092
1.59	.9998	−.0192	−52.067	−.0192
1.60	.9996	−.0292	−34.233	−.0292

TABLE XIVa. DEGREES TO RADIANS

	Radians	Tenths	Hundredths	Thousandths	Ten-thousandths
1	57°17′44″.8	5°43′46″.5	0°34′22″.6	0° 3′26″.3	0°0′20″.6
2	114°35′29″.6	11°27′33″.0	1° 8′45″.′3	0° 6′52″.5	0°0′41″.3
3	171°53′14″.4	17°11′19″.4	1°43′07″.9	0°10′18″.8	0°1′01″.9
4	229°10′59″.2	22°55′05″.9	2°17′30″.6	0°13′45″.1	0°1′22″.5
5	286°28′44″.0	28°38′52″.4	2°51′53″.2	0°17′11″.3	0°1′43″.1
6	343°46′28″.8	34°22′38″.9	3°26′15″.9	0°20′37″.6	0°2′03″.8
7	401° 4′13″.6	40° 6′25″.4	4° 0′38″.5	0°24′03″.9	0°2′24″.4
8	458°21′58″.4	45°50′11″.8	4°35′01″.2	0°27′30″.1	0°2′45″.0
9	515°39′43″.3	51°33′58″.3	5° 9′23″.8	0°30′56″.4	0°3′05″.6

TABLE XIV B. MINUTES IN DECIMALS OF A DEGREE

′	0″	10″	15″	20″	30″	40″	45″	50″	′
0	.00000	.00278	.00417	.00556	.00833	.01111	.01250	.01389	0
1	.01667	.01944	.02083	.02222	.02500	.02778	.02917	.03056	1
2	.03333	.03611	.03750	.03889	.04167	.04444	.04583	.04722	2
3	.05000	.05278	.05417	.05556	.05833	.06111	.06250	.06389	3
4	.06667	.06944	.07083	.07222	.07500	.07778	.07917	.08056	4
5	.08333	.08611	.08750	.08889	.09167	.09444	.09583	.09722	5
6	.10000	.10278	.10417	.10556	.10833	.11111	.11250	.11389	6
7	.11667	.11944	.12083	.12222	.12500	.12778	.12917	.13056	7
8	.13333	.13611	.13750	.13889	.14167	.14444	.14583	.14722	8
9	.15000	.15278	.15417	.15556	.15833	.16111	.16250	.16389	9
10	.16667	.16944	.17083	.17222	.17500	.17778	.17917	.18056	10
11	.18333	.18611	.18750	.18889	.19167	.19444	.19583	.19722	11
12	.20000	.20278	.20417	.20556	.20833	.21111	.21250	.21389	12
13	.21667	.21944	.22083	.22222	.22500	.22778	.22917	.23056	13
14	.23333	.23611	.23750	.23889	.24167	.24444	.24583	.24722	14
15	.25000	.25278	.25417	.25556	.25833	.26111	.26250	.26389	15
16	.26667	.26944	.27083	.27222	.27500	.27778	.27917	.28056	16
17	.28333	.28611	.28750	.28889	.29167	.29444	.29583	.29722	17
18	.30000	.30278	.30417	.30556	.30833	.31111	.31250	.31389	18
19	.31667	.31944	.32083	.32222	.32500	.32778	.32917	.33056	19
20	.33333	.33611	.33750	.33889	.34167	.34444	.34583	.34722	20
21	.35000	.35278	.35417	.35556	.35833	.36111	.36250	.36389	21
22	.36667	.36944	.37083	.37222	.37500	.37778	.37917	.38056	22
23	.38333	.38611	.38750	.38889	.39167	.39444	.39583	.39722	23
24	.40000	.40278	.40417	.40556	.40833	.41111	.41250	.41389	24
25	.41667	.41944	.42083	.42222	.42500	.42778	.42917	.43056	25
26	.43333	.43611	.43750	.43889	.44167	.44444	.44583	.44722	26
27	.45000	.45278	.45417	.45556	.45833	.46111	.46250	.46389	27
28	.46667	.46944	.47083	.47222	.47500	.47778	.47917	.48056	28
29	.48333	.48611	.48750	.48889	.49167	.49444	.49583	.49722	29
30	.50000	.50278	.50417	.50556	.50833	.51111	.51250	.51389	30
31	.51667	.51944	.52083	.52222	.52500	.52778	.52917	.53056	31
32	.53333	.53611	.53750	.53889	.54167	.54444	.54583	.54722	32
33	.55000	.55278	.55417	.55556	.55833	.56111	.56250	.56389	33
34	.56667	.56944	.57083	.57222	.57500	.57778	.57917	.58056	34
35	.58333	.58611	.58750	.58889	.59167	.59444	.59583	.59722	35
36	.60000	.60278	.60417	.60556	.60833	.61111	.61250	.61389	36
37	.61667	.61944	.62083	.62222	.62500	.62778	.62917	.63056	37
38	.63333	.63611	.63750	.63889	.64167	.64444	.64583	.64722	38
39	.65000	.65278	.65417	.65556	.65833	.66111	.66250	.66389	39
40	.66667	.66944	.67083	.67222	.67500	.67778	.67917	.68056	40
41	.68333	.68611	.68750	.68889	.69167	.69444	.69583	.69722	41
42	.70000	.70278	.70417	.70556	.70833	.71111	.71250	.71389	42
43	.71667	.71944	.72083	.72222	.72500	.72778	.72917	.73056	43
44	.73333	.73611	.73750	.73889	.74167	.74444	.74583	.74722	44
45	.75000	.75278	.75417	.75556	.75833	.76111	.76250	.76389	45
46	.76667	.76944	.77083	.77222	.77500	.77778	.77917	.78056	46
47	.78333	.78611	.78750	.78889	.79167	.79444	.79583	.79722	47
48	.80000	.80278	.80417	.80556	.80833	.81111	.81250	.81389	48
49	.81667	.81944	.82083	.82222	.82500	.82778	.82917	.83056	49
50	.83333	.83611	.83750	.83889	.84167	.84444	.84583	.84722	50
51	.85000	.85278	.85417	.85556	.85833	.86111	.86250	.86389	51
52	.86667	.86944	.87083	.87222	.87500	.87778	.87917	.88056	52
53	.88333	.88611	.88750	.88889	.89167	.89444	.89583	.89722	53
54	.90000	.90278	.90417	.90556	.90833	.91111	.91250	.91389	54
55	.91667	.91944	.92083	.92222	.92500	.92778	.92917	.93056	55
56	.93333	.93611	.93750	.93889	.94167	.94444	.94583	.94722	56
57	.95000	.95278	.95417	.95556	.95833	.96111	.96250	.96389	57
58	.96667	.96944	.97083	.97222	.97500	.97778	.97917	.98056	58
59	.98333	.98611	.98750	.98889	.99167	.99444	.99583	.99722	59
′	0″	10″	15″	20″	30″	40″	45″	50″	′

PART II. TABLES XV–XIX

EXPLANATION OF TABLES

11. Table XV. Powers, roots, and reciprocals. This table gives the squares, cubes, square roots, cube roots, and reciprocals of numbers which have three significant digits, directly to five decimal places. The use of the table for finding the squares, cubes, and reciprocals of numbers in which one digit precedes the decimal point needs no explanation. The digits for the squares, cubes, and reciprocals of all other three-digit numbers may also be obtained from the table, and the position of the decimal point in the result may easily be determined by inspection.

The columns headed $\sqrt{n}$ and $\sqrt{10\,n}$ give the square roots of numbers with three significant digits for which respectively one and two digits precede the decimal point. The square roots of all other three-digit numbers may be obtained from the tabulated values by making the proper change in the position of the decimal point. Thus,

$$\sqrt{2.37} = 1.53948;$$
$$\sqrt{23.7} = 4.86826;$$
$$\sqrt{237} = 10\sqrt{2.37} = 15.3948;$$
$$\sqrt{2370} = 10\sqrt{23.7} = 48.6826;$$
$$\sqrt{.237} = (.1)\sqrt{23.7} = .486826;$$
$$\sqrt{.0237} = (0.1)\sqrt{2.37} = .153948.$$

The columns headed $\sqrt[3]{n}$, $\sqrt[3]{10\,n}$ and $\sqrt[3]{100\,n}$ give the cube roots of numbers with three significant figures for which respectively one, two, and three significant digits precede the decimal point. The cube roots of all other numbers with three significant figures may be obtained from the tabulated values in a manner similar to that which has been illustrated for finding square roots.

12. Table XVI. Natural logarithms. This table gives the logarithms of N, $10\,N$, and $100\,N$ to the base e ($= 2.71828$) for equidistant values of N from 0.00 to 10.00. For all values of N and $10\,N$ which are less than unity 10 must be subtracted from the tabulated values of the logarithms. The differences between logarithms of consecutive values of $100\,N$ are given in the column headed d. The differences between the logarithms for consecutive values of N and $10\,N$ are the same or approximately the same as the differences for the logarithms of $100\,N$. It should be noted that when

171

the decimal point of a number is moved one place to the right the natural logarithm of the resulting number may be obtained by adding $\log_e 10 = 2.302585$ to the logarithm of the original number. Likewise, 2.302585 should be subtracted when the decimal point is moved one place to the left.

13. Table XVII. Exponential and hyperbolic functions. This table gives the values of e^x, e^{-x}, sinh x, cosh x, and the common logarithms of e^x, sinh x, and cosh x for equidistant values of x from 0.00 to 3.00 and for selected values of x from 3.00 to 10.00. The common logarithm of e^{-x} may be obtained by taking the cologarithm of e^x.

14. Table XVIII. Multiples of M and of 1/M. We have

$$\log_{10} N = M \log_e N,$$

and $$\log_e N = (1/M) \log_{10} N,$$

where $1/M = 2.302585$. The purpose of this table is to facilitate the multiplication by M and $1/M$.

Example. Find $\log_e 68.35$ from the common logarithm of 68.35.
We have
$$\log_e 68.35 = (1/M) \log_{10} 68.35$$
$$= (1/M)(1.83474)$$
$$= 1/M + (1/M)(.83) + (1/M)(.0047) + (1/M)(.00004).$$

By means of the table we may obtain the products in the last expression. Therefore,

$$
\begin{aligned}
\log_e 68.35 &= 2.302585 \\
&+ 1.911146 \\
&+ \quad\ 10822 \\
&+ \quad\quad\ \ 9 \\
\hline
&= \ 4.22456
\end{aligned}
$$

15. Table XIX. Ten-place logarithms of prime numbers. This table is self explanatory.

TABLES XV–XIX

TABLE XV. POWERS, ROOTS AND RECIPROCALS

n	$1/n$	n^2	n^3	$\sqrt{n}$	$\sqrt{10n}$	$\sqrt[3]{n}$	$\sqrt[3]{10n}$	$\sqrt[3]{100n}$
1.00	1.00000	1.0000	1.00000	1.00000	3.16228	1.00000	2.15443	4.64159
.01	.990099	1.0201	1.03030	.00499	.17805	.00332	.16159	.65701
.02	.980392	1.0404	1.06121	.00995	.19374	.00662	.16870	.67233
.03	.970874	1.0609	1.09273	.01489	.20936	.00990	.17577	.68755
.04	.961538	1.0816	1.12486	.01980	.22490	.01316	.18279	.70267
1.05	.952381	1.1025	1.15762	1.02470	3.24037	1.01640	2.18976	4.71769
.06	.943396	1.1236	1.19102	.02956	.25576	.01961	.19669	.73262
.07	.934579	1.1449	1.22504	.03441	.27109	.02281	.20358	.74746
.08	.925926	1.1664	1.25971	.03923	.28634	.02599	.21042	.76220
.09	.917431	1.1881	1.29503	.04403	.30151	.02914	.21722	.77686
1.10	.909091	1.2100	1.33100	1.04881	3.31662	1.03228	2.22398	4.79142
.11	.900901	1.2321	1.36763	.05357	.33167	.03540	.23070	.80590
.12	.892857	1.2544	1.40493	.05830	.34664	.03850	.23738	.82028
.13	.884956	1.2769	1.44290	.06301	.36155	.04158	.24402	.83459
.14	.877193	1.2996	1.48154	.06771	.37639	.04464	.25062	.84881
1.15	.869565	1.3225	1.52088	1.07238	3.39116	1.04769	2.25718	4.86294
.16	.862069	1.3456	1.56090	.07703	.40588	.05072	.26370	.87700
.17	.854701	1.3689	1.60161	.08167	.42053	.05373	.27019	.89097
.18	.847458	1.3924	1.64303	.08628	.43511	.05672	.27664	.90487
.19	.840336	1.4161	1.68516	.09087	.44964	.05970	.28305	.91868
1.20	.833333	1.4400	1.72800	1.09545	3.46410	1.06266	2.28943	4.93242
.21	.826446	1.4641	1.77156	.10000	.47851	.06560	.29577	.94609
.22	.819672	1.4884	1.81585	.10454	.49285	.06853	.30208	.95968
.23	.813008	1.5129	1.86087	.10905	.50714	.07144	.30835	.97319
.24	.806452	1.5376	1.90662	.11355	.52136	.07434	.31459	.98663
1.25	.800000	1.5625	1.95312	1.11803	3.53553	1.07722	2.32079	5.00000
.26	.793651	1.5876	2.00038	.12250	.54965	.08008	.32697	.01330
.27	.787402	1.6129	2.04838	.12694	.56371	.08293	.33311	.02653
.28	.781250	1.6384	2.09715	.13137	.57771	.08577	.33921	.03968
.29	.775194	1.6641	2.14669	.13578	.59166	.08859	.34529	.05277
1.30	.769231	1.6900	2.19700	1.14018	3.60555	1.09139	2.35133	5.06580
.31	.763359	1.7161	2.24809	.14455	.61939	.09418	.35735	.07875
.32	.757576	1.7424	2.29997	.14891	.63318	.09696	.36333	.09164
.33	.751880	1.7689	2.35264	.15326	.64692	.09972	.36928	.10447
.34	.746269	1.7956	2.40610	.15758	.66060	.10247	.37521	.11723
1.35	.740741	1.8225	2.46038	1.16190	3.67423	1.10521	2.38110	5.12993
.36	.735294	1.8496	2.51546	.16619	.68782	.10793	.38697	.14256
.37	.729927	1.8769	2.57135	.17047	.70135	.11064	.39280	.15514
.38	.724638	1.9044	2.62807	.17473	.71484	.11334	.39861	.16765
.39	.719424	1.9321	2.68562	.17898	.72827	.11602	.40439	.18010
1.40	.714286	1.9600	2.74400	1.18322	3.74166	1.11869	2.41014	5.19249
.41	.709220	1.9881	2.80322	.18743	.75500	.12135	.41587	.20483
.42	.704225	2.0164	2.86329	.19164	.76829	.12399	.42156	.21710
.43	.699301	2.0449	2.92421	.19583	.78153	.12662	.42724	.22932
.44	.694444	2.0736	2.98598	.20000	.79473	.12924	.43288	.24148
1.45	.689655	2.1025	3.04862	1.20416	3.80789	1.13185	2.43850	5.25359
.46	.684932	2.1316	3.11214	.20830	.82099	.13445	.44409	.26564
.47	.680272	2.1609	3.17652	.21244	.83406	.13703	.44966	.27763
.48	.675676	2.1904	3.24179	.21655	.84708	.13960	.45520	.28957
.49	.671141	2.2201	3.30795	.22066	.86005	.14216	.46072	.30146
1.50	.666667	2.2500	3.37500	1.22474	3.87298	1.14471	2.46621	5.31329
n	$1/n$	n^2	n^3	$\sqrt{n}$	$\sqrt{10n}$	$\sqrt[3]{n}$	$\sqrt[3]{10n}$	$\sqrt[3]{100n}$

n	$1/n$	n^2	n^3	$\sqrt{n}$	$\sqrt{10n}$	$\sqrt[3]{n}$	$\sqrt[3]{10n}$	$\sqrt[3]{100n}$
1.50	.666667	2.2500	3.37500	1.22474	3.87298	1.14471	2.46621	5.31329
.51	.662252	2.2801	3.44295	.22882	.88587	.14725	.47168	.32507
.52	.657895	2.3104	3.51181	.23288	.89872	.14978	.47712	.33680
.53	.653595	2.3409	3.58158	.23693	.91152	.15230	.48255	.34848
.54	.649351	2.3716	3.65226	.24097	.92428	.15480	.48794	.36011
1.55	.645161	2.4025	3.72388	1.24499	3.93700	1.15729	2.49332	5.37169
.56	.641026	2.4336	3.79642	.24900	.94968	.15978	.49867	.38321
.57	.636943	2.4649	3.86989	.25300	.96232	.16225	.50399	.39469
.58	.632911	2.4964	3.94431	.25698	.97492	.16471	.50930	.40612
.59	.628931	2.5281	4.01968	.26095	.98748	.16717	.51458	.41750
1.60	.625000	2.5600	4.09600	1.26491	4.00000	1.16961	2.51984	5.42884
.61	.621118	2.5921	4.17328	.26886	.01248	.17204	.52508	.44012
.62	.617284	2.6244	4.25153	.27279	.02492	.17446	.53030	.45136
.63	.613497	2.6569	4.33075	.27671	.03733	.17687	.53549	.46256
.64	.609756	2.6896	4.41094	.28062	.04969	.17927	.54067	.47370
1.65	.606061	2.7225	4.49212	1.28452	4.06202	1.18167	2.54582	5.48481
.66	.602410	2.7556	4.57430	.28841	.07431	.18405	.55095	.49586
.67	.598802	2.7889	4.65746	.29228	.08656	.18642	.55607	.50688
.68	.595238	2.8224	4.74163	.29615	.09878	.18878	.56116	.51785
.69	.591716	2.8561	4.82681	.30000	.11096	.19114	.56623	.52877
1.70	.588235	2.8900	4.91300	1.30384	4.12311	1.19348	2.57128	5.53966
.71	.584795	2.9241	5.00021	.30767	.13521	.19582	.57631	.55050
.72	.581395	2.9584	5.08845	.31149	.14729	.19815	.58133	.56130
.73	.578035	2.9929	5.17772	.31529	.15933	.20046	.58632	.57205
.74	.574713	3.0276	5.26802	.31909	.17133	.20277	.59129	.58277
1.75	.571429	3.0625	5.35938	1.32288	4.18330	1.20507	2.59625	5.59344
.76	.568182	3.0976	5.45178	.32665	.19524	.20736	.60118	.60408
.77	.564972	3.1329	5.54523	.33041	.20714	.20964	.60610	.61467
.78	.561798	3.1684	5.63975	.33417	.21900	.21192	.61100	.62523
.79	.558659	3.2041	5.73534	.33791	.23084	.21418	.61588	.63574
1.80	.555556	3.2400	5.83200	1.34164	4.24264	1.21644	2.62074	5.64622
.81	.552486	3.2761	5.92974	.34536	.25441	.21869	.62559	.65665
.82	.549451	3.3124	6.02857	.34907	.26615	.22093	.63041	.66705
.83	.546448	3.3489	6.12849	.35277	.27785	.22316	.63522	.67741
.84	.543478	3.3856	6.22950	.35647	.28952	.22539	.64001	.68773
1.85	.540541	3.4225	6.33162	1.36015	4.30116	1.22760	2.64479	5.69802
.86	.537634	3.4596	6.43486	.36382	.31277	.22981	.64954	.70827
.87	.534759	3.4969	6.53920	.36748	.32435	.23201	.65428	.71848
.88	.531915	3.5344	6.64467	.37113	.33590	.23420	.65901	.72865
.89	.529101	3.5721	6.75127	.37477	.34741	.23639	.66371	.73879
1.90	.526316	3.6100	6.85900	1.37840	4.35890	1.23856	2.66840	5.74890
.91	.523560	3.6481	6.96787	.38203	.37035	.24073	.67307	.75897
.92	.520833	3.6864	7.07789	.38564	.38178	.24289	.67773	.76900
.93	.518135	3.7249	7.18906	.38924	.39318	.24505	.68237	.77900
.94	.515464	3.7636	7.30138	.39284	.40454	.24719	.68700	.78896
1.95	.512821	3.8025	7.41488	1.39642	4.41588	1.24933	2.69161	5.79889
.96	.510204	3.8416	7.52954	.40000	.42719	.25146	.69620	.80879
.97	.507614	3.8809	7.64537	.40357	.43847	.25359	.70078	.81865
.98	.505051	3.9204	7.76239	.40712	.44972	.25571	.70534	.82848
.99	.502513	3.9601	7.88060	.41067	.46094	.25782	.70989	.83827
2.00	.500000	4.0000	8.00000	1.41421	4.47214	1.25992	2.71442	5.84804
n	$1/n$	n^2	n^3	$\sqrt{n}$	$\sqrt{10n}$	$\sqrt[3]{n}$	$\sqrt[3]{10n}$	$\sqrt[3]{100n}$

n	$1/n$	n^2	n^3	$\sqrt{n}$	$\sqrt{10n}$	$\sqrt[3]{n}$	$\sqrt[3]{10n}$	$\sqrt[3]{100n}$
2.00	.500000	4.0000	8.00000	1.41421	4.47214	1.25992	2.71442	5.84804
.01	.497512	4.0401	8.12060	.41774	.48330	.26202	.71893	.85777
.02	.495050	4.0804	8.24241	.42127	.49444	.26411	.72344	.86746
.03	.492611	4.1209	8.36543	.42478	.50555	.26619	.72792	.87713
.04	.490196	4.1616	8.48966	.42829	.51664	.26827	.73239	.88677
2.05	.487805	4.2025	8.61512	1.43178	4.52769	1.27033	2.73685	5.89637
.06	.485437	4.2436	8.74182	.43527	.53872	.27240	.74129	.90594
.07	.483092	4.2849	8.86974	.43875	.54973	.27445	.74572	.91548
.08	.480769	4.3264	8.99891	.44222	.56070	.27650	.75014	.92499
.09	.478469	4.3681	9.12933	.44568	.57165	.27854	.75454	.93447
2.10	.476190	4.4100	9.26100	1.44914	4.58258	1.28058	2.75892	5.94392
.11	.473934	4.4521	9.39393	.45258	.59347	.28261	.76330	.95334
.12	.471698	4.4944	9.52813	.45602	.60435	.28463	.76766	.96273
.13	.469484	4.5369	9.66360	.45945	.61519	.28665	.77200	.97209
.14	.467290	4.5796	9.80034	.46287	.62601	.28866	.77633	.98142
2.15	.465116	4.6225	9.93838	1.46629	4.63681	1.29066	2.78065	5.99073
.16	.462963	4.6656	10.0777	.46969	.64758	.29266	.78495	6.00000
.17	.460829	4.7089	10.2183	.47309	.65833	.29465	.78924	.00925
.18	.458716	4.7524	10.3602	.47648	.66905	.29664	.79352	.01846
.19	.456621	4.7961	10.5035	.47986	.67974	.29862	.79779	.02765
2.20	.454545	4.8400	10.6480	1.48324	4.69042	1.30059	2.80204	6.03681
.21	.452489	4.8841	10.7939	.48661	.70106	.30256	.80628	.04594
.22	.450450	4.9284	10.9410	.48997	.71169	.30452	.81050	.05505
.23	.448430	4.9729	11.0896	.49332	.72229	.30648	.81472	.06413
.24	.446429	5.0176	11.2394	.49666	.73286	.30843	.81892	.07318
2.25	.444444	5.0625	11.3906	1.50000	4.74342	1.31037	2.82311	6.08220
.26	.442478	5.1076	11.5432	.50333	.75395	.31231	.82728	.09120
.27	.440529	5.1529	11.6971	.50665	.76445	.31424	.83145	.10017
.28	.438596	5.1984	11.8524	.50997	.77493	.31617	.83560	.10911
.29	.436681	5.2441	12.0090	.51327	.78539	.31809	.83974	.11803
2.30	.434783	5.2900	12.1670	1.51658	4.79583	1.32001	2.84387	6.12693
.31	.432900	5.3361	12.3264	.51987	.80625	.32192	.84798	.13579
.32	.431034	5.3824	12.4872	.52315	.81664	.32382	.85209	.14463
.33	.429185	5.4289	12.6493	.52643	.82701	.32572	.85618	.15345
.34	.427350	5.4756	12.8129	.52971	.83735	.32761	.86026	.16224
2.35	.425532	5.5225	12.9779	1.53297	4.84768	1.32950	2.86433	6.17101
.36	.423729	5.5696	13.1443	.53623	.85798	.33139	.86838	.17975
.37	.421941	5.6169	13.3121	.53948	.86826	.33326	.87243	.18846
.38	.420168	5.6644	13.4813	.54272	.87852	.33514	.87646	.19715
.39	.418410	5.7121	13.6519	.54596	.88876	.33700	.88049	.20582
2.40	.416667	5.7600	13.8240	1.54919	4.89898	1.33887	2.88450	6.21447
.41	.414938	5.8081	13.9975	.55242	.90918	.34072	.88850	.22308
.42	.413223	5.8564	14.1725	.55563	.91935	.34257	.89249	.23168
.43	.411523	5.9049	14.3489	.55885	.92950	.34442	.89647	.24025
.44	.409836	5.9536	14.5268	.56205	.93964	.34626	.90044	.24880
2.45	.408163	6.0025	14.7061	1.56525	4.94975	1.34810	2.90439	6.25732
.46	.406504	6.0516	14.8869	.56844	.95984	.34993	.90834	.26583
.47	.404858	6.1009	15.0692	.57162	.96991	.35176	.91227	.27431
.48	.403226	6.1504	15.2530	.57480	.97996	.35358	.91620	.28276
.49	.401606	6.2001	15,4382	.57797	.98999	.35540	.92011	.29119
2.50	.400000	6.2500	15,6250	1.58114	5.00000	1.35721	2.92402	6.29961
n	$1/n$	n^2	n^3	$\sqrt{n}$	$\sqrt{10n}$	$\sqrt[3]{n}$	$\sqrt[3]{10n}$	$\sqrt[3]{100n}$

n	$1/n$	n^2	n^3	$\sqrt{n}$	$\sqrt{10n}$	$\sqrt[3]{n}$	$\sqrt[3]{10n}$	$\sqrt[3]{100n}$
2.50	.400000	6.2500	15.6250	1.58114	5.00000	1.35721	2.92402	6.29961
.51	.398406	6.3001	15.8133	.58430	.00999	.35902	.92791	.30799
.52	.396825	6.3504	16.0030	.58745	.01996	.36082	.93179	.31636
.53	.395257	6.4009	16.1943	.59060	.02991	.36262	.93567	.32470
.54	.393701	6.4516	16.3871	.59374	.03984	.36441	.93953	.33303
2.55	.392157	6.5025	16.5814	1.59687	5.04975	1.36620	2.94338	6.34133
.56	.390625	6.5536	16.7772	.60000	.05964	.36798	.94723	.34960
.57	.389105	6.6049	16.9746	.60312	.06952	.36976	.95106	.35786
.58	.387597	6.6564	17.1735	.60624	.07937	.37153	.95488	.36610
.59	.386100	6.7081	17.3740	.60935	.08920	.37330	.95869	.37431
2.60	.384615	6.7600	17.5760	1.61245	5.09902	1.37507	2.96250	6.38250
.61	.383142	6.8121	17.7796	.61555	.10882	.37683	.96629	.39068
.62	.381679	6.8644	17.9847	.61864	.11859	.37859	.97007	.39883
.63	.380228	6.9169	18.1914	.62173	.12835	.38034	.97385	.40696
.64	.378788	6.9696	18.3997	.62481	.13809	.38208	.97761	.41507
2.65	.377358	7.0225	18.6096	1.62788	5.14782	1.38383	2.98137	6.42316
.66	.375940	7.0756	18.8211	.63095	.15752	.38557	.98511	.43123
.67	.374532	7.1289	19.0342	.63401	.16720	.38730	.98885	.43928
.68	.373134	7.1824	19.2488	.63707	.17687	.38903	.99257	.44731
.69	.371747	7.2361	19.4651	.64012	.18652	.39076	.99629	.45531
2.70	.370370	7.2900	19.6830	1.64317	5.19615	1.39248	3.00000	6.46330
.71	.369004	7.3441	19.9025	.64621	.20577	.39419	.00370	.47127
.72	.367647	7.3984	20.1236	.64924	.21536	.39591	.00739	.47922
.73	.366300	7.4529	20.3464	.65227	.22494	.39761	.01107	.48715
.74	.364964	7.5076	20.5708	.65529	.23450	.39932	.01474	.49507
2.75	.363636	7.5625	20.7969	1.65831	5.24404	1.40102	3.01841	6.50296
.76	.362319	7.6176	21.0246	.66132	.25357	.40272	.02206	.51083
.77	.361011	7.6729	21.2539	.66433	.26308	.40441	.02570	.51868
.78	.359712	7.7284	21.4850	.66733	.27257	.40610	.02934	.52652
.79	.358423	7.7841	21.7176	.67033	.28205	.40778	.03297	.53434
2.80	.357143	7.8400	21.9520	1.67332	5.29150	1.40946	3.03659	6.54213
.81	.355872	7.8961	22.1880	.67631	.30094	.41114	.04020	.54991
.82	.354610	7.9524	22.4258	.67929	.31037	.41281	.04380	.55767
.83	.353357	8.0089	22.6652	.68226	.31977	.41448	.04740	.56541
.84	.352113	8.0656	22.9063	.68523	.32917	.41614	.05098	.57314
2.85	.350877	8.1225	23.1491	1.68819	5.33854	1.41780	3.05456	6.58084
.86	.349650	8.1796	23.3937	.69115	.34790	.41946	.05813	.58853
.87	.348432	8.2369	23.6399	.69411	.35724	.42111	.06169	.59620
.88	.347222	8.2944	23.8879	.69706	.36656	.42276	.06524	.60385
.89	.346021	8.3521	24.1376	.70000	.37587	.42440	.06878	.61149
2.90	.344828	8.4100	24.3890	1.70294	5.38516	1.42604	3.07232	6.61911
.91	.343643	8.4681	24.6422	.70587	.39444	.42768	.07584	.62671
.92	.342466	8.5264	24.8971	.70880	.40370	.42931	.07936	.63429
.93	.341297	8.5849	25.1538	.71172	.41295	.43094	.08287	.64185
.94	.340136	8.6436	25.4122	.71464	.42218	.43257	.08638	.64940
2.95	.338983	8.7025	25.6724	1.71756	5.43139	1.43419	3.08987	6.65693
.96	.337838	8.7616	25.9343	.72047	.44059	.43581	.09336	.66444
.97	.336700	8.8209	26.1981	.72337	.44977	.43743	.09684	.67194
.98	.335570	8.8804	26.4636	.72627	.45894	.43904	.10031	.67942
.99	.334448	8.9401	26.7309	.72916	.46809	.44065	.10378	.68688
3.00	.333333	9.0000	27.0000	1.73205	5.47723	1.44225	3.10723	6.69433
n	$1/n$	n^2	n^3	$\sqrt{n}$	$\sqrt{10n}$	$\sqrt[3]{n}$	$\sqrt[3]{10n}$	$\sqrt[3]{100n}$

n	$1/n$	n^2	n^3	$\sqrt{n}$	$\sqrt{10n}$	$\sqrt[3]{n}$	$\sqrt[3]{10n}$	$\sqrt[3]{100n}$
3.00	.333333	9.0000	27.0000	1.73205	5.47723	1.44225	3.10723	6.69433
.01	.332226	9.0601	27.2709	.73494	.48635	.44385	.11068	.70176
.02	.331126	9.1204	27.5436	.73781	.49545	.44545	.11412	.70917
.03	.330033	9.1809	27.8181	.74069	.50454	.44704	.11756	.71657
.04	.328947	9.2416	28.0945	.74356	.51362	.44863	.12098	.72395
3.05	.327869	9.3025	28.3726	1.74642	5.52268	1.45022	3.12440	6.73132
.06	.326797	9.3636	28.6526	.74929	.53173	.45180	.12781	.73866
.07	.325733	9.4249	28.9344	.75214	.54076	.45338	.13121	.74600
.08	.324675	9.4864	29.2181	.75499	.54977	.45496	.13461	.75331
.09	.323625	9.5481	29.5036	.75784	.55878	.45653	.13800	.76061
3.10	.322581	9.6100	29.7910	1.76068	5.56776	1.45810	3.14138	6.76790
.11	.321543	9.6721	30.0802	.76352	.57674	.45967	.14475	.77517
.12	.320513	9.7344	30.3713	.76635	.58570	.46123	.14812	.78242
.13	.319489	9.7969	30.6643	.76918	.59464	.46279	.15148	.78966
.14	.318471	9.8596	30.9591	.77200	.60357	.46434	.15483	.79688
3.15	.317460	9.9225	31.2559	1.77482	5.61249	1.46590	3.15818	6.80409
.16	.316456	9.9856	31.5545	.77764	.62139	.46745'	.16152	.81128
.17	.315457	10.0489	31.8550	.78045	.63028	.46899	.16485	.81846
.18	.314465	10.1124	32.1574	.78326	.63915	.47054	.16817	.82562
.19	.313480	10.1761	32.4618	.78606	.64801	.47208	.17149	.83277
3.20	.312500	10.2400	32.7680	1.78885	5.65685	1.47361	3.17480	6.83990
.21	.311526	10.3041	33.0762	.79165	.66569	.47515	.17811	.84702
.22	.310559	10.3684	33.3862	.79444	.67450	.47668	.18140	.85412
.23	.309598	10.4329	33.6983	.79722	.68331	.47820	.18469	.86121
.24	.308642	10.4976	34.0122	.80000	.69210	.47973	.18798	.86829
3.25	.307692	10.5625	34.3281	1.80278	5.70088	1.48125	3.19125	6.87534
.26	.306748	10.6276	34.6460	.80555	.70964	.48277	.19452	.88239
.27	.305810	10.6929	34.9658	.80831	.71839	.48428	.19778	.88942
.28	.304878	10.7584	35.2876	.81108	.72713	.48579	.20104	.89643
.29	.303951	10.8241	35.6113	.81384	.73585	.48730	.20429	.90344
3.30	.303030	10.8900	35.9370	1.81659	5.74456	1.48881	3.20753	6.91042
.31	.302115	10.9561	36.2647	.81934	.75326	.49031	.21077	.91740
.32	.301205	11.0224	36.5944	.82209	.76194	.49181	.21400	.92436
.33	.300300	11.0889	36.9260	.82483	.77062	.49330	.21722	.93130
.34	.299401	11.1556	37.2597	.82757	.77927	.49480	.22044	.93823
3.35	.298507	11.2225	37.5954	1.83030	5.78792	1.49629	3.22365	6.94515
.36	.297619	11.2896	37.9331	.83303	.79655	.49777	.22686	.95205
.37	.296736	11.3569	38.2728	.83576	.80517	.49926	.23006	.95894
.38	.295858	11.4244	38.6145	.83848	.81378	.50074	.23325	.96582
.39	.294985	11.4921	38.9582	.84120	.82237	.50222	.23643	.97268
3.40	.294118	11.5600	39.3040	1.84391	5.83095	1.50369	3.23961	6.97953
.41	.293255	11.6281	39.6518	.84662	.83952	.50517	.24278	.98637
.42	.292398	11.6964	40.0017	.84932	.84808	.50664	.24595	.99319
.43	.291545	11.7649	40.3536	.85203	.85662	.50810	.24911	7.00000
.44	.290698	11.8336	40.7076	.85472	.86515	.50957	.25227	.00680
3.45	.289855	11.9025	41.0636	1.85742	5.87367	1.51103	3.25542	7.01358
.46	.289017	11.9716	41.4217	.86011	.88218	.51249	.25856	.02035
.47	.288184	12.0409	41.7819	.86279	.89067	.51394	.26169	.02711
.48	.287356	12.1104	42.1442	.86548	.89915	.51540	.26482	.03385
.49	.286533	12.1801	42.5085	.86815	.90762	.51685	.26795	.04058
3.50	.285714	12.2500	42.8750	1.87083	5.91608	1.51829	3.27107	7.04730
n	$1/n$	n^2	n^3	$\sqrt{n}$	$\sqrt{10n}$	$\sqrt[3]{n}$	$\sqrt[3]{10n}$	$\sqrt[3]{100n}$

n	$1/n$	n^2	n^3	$\sqrt{n}$	$\sqrt{10n}$	$\sqrt[3]{n}$	$\sqrt[3]{10n}$	$\sqrt[3]{100n}$
3.50	.285714	12.2500	42.8750	1.87083	5.91608	1.51829	3.27107	7.04730
.51	.284900	12.3201	43.2436	.87350	.92453	.51974	.27418	.05400
.52	.284091	12.3904	43.6142	.87617	.93296	.52118	.27729	.06070
.53	.283286	12.4609	43.9870	.87883	.94138	.52262	.28039	.06738
.54	.282486	12.5316	44.3619	.88149	.94979	.52406	.28348	.07404
3.55	.281690	12.6025	44.7389	1.88414	5.95819	1.52549	3.28657	7.08070
.56	.280899	12.6736	45.1180	.88680	.96657	.52692	.28965	.08734
.57	.280112	12.7449	45.4993	.88944	.97495	.52835	.29273	.09397
.58	.279330	12.8164	45.8827	.89209	.98331	.52978	.29580	.10059
.59	.278552	12.8881	46.2683	.89473	.99166	.53120	.29887	.10719
3.60	.277778	12.9600	46.6560	1.89737	6.00000	1.53262	3.30193	7.11379
.61	.277008	13.0321	47.0459	.90000	.00833	.53404	.30498	.12037
.62	.276243	13.1044	47.4379	.90263	.01664	.53545	.30803	.12694
.63	.275482	13.1769	47.8321	.90526	.02495	.53686	.31107	.13349
.64	.274725	13.2496	48.2285	.90788	.03324	.53827	.31411	.14004
3.65	.273973	13.3225	48.6271	1.91050	6.04152	1.53968	3.31714	7.14657
.66	.273224	13.3956	49.0279	.91311	.04979	.54109	.32017	.15309
.67	.272480	13.4689	49.4309	.91572	.05805	.54249	.32319	.15960
.68	.271739	13.5424	49.8360	.91833	.06630	.54389	.32621	.16610
.69	.271003	13.6161	50.2434	.92094	.07454	.54529	.32922	.17258
3.70	.270270	13.6900	50.6530	1.92354	6.08276	1.54668	3.33222	7.17905
.71	.269542	13.7641	51.0648	.92614	.09098	.54807	.33522	.18552
.72	.268817	13.8384	51.4788	.92873	.09918	.54946	.33822	.19197
.73	.268097	13.9129	51.8951	.93132	.10737	.55085	.34120	.19840
.74	.267380	13.9876	52.3136	.93391	.11555	.55223	.34419	.20483
3.75	.266667	14.0625	52.7344	1.93649	6.12372	1.55362	3.34716	7.21125
.76	.265957	14.1376	53.1574	.93907	.13188	.55500	.35014	.21765
.77	.265252	14.2129	53.5826	.94165	.14003	.55637	.35310	.22405
.78	.264550	14.2884	54.0102	.94422	.14817	.55775	.35607	.23043
.79	.263852	14.3641	54.4399	.94679	.15630	.55912	.35902	.23680
3.80	.263158	14.4400	54.8720	1.94936	6.16441	1.56049	3.36198	7.24316
.81	.262467	14.5161	55.3063	.95192	.17252	.56186	.36492	.24950
.82	.261780	14.5924	55.7430	.95448	.18061	.56322	.36786	.25584
.83	.261097	14.6689	56.1819	.95704	.18870	.56459	.37080	.26217
.84	.260417	14.7456	56.6231	.95959	.19677	.56595	.37373	.26848
3.85	.259740	14.8225	57.0666	1.96214	6.20484	1.56731	3.37666	7.27479
.86	.259067	14.8996	57.5125	.96469	.21289	.56866	.37958	.28108
.87	.258398	14.9769	57.9606	.96723	.22093	.57001	.38249	.28736
.88	.257732	15.0544	58.4111	.96977	.22896	.57137	.38540	.29363
.89	.257069	15.1321	58.8639	.97231	.23699	.57271	.38831	.29989
3.90	.256410	15.2100	59.3190	1.97484	6.24500	1.57406	3.39121	7.30614
.91	.255754	15.2881	59.7765	.97737	.25300	.57541	.39411	.31238
.92	.255102	15.3664	60.2363	.97990	.26099	.57675	.39700	.31861
.93	.254453	15.4449	60.6985	.98242	.26897	.57809	.39988	.32483
.94	.253807	15.5236	61.1630	.98494	.27694	.57942	.40277	.33104
3.95	.253165	15.6025	61.6299	1.98746	6.28490	1.58076	3.40564	7.33723
.96	.252525	15.6816	62.0991	.98997	.29285	.58209	.40851	.34342
.97	.251889	15.7609	62.5708	.99249	.30079	.58342	.41138	.34960
.98	.251256	15.8404	63.0448	.99499	.30872	.58475	.41424	.35576
.99	.250627	15.9201	63.5212	.99750	.31664	.58608	.41710	.36192
4.00	.250000	16.0000	64.0000	2.00000	6.32456	1.58740	3.41995	7.36806
n	$1/n$	n^2	n^3	$\sqrt{n}$	$\sqrt{10n}$	$\sqrt[3]{n}$	$\sqrt[3]{10n}$	$\sqrt[3]{100n}$

n	$1/n$	n^2	n^3	$\sqrt{n}$	$\sqrt{10n}$	$\sqrt[3]{n}$	$\sqrt[3]{10n}$	$\sqrt[3]{100n}$
4.00	.250000	16.0000	64.0000	2.00000	6.32456	1.58740	3.41995	7.36806
.01	.249377	16.0801	64.4812	.00250	.33246	.58872	.42280	.37420
.02	.248756	16.1604	64.9648	.00499	.34035	.59004	.42564	.38032
.03	.248139	16.2409	65.4508	.00749	.34823	.59136	.42848	.38644
.04	.247525	16.3216	65.9393	.00998	.35610	.59267	.43131	.39254
4.05	.246914	16.4025	66.4301	2.01246	6.36396	1.59399	3.43414	7.39864
.06	.246305	16.4836	66.9234	.01494	.37181	.59530	.43697	.40472
.07	.245700	16.5649	67.4191	01742	.37966	.59661	.43979	.41080
.08	.245098	16.6464	67.9173	.01990	.38749	.59791	.44260	.41686
.09	.244499	16.7281	68.4179	.02237	.39531	.59922	.44541	.42291
4.10	.243902	16.8100	68.9210	2.02485	6.40312	1.60052	3.44822	7.42896
.11	.243309	16.8921	69.4265	.02731	.41093	.60182	.45102	.43499
.12	.242718	16.9744	69.9345	.02978	.41872	.60312	.45382	.44102
.13	.242131	17.0569	70.4450	.03224	.42651	.60441	.45661	.44703
.14	.241546	17.1396	70.9579	.03470	.43428	.60571	.45939	.45304
4.15	.240964	17.2225	71.4734	2.03715	6.44205	1.60700	3.46218	7.45904
.16	.240385	17.3056	71.9913	.03961	.44981	.60829	.46496	.46502
.17	.239808	17.3889	72.5117	.04206	.45755	.60958	.46773	.47100
.18	.239234	17.4724	73.0346	.04450	.46529	.61086	.47050	.47697
.19	.238663	17.5561	73.5601	.04695	.47302	.61215	.47327	.48292
4.20	.238095	17.6400	74.0880	2.04939	6.48074	1.61343	3.47603	7.48887
.21	.237530	17.7241	74.6185	.05183	.48845	.61471	.47878	.49481
.22	.236967	17.8084	75.1514	.05426	.49615	.61599	.48154	.50074
.23	.236407	17.8929	75.6870	.05670	.50384	.61726	.48428	.50666
.24	.235849	17.9776	76.2250	.05913	.51153	.61853	.48703	.51257
4.25	.235294	18.0625	76.7656	2.06155	6.51920	1.61981	3.48977	7.51847
.26	.234742	18.1476	77.3088	.06398	.52687	.62108	.49250	.52437
.27	.234192	18.2329	77.8545	.06640	.53452	.62234	.49523	.53025
.28	.233645	18.3184	78.4028	.06882	.54217	.62361	.49796	.53612
.29	.233100	18.4041	78.9536	.07123	.54981	.62487	.50068	.54199
4.30	.232558	18.4900	79.5070	2.07364	6.55744	1.62613	3.50340	7.54784
.31	.232019	18.5761	80.0630	.07605	.56506	.62739	.50611	.55369
.32	.231481	18.6624	80.6216	.07846	.57267	.62865	.50882	.55953
.33	.230947	18.7489	81.1827	.08087	.58027	.62991	.51153	.56535
.34	.230415	18.8356	81.7465	.08327	.58787	.63116	.51423	.57117
4.35	.229885	18.9225	82.3129	2.08567	6.59545	1.63241	3.51692	7.57698
.36	.229358	19.0096	82.8819	.08806	.60303	.63366	.51962	.58279
.37	.228833	19.0969	83.4535	.09045	.61060	.63491	.52231	.58858
.38	.228311	19.1844	84.0277	.09284	.61816	.63619	.52499	.59436
.39	.227790	19.2721	84.6045	.09523	.62571	.63740	.52767	.60014
4.40	.227273	19.3600	85.1840	2.09762	6.63325	1.63864	3.53035	7.60590
.41	.226757	19.4481	85.7661	.10000	.64078	.63988	.53302	.61166
.42	.226244	19.5364	86.3509	.10238	.64831	.64112	.53569	.61741
.43	.225734	19.6249	86.9383	.10476	.65582	.64236	.53835	.62315
.44	.225225	19.7136	87.5284	.10713	.66333	.64359	.54101	.62888
4.45	.224719	19.8025	88.1211	2.10950	6.67083	1.64483	3.54367	7.63461
.46	.224215	19.8916	88.7165	.11187	.67832	.64606	.54632	.64032
.47	.223714	19.9809	89.3146	.11424	.68581	.64729	.54897	.64603
.48	.223214	20.0704	89.9154	.11660	.69328	.64851	.55162	.65172
.49	.222717	20.1601	90.5188	.11896	.70075	.64974	.55426	.65741
4.50	.222222	20.2500	91.1250	2.12132	6.70820	1.65096	3.55689	7.66309
n	$1/n$	n^2	n^3	$\sqrt{n}$	$\sqrt{10n}$	$\sqrt[3]{n}$	$\sqrt[3]{10n}$	$\sqrt[3]{100n}$

n	$1/n$	n^2	n^3	$\sqrt{n}$	$\sqrt{10n}$	$\sqrt[3]{n}$	$\sqrt[3]{10n}$	$\sqrt[3]{100n}$
4.50	.222222	20.2500	91.1250	2.12132	6.70820	1.65096	3.55689	7.66309
.51	.221729	20.3401	91.7339	.12368	.71565	.65219	.55953	.66877
.52	.221239	20.4304	92.3454	.12603	.72309	.65341	.56215	.67443
.53	.220751	20.5209	92.9597	.12838	.73053	.65462	.56478	.68009
.54	.220264	20.6116	93.5767	.13073	.73795	.65584	.56740	.68573
4.55	.219780	20.7025	94.1964	2.13307	6.74537	1.65706	3.57002	7.69137
.56	.219298	20.7936	94.8188	.13542	.75278	.65827	.57263	.69700
.57	.218818	20.8849	95.4440	.13776	.76018	.65948	.57524	.70262
.58	.218341	20.9764	96.0719	.14009	.76757	.66069	.57785	.70824
.59	.217865	21.0681	96.7026	.14243	.77495	.66190	.58045	.71384
4.60	.217391	21.1600	97.3360	2.14476	6.78233	1.66310	3.58305	7.71944
.61	.216920	21.2521	97.9722	.14709	.78970	.66431	.58564	.72503
.62	.216450	21.3444	98.6111	.14942	.79706	.66551	.58823	.73061
.63	.215983	21.4369	99.2528	.15174	.80441	.66671	.59082	.73619
.64	.215517	21.5296	99.8973	.15407	.81175	.66791	.59340	.74175
4.65	.215054	21.6225	100.545	2.15639	6.81909	1.66911	3.59598	7.74731
.66	.214592	21.7156	101.195	.15870	.82642	.67030	.59856	.75286
.67	.214133	21.8089	101.848	.16102	.83374	.67150	.60113	.75840
.68	.213675	21.9024	102.503	.16333	.84105	.67269	.60370	.76394
.69	.213220	21.9961	103.162	.16564	.84836	.67388	.60626	.76946
4.70	.212766	22.0900	103.823	2.16795	6.85565	1.67507	3.60883	7.77498
.71	.212314	22.1841	104.487	.17025	.86294	.67626	.61138	.78049
.72	.211864	22.2784	105.154	.17256	.87023	.67744	.61394	.78599
.73	.211416	22.3729	105.824	.17486	.87750	.67863	.61649	.79149
.74	.210970	22.4676	106.496	.17715	.88477	.67981	.61903	.79697
4.75	.210526	22.5625	107.172	2.17945	6.89202	1.68099	3.62158	7.80245
.76	.210084	22.6576	107.850	.18174	.89928	.68217	.62412	.80793
.77	.209644	22.7529	108.531	.18403	.90652	.68334	.62665	.81339
.78	.209205	22.8484	109.215	.18632	.91375	.68452	.62919	.81885
.79	.208768	22.9441	109.902	.18861	.92098	.68569	.63172	.82429
4.80	.208333	23.0400	110.592	2.19089	6.92820	1.68687	3.63424	7.82974
.81	.207900	23.1361	111.285	.19317	.93542	.68804	.63676	.83517
.82	.207469	23.2324	111.980	.19545	.94262	.68920	.63928	.84059
.83	.207039	23.3289	112.679	.19773	.94982	.69037	.64180	.84601
.84	.206612	23.4256	113.380	2.20000	.95701	.69154	.64431	.85142
4.85	.206186	23.5225	114.084	2.20227	6.96419	1.69270	3.64682	7.85683
.86	.205761	23.6196	114.791	.20454	.97137	.69386	.64932	.86222
.87	.205339	23.7169	115.501	.20681	.97854	.69503	.65182	.86761
.88	.204918	23.8144	116.214	.20907	.98570	.69619	.65432	.87299
.89	.204499	23.9121	116.930	.21133	.99285	.69734	.65681	.87837
4.90	.204082	24.0100	117.649	2.21359	7.00000	1.69850	3.65931	7.88374
.91	.203666	24.1081	118.371	.21585	.00714	.69965	.66179	.88909
.92	.203252	24.2064	119.095	.21811	.01427	.70081	.66428	.89445
.93	.202840	24.3049	119.823	.22036	.02140	.70196	.66676	.89979
.94	.202429	24.4036	120.554	.22261	.02851	.70311	.66924	.90513
4.95	.202020	24.5025	121.287	2.22486	7.03562	1.70426	3.67171	7.91046
.96	.201613	24.6016	122.024	.22711	.04273	.70540	.67418	.91578
.97	.201207	24.7009	122.763	.22935	.04982	.70655	.67665	.92110
.98	.200803	24.8004	123.506	.23159	.05691	.70769	.67911	.92641
.99	.200401	24.9001	124.251	.23383	.06399	.70884	.68157	.93171
5.00	.200000	25.0000	125.000	2.23607	7.07107	1.70998	3.68403	7.93701
n	$1/n$	n^2	n^3	$\sqrt{n}$	$\sqrt{10n}$	$\sqrt[3]{n}$	$\sqrt[3]{10n}$	$\sqrt[3]{100n}$

n	$1/n$	n^2	n^3	$\sqrt{n}$	$\sqrt{10n}$	$\sqrt[3]{n}$	$\sqrt[3]{10n}$	$\sqrt[3]{100n}$
5.00	.200000	25.0000	125.000	2.23607	7.07107	1.70998	3.68403	7.93701
.01	.199601	25.1001	125.752	.23830	.07814	.71112	.68649	.94229
.02	.199203	25.2004	126.506	.24054	.08520	.71225	.68894	.94757
.03	.198807	25.3009	127.264	.24277	.09225	.71339	.69138	.95285
.04	.198413	25.4016	128.024	.24499	.09930	.71452	.69383	.95811
5.05	.198020	25.5025	128.788	2.24722	7.10634	1.71566	3.69627	7.96337
.06	.197628	25.6036	129.554	.24944	.11337	.71679	.69871	.96863
.07	.197239	25.7049	130.324	.25167	12039	.71792	.70114	.97387
.08	.196850	25.8064	131.097	.25389	.12741	.71905	.70357	.97911
.09	.196464	25.9081	131.872	.25610	.13442	.72017	.70600	.98434
5.10	.196078	26.0100	132.651	2.25832	7.14143	1.72130	3.70843	7.98957
.11	.195695	26.1121	133.433	.26053	.14843	.72242	.71085	.99479
.12	.195312	26.2144	134.218	.26274	.15542	.72355	.71327	8.00000
.13	.194932	26.3169	135.006	.26495	.16240	.72467	.71569	.00520
.14	.194553	26.4196	135.797	.26716	.16938	.72579	.71810	.01040
.15	.194175	26.5225	136.591	2.26936	7.17635	1.72691	3.72051	8.01559
.16	.193798	26.6256	137.388	.27156	.18331	.72802	.72292	.02078
.17	.193424	26.7289	138.188	.27376	.19027	.72914	.72532	.02596
.18	.193050	26.8324	138.992	.27596	.19722	.73025	.72772	.03113
.19	.192678	26.9361	139.798	.27816	.20417	.73137	.73012	.03629
5.20	.192308	27.0400	140.608	2.28035	7.21110	1.73248	3.73251	8.04145
.21	.191939	27.1441	141.421	.28254	.21803	.73359	.73490	.04660
.22	.191571	27.2484	142 237	.28473	.22496	.73470	.73729	.05175
.23	.191205	27.3529	143.056	.28692	.23187	.73580	.73968	.05689
.24	.190840	27.4576	143.878	.28910	.23878	.73691	.74206	.06202
5.25	.190476	27.5625	144.703	2.29129	7.24569	1.73801	3.74443	8.06714
.26	.190114	27.6676	145.532	.29347	.25259	.73912	.74681	.07226
.27	.189753	27.7729	146.363	.29565	.25948	.74022	.74918	.07737
.28	.189394	27.8784	147.198	.29783	.26636	.74132	.75155	.08248
.29	.189036	27.9841	148.036	.30000	.27324	.74242	.75392	.08758
5.30	.188679	28.0900	148.877	2.30217	7.28011	1.74351	3.75629	8.09267
.31	.188324	28.1961	149.721	.30434	.28697	.74461	.75865	.09776
.32	.187970	28.3024	150.569	.30651	.29383	.74570	.76101	.10284
.33	.187617	28.4089	151.419	.30868	.30068	.74680	.76336	.10791
.34	.187266	28.5156	152.273	.31084	.30753	.74789	.76571	.11298
5.35	.186916	28.6225	153.130	2.31301	7.31437	1.74898	3.76806	8.11804
.36	.186567	28.7296	153.991	.31517	.32120	.75007	.77041	.12310
.37	.186220	28.8369	154.854	.31733	.32803	.75116	.77275	.12814
.38	.185874	28.9444	155.721	.31948	.33485	.75224	.77509	.13319
.39	.185529	29.0521	156.591	.32164	.34166	.75333	.77743	.13822
5.40	.185185	29.1600	157.464	2.32379	7.34847	1.75441	3.77976	8.14325
.41	.184843	29.2681	158.340	.32594	.35527	.75549	.78209	.14828
.42	.184502	29.3764	159.220	.32809	.36206	.75657	.78442	.15329
.43	.184162	29.4849	160.103	.33024	.36885	.75765	.78675	.15831
.44	.183824	29.5936	160.989	.33238	.37564	.75873	.78907	.16331
5.45	.183486	29.7025	161.879	2.33452	7.38241	1.75981	3.79139	8.16831
.46	.183150	29.8116	162.771	.33666	.38918	.76088	.79371	.17330
.47	.182815	29.9209	163.667	.33880	.39594	.76196	.79603	.17829
.48	.182482	30.0304	164.567	.34094	.40270	.76303	.79834	.18327
.49	.182149	30.1401	165.469	.34307	.40945	.76410	.80065	.18824
5.50	.181818	30.2500	166.375	2.34521	7.41620	1.76517	3.80295	8.19321
n	$1/n$	n^2	n^3	$\sqrt{n}$	$\sqrt{10n}$	$\sqrt[3]{n}$	$\sqrt[3]{10n}$	$\sqrt[3]{100n}$

n	$1/n$	n^2	n^3	$\sqrt{n}$	$\sqrt{10n}$	$\sqrt[3]{n}$	$\sqrt[3]{10n}$	$\sqrt[3]{100n}$
5.50	.181818	30.2500	166.375	2.34521	7.41620	1.76517	3.80295	8.19321
.51	.181488	30.3601	167.284	.34734	.42294	.76624	.80526	.19818
.52	.181159	30.4704	168.197	.34947	.42967	.76731	.80756	.20313
.53	.180832	30.5809	169.112	.35160	.43640	.76838	.80985	.20808
.54	.180505	30.6916	170,031	.35372	.44312	.76944	.81215	.21303
5.55	.180180	30.8025	170.954	2.35584	7.44983	1.77051	3.81444	8.21797
.56	.179856	30.9136	171.880	.35797	.45654	.77157	.81673	.22290
.57	.179533	31.0249	172.809	.36008	.46324	.77263	.81902	.22783
.58	.179211	31.1364	173.741	.36220	.46994	.77369	.82130	.23275
.59	.178891	31.2481	174.677	.36432	.47663	.77475	.82358	.23766
5.60	.178571	31.3600	175.616	2.36643	7.48331	1.77581	3.82586	8.24257
.61	.178253	31.4721	176.558	.36854	.48999	.77686	.82814	.24747
.62	.177936	31.5844	177.504	.37065	.49667	.77792	.83041	.25237
.63	.177620	31.6969	178.454	.37276	.50333	.77897	.83268	.25726
.64	.177305	31.8096	179.406	.37487	.50999	.78003	.83495	.26215
5.65	.176991	31.9225	180.362	2.37697	7.51665	1.78108	3.83722	8.26703
.66	.176678	32.0356	181.321	.37908	.52330	.78213	.83948	.27190
.67	.176367	32.1489	182.284	.38118	.52994	.78318	.84174	.27677
.68	.176056	32.2624	183.250	.38328	.53658	.78422	.84399	.28164
.69	.175747	32.3761	184.220	.38537	.54321	.78527	.84625	.28649
5.70	.175439	32.4900	185.193	2.38747	7.54983	1.78632	3.84850	8.29134
.71	.175131	32.6041	186.169	.38956	.55645	.78736	.85075	.29619
.72	.174825	32.7184	187.149	.39165	.56307	.78840	.85300	.30103
.73	.174520	32.8329	188.133	.39374	.56968	.78944	.85524	.30587
.74	.174216	32.9476	189.119	.39583	.57628	.79048	.85748	.31069
5.75	.173913	33.0625	190.109	2.39792	7.58288	1.79152	3.85972	8.31552
.76	.173611	33.1776	191.103	.40000	.58947	.79256	.86196	.32034
.77	.173310	33.2929	192.100	.40208	.59605	.79360	.86419	.32515
.78	.173010	33.4084	193.101	.40416	.60263	.79463	.86642	.32995
.79	.172712	33.5241	194.105	.40624	.60920	.79567	.86865	.33476
5.80	.172414	33.6400	195.112	2.40832	7.61577	1.79670	3.87088	8.33955
.81	.172117	33.7561	196.123	.41039	.62234	.79773	.87310	.34434
.82	.171821	33.8724	197.137	.41247	.62889	.79876	.87532	.34913
.83	.171527	33.9889	198.155	.41454	.63544	.79979	.87754	.35390
.84	.171233	34.1056	199.177	.41661	.64199	.80082	.87975	.35868
5.85	.170940	34.2225	200.202	2.41868	7.64853	1.80185	3.88197	8.36345
.86	.170649	34.3396	201.230	.42074	.65506	.80288	.88418	.36821
.87	.170358	34.4569	202.262	.42281	.66159	.80390	.88639	.37297
.88	.170068	34.5744	203.297	.42487	.66812	.80492	.88859	.37772
.89	.169779	34.6921	204.336	.42693	.67463	.80595	.89080	.38247
5.90	.169492	34.8100	205.379	2.42899	7.68115	1.80697	3.89300	8.38721
.91	.169205	34.9281	206.425	.43105	.68765	.80799	.89519	.39194
.92	.168919	35.0464	207.475	.43311	.69415	.80901	.89739	.39667
.93	.168634	35.1649	208.528	.43516	.70065	.81003	.89958	.40140
.94	.168350	35.2836	209.585	.43721	.70714	.81104	.90177	.40612
5.95	.168067	35.4025	210.645	2.43926	7.71362	1.81206	3.90396	8.41083
.96	.167785	35.5216	211.709	.44131	.72010	.81307	.90615	.41554
.97	.167504	35.6409	212.776	.44336	.72658	.81409	.90833	.42025
.98	.167224	35.7604	213.847	.44540	.73305	.81510	.91051	.42494
.99	.166945	35.8801	214.922	.44745	.73951	.81611	.91269	.42964
6.00	.166667	36.0000	216.000	2.44949	7.74597	1.81712	3.91487	8.43433
n	$1/n$	n^2	n^3	$\sqrt{n}$	$\sqrt{10n}$	$\sqrt[3]{n}$	$\sqrt[3]{10n}$	$\sqrt[3]{100n}$

n	$1/n$	n^2	n^3	$\sqrt{n}$	$\sqrt{10n}$	$\sqrt[3]{n}$	$\sqrt[3]{10n}$	$\sqrt[3]{100n}$
6.00	.166667	36.0000	216.000	2.44949	7.74597	1.81712	3.91487	8.43433
.01	.166389	36.1201	217.082	.45153	.75242	.81813	.91704	.43901
.02	.166113	36.2404	218.167	.45357	.75887	.81914	.91921	.44369
.03	.165837	36.3609	219.256	.45561	.76531	.82014	.92138	.44836
.04	.165563	36.4816	220.349	.45764	.77174	.82115	.92355	.45303
6.05	.165289	36.6025	221.445	2.45967	7.77817	1.82215	3.92571	8.45769
.06	.165017	36.7236	222.545	46171	.78460	.82316	.92787	.46235
.07	.164745	36.8449	223.649	.46374	.79102	.82416	.93003	.46700
.08	.164474	36.9664	224.756	.46577	.79744	.82516	.93219	.47165
.09	.164204	37.0881	225.867	.46779	.80385	.82616	.93434	.47629
6.10	.163934	37.2100	226.981	2.46982	7.81025	1.82716	3.93650	8.48093
.11	.163666	37.3321	228.099	.47184	.81665	.82816	.93865	.48556
.12	.163399	37.4544	229.221	.47386	.82304	.82915	.94079	.49018
.13	.163132	37.5769	230.346	.47588	.82943	.83015	.94294	.49481
.14	.162866	37.6996	231.476	.47790	.83582	.83115	.94508	.49942
6.15	.162602	37.8225	232.608	2.47992	7.84219	1.83214	3.94722	8.50403
.16	.162338	37.9456	233.745	.48193	.84857	.83313	.94936	.50864
.17	.162075	38.0689	234.885	.48395	.85493	.83412	.95150	.51324
.18	.161812	38.1924	236.029	.48596	.86130	.83511	.95363	.51784
.19	.161551	38.3161	237.177	.48797	.86766	.83610	.95576	.52243
6.20	.161290	38.4400	238.328	2.48998	7.87401	1.83709	3.95789	8.52702
.21	.161031	38.5641	239.483	.49199	.88036	.83808	.96002	.53160
.22	.160772	38.6884	240.642	.49399	.88670	.83906	.96214	.53618
.23	.160514	38.8129	241.804	.49600	.89303	.84005	.96427	.54075
.24	.160256	38.9376	242.971	.49800	.89937	.84103	.96638	.54532
6.25	.160000	39.0625	244.141	2.50000	7.90569	1.84202	3.96850	8.54988
.26	.159744	39.1876	245.314	.50200	.91202	.84300	.97062	.55444
.27	.159490	39.3129	246.492	.50400	.91833	.84398	.97273	.55899
.28	.159236	39.4384	247.673	.50599	.92465	.84496	.97484	.56354
.29	.158983	39.5641	248.858	.50799	.93095	.84594	.97695	.56808
6.30	.158730	39.6900	250.047	2.50998	7.93725	1.84691	3.97906	8.57262
.31	.158479	39.8161	251.240	.51197	.94355	.84789	.98116	.57715
.32	.158228	39.9424	252.436	.51396	.94984	.84887	.98326	.58168
.33	.157978	40.0689	253.636	.51595	.95613	.84984	.98536	.58620
.34	.157729	40.1956	254.840	.51794	.96241	.85082	.98746	.59072
6.35	.157480	40.3225	256.048	2.51992	7.96869	1.85179	3.98956	8.59524
.36	.157233	40.4496	257.259	.52190	.97496	.85276	.99165	.59975
.37	.156986	40.5769	258.475	.52389	.98123	.85373	.99374	.60425
.38	.156740	40.7044	259.694	.52587	.98749	.85470	.99583	.60875
.39	.156495	40.8321	260.917	.52784	.99375	.85567	.99792	.61325
6.40	.156250	40.9600	262.144	2.52982	8.00000	1.85664	4.00000	8.61774
.41	.156006	41.0881	263.375	.53180	.00625	.85760	.00208	.62222
.42	.155763	41.2164	264.609	.53377	.01249	.85857	.00416	.62671
.43	.155521	41.3449	265.848	.53574	.01873	.85953	.00624	.63118
.44	.155280	41.4736	267.090	.53772	.02496	.86050	.00832	.63566
6.45	.155039	41.6025	268.336	2.53969	8.03119	1.86146	4.01039	8.64012
.46	.154799	41.7316	269.586	.54165	.03741	.86242	.01246	.64459
.47	.154560	41.8609	270.840	.54362	.04363	.86338	.01453	.64904
.48	.154321	41.9904	272.098	.54558	.04984	.86434	.01660	.65350
.49	.154083	42.1201	273.359	.54755	.05605	.86530	.01866	.65795
6.50	.153846	42.2500	274.625	2.54951	8.06226	1.86626	4.02073	8.66239
n	$1/n$	n^2	n^3	$\sqrt{n}$	$\sqrt{10n}$	$\sqrt[3]{n}$	$\sqrt[3]{10n}$	$\sqrt[3]{100n}$

n	$1/n$	n^2	n^3	$\sqrt{n}$	$\sqrt{10n}$	$\sqrt[3]{n}$	$\sqrt[3]{10n}$	$\sqrt[3]{100n}$
6.50	.153846	42.2500	274.625	2.54951	8.06226	1.86626	4.02073	8.66239
.51	.153610	42.3801	275.894	.55147	.06846	.86721	.02279	.66683
.52	.153374	42.5104	277.168	.55343	.07465	.86817	.02485	.67127
.53	.153139	42.6409	278.445	.55539	.08084	.86912	.02690	.67570
.54	.152905	42.7716	279.726	.55734	.08703	.87008	.02896	.68012
6.55	.152672	42.9025	281.011	2.55930	8.09321	1.87103	4.03101	8.68455
.56	.152439	43.0336	282.300	.56125	.09938	.87198	.03306	.68896
.57	.152207	43.1649	283.593	.56320	.10555	.87293	.03511	.69338
.58	.151976	43.2964	284.890	.56515	.11172	.87388	.03715	.69778
.59	.151745	43.4281	286.191	.56710	.11788	.87483	.03920	.70219
6.60	.151515	43.5600	287.496	2.56905	8.12404	1.87578	4.04124	8.70659
.61	.151286	43.6921	288.805	.57099	.13019	.87672	.04328	.71098
.62	.151057	43.8244	290.118	.57294	.13634	.87767	.04532	.71537
.63	.150830	43.9569	291.434	.57488	.14248	.87862	.04735	.71976
.64	.150602	44.0896	292.755	.57682	.14862	.87956	.04939	.72414
6.65	.150376	44.2225	294.080	2.57876	8.15475	1.88050	4.05142	8.72852
.66	.150150	44.3556	295.408	.58070	.16088	.88144	.05345	.73289
.67	.149925	44.4889	296.741	.58263	.16701	.88239	.05548	.73726
.68	.149701	44.6224	298.078	.58457	.17313	.88333	.05750	.74162
.69	.149477	44.7561	299.418	.58650	.17924	.88427	.05953	.74598
6.70	.149254	44.8900	300.763	2.58844	8.18535	1.88520	4.06155	8.75034
.71	.149031	45.0241	302.112	.59037	.19146	.88614	.06357	.75469
.72	.148810	45.1584	303.464	.59230	.19756	.88708	.06559	.75904
.73	.148588	45.2929	304.821	.59422	.20366	.88801	.06760	.76338
.74	.148368	45.4276	306.182	.59615	.20975	.88895	.06961	.76772
6.75	.148148	45.5625	307.547	2.59808	8.21584	1.88988	4.07163	8.77205
.76	.147929	45.6976	308.916	.60000	.22192	.89081	.07364	.77638
.77	.147710	45.8329	310.289	.60192	.22800	.89175	.07564	.78071
.78	.147493	45.9684	311.666	.60384	.23408	.89268	.07765	.78503
.79	.147275	46.1041	313.047	.60576	.24015	.89361	.07965	.78935
6.80	.147059	46.2400	314.432	2.60768	8.24621	1.89454	4.08166	8.79366
.81	.146843	46.3761	315.821	.60960	.25227	.89546	.08365	.79797
.82	.146628	46.5124	317.215	.61151	.25833	.89639	.08565	.80227
.83	.146413	46.6489	318.612	.61343	.26438	.89732	.08765	.80657
.84	.146199	46.7856	320.014	.61534	.27043	.89824	.08964	.81087
6.85	.145985	46.9225	321.419	2.61725	8.27647	1.89917	4.09163	8.81516
.86	.145773	47.0596	322.829	.61916	.28251	.90009	.09362	.81945
.87	.145560	47.1969	324.243	.62107	.28855	.90102	.09561	.82373
.88	.145349	47.3344	325.661	.62298	.29458	.90194	.09760	.82801
.89	.145138	47.4721	327.083	.62488	.30060	.90286	.09958	.83228
6.90	.144928	47.6100	328.509	2.62679	8.30662	1.90378	4.10157	8.83656
.91	.144718	47.7481	329.939	.62869	.31264	.90470	.10355	.84082
.92	.144509	47.8864	331.374	.63059	.31865	.90562	.10552	.84509
.93	.144300	48.0249	332.813	.63249	.32466	.90653	.10750	.84934
.94	.144092	48.1636	334.255	.63439	.33067	.90745	.10948	.85360
6.95	.143885	48.3025	335.702	2.63629	8.33667	1.90837	4.11145	8.85785
.96	.143678	48.4416	337.154	.63818	.34266	.90928	.11342	.86210
.97	.143472	48.5809	338.609	.64008	.34865	.91019	.11539	.86634
.98	.143266	48.7204	340.068	.64197	.35464	.91111	.11736	.87058
.99	.143062	48.8601	341.532	.64386	.36062	.91202	.11932	.87481
7.00	.142857	49.0000	343.000	2.64575	8.36660	1.91293	4.12129	8.87904
n	$1/n$	n^2	n^3	$\sqrt{n}$	$\sqrt{10n}$	$\sqrt[3]{n}$	$\sqrt[3]{10n}$	$\sqrt[3]{100n}$

n	$1/n$	n^2	n^3	$\sqrt{n}$	$\sqrt{10n}$	$\sqrt[3]{n}$	$\sqrt[3]{10n}$	$\sqrt[3]{100n}$
7.00	.142857	49.0000	343.000	2.64575	8.36660	1.91293	4.12129	8.87904
.01	.142653	49.1401	344.472	.64764	.37257	.91384	.12325	.88327
.02	.142450	49.2804	345.948	.64953	.37854	.91475	.12521	.88749
.03	.142248	49.4209	347.429	.65141	.38451	.91566	.12716	.89171
.04	.142045	49.5616	348.914	.65330	.39047	.91657	.12912	.89592
7.05	.141844	49.7025	350.403	2.65518	8.39643	1.91747	4.13107	8.90013
.06	.141643	49.8436	351.896	.65707	.40238	.91838	.13303	.90434
.07	.141443	49.9849	353.393	.65895	.40833	.91929	.13498	.90854
.08	.141243	50.1264	354.895	.66083	.41427	.92019	.13693	.91274
.09	.141044	50.2681	356.401	.66271	.42021	.92109	.13887	.91693
7.10	.140845	50.4100	357.911	2.66458	8.42615	1.92200	4.14082	8.92112
.11	.140647	50.5521	359.425	.66646	.43208	.92290	.14276	.92531
.12	.140449	50.6944	360.944	.66833	.43801	.92380	.14470	.92949
.13	.140252	50.8369	362.467	.67021	.44393	.92470	.14664	.93367
.14	.140056	50.9796	363.994	.67208	.44985	.92560	.14858	.93784
7.15	.139860	51.1225	365.526	2.67395	8.45577	1.92650	4.15052	8.94201
.16	.139665	51.2656	367.062	.67582	.46168	.92740	.15245	.94618
.17	.139470	51.4089	368.602	.67769	.46759	.92829	.15438	.95034
.18	.139276	51.5524	370.146	.67955	.47349	.92919	.15631	.95450
.19	.139082	51.6961	371.695	.68142	.47939	.93008	.15824	.95866
7.20	.138889	51.8400	373.248	2.68328	8.48528	1.93098	4.16017	8.96281
.21	.138696	51.9841	374.805	.68514	.49117	.93187	.16209	.96696
.22	.138504	52.1284	376.367	.68701	.49706	.93277	.16402	.97110
.23	.138313	52.2729	377.933	.68887	.50294	.93366	.16594	.97524
.24	.138122	52.4176	379.503	.69072	.50882	.93455	.16786	.97938
7.25	.137931	52.5625	381.078	2.69258	8.51469	1.93544	4.16978	8.98351
.26	.137741	52.7076	382.657	.69444	.52056	.93633	.17169	.98764
.27	.137552	52.8529	384.241	.69629	.52643	.93722	.17361	.99176
.28	.137363	52.9984	385.828	.69815	.53229	.93810	.17552	.99588
.29	.137174	53.1441	387.420	.70000	.53815	.93899	.17743	9.00000
7.30	.136986	53.2900	389.017	2.70185	8.54400	1.93988	4.17934	9.00411
.31	.136799	53.4361	390.618	.70370	.54985	.94076	.18125	.00822
.32	.136612	53.5824	392.223	.70555	.55570	.94165	.18315	.01233
.33	.136426	53.7289	393.833	.70740	.56154	.94253	.18506	.01643
.34	.136240	53.8756	395.447	.70924	.56738	.94341	.18696	.02053
7.35	.136054	54.0225	397.065	2.71109	8.57321	1.94430	4.18886	9.02462
.36	.135870	54.1696	398.688	.71293	.57904	.94518	.19076	.02871
.37	.135685	54.3169	400.316	.71477	.58487	.94606	.19266	.03280
.38	.135501	54.4644	401.947	.71662	.59069	.94694	.19455	.03689
.39	.135318	54.6121	403.583	.71846	.59651	.94782	.19644	.04097
7.40	.135135	54.7600	405.224	2.72029	8.60233	1.94870	4.19834	9.04504
.41	.134953	54.9081	406.869	.72213	.60814	.94957	.20023	.04911
.42	.134771	55.0564	408.518	.72397	.61394	.95045	.20212	.05318
.43	.134590	55.2049	410.172	.72580	.61974	.95132	.20400	.05725
.44	.134409	55.3536	411.831	.72764	.62554	.95220	.20589	.06131
7.45	.134228	55.5025	413.494	2.72947	8.63134	1.95307	4.20777	9.06537
.46	.134048	55.6516	415.161	.73130	.63713	.95395	.20965	.06942
.47	.133869	55.8009	416.833	.73313	.64292	.95482	.21153	.07347
.48	.133690	55.9504	418.509	.73496	.64870	.95569	.21341	.07752
.49	.133511	56.1001	420.190	.73679	.65448	.95656	.21529	.08156
7.50	.133333	56.2500	421.875	2.73861	8.66025	1.95743	4.21716	9.08560
n	$1/n$	n^2	n^3	$\sqrt{n}$	$\sqrt{10n}$	$\sqrt[3]{n}$	$\sqrt[3]{10n}$	$\sqrt[3]{100n}$

n	$1/n$	n^2	n^3	$\sqrt{n}$	$\sqrt{10n}$	$\sqrt[3]{n}$	$\sqrt[3]{10n}$	$\sqrt[3]{100n}$
7.50	.133333	56.2500	421.875	2.73861	8.66025	1.95743	4.21716	9.08560
.51	.133156	56.4001	423.565	.74044	.66603	.95830	.21904	.08964
.52	.132979	55.5504	425.259	.74226	.67179	.95917	.22091	.09367
.53	.132802	56.7009	426.958	.74408	.67756	.96004	.22278	.09770
.54	.132626	56.8516	428.661	.74591	.68332	.96091	.22465	.10173
7.55	.132450	57.0025	430.369	2.74773	8.68907	1.96177	4.22651	9.10575
.56	.132275	57.1536	432.081	.74955	.69483	.96264	.22838	.10977
.57	.132100	57.3049	433.798	.75136	.70057	.96350	.23024	.11378
.58	.131926	57.4564	435.520	.75318	.70632	.96437	.23210	.11779
.59	.131752	57.6081	437.245	.75500	.71206	.96523	.23396	.12180
7.60	.131579	57.7600	438.976	2.75681	8.71780	1.96610	4.23582	9.12581
.61	.131406	57.9121	440.711	.75862	.72353	.96696	.23768	.12981
.62	.131234	58.0644	442.451	.76043	.72926	.96782	.23954	.13380
.63	.131062	58.2169	444.195	.76225	.73499	.96868	.24139	.13780
.64	.130890	58.3696	445.944	.76405	.74071	.96954	.24324	.14179
7.65	.130719	58.5225	447.697	2.76586	8.74643	1.97040	4.24509	9.14577
.66	.130548	58.6756	449.455	.76767	.75214	.97126	.24694	.14976
.67	.130378	58.8289	451.218	.76948	.75785	.97211	.24879	.15374
.68	.130208	58.9824	452.985	.77128	.76356	.97297	.25063	.15771
.69	.130039	59.1361	454.757	.77308	.76926	.97383	.25248	.16169
7.70	.129870	59.2900	456.533	2.77489	8.77496	1.97468	4.25432	9.16566
.71	.129702	59.4441	458.314	.77669	.78066	.97554	.25616	.16962
.72	.129534	59.5984	460.100	.77849	.78635	.97639	.25800	.17359
.73	.129366	59.7529	461.890	.78029	.79204	.97724	.25984	.17754
.74	.129199	59.9076	463.685	.78209	.79773	.97809	.26167	.18150
7.75	.129032	60.0625	465.484	2.78388	8.80341	1.97895	4.26351	9.18545
.76	.128866	60.2176	467.289	.78568	.80909	.97980	.26534	.18940
.77	.128700	60.3729	469.097	.78747	.81476	.98065	.26717	.19335
.78	.128535	60.5284	470.911	.78927	.82043	.98150	.26900	.19729
.79	.128370	60.6841	472.729	.79106	.82610	.98234	.27083	.20123
7.80	.128205	60.8400	474.552	2.79285	8.83176	1.98319	4.27266	9.20516
.81	.128041	60.9961	476.380	.79464	.83742	.98404	.27448	.20910
.82	.127877	61.1524	478.212	.79643	.84308	.98489	.27631	.21302
.83	.127714	61.3089	480.049	.79821	.84873	.98573	.27813	.21695
.84	.127551	61.4656	481.890	.80000	.85438	.98658	.27995	.22087
7.85	.127389	61.6225	483.737	2.80179	8.86002	1.98742	4.28177	9.22479
.86	.127226	61.7796	485.588	.80357	.86566	.98826	.28359	.22871
.87	.127065	61.9369	487.443	.80535	.87130	.98911	.28540	.23262
.88	.126904	62.0944	489.304	.80713	.87694	.98995	.28722	.23653
.89	.126743	62.2521	491.169	.80891	.88257	.99079	.28903	.24043
7.90	.126582	62.4100	493.039	2.81069	8.88819	1.99163	4.29084	9.24434
.91	.126422	62.5681	494.914	.81247	.89382	.99247	.29265	.24823
.92	.126263	62.7264	496.793	.81425	.89944	.99331	.29446	.25213
.93	.126103	62.8849	498.677	.81603	.90505	.99415	.29627	.25602
.94	.125945	63.0436	500.566	.81780	.91067	.99499	.29807	.25991
7.95	.125786	63.2025	502.460	2.81957	8.91628	1.99582	4.29987	9.26380
.96	.125628	63.3616	504.358	.82135	.92188	.99666	.30168	.26768
.97	.125471	63.5209	506.262	.82312	.92749	.99750	.30348	.27156
.98	.125313	63.6804	508.170	.82489	.93308	.99833	.30528	.27544
.99	.125156	63.8401	510.082	.82666	.93868	.99917	.30707	.27931
8.00	.125000	64.0000	512.000	2.82843	8.94427	2.00000	4.30887	9.28318
n	$1/n$	n^2	n^3	$\sqrt{n}$	$\sqrt{10n}$	$\sqrt[3]{n}$	$\sqrt[3]{10n}$	$\sqrt[3]{100n}$

n	$1/n$	n^2	n^3	$\sqrt{n}$	$\sqrt{10n}$	$\sqrt[3]{n}$	$\sqrt[3]{10n}$	$\sqrt[3]{100n}$
8.00	.125000	64.0000	512.000	2.82843	8.94427	2.00000	4.30887	9.28318
.01	.124844	64.1601	513.922	.83019	.94986	.00083	.31066	.28704
.02	.124688	64.3204	515.850	.83196	.95545	.00167	.31246	.29091
.03	.124533	64.4809	517.782	.83373	.96103	.00250	.31425	.29477
.04	.124378	64.6416	519.718	.83549	.96660	.00333	.31604	.29862
8.05	.124224	64.8025	521.660	2.83725	8.97218	2.00416	4.31783	9.30248
.06	.124069	64.9636	523.607	.83901	.97775	.00499	.31961	.30633
.07	.123916	65.1249	525.558	.84077	.98332	.00582	.32140	.31018
.08	.123762	65.2864	527.514	.84253	.98888	.00664	.32318	.31402
.09	.123609	65.4481	529.475	.84429	.99444	.00747	.32497	.31786
8.10	.123457	65.6100	531.441	2.84605	9.00000	2.00830	4.32675	9.32170
.11	.123305	65.7721	533.412	.84781	.00555	.00912	.32853	.32553
.12	.123153	65.9344	535.387	.84956	.01110	.00995	.33031	.32936
.13	.123001	66.0969	537.368	.85132	.01665	.01078	.33208	.33319
.14	.122850	66.2596	539.353	.85307	.02219	.01160	.33386	.33702
8.15	.122699	66.4225	541.343	2.85482	9.02774	2.01242	4.33563	9.34084
.16	.122549	66.5856	543.338	.85657	.03327	.01325	.33741	.34466
.17	.122399	66.7489	545.339	.85832	.03881	.01407	.33918	.34847
.18	.122249	66.9124	547.343	.86007	.04434	.01489	.34095	.35229
.19	.122100	67.0761	549.353	.86182	.04986	.01571	.34271	.35610
8.20	.121951	67.2400	551.368	2.86356	9.05539	2.01653	4.34448	9.35990
.21	.121803	67.4041	553.388	.86531	.06091	.01735	.34625	.36370
.22	.121655	67.5684	555.412	.86705	.06642	.01817	.34801	.36751
.23	.121507	67.7329	557.442	.86880	.07193	.01899	.34977	.37130
.24	.121359	67.8976	559.476	.87054	.07744	.01980	.35153	.37510
8.25	.121212	68.0625	561.516	2.87228	9.08295	2.02062	4.35329	9.37889
.26	.121065	68.2276	563.560	.87402	.08845	.02144	.35505	.38268
.27	.120919	68.3929	565.609	.87576	.09395	.02225	.35681	.38646
.28	.120773	68.5584	567.664	.87750	.09945	.02307	.35856	.39024
.29	.120627	68.7241	569.723	.87924	.10494	.02388	.36032	.39402
8.30	.120482	68.8900	571.787	2.88097	9.11043	2.02469	4.36207	9.39780
.31	.120337	69.0561	573.856	.88271	.11592	.02551	.36382	.40157
.32	.120192	69.2224	575.930	.88444	.12140	.02632	.36557	.40534
.33	.120048	69.3889	578.010	.88617	.12688	.02713	.36732	.40911
.34	.119904	69.5556	580.094	.88791	.13236	.02794	.36907	.41287
8.35	.119760	69.7225	582.183	2.88964	9.13783	2.02875	4.37081	9.41663
.36	.119617	69.8896	584.277	.89137	.14330	.02956	.37256	.42039
.37	.119474	70.0569	586.376	.89310	.14877	.03037	.37430	.42414
.38	.119332	70.2244	588.480	.89482	.15423	.03118	.37604	.42789
.39	.119190	70.3921	590.590	.89655	.15969	.03199	.37778	.43164
8.40	.119048	70.5600	592.704	2.89828	9.16515	2.03279	4.37952	9.43539
.41	.118906	70.7281	594.823	.90000	.17061	.03360	.38126	.43913
.42	.118765	70.8964	596.948	.90172	.17606	.03440	.38299	.44287
.43	.118624	71.0649	599.077	.90345	.18150	.03521	.38473	.44661
.44	.118483	71.2336	601.212	.90517	.18695	.03601	.38646	.45034
8.45	.118343	71.4025	603.351	2.90689	9.19239	2.03682	4.38819	9.45407
.46	.118203	71.5716	605.496	.90861	.19783	.03762	.38992	.45780
.47	.118064	71.7409	607.645	.91033	.20326	.03842	.39165	.46152
.48	.117925	71.9104	609.800	.91204	.20869	.03923	.39338	.46525
.49	.117786	72.0801	611.960	.91376	.21412	.04003	.39510	.46897
8.50	.117647	72.2500	614.125	2.91548	9.21954	2.04083	4.39683	9.47268

n	$1/n$	n^2	n^3	$\sqrt{n}$	$\sqrt{10n}$	$\sqrt[3]{n}$	$\sqrt[3]{10n}$	$\sqrt[3]{100n}$

n	$1/n$	n^2	n^3	$\sqrt{n}$	$\sqrt{10n}$	$\sqrt[3]{n}$	$\sqrt[3]{10n}$	$\sqrt[3]{100n}$
8.50	.117647	72.2500	614.125	2.91548	9.21954	2.04083	4.39683	9.47268
.51	.117509	72.4201	616.295	.91719	.22497	.04163	.39855	.47640
.52	.117371	72.5904	618.470	.91890	.23038	.04243	.40028	.48011
.53	.117233	72.7609	620.650	.92062	.23580	.04323	.40200	.48381
.54	.117096	72.9316	622.836	.92233	.24121	.04402	.40372	.48752
8.55	.116959	73.1025	625.026	2.92404	9.24662	2.04482	4.40543	9.49122
.56	.116822	73.2736	627.222	.92575	.25203	.04562	.40715	.49492
.57	.116686	73.4449	629.423	.92746	.25743	.04641	.40887	.49861
.58	.116550	73.6164	631.629	.92916	.26283	.04721	.41058	.50231
.59	.116414	73.7881	633.840	.93087	.26823	.04801	.41229	.50600
8.60	.116279	73.9600	636.056	2.93258	9.27362	2.04880	4.41400	9.50969
.61	.116144	74.1321	638.277	.93428	.27901	.04959	.41571	.51337
.62	.116009	74.3044	640.504	.93598	.28440	.05039	.41742	.51705
.63	.115875	74.4769	642.736	.93769	.28978	.05118	.41913	.52073
.64	.115741	74.6496	644.973	.93939	.29516	.05197	.42084	.52441
8.65	.115607	74.8225	647.215	2.94109	9.30054	2.05276	4.42254	9.52808
.66	.115473	74.9956	649.462	.94279	.30591	.05355	.42425	.53175
.67	.115340	75.1689	651.714	.94449	.31128	.05434	.42595	.53542
.68	.115207	75.3424	653.972	.94618	.31665	.05513	.42765	.53908
.69	.115075	75.5161	656.235	.94788	.32202	.05592	.42935	.54274
8.70	.114943	75.6900	658.503	2.94958	9.32738	2.05671	4.43105	9.54640
.71	.114811	75.8641	660.776	.95127	.33274	.05750	.43274	.55006
.72	.114679	76.0384	663.055	.95296	.33809	.05828	.43444	.55371
.73	.114548	76.2129	665.339	.95466	.34345	.05907	.43613	.55736
.74	.114416	76.3876	667.628	.95635	.34880	.05986	.43783	.56101
8.75	.114286	76.5625	669.922	2.95804	9.35414	2.06064	4.43952	9.56466
.76	.114155	76.7376	672.221	.95973	.35949	.06143	.44121	.56830
.77	.114025	76.9129	674.526	.96142	.36483	.06221	.44290	.57194
.78	.113895	77.0884	676.836	.96311	.37017	.06299	.44459	.57557
.79	.113766	77.2641	679.151	.96479	.37550	.06378	.44627	.57921
8.80	.113636	77.4400	681.472	2.96648	9.38083	2.06456	4.44796	9.58284
.81	.113507	77.6161	683.798	.96816	.38616	.06534	.44964	.58647
.82	.113379	77.7924	686.129	.96985	.39149	.06612	.45133	.59009
.83	.113250	77.9689	688.465	.97153	.39681	.06690	.45301	.59372
.84	.113122	78.1456	690.807	.97321	.40213	.06768	.45469	.59734
8.85	.112994	78.3225	693.154	2.97489	9.40744	2.06846	4.45637	9.60095
.86	.112867	78.4996	695.506	.97658	.41276	.06924	.45805	.60457
.87	.112740	78.6769	697.864	.97825	.41807	.07002	.45972	.60818
.88	.112613	78.8544	700.227	.97993	.42338	.07080	.46140	.61179
.89	.112486	79.0321	702.595	.98161	.42868	.07157	.46307	.61540
8.90	.112360	79.2100	704.969	2.98329	9.43398	2.07235	4.46475	9.61900
.91	.112233	79.3881	707.348	.98496	.43928	.07313	.46642	.62260
.92	.112108	79.5664	709.732	.98664	.44458	.07390	.46809	.62620
.93	.111982	79.7449	712.122	.98831	.44987	.07468	.46976	.62980
.94	.111857	79.9236	714.517	.98998	.45516	.07545	.47142	.63339
8.95	.111732	80.1025	716.917	2.99166	9.46044	2.07622	4.47309	9.63698
.96	.111607	80.2816	719.323	.99333	.46573	.07700	.47476	.64057
.97	.111483	80.4609	721.734	.99500	.47101	.07777	.47642	.64415
.98	.111359	80.6404	724.151	.99666	.47629	.07854	.47808	.64774
.99	.111235	80.8201	726.573	.99833	.48156	.07931	.47974	.65132
9.00	.111111	81.0000	729.000	3.00000	9.48683	2.08008	4.48140	9.65489
n	$1/n$	n^2	n^3	$\sqrt{n}$	$\sqrt{10n}$	$\sqrt[3]{n}$	$\sqrt[3]{10n}$	$\sqrt[3]{100n}$

n	$1/n$	n^2	n^3	$\sqrt{n}$	$\sqrt{10n}$	$\sqrt[3]{n}$	$\sqrt[3]{10n}$	$\sqrt[3]{100n}$
9.00	.111111	81.0000	729.000	3.00000	9.48683	2.08008	4.48140	9.65489
.01	.110988	81.1801	731.433	.00167	.49210	.08085	.48306	.65847
.02	.110865	81.3604	733.871	.00333	.49737	.08162	.48472	.66204
.03	.110742	81.5409	736.314	.00500	.50263	.08239	.48638	.66561
.04	.110619	81.7216	738.763	.00666	.50789	.08316	.48803	.66918
9.05	.110497	81.9025	741.218	3.00832	9.51315	2.08393	4.48969	9.67274
.06	.110375	82.0836	743.677	.00998	.51840	.08470	.49134	.67630
.07	.110254	82.2649	746.143	.01164	.52365	.08546	.49299	.67986
.08	.110132	82.4464	748.613	.01330	.52890	.08623	.49464	.68342
.09	.110011	82.6281	751.089	.01496	.53415	.08699	.49629	.68697
9.10	.109890	82.8100	753.571	3.01662	9.53939	2.08776	4.49794	9.69052
.11	.109769	82.9921	756.058	.01828	.54463	.08852	.49959	.69407
.12	.109649	83.1744	758.551	.01993	.54987	.08929	.50123	.69762
.13	.109529	83.3569	761.048	.02159	.55510	.09005	.50288	.70116
.14	.109409	83.5396	763.552	.02324	.56033	.09081	.50452	.70470
9.15	.109290	83.7225	766.061	3.02490	9.56556	2.09158	4.50616	9.70824
.16	.109170	83.9056	768.575	.02655	.57079	.09234	.50781	.71177
.17	.109051	84.0889	771.095	.02820	.57601	.09310	.50945	.71531
.18	.108932	84.2724	773.621	.02985	.58123	.09386	.51108	.71884
.19	.108814	84.4561	776.152	.03150	.58645	.09462	.51272	.72236
9.20	.108696	84.6400	778.688	3.03315	9.59166	2.09538	4.51436	9.72589
.21	.108578	84.8241	781.230	.03480	.59687	.09614	.51599	.72941
.22	.108460	85.0084	783.777	.03645	.60208	.09690	.51763	.73293
.23	.108342	85.1929	786.330	.03809	.60729	.09765	.51926	.73645
.24	.108225	85.3776	788.889	.03974	.61249	.09841	.52089	.73996
9.25	.108108	85.5625	791.453	3.04138	9.61769	2.09917	4.52252	9.74348
.26	.107991	85.7476	794.023	.04302	.62289	.09992	.52415	.74699
.27	.107875	85.9329	796.598	.04467	.62808	.10068	.52578	.75049
.28	.107759	86.1184	799.179	.04631	.63328	.10144	.52740	.75400
.29	.107643	86.3041	801.765	.04795	.63846	.10219	.52903	.75750
9.30	.107527	86.4900	804.357	3.04959	9.64365	2.10294	4.53065	9.76100
.31	.107411	86.6761	806.954	.05123	.64883	.10370	.53228	.76450
.32	.107296	86.8624	809.558	.05287	.65401	.10445	.53390	.76799
.33	.107181	87.0489	812.166	.05450	.65919	.10520	.53552	.77148
.34	.107066	87.2356	814.781	.05614	.66437	.10595	.53714	.77497
9.35	.106952	87.4225	817.400	3.05778	9.66954	2.10671	4.53876	9.77846
.36	.106838	87.6096	820.026	.05941	.67471	.10746	.54038	.78195
.37	.106724	87.7969	822.657	.06105	.67988	.10821	.54199	.78543
.38	.106610	87.9844	825.294	.06268	.68504	.10896	.54361	.78891
.39	.106496	88.1721	827.936	.06431	.69020	.10971	.54522	.79239
9.40	.106383	88.3600	830.584	3.06594	9.69536	2.11045	4.54684	9.79586
.41	.106270	88.5481	833.238	.06757	.70052	.11120	.54845	.79933
.42	.106157	88.7364	835.897	.06920	.70567	.11195	.55006	.80280
.43	.106045	88.9249	838.562	.07083	.71082	.11270	.55167	.80627
.44	.105932	89.1136	841.232	.07246	.71597	.11344	.55328	.80974
9.45	.105820	89.3025	843.909	3.07409	9.72111	2.11419	4.55488	9.81320
.46	.105708	89.4916	846.591	.07571	.72625	.11494	.55649	.81666
.47	.105597	89.6809	849.278	.07734	.73139	.11568	.55809	.82012
.48	.105485	89.8704	851.971	.07896	.73653	.11642	.55970	.82357
.49	.105374	90.0601	854.670	.08058	.74166	.11717	.56130	.82703
9.50	.105263	90.2500	857.375	3.08221	9.74679	2.11791	4.56290	9.83048
n	$1/n$	n^2	n^3	$\sqrt{n}$	$\sqrt{10n}$	$\sqrt[3]{n}$	$\sqrt[3]{10n}$	$\sqrt[3]{100n}$

n	$1/n$	n^2	n^3	$\sqrt{n}$	$\sqrt{10n}$	$\sqrt[3]{n}$	$\sqrt[3]{10n}$	$\sqrt[3]{100n}$
9.50	.105263	90.2500	857.375	3.08221	9.74679	2.11791	4.56290	9.83048
.51	.105152	90.4401	860.085	.08383	.75192	.11865	.56450	.83392
.52	.105042	90.6304	862.801	.08545	.75705	.11940	.56610	.83737
.53	.104932	90.8209	865.523	.08707	.76217	.12014	.56770	.84081
.54	.104822	91.0116	868.251	.08869	.76729	.12088	.56930	.84425
9.55	.104712	91.2025	870.984	3.09031	9.77241	2.12162	4.57089	9.84769
.56	.104603	91.3936	873.723	.09192	.77753	.12236	.57249	.85113
.57	.104493	91.5849	876.467	.09354	.78264	.12310	.57408	.85456
.58	.104384	91.7764	879.218	.09516	.78775	.12384	.57567	.85799
.59	.104275	91.9681	881.974	.09677	.79285	.12458	.57727	.86142
9.60	.104167	92.1600	884.736	3.09839	9.79796	2.12532	4.57886	9.86485
.61	.104058	92.3521	887.504	.10000	.80306	.12605	.58045	.86827
.62	.103950	92.5444	890.277	.10161	.80816	.12679	.58204	.87169
.63	.103842	92.7369	893.056	.10322	.81326	.12753	.58362	.87511
.64	.103734	92.9296	895.841	.10483	.81835	.12826	.58521	.87853
9.65	.103627	93.1225	898.632	3.10644	9.82344	2.12900	4.58679	9.88195
.66	.103520	93.3156	901.429	.10805	.82853	.12974	.58838	.88536
.67	.103413	93.5089	904.231	.10966	.83362	.13047	.58996	.88877
.68	.103306	93.7024	907.039	.11127	.83870	.13120	.59154	.89217
.69	.103199	93.8961	909.853	.11288	.84378	.13194	.59312	.89558
9.70	.103093	94.0900	912.673	3.11448	9.84886	2.13267	4.59470	9.89898
.71	.102987	94.2841	915.499	.11609	.85393	.13340	.59628	.90238
.72	.102881	94.4784	918.330	.11769	.85901	.13414	.59786	.90578
.73	.102775	94.6729	921.167	.11929	.86408	.13487	.59943	.90918
.74	.102669	94.8676	924.010	.12090	.86914	.13560	.60101	.91257
9.75	.102564	95.0625	926.859	3.12250	9.87421	2.13633	4.60258	9.91596
.76	.102459	95.2576	929.714	.12410	.87927	.13706	.60416	.91935
.77	.102354	95.4529	932.575	.12570	.88433	.13779	.60573	.92274
.78	.102249	95.6484	935.441	.12730	.88939	.13852	.60730	.92612
.79	.102145	95.8441	938.314	.12890	.89444	.13925	.60887	.92950
9.80	.102041	96.0400	941.192	3.13050	9.89949	2.13997	4.61044	9.93288
.81	.101937	96.2361	944.076	.13209	.90454	.14070	.61200	.93626
.82	.101833	96.4324	946.966	.13369	.90959	.14143	.61357	.93964
.83	.101729	96.6289	949.862	.13528	.91464	.14216	.61514	.94301
.84	.101626	96.8256	952.764	.13688	.91968	.14288	.61670	.94638
9.85	.101523	97.0225	955.672	3.13847	9.92472	2.14361	4.61826	9.94975
.86	.101420	97.2196	958.585	.14006	.92975	.14433	.61983	.95311
.87	.101317	97.4169	961.505	.14166	.93479	.14506	.62139	.95648
.88	.101215	97.6144	964.430	.14325	.93982	.14578	.62295	.95984
.89	.101112	97.8121	967.362	.14484	.94485	.14651	.62451	.96320
9.90	.101010	98.0100	970.299	3.14643	9.94987	2.14723	4.62607	9.96655
.91	.100908	98.2081	973.242	.14802	.95490	.14795	.62762	.96991
.92	.100806	98.4064	976.191	.14960	.95992	.14867	.62918	.97326
.93	.100705	98.6049	979.147	.15119	.96494	.14940	.63073	.97661
.94	.100604	98.8036	982.108	.15278	.96995	.15012	.63229	.97996
9.95	.100503	99.0025	985.075	3.15436	9.97497	2.15084	4.63384	9.98331
.96	.100402	99.2016	988.048	.15595	.97998	.15156	.63539	.98665
.97	.100301	99.4009	991.027	.15753	.98499	.15228	.63694	.98999
.98	.100200	99.6004	994.012	.15911	.98999	.15300	.63849	.99333
.99	.100100	99.8001	997.003	.16070	.99500	.15372	.64004	.99667
10.00	.100000	100.000	1000.00	3.16228	10.0000	2.15443	4.64159	10.0000
n	$1/n$	n^2	n^3	$\sqrt{n}$	$\sqrt{10n}$	$\sqrt[3]{n}$	$\sqrt[3]{10n}$	$\sqrt[3]{100n}$

TABLE XVI. NATURAL LOGARITHMS

N	Log N	Log 10N	Log 100N	d	N	Log N	Log 10N	Log 100N	d
.00	....	....	....		.50	9.30685	1.60944	3.91202	1981
.01	5.39483	7.69741	0.00000	69315	.51	.32666	.62924	.93183	1941
.02	6.08798	8.39056	.69315	40546	.52	.34607	.64866	.95124	1905
.03	.49344	.79603	1.09861	28768	.53	.36512	.66771	.97029	1869
.04	.78112	9.08371	.38629	22315	.54	.38381	.68640	.98898	1835
.05	7.00427	9.30685	1.60944	18232	.55	9.40216	1.70475	4.00733	1802
.06	.18659	.48917	.79176	15415	.56	.42018	.72277	.02535	1770
.07	.34074	.64333	.94591	13353	.57	.43788	.74047	.04305	1739
.08	.47427	.77686	2.07944	11778	.58	.45527	.75786	.06044	1710
.09	.59205	.89464	.19722	10537	.59	.47237	.77495	.07754	1680
.10	7.69741	0.00000	2.30259	9531	.60	9.48917	1.79176	4.09434	1653
.11	.79273	.09531	.39790	8701	.61	.50570	.80829	.11087	1626
.12	.87974	.18232	.48491	8004	.62	.52196	.82455	.12713	1600
.13	.95978	.26236	.56495	7411	.63	.53796	.84055	.14313	1575
.14	8.03389	.33647	.63906	6899	.64	.55371	.85630	.15888	1551
.15	8.10288	0.40547	2.70805	6454	.65	9.56922	1.87180	4.17439	1526
.16	.16742	.47000	.77259	6062	.66	.58448	.88707	.18965	1504
.17	.22804	.53063	.83321	5716	.67	.59952	.90211	.20469	1482
.18	.28520	.58779	.89037	5407	.68	.61434	.91692	.21951	1460
.19	.33927	.64185	.94444	5129	.69	.62894	.93152	.23411	1439
.20	8.39056	0.69315	2.99573	4879	.70	9.64333	1.94591	4.24850	1418
.21	.43935	.74194	3.04452	4652	.71	.65751	.96009	.26268	1399
.22	.48587	.78846	.09104	4445	.72	.67150	.97408	.27667	1379
.23	.53032	.83291	.13549	4256	.73	.68529	.98787	.29046	1361
.24	.57288	.87547	.17805	4083	.74	.69889	2.00148	.30407	1342
.25	8.61371	0.91629	3.21888	3922	.75	9.71232	2.01490	4.31749	1324
.26	.65293	.95551	.25810	3774	.76	.72556	.02815	.33073	1308
.27	.69067	.99325	.29584	3636	.77	.73864	.04122	.34381	1290
.28	.72703	1.02962	.33220	3510	.78	.75154	.05412	.35671	1274
.29	.76213	.06471	.36730	3390	.79	.76428	.06686	.36945	1258
.30	8.79603	1.09861	3.40120	3279	.80	9.77686	2.07944	4.38203	1242
.31	.82882	.13140	.43399	3175	.81	.78928	.09186	.39445	1227
.32	.86057	.16315	.46574	3077	.82	.80155	.10413	.40672	1212
.33	.89134	.19392	.49651	2985	.83	.81367	.11626	.41884	1198
.34	.92119	.22378	.52636	2899	.84	.82565	.12823	.43082	1183
.35	8.95018	1.25276	3.55535	2817	.85	9.83748	2.14007	4.44265	1170
.36	.97835	.28093	.58352	2740	.86	.84918	.15176	.45435	1156
.37	9.00575	.30833	.61092	2667	.87	.86074	.16332	.46591	1143
.38	.03242	.33500	.63759	2597	.88	.87217	.17475	.47734	1130
.39	.05839	.36098	.66356	2532	.89	.88347	.18605	.48864	1117
.40	9.08371	1.38629	3.68888	2469	.90	9.89464	2.19722	4.49981	1105
.41	.10840	.41099	.71357	2410	.91	.90569	.20827	.51086	1093
.42	.13250	.43508	.73767	2353	.92	.91662	.21920	.52179	1081
.43	.15603	.45862	.76120	2299	.93	.92743	.23001	.53260	1069
.44	.17902	.48160	.78419	2247	.94	.93812	.24071	.54329	1059
.45	9.20149	1.50408	3.80666	2198	.95	9.94871	2.25129	4.55388	1047
.46	.22347	.52606	.82864	2151	.96	.95918	.26176	.56435	1036
.47	.24498	.54756	.85015	2105	.97	.96954	.27213	.57471	1026
.48	.26603	.56862	.87120	2062	.98	.97980	.28238	.58497	1015
.49	.28665	.58924	.89182	2020	.99	.98995	.29253	.59512	1005
.50	9.30685	1.60944	3.91202		1.00	0.00000	2.30259	4.60517	

N	Log N	Log 10N	Log 100N	d	N	Log N	Log 10N	Log 100N	d
1.00	0.00000	2.30259	4.60517		**1.50**	0.40547	2.70805	5.01064	
1 01	.00995	.31254	.61512	995	1.51	.41211	.71469	.01728	664
1.02	.01980	.32239	.62497	985	1.52	.41871	.72130	.02388	660
1.03	.02956	.33214	.63473	976	1.53	.42527	.72785	.03044	656
1.04	.03922	.34181	.64439	966	1.54	.43178	.73437	.03695	651
				957					648
1.05	0.04879	2.35138	4.65396		**1.55**	0.43825	2.74084	5.04343	
1.06	.05827	.36085	.66344	948	1.56	.44469	.74727	.04986	643
1.07	.06766	.37024	.67283	939	1.57	.45108	.75366	.05625	639
1.08	.07696	.37955	.68213	930	1.58	.45742	.76001	.06260	635
1.09	.08618	.38876	.69135	922	1.59	.46373	.76632	.06890	630
				913					627
1.10	0.09531	2.39790	4.70048		**1.60**	0.47000	2.77259	5.07517	
1.11	.10436	.40695	.70953	905	1.61	.47623	.77882	.08140	623
1.12	.11333	.41591	.71850	897	1.62	.48243	.78501	.08760	620
1.13	.12222	.42480	.72739	889	1.63	.48858	.79117	.09375	615
1.14	.13103	.43361	.73620	881	1.64	.49470	.79728	.09987	612
				873					608
1.15	0.13976	2.44235	4.74493		**1.65**	0.50078	2.80336	5.10595	
1.16	.14842	.45101	.75359	866	1.66	.50682	.80940	.11199	604
1.17	.15700	.45959	.76217	858	1.67	.51282	.81541	.11799	600
1.18	.16551	.46810	.77068	851	1.68	.51879	.82138	.12396	597
1.19	.17395	.47654	.77912	844	1.69	.52473	.82731	.12990	594
				837					590
1.20	0.18232	2.48491	4.78749		**1.70**	0.53063	2.83321	5.13580	
1.21	.19062	.49321	.79579	830	1.71	.53649	.83908	.14166	586
1.22	.19885	.50144	.80402	823	1.72	.54232	.84491	.14749	583
1.23	.20701	.50960	.81218	816	1.73	.54812	.85071	.15329	580
1.24	.21511	.51770	.82028	810	1.74	.55389	.85647	.15906	577
				803					573
1.25	0.22314	2.52573	4.82831		**1.75**	0.55962	2.86220	5.16479	
1.26	.23111	.53370	.83628	797	1.76	.56531	.86790	.17048	569
1.27	.23902	.54160	.84419	791	1.77	.57098	.87356	.17615	567
1.28	.24686	.54945	.85203	784	1.78	.57661	.87920	.18178	563
1.29	.25464	.55723	.85981	778	1.79	.58222	.88480	.18739	561
				772					557
1.30	0.26236	2.56495	4.86753		**1.80**	0.58779	2.89037	5.19296	
1.31	.27003	.57261	.87520	767	1.81	.59333	.89591	.19850	554
1.32	.27763	.58022	.88280	760	1.82	.59884	.90142	.20401	551
1.33	.28518	.58776	.89035	755	1.83	.60432	.90690	.20949	548
1.34	.29267	.59525	.89784	749	1.84	.60977	.91235	.21494	545
				743					542
1.35	0.30010	2.60269	4.90527		**1.85**	0.61519	2.91777	5.22036	
1.36	.30748	.61007	.91265	738	1.86	.62058	.92316	.22575	539
1.37	.31481	.61740	.91998	733	1.87	.62594	.92852	.23111	536
1.38	.32208	.62467	.92725	727	1.88	.63127	.93386	.23644	533
1.39	.32930	.63189	.93447	722	1.89	.63658	.93916	.24175	531
				717					527
1.40	0.33647	2.63906	4.94164		**1.90**	0.64185	2.94444	5.24702	
1.41	.34359	.64617	.94876	712	1.91	.64710	.94969	.25227	525
1.42	.35066	.65324	.95583	707	1.92	.65233	.95491	.25750	523
1.43	.35767	.66026	.96284	701	1.93	.65752	.96011	.26269	519
1.44	.36464	.66723	.96981	697	1.94	.66269	.96527	.26786	517
				692					514
1.45	0.37156	2.67415	4.97673		**1.95**	0.66783	2.97041	5.27300	
1.46	.37844	.68102	.98361	688	1.96	.67294	.97553	.27811	511
1.47	.38526	.68785	.99043	682	1.97	.67803	.98062	.28320	509
1.48	.39204	.69463	.99721	678	1.98	.68310	.98568	.28827	507
1.49	.39878	.70136	5.00395	674	1.99	.68813	.99072	.29330	503
				669					502
1.50	0.40547	2.70805	5.01064		**2.00**	0.69315	2.99573	5.29832	

N	Log N	Log 10N	Log 100N	d	N	Log N	Log 10N	Log 100N	d
2.00	0.69315	2.99573	5.29832	498	**2.50**	0.91629	3.21888	5.52146	399
2.01	.69813	3.00072	.30330	497	2.51	.92028	.22287	.52545	398
2.02	.70310	.00568	.30827	494	2.52	.92426	.22648	.52943	396
2.03	.70804	.01062	.31321	491	2.53	.92822	.23080	.53339	394
2.04	.71295	.01553	.31812	489	2.54	.93216	.23475	.53733	393
2.05	0.71784	3.02042	5.32301	487	**2.55**	0.93609	3.23868	5.54126	392
2.06	.72271	.02529	.32788	484	2.56	.94001	.24259	.54518	390
2.07	.72755	.03013	.33272	482	2.57	.94391	.24649	.54908	388
2.08	.73237	.03495	.33754	479	2.58	.94779	.25037	.55296	387
2.09	.73716	.03975	.34233	478	2.59	.95166	.25424	.55683	385
2.10	0.74194	3.04452	5.34711	475	**2.60**	0.95551	3.25810	5.56068	384
2.11	.74669	.04927	.35186	473	2.61	.95935	.26194	.56452	382
2.12	.75142	.05400	.35659	470	2.62	.96317	.26576	.56834	381
2.13	.75612	.05871	.36129	469	2.63	.96698	.26957	.57215	380
2.14	.76081	.06339	.36598	466	2.64	.97078	.27336	.57595	378
2.15	0.76547	3.06805	5.37064	464	**2.65**	0.97456	3.27714	5.57973	377
2.16	.77011	.07269	.37528	462	2.66	.97833	.28091	.58350	375
2.17	.77473	.07731	.37990	460	2.67	.98208	.28466	.58725	374
2.18	.77932	.08191	.38450	457	2.68	.98582	.28840	.59099	372
2.19	.78390	.08649	.38907	456	2.69	.98954	.29213	.59471	371
2.20	0.78846	3.09104	5.39363	453	**2.70**	0.99325	3.29584	5.59842	370
2.21	.79299	.09558	.39816	452	2.71	.99695	.29953	.60212	368
2.22	.79751	.10009	.40268	449	2.72	1.00063	.30322	.60580	367
2.23	.80200	.10459	.40717	448	2.73	.00430	.30689	.60947	366
2.24	.80648	.10906	.41165	445	2.74	.00796	.31054	.61313	364
2.25	0.81093	3.11352	5.41610	443	**2.75**	1.01160	3.31419	5.61677	363
2.26	.81536	.11795	.42053	442	2.76	.01523	.31772	.62040	362
2.27	.81978	.12236	.42495	440	2.77	.01885	.32143	.62402	360
2.28	.82418	.12676	.42935	437	2.78	.02245	.32504	.62762	359
2.29	.82855	.13114	.43372	436	2.79	.02604	.32863	.63121	358
2.30	0.83291	3.13549	5.43808	434	**2.80**	1.02962	3.33220	5.63479	356
2.31	.83725	.13983	.44242	432	2.81	.03318	.33577	.63835	356
2.32	.84157	.14415	.44674	430	2.82	.03674	.33932	.64191	354
2.33	.84587	.14845	.45104	428	2.83	.04028	.34286	.64545	352
2.34	.85015	.15274	.45532	427	2.84	.04380	.34639	.64897	352
2.35	0.85442	3.15700	5.45959	424	**2.85**	1.04732	3.34990	5.65249	350
2.36	.85866	.16125	.46383	423	2.86	.05082	.35341	.65599	349
2.37	.86289	.16548	.46806	421	2.87	.05431	.35690	.65948	348
2.38	.86710	.16969	.47227	419	2.88	.05779	.36038	.66296	347
2.39	.87129	.17388	.47646	418	2.89	.06126	.36384	.66643	345
2.40	0.87547	3.17805	5.48064	416	**2.90**	1.06471	3.36730	5.66988	344
2.41	.87963	.18221	.48480	414	2.91	.06815	.37074	.67332	343
2.42	.88377	.18635	.48894	412	2.92	.07158	.37417	.67675	342
2.43	.88789	.19048	.49306	411	2.93	.07500	.37759	.68017	341
2.44	.89200	.19458	.49717	409	2.94	.07841	.38099	.68358	340
2.45	0.89609	3.19867	5.50126	407	**2.95**	1.08181	3.38439	5.68698	338
2.46	.90016	.20275	.50533	406	2.96	.08519	.38777	.69036	337
2.47	.90422	.20680	.50939	404	2.97	.08856	.39115	.69373	336
2.48	.90826	.21084	.51343	402	2.98	.09192	.39451	.69709	335
2.49	.91228	.21487	.51745	401	2.99	.09527	.39786	.70044	334
2.50	0.91629	3.21888	5.52146		**3.00**	1.09861	3.40120	5.70378	

N	Log N	Log 10N	Log 100N	d	N	Log N	Log 10N	Log 100N	d
3.00	1.09861	3.40120	5.70378	333	**3.50**	1.25276	3.55535	5.85793	286
3.01	.10194	.40453	.70711	332	3.51	.25562	.55820	.86079	284
3.02	.10526	.40784	.71043	330	3.52	.25846	.56105	.86363	284
3.03	.10856	.41115	.71373	330	3.53	.26130	.56388	.86647	283
3.04	.11186	.41444	.71703	328	3.54	.26413	.56671	.86930	282
3.05	1.11514	3.41773	5.72031	328	**3.55**	1.26695	3.56953	5.87212	281
3.06	.11841	.42100	.72359	326	3.56	.26976	.57235	.87493	281
3.07	.12168	.42426	.72685	325	3.57	.27257	.57515	.87774	279
3.08	.12493	.42751	.73010	324	3.58	.27536	.57795	.88053	279
3.09	.12817	.43076	.73334	323	3.59	.27815	.58074	.88332	278
3.10	1.13140	3.43399	5.73657	322	**3.60**	1.28093	3.58352	5.88610	278
3.11	.13462	.43721	.73979	321	3.61	.28371	.58629	.88888	276
3.12	.13783	.44042	.74300	320	3.62	.28647	.58906	.89164	276
3.13	.14103	.44362	.74620	319	3.63	.28923	.59182	.89440	275
3.14	.14422	.44681	.74939	318	3.64	.29198	.59457	.89715	275
3.15	1.14740	3.44999	5.75257	317	**3.65**	1.29473	3.59731	5.89990	273
3.16	.15057	.45316	.75574	316	3.66	.29746	.60005	.90263	273
3.17	.15373	.45632	.75890	315	3.67	.30019	.60278	.90536	272
3.18	.15688	.45947	.76205	314	3.68	.30291	.60550	.90808	272
3.19	.16002	.46261	.76519	313	3.69	.30563	.60821	.91080	270
3.20	1.16315	3.46574	5.76832	312	**3.70**	1.30833	3.61092	5.91350	270
3.21	.16627	.46886	.77144	311	3.71	.31103	.61362	.91620	269
3.22	.16938	.47197	.77455	310	3.72	.31372	.61631	.91889	269
3.23	.17248	.47507	.77765	309	3.73	.31641	.61899	.92158	268
3.24	.17557	.47816	.78074	309	3.74	.31909	.62167	.92426	267
3.25	1.17865	3.48124	5.78383	307	**3.75**	1.32176	3.62434	5.92693	266
3.26	.18173	.48431	.78690	306	3.76	.32442	.62700	.92959	266
3.27	.18479	.48738	.78996	305	3.77	.32708	.62966	.93225	264
3.28	.18784	.49043	.79301	305	3.78	.32972	.63231	.93489	265
3.29	.19089	.49347	.79606	303	3.79	.33237	.63495	.93754	263
3.30	1.19392	3.49651	5.79909	303	**3.80**	1.33500	3.63759	5.94017	263
3.31	.19695	.49953	.80212	301	3.81	.33763	.64021	.94280	262
3.32	.19996	.50255	.80513	301	3.82	.34025	.64284	.94542	261
3.33	.20297	.50556	.80814	300	3.83	.34286	.64545	.94803	261
3.34	.20597	.50856	.81114	299	3.84	.34547	.64806	.95064	260
3.35	1.20896	3.51155	5.81413	298	**3.85**	1.34807	3.65066	5.95324	260
3.36	.21194	.51453	.81711	297	3.86	.35067	.65325	.95584	258
3.37	.21491	.51750	.82008	297	3.87	.35325	.65584	.95842	259
3.38	.21788	.52046	.82305	295	3.88	.35584	.65842	.96101	257
3.39	.22083	.52342	.82600	295	3.89	.35841	.66099	.96358	257
3.40	1.22378	3.52636	5.82895	293	**3.90**	1.36098	3.66356	5.96615	256
3.41	.22671	.52930	.83188	293	3.91	.36354	.66612	.96871	255
3.42	.22964	.53223	.83481	292	3.92	.36609	.66868	.97126	255
3.43	.23256	.53515	.83773	291	3.93	.36864	.67122	.97381	254
3.44	.23547	.53806	.84064	290	3.94	.37118	.67377	.97635	254
3.45	1.23837	3.54096	5.84354	290	**3.95**	1.37372	3.67630	5.97889	252
3.46	.24127	.54385	.84644	288	3.96	.37624	.67883	.98141	253
3.47	.24415	.54674	.84932	288	3.97	.37877	.68135	.98394	251
3.48	.24703	.54962	.85220	287	3.98	.38128	.68387	.98645	251
3.49	.24990	.55249	.85507	286	3.99	.38379	.68638	.98896	250
3.50	1.25276	3.55535	5.85793		**4.00**	1.38629	3.68888	5.99146	

N	Log N	Log 10N	Log 100N	d	N	Log N	Log 10N	Log 100N	d
4.00	1.38629	3.68888	5.99146	250	**4.50**	1.50408	3.80666	6.10925	222
4.01	.38879	.69138	.99396	249	4.51	.50630	.80888	.11147	221
4.02	.39128	.69387	.99645	249	4.52	.50851	.81110	.11368	221
4.03	.39377	.69635	.99894	247	4.53	.51072	.81331	.11589	221
4.04	.39624	.69883	6.00141	248	4.54	.51293	.81551	.11810	220
4.05	1.39872	3.70130	6.00389	246	**4.55**	1.51513	3.81771	6.12030	219
4.06	.40118	.70377	.00635	246	4.56	.51732	.81991	.12249	219
4.07	.40364	.70623	.00881	246	4.57	.51951	.82210	.12468	219
4.08	.40610	.70868	.01127	245	4.58	.52170	.82428	.12687	218
4.09	.40854	.71113	.01372	244	4.59	.52388	.82647	.12905	218
4.10	1.41099	3.71357	6.01616	243	**4.60**	1.52606	3.82864	6.13123	217
4.11	.41342	.71601	.01859	243	4.61	.52823	.83081	.13340	216
4.12	.41585	.71844	.02102	243	4.62	.53039	.83298	.13556	217
4.13	.41828	.72086	.02345	242	4.63	.53256	.83514	.13773	215
4.14	.42070	.72328	.02587	241	4.64	.53471	.83730	.13988	216
4.15	1.42311	3.72569	6.02828	241	**4.65**	1.53687	3.83945	6.14204	215
4.16	.42552	.72810	.03069	240	4.66	.53902	.84160	.14419	214
4.17	.42792	.73050	.03309	239	4.67	.54116	.84374	.14633	214
4.18	.43031	.73290	.03548	239	4.68	.54330	.84588	.14847	213
4.19	.43270	.73529	.03787	238	4.69	.54543	.84802	.15060	213
4.20	1.43508	3.73767	6.04025	238	**4.70**	1.54756	3.85015	6.15273	213
4.21	.43746	.74005	.04263	238	4.71	.54969	.85227	.15486	212
4.22	.43984	.74242	.04501	236	4.72	.55181	.85439	.15698	212
4.23	.44220	.74479	.04737	236	4.73	.55393	.85651	.15910	211
4.24	.44456	.74715	.04973	236	4.74	.55604	.85862	.16121	210
4.25	1.44692	3.74950	6.05209	235	**4.75**	1.55814	3.86073	6.16331	211
4.26	.44927	.75185	.05444	234	4.76	.56025	.86283	.16542	210
4.27	.45161	.75420	.05678	234	4.77	.56235	.86493	.16752	209
4.28	.45395	.75654	.05912	234	4.78	.56444	.86703	.16961	209
4.29	.45629	.75887	.06146	233	4.79	.56653	.86912	.17170	209
4.30	1.45862	3.76120	6.06379	232	4.80	1.56862	3.87120	6.17379	208
4.31	.46094	.76352	.06611	232	4.81	.57070	.87328	.17587	207
4.32	.46326	.76584	.06843	231	4.82	.57277	.87536	.17794	208
4.33	.46557	.76815	.07074	230	4.83	.57485	.87743	.18002	206
4.34	.46787	.77046	.07304	231	4.84	.57691	.87950	.18208	207
4.35	1.47018	3.77276	6.07535	229	**4.85**	1.57898	3.88156	6.18415	206
4.36	.47247	.77506	.07764	229	4.86	.58104	.88362	.18621	205
4.37	.47476	.77735	.07993	229	4.87	.58309	.88568	.18826	206
4.38	.47705	.77963	.08222	228	4.88	.58515	.88773	.19032	204
4.39	.47933	.78191	.08450	227	4.89	.58719	.88978	.19236	205
4.40	1.48160	3.78419	6.08677	227	**4.90**	1.58924	3.89182	6.19441	203
4.41	.48387	.78646	.08904	227	4.91	.59127	.89386	.19644	204
4.42	.48614	.78872	.09131	226	4.92	.59331	.89589	.19848	203
4.43	.48840	.79098	.09357	225	4.93	.59534	.89792	.20051	203
4.44	.49065	.79324	.09582	225	4.94	.59737	.89995	.20254	202
4.45	1.49290	3.79549	6.09807	225	**4.95**	1.59939	3.90197	6.20456	202
4.46	.49515	.79773	.10032	224	4.96	.60141	.90399	.20658	201
4.47	.49739	.79997	.10256	223	4.97	.60342	.90600	.20859	201
4.48	.49962	.80221	.10479	223	4.98	.60543	.90801	.21060	201
4.49	.50185	.80444	.10702	223	4.99	.60744	.91002	.21261	200
4.50	1.50408	3.80666	6.10925		**5.00**	1.60944	3.91202	6.21461	

N	Log N	Log 10N	Log 100N	d	N	Log N	Log 10N	Log 100N	d
5.00	1.60944	3.91202	6.21461		**5.50**	1.70475	4.00733	6.30992	
5.01	.61144	.91402	.21661	200	5.51	.70656	.00915	.31173	181
5.02	.61343	.91602	.21860	199	5.52	.70838	.01096	.31355	182
5.03	.61542	.91801	.22059	199	5.53	.71019	.01277	.31536	181
5.04	.61741	.91999	.22258	199	5.54	.71199	.01458	.31716	180
				198					181
5.05	1.61939	3.92197	6.22456		**5.55**	1.71380	4.01638	6.31897	
5.06	.62137	.92395	.22654	198	5.56	.71560	.01818	.32077	180
5.07	.62334	.92593	.22851	197	5.57	.71740	.01998	.32257	180
5.08	.62531	.92790	.23048	197	5.58	.71919	.02177	.32436	179
5.09	.62728	.92986	.23245	197	5.59	.72098	.02356	.32615	179
				196					179
5.10	1.62924	3.93183	6.23441		**5.60**	1.72277	4.02535	6.32794	
5.11	.63120	.93378	.23637	196	5.61	.72455	.02714	.32972	178
5.12	.63315	.93574	.23832	195	5.62	.72633	.02892	.33150	178
5.13	.63511	.93769	.24028	196	5.63	.72811	.03069	.33328	178
5.14	.63705	.93964	.24222	194	5.64	.72988	.03247	.33505	177
				195					178
5.15	1.63900	3.94158	6.24417		**5.65**	1.73166	4.03424	6.33683	
5.16	.64094	.94352	.24611	194	5.66	.73342	.03601	.33859	176
5.17	.64287	.94546	.24804	193	5.67	.73519	.03777	.34036	177
5.18	.64481	.94739	.24998	194	5.68	.73695	.03954	.34212	176
5.19	.64673	.94932	.25190	192	5.69	.73871	.04130	.34388	176
				193					176
5.20	1.64866	3.95124	6.25383		**5.70**	1.74047	4.04305	6.34564	
5.21	.65058	.95316	.25575	192	5.71	.74222	.04480	.34739	175
5.22	.65250	.95508	.25767	192	5.72	.74397	.04655	.34914	175
5.23	.65441	.95700	.25958	191	5.73	.74572	.04830	.35089	175
5.24	.65632	.95891	.26149	191	5.74	.74746	.05004	.35263	174
				191					174
5.25	1.65823	3.96081	6.26340		**5.75**	1.74920	4.05178	6.35437	
5.26	.66013	.96272	.26530	190	5.76	.75094	.05352	.35611	174
5.27	.66203	.96462	.26720	190	5.77	.75267	.05526	.35784	173
5.28	.66393	.96651	.26910	190	5.78	.75440	.05699	.35957	173
5.29	.66582	.96840	.27099	189	5.79	.75613	.05872	.36130	173
				189					173
5.30	1.66771	3.97029	6.27288		**5.80**	1.75786	4.06044	6.36303	
5.31	.66959	.97218	.27476	188	5.81	.75958	.06217	.36475	172
5.32	.67147	.97406	.27664	188	5.82	.76130	.06389	.36647	172
5.33	.67335	.97594	.27852	188	5.83	.76302	.06560	.36819	172
5.34	.67523	.97781	.28040	188	5.84	.76473	.06732	.36990	171
				187					171
5.35	1.67710	3.97968	6.28227		**5.85**	1.76644	4.06903	6.37161	
5.36	.67896	.98155	.28413	186	5.86	.76815	.07073	.37332	171
5.37	.68083	.98341	.28600	187	5.87	.76985	.07244	.37502	170
5.38	.68269	.98527	.28786	186	5.88	.77156	.07414	.37673	171
5.39	.68455	.98713	.28972	186	5.89	.77326	.07584	.37843	170
				185					169
5.40	1.68640	3.98898	6.29157		**5.90**	1.77495	4.07754	6.38012	
5.41	.68825	.99083	.29342	185	5.91	.77665	.07923	.38182	170
5.42	.69010	.99268	.29527	185	5.92	.77834	.08092	.38351	169
5.43	.69194	.99452	.29711	184	5.93	.78002	.08261	.38519	168
5.44	.69378	.99636	.29895	184	5.94	.78171	.08429	.38688	169
				184					168
5.45	1.69562	3.99820	6.30079		**5.95**	1.78339	4.08598	6.38856	
5.46	.69745	4.00003	.30262	183	5.96	.78507	.08766	.39024	168
5.47	.69928	.00186	.30445	183	5.97	.78675	.08933	.39192	168
5.48	.70111	.00369	.30628	183	5.98	.78842	.09101	.39359	167
5.49	.70293	.00551	.30810	182	5.99	.79009	.09268	.39526	167
				182					167
5.50	1.70475	4.00733	6.30992		**6.00**	1.79176	4.09434	6.39693	

N	Log N	Log 10N	Log 100N	d	N	Log N	Log 10N	Log 100N	d
6.00	1.79176	4.09434	6.39693		**6.50**	1.87180	4.17439	6.47697	
6.01	.79342	.09601	.39859	166	6.51	.87334	.17592	.47851	154
6.02	.79509	.09767	.40026	167	6.52	.87487	.17746	.48004	153
6.03	.79675	.09933	.40192	166	6.53	.87641	.17899	.48158	154
6.04	.79840	.10099	.40357	165	6.54	.87794	.18052	.48311	153
				166					153
6.05	1.80006	4.10264	6.40523		**6.55**	1.87947	4.18205	6.48464	
6.06	.80171	.10429	.40688	165	6.56	.88099	.18358	.48616	152
6.07	.80336	.10594	.40853	165	6.57	.88251	.18510	.48768	152
6.08	.80500	.10759	.41017	164	6.58	.88403	.18662	.48920	152
6.09	.80655	.10923	.41182	165	6.59	.88555	.18814	.49072	152
				164					152
6.10	1.80829	4.11087	6.41346		**6.60**	1.88707	4.18965	6.49224	
6.11	.80993	.11251	.41510	164	6.61	.88858	.19117	.49375	151
6.12	.81156	.11415	.41673	163	6.62	.89010	.19268	.49527	152
6.13	.81319	.11578	.41836	163	6.63	.89160	.19419	.49677	150
6.14	.81482	.11741	.41999	163	6.64	.89311	.19570	.49828	151
				163					151
6.15	1.81645	4.11904	6.42162		**6.65**	1.89462	4.19720	6.49979	
6.16	.81808	.12066	.42325	163	6.66	.89612	.19870	.50129	150
6.17	.81970	.12228	.42487	162	6.67	.89762	.20020	.50279	150
6.18	.82132	.12390	.42649	162	6.68	.89912	.20170	.50429	150
6.19	.82294	.12552	.42811	162	6.69	.90061	.20320	.50578	149
				161					150
6.20	1.82455	4.12713	6.42972		**6.70**	1.90211	4.20469	6.50728	
6.21	.82616	.12875	.43133	161	6.71	.90360	.20618	.50877	149
6.22	.82777	.13036	.43294	161	6.72	.90509	.20767	.51026	149
6.23	.82938	.13196	.43455	161	6.73	.90658	.20916	.51175	149
6.24	.83098	.13357	.43615	160	6.74	.90806	.21065	.51323	148
				160					148
6.25	1.83258	4.13517	6.43775		**6.75**	1.90954	4.21213	6.51471	
6.26	.83418	.13677	.43935	160	6.76	.91102	.21361	.51619	148
6.27	.83578	.13836	.44095	160	6.77	.91250	.21509	.51767	148
6.28	.83737	.13996	.44254	159	6.78	.91398	.21656	.51915	148
6.29	.83896	.14155	.44413	159	6.79	.91545	.21804	.52062	147
				159					147
6.30	1.84055	4.14313	6.44572		**6.80**	1.91692	4.21951	6.52209	
6.31	.84214	.14472	.44731	159	6.81	.91839	.22098	.52356	147
6.32	.84372	.14630	.44889	158	6.82	.91986	.22244	.52503	147
6.33	.84530	.14789	.45047	158	6.83	.92132	.22391	.52649	146
6.34	.84688	.14946	.45205	158	6.84	.92279	.22537	.52796	147
				157					146
6.35	1.84845	4.15104	6.45362		**6.85**	1.92425	4.22683	6.52942	
6.36	.85003	.15261	.45520	158	6.86	.92571	.22829	.53088	146
6.37	.85160	.15418	.45677	157	6.87	.92716	.22975	.53233	145
6.38	.85317	.15575	.45834	157	6.88	.92862	.23120	.53379	146
6.39	.85473	.15732	.45990	156	6.89	.93007	.23266	.53524	145
				157					145
6.40	1.85630	4.15888	6.46147		**6.90**	1.93152	4.23411	6.53669	
6.41	.85786	.16044	.46303	156	6.91	.93297	.23555	.53814	145
6.42	.85942	.16200	.46459	156	6.92	.93442	.23700	.53959	145
6.43	.86097	.16356	.46614	155	6.93	.93586	.23844	.54103	144
6.44	.86253	.16511	.46770	156	6.94	.93730	.23989	.54247	144
				155					144
6.45	1.86408	4.16667	6.46925		**6.95**	1.93874	4.24133	6.54391	
6.46	.86563	.16821	.47080	155	6.96	.94018	.24276	.54535	144
6.47	.86718	.16976	.47235	155	6.97	.94162	.24420	.54679	144
6.48	.86872	.17131	.47389	154	6.98	.94305	.24563	.54822	143
6.49	.87026	.17285	.47543	154	6.99	.94448	.24707	.54965	143
				154					143
6.50	1.87180	4.17439	6.47697		**7.00**	1.94591	4.24850	6.55108	

N	Log N	Log 10N	Log 100N	d	N	Log N	Log 10N	Log 100N	d
7.00	1.94591	4.24950	6.55108		**7.50**	2.01490	4.31749	6.62007	
7.01	.94734	.24992	.55251	143	7.51	.01624	.31882	.62141	134
7.02	.94876	.25135	.55393	142	7.52	.01757	.32015	.62274	133
7.03	.95019	.25277	.55536	143	7.53	.01890	.32148	.62407	133
7.04	.95161	.25419	.55678	142	7.54	.02022	.32281	.62539	132
				142					133
7.05	1.95303	4.25561	6.55820		**7.55**	2.02155	4.32413	6.62672	
7.06	.95445	.25703	.55962	142	7.56	.02287	.32546	.62804	132
7.07	.95586	.25845	.56103	141	7.57	.02419	.32678	.62936	132
7.08	.95727	.25986	.56244	141	7.58	.02551	.32810	.63068	132
7.09	.95869	.26127	.56386	142	7.59	.02683	.32942	.63200	132
				140					132
7.10	1.96009	4.26268	6.56526		**7.60**	2.02815	4.33073	6.63332	
7.11	.96150	.26409	.56667	141	7.61	.02946	.33205	.63463	131
7.12	.96291	.26549	.56808	141	7.62	.03078	.33336	.63595	132
7.13	.96431	.26690	.56948	140	7.63	.03209	.33467	.63726	131
7.14	.96571	.26830	.57088	140	7.64	.03340	.33598	.63857	131
				140					131
7.15	1.96711	4.26970	6.57228		**7.65**	2.03471	4.33729	6.63988	
7.16	.96851	.27110	.57368	140	7.66	.03601	.33860	.64118	130
7.17	.96991	.27249	.57508	140	7.67	.03732	.33990	.64249	131
7.18	.97130	.27388	.57647	139	7.68	.03862	.34120	.64379	130
7.19	.97269	.27528	.57786	139	7.69	.03992	.34251	.64509	130
				139					130
7.20	1.97408	4.27667	6.57925		**7.70**	2.04122	4.34381	6.64639	
7.21	.97547	.27805	.58064	139	7.71	.04252	.34510	.64769	130
7.22	.97685	.27944	.58203	139	7.72	.04381	.34640	.64898	129
7.23	.97824	.28082	.58341	138	7.73	.04511	.34769	.65028	130
7.24	.97962	.28221	.58479	138	7.74	.04640	.34899	.65157	129
				138					129
7.25	1.98100	4.28359	6.58617		**7.75**	2.04769	4.35028	6.65286	
7.26	.98238	.28496	.58755	138	7.76	.04898	.35157	.65415	129
8.27	.98376	.28634	.58893	138	7.77	.05027	.35286	.65544	129
7.28	.98513	.28772	.59030	137	7.78	.05156	.35414	.65673	129
7.29	.98650	.28909	.59167	137	7.79	.05284	.35543	.65801	128
				137					128
7.30	1.98787	4.29046	6.59304		**7.80**	2.05412	4.35671	6.65929	
7.31	.98924	.29183	.59441	137	7.81	.05540	.35799	.66058	129
7.32	.99061	.29320	.59578	137	7.82	.05668	.35927	.66185	127
7.33	.99198	.29456	.59715	137	7.83	.05796	.36055	.66313	128
7.34	.99334	.29592	.59851	136	7.84	.05924	.36182	.66441	128
				136					127
7.35	1.99470	4.29729	6.59987		**7.85**	2.06051	4.36310	6.66568	
7.36	.99606	.29865	.60123	136	7.86	.06179	.36437	.66696	128
7.37	.99742	.30000	.60259	136	7.87	.06306	.36564	.66823	127
7.38	.99877	.30136	.60394	135	7.88	.06433	.36691	.66950	127
7.39	2.00013	.30271	.60530	136	7.89	.06560	.36818	.67077	127
				135					126
7.40	2.00148	4.30407	6.60665		**7.90**	2.06686	4.36945	6.67203	
7.41	.00283	.30542	.60800	135	7.91	.06813	.37071	.67330	127
7.42	.00418	.30676	.60935	135	7.92	.06939	.37198	.67456	126
7.43	.00553	.30811	.61070	135	7.93	.07065	.37324	.67582	126
7.44	.00687	.30946	.61204	134	7.94	.07191	.37450	.67708	126
				134					126
7.45	2.00821	4.31080	6.61338		**7.95**	2.07317	4.37576	6.67834	
7.46	.00956	.31214	.61473	135	7.96	.07443	.37701	.67960	126
7.47	.01089	.31348	.61607	134	7.97	.07568	.37827	.68085	125
7.48	.01223	.31482	.61740	133	7.98	.07694	.37952	.68211	126
7.49	.01357	.31615	.61874	134	7.99	.07819	.38178	.68336	125
				133					125
7.50	2.01490	4.31749	6.62007		**8.00**	2.07944	4.38203	6.68461	

N	Log N	Log 10N	Log 100N	d	N	Log N	Log 10N	Log 100N	d
8.00	2.07944	4.38203	6.68461		**8.50**	2.14007	4.44265	6.74524	
8.01	.08069	.38328	.68586	125	8.51	.14124	.44383	.74641	117
8.02	.08194	.38452	.68711	125	8.52	.14242	.44500	.74759	118
8.03	.08318	.38577	.68835	124	8.53	.14359	.44617	.74876	117
8.04	.08443	.38701	.68960	125	8.54	.14476	.44735	.74993	117
				124					117
8.05	2.08567	4.38826	6.69084		**8.55**	2.14593	4.44852	6.75110	
8.06	.08691	.38950	.69208	124	8.56	.14710	.44969	.75227	117
8.07	.08815	.39074	.69332	124	8.57	.14827	.45085	.75344	117
8.08	.08939	.39198	.69456	124	8.58	.14943	.45202	.75460	116
8.09	.09063	.39321	.69580	124	8.59	.15060	.45318	.75577	117
				123					116
8.10	2.09186	4.39445	6.69703		**8.60**	2.15176	4.45435	6.75693	
8.11	.09310	.39568	.69827	124	8.61	.15292	.45551	.75809	116
8.12	.09433	.39692	.69950	123	8.62	.15409	.45667	.75926	117
8.13	.09556	.39815	.70073	123	8.63	.15524	.45783	.76041	115
8.14	.09679	.39938	.70196	123	8.64	.15640	.45899	.76157	116
				123					116
8.15	2.09802	4.40060	6.70319		**8.65**	2.15756	4.46014	6.76273	
8.16	.09924	.40183	.70441	122	8.66	.15871	.46130	.76388	115
8.17	.10047	.40305	.70564	123	8.67	.15987	.46245	.76504	116
8.18	.10169	.40428	.70686	122	8.68	.16102	.46361	.76619	115
8.19	.10291	.40550	.70808	122	8.69	.16217	.46476	.76734	115
				122					115
8.20	2.10413	4.40672	6.70930		**8.70**	2.16332	4.46591	6.76849	
8.21	.10535	.40794	.71052	122	8.71	.16447	.46706	.76964	115
8.22	.10657	.40916	.71174	122	8.72	.16562	.46820	.77079	115
8.23	.10779	.41037	.71296	122	8.73	.16677	.46935	.77194	115
8.24	.10900	.41159	.71417	121	8.74	.16791	.47050	.77308	114
				121					114
8.25	2.11021	4.41280	6.71538		**8.75**	2.16905	4.47164	6.77422	
8.26	.11142	.41401	.71659	121	8.76	.17020	.47278	.77537	115
8.27	.11263	.41522	.71780	121	8.77	.17134	.47392	.77651	114
8.28	.11384	.41643	.71901	121	8.78	.17248	.47506	.77765	114
8.29	.11505	.41764	.72022	121	8.79	.17361	.47620	.77878	113
				121					114
8.30	2.11626	4.41884	6.72143		**8.80**	2.17475	4.47734	6.77992	
8.31	.11746	.42004	.72263	120	8.81	.17589	.47847	.78106	114
8.32	.11866	.42125	.72383	120	8.82	.17702	.47961	.78219	113
8.33	.11986	.42245	.72503	120	8.83	.17816	.48074	.78333	114
8.34	.12106	.42365	.72623	120	8.84	.17929	.48187	.78446	113
				120					113
8.35	2.12226	4.42485	6.72743		**8.85**	2.18042	4.48300	6.78559	
8.36	.12346	.42604	.72863	120	8.86	.18155	.48413	.78672	113
8.37	.12465	.42724	.72982	119	8.87	.18267	.48526	.78784	112
8.38	.12585	.42843	.73102	120	8.88	.18380	.48639	.78897	113
8.39	.12704	.42963	.73221	119	8.89	.18493	.48751	.79010	113
				119					112
8.40	2.12823	4.43082	6.73340		**8.90**	2.18605	4.48864	6.79122	
8.41	.12942	.43201	.73459	119	8.91	.18717	.48976	.79234	112
8.42	.13061	.43319	.73578	119	8.92	.18830	.49088	.79347	113
8.43	.13180	.43438	.73697	119	8.93	.18942	.49200	.79459	112
8.44	.13298	.43557	.73815	118	8.94	.19054	.49312	.79571	112
				119					111
8.45	2 13417	4.43675	6.73934		**8.95**	2.19165	4.49424	6.79682	
8.46	.13535	.43793	.74052	118	8.96	.19277	.49536	.79794	112
8.47	.13653	.43912	.74170	118	8.97	.19389	.49647	.79906	112
8.48	.13771	.44030	.74288	118	8.98	.19500	.49758	.80017	111
8.49	.13889	.44147	.74406	118	8.99	.19611	.49870	.80128	111
				118					111
8.50	2.14007	4.44265	6.74524		**9.00**	2.19722	4.49981	6.80239	

N	Log N	Log 10N	Log 100N	d	N	Log N	Log 10N	Log 100N	d
9.00	2.19722	4.49981	6.80239	112	**9.50**	2.25129	4.55388	6.85646	105
9.01	.19834	.50092	.80351	110	9.51	.25234	.55493	.85751	106
9.02	.19944	.50203	.80461	111	9.52	.25339	.55598	.85857	104
9.03	.20055	.50314	.80572	111	9.53	.25444	.55703	.85961	105
9.04	.20166	.50424	.80683	110	9.54	.25549	.55808	.86066	105
9.05	2.20276	4.50535	6.80793	111	**9.55**	2.25654	4.55913	6.86171	105
9.06	.20387	.50645	.80904	110	9.56	.25759	.56017	.86276	104
9.07	.20497	.50756	.81014	110	9.57	.25863	.56122	.86380	105
9.08	.20607	.50866	.81124	111	9.58	.25968	.56226	.86485	104
9.09	.20717	.50976	.81235	109	9.59	.26072	.56331	.86589	104
9.10	2.20827	4.51086	6.81344	110	**9.60**	2.26176	4.56435	6.86693	104
9.11	.20937	.51196	.81454	110	9.61	.26280	.56539	.86797	104
9.12	.21047	.51305	.81564	110	9.62	.26384	.56643	.86901	104
9.13	.21157	.51415	.81674	109	9.63	.26488	.56747	.87005	104
9.14	.21266	.51525	.81783	109	9.64	.26592	.56851	.87109	104
9.15	2.21375	4.51634	6.81892	110	**9.65**	2.26696	4.56954	6.87213	103
9.16	.21485	.51743	.82002	109	9.66	.26799	.57058	.87316	104
9.17	.21594	.51852	.82111	109	9.67	.26903	.57161	.87420	103
9.18	.21703	.51961	.82220	109	9.68	.27006	.57265	.87523	103
9.19	.21812	.52070	.82329	108	9.69	.27109	.57368	.87626	104
9.20	2.21920	4.52179	6.82437	109	**9.70**	2.27213	4.57471	6.87730	103
9.21	.22029	.52287	.82546	109	9.71	.27316	.57574	.87833	103
9.22	.22138	.52396	.82655	108	9.72	.27419	.57677	.87936	102
9.23	.22246	.52504	.82763	108	9.73	.27521	.57780	.88038	103
9.24	.22354	.52613	.82871	108	9.74	.27624	.57883	.88141	103
9.25	2.22462	4.52721	6.82979	108	**9.75**	2.27727	4.57985	6.88244	102
9.26	.22570	.52829	.83087	108	9.76	.27829	.58088	.88346	103
9.27	.22678	.52937	.83195	108	9.77	.27932	.58190	.88449	102
9.28	.22786	.53045	.83303	108	9.78	.28034	.58292	.88551	102
9.29	.22894	.53152	.83411	107	9.79	.28136	.58395	.88653	102
9.30	2.23001	4.53260	6.83518	108	**9.80**	2.28238	4.58497	6.88755	102
9.31	.23109	.53367	.83626	107	9.81	.28340	.58599	.88857	102
9.32	.23216	.53475	.83733	108	9.82	.28442	.58701	.88959	102
9.33	.23324	.53582	.83841	107	9.83	.28544	.58802	.89061	102
9.34	.23431	.53689	.83948	107	9.84	.28646	.58904	.89163	101
9.35	2.23538	4.53796	6.84055	107	**9.85**	2.28747	4.59006	6.89264	102
9.36	.23645	.53903	.84162	106	9.86	.28849	.59107	.89366	101
9.37	.23751	.54010	.84268	107	9.87	.28950	.59208	.89467	101
9.38	.23858	.54116	.84375	107	9.88	.29051	.59310	.89568	101
9.39	.23965	.54223	.84482	106	9.89	.29152	.59411	.89669	101
9.40	2.24071	4.54329	6.84588	106	**9.90**	2.29253	4.59512	6.89770	101
9.41	.24177	.54436	.84694	107	9.91	.29354	.59613	.89871	101
9.42	.24284	.54542	.84801	106	9.92	.29455	.59714	.89972	101
9.43	.24390	.54648	.84907	106	9.93	.29556	.59815	.90073	101
9.44	.24496	.54754	.85013	105	9.94	.29657	.59915	.90174	100
9.45	2.24601	4.54860	6.85118	106	**9.95**	2.29757	4.60016	6.90274	101
9.46	.24707	.54966	.85224	106	9.96	.29858	.60116	.90375	100
9.47	.24813	.55071	.85330	105	9.97	.29958	.60217	.90475	100
9.48	.24918	.55177	.85435	106	9.98	.30058	.60317	.90575	100
9.49	.25024	.55282	.85541	105	9.99	.30158	.60417	.90675	101
9.50	2.25129	4.55388	6.85646		**10.00**	2.30259	4.60517	6.90776	

TABLE XVII. EXPONENTIAL AND HYPERBOLIC FUNCTIONS

x	e^x	e^{-x}	Sinh x	Cosh x	L e^x	L Sinh x	L Cosh x
0.00	1.0000	1.0000	0.0000	1.0000	.00000	$-\infty$	.00000
.01	.0101	.99005	.0100	.0001	.00434	.00001	.00002
.02	.0202	.98020	.0200	.0002	.00869	.30106	.00009
.03	.0305	.97045	.0300	.0005	.01303	.47719	.00020
.04	.0408	.96079	.0400	.0008	.01737	.60218	.00035
0.05	1.0513	.95123	0.0500	1.0013	.02171	.69915	.00054
.06	.0618	.94176	.0600	.0018	.02606	.77841	.00078
.07	.0725	.93239	.0701	.0025	.03040	.84545	.00106
.08	.0833	.92312	.0801	.0032	.03474	.90355	.00139
.09	.0942	.91393	.0901	.0041	.03909	.95483	.00176
0.10	1.1052	.90484	0.1002	1.0050	.04343	.00072	.00217
.11	.1163	.89583	.1102	.0061	.04777	.04227	.00262
.12	.1275	.88692	.1203	.0072	.05212	.08022	.00312
.13	.1388	.87810	.1304	.0085	.05646	.11517	.00366
.14	.1503	.86936	.1405	.0098	.06080	.14755	.00424
0.15	1.1618	.86071	0.1506	1.0113	.06514	.17772	.00487
.16	.1735	.85214	.1607	.0128	.06949	.20597	.00554
.17	.1853	.84366	.1708	.0145	.07383	.23254	.00625
.18	.1972	.83527	.1810	.0162	.07817	.25762	.00700
.19	.2092	.82696	.1911	.0181	.08252	.28136	.00779
0.20	1.2214	.81873	0.2013	1.0201	.08686	.30392	.00863
.21	.2337	.81058	.2115	.0221	.09120	.32541	.00951
.22	.2461	.80252	.2218	.0243	.09554	.34592	.01043
.23	.2586	.79453	.2320	.0266	.09989	.36555	.01139
.24	.2712	.78663	.2423	.0289	.10423	.38437	.01239
0.25	1.2840	.77880	0.2526	1.0314	.10857	.40245	.01343
.26	.2969	.77105	.2629	.0340	.11292	.41986	.01452
.27	.3100	.76338	.2733	.0367	.11726	.43663	.01564
.28	.3231	.75578	.2837	.0395	.12160	.45282	.01681
.29	.3364	.74826	.2941	.0423	.12595	.46847	.01801
0.30	1.3499	.74082	0.3045	1.0453	.13029	.48362	.01926
.31	.3634	.73345	.3150	.0484	.13463	.49830	.02054
.32	.3771	.72615	.3255	.0516	.13897	.51254	.02187
.33	.3910	.71892	.3360	.0549	.14332	.52637	.02323
.34	.4049	.71177	.3466	.0584	.14766	.53981	.02463
0.35	1.4191	.70469	0.3572	1.0619	.15200	.55290	.02607
.36	.4333	.69768	.3678	.0655	.15635	.56564	.02755
.37	.4477	.69073	.3785	.0692	.16069	.57807	.02907
.38	.4623	.68386	.3892	.0731	.16503	.59019	.03063
.39	.4770	.67706	.4000	.0770	.16937	.60202	.03222
0.40	1.4918	.67032	0.4108	1.0811	.17372	.61358	.03385
.41	.5068	.66365	.4216	.0852	.17806	.62488	.03552
.42	.5220	.65705	.4325	.0895	.18240	.63594	.03723
.43	.5373	.65051	.4434	.0939	.18675	.64677	.03897
.44	.5527	.64404	.4543	.0984	.19109	.65738	.04075
0.45	1.5683	.63763	0.4653	1.1030	.19543	.66777	.04256
.46	.5841	.63128	.4764	.1077	.19978	.67797	.04441
.47	.6000	.62500	.4875	.1125	.20412	.68797	.04630
.48	.6161	.61878	.4986	.1174	.20846	.69779	.04822
.49	.6323	.61263	.5098	.1225	.21280	.70744	.05018
0.50	1.6487	.60653	0.5211	1.1276	.21715	.71692	.05217

x	e^x	e^{-x}	Sinh x	Cosh x	L e^x	L Sinh x	L Cosh x
0.50	1.6487	.60653	0.5211	1.1276	.21715	.71692	.05217
.51	.6653	.60050	.5324	.1329	.22149	.72624	.05419
.52	.6820	.59452	.5438	.1383	.22583	.73540	.05625
.53	.6989	.58860	.5552	.1438	.23018	.74442	.05834
.54	.7160	.58275	.5666	.1494	.23452	.75330	.06046
0.55	1.7333	.57695	0.5782	1.1551	.23886	.76204	.06262
.56	.7507	.57121	.5897	.1609	.24320	.77065	.06481
.57	.7683	.56553	.6014	.1669	.24755	.77914	.06703
.58	.7860	.55990	.6131	.1730	.25189	.78751	.06929
.59	.8040	.55433	.6248	.1792	.25623	.79576	.07157
0.60	1.8221	.54881	0.6367	1.1855	.26058	.80390	.07389
.61	.8404	.54335	.6485	.1919	.26492	.81194	.07624
.62	.8589	.53794	.6605	.1984	.26926	.81987	.07861
.63	.8776	.53259	.6725	.2051	.27361	.82770	.08102
.64	.8965	.52729	.6846	.2119	.27795	.83543	.08346
0.65	1.9155	.52205	0.6967	1.2188	.28229	.84308	.08593
.66	.9348	.51685	.7090	.2258	.28663	.85063	.08843
.67	.9542	.51171	.7213	.2330	.29098	.85809	.09095
.68	.9739	.50662	.7336	.2402	.29532	.86548	.09351
.69	.9937	.50158	.7461	.2476	.29966	.87278	.09609
0.70	2.0138	.49659	0.7586	1.2552	.30401	.88000	.09870
.71	.0340	.49164	.7712	.2628	.30835	.88715	.10134
.72	.0544	.48675	.7838	.2706	.31269	.89423	.10401
.73	.0751	.48191	.7966	.2785	.31703	.90123	.10670
.74	.0959	.47711	.8094	.2865	.32138	.90817	.10942
0.75	2.1170	.47237	0.8223	1.2947	.32572	.91504	.11216
.76	.1383	.46767	.8353	.3030	.33006	.92185	.11493
.77	.1598	.46301	.8484	.3114	.33441	.92859	.11773
.78	.1815	.45841	.8615	.3199	.33875	.93527	.12055
.79	.2034	.45384	.8748	.3286	.34309	.94190	.12340
0.80	2.2255	.44933	0.8881	1.3374	.34744	.94846	.12627
.81	.2479	.44486	.9015	.3464	.35178	.95498	.12917
.82	.2705	.44043	.9150	.3555	.35612	.96144	.13209
.83	.2933	.43605	.9286	.3647	.36046	.96784	.13503
.84	.3164	.43171	.9423	.3740	.36481	.97420	.13800
0.85	2.3396	.42741	0.9561	1.3835	.36915	.98051	.14099
.86	.3632	.42316	.9700	.3932	.37349	.98677	.14400
.87	.3869	.41895	.9840	.4029	.37784	.99299	.14704
.88	.4109	.41478	.9981	.4128	.38218	.99916	.15009
.89	.4351	.41066	1.0122	.4229	.38652	.00528	.15317
0.90	2.4596	.40657	1.0265	1.4331	.39087	.01137	.15627
.91	.4843	.40252	.0409	.4434	.39521	.01741	.15939
.92	.5093	.39852	.0554	.4539	.39955	.02341	.16254
.93	.5345	.39455	.0700	.4645	.40389	.02937	.16570
.94	.5600	.39063	.0847	.4753	.40824	.03530	.16888
0.95	2.5857	.38674	1.0995	1.4862	.41258	.04119	.17208
.96	.6117	.38289	.1144	.4973	.41692	.04704	.17531
.97	.6379	.37908	.1294	.5085	.42127	.05286	.17855
.98	.6645	.37531	.1446	.5199	.42561	.05864	.18181
.99	.6912	.37158	.1598	.5314	.42995	.06439	.18509
1.00	2.7183	.36788	1.1752	1.5431	.43429	.07011	.18839

x	e^x	e^{-x}	Sinh x	Cosh x	L e^x	L Sinh x	L Cosh x
1.00	2.7183	.36788	1.1752	1.5431	.43429	.07011	.18839
.01	.7456	.36422	.1907	.5549	.43864	.07580	.19171
.02	.7732	.36059	.2063	.5669	.44298	.08146	.19504
.03	.8011	.35701	.2220	.5790	.44732	.08708	.19839
.04	.8292	.35345	.2379	.5913	.45167	.09268	.20176
1.05	2.8577	.34994	1.2539	1.6038	.45601	.09825	.20515
.06	.8864	.34646	.2700	.6164	.46035	.10379	.20855
.07	.9154	.34301	.2862	.6292	.46470	.10930	.21197
.08	.9447	.33960	.3025	.6421	.46904	.11479	.21541
.09	.9743	.33622	.3190	.6552	.47338	.12025	.21886
1.10	3.0042	.33287	1.3356	1.6685	.47772	.12569	.22233
.11	.0344	.32956	.3524	.6820	.48207	.13111	.22582
.12	.0649	.32628	.3693	.6956	.48641	.13649	.22931
.13	.0957	.32303	.3863	.7093	.49075	.14186	.23283
.14	.1268	.31982	.4035	.7233	.49510	.14720	.23636
1.15	3.1582	.31664	1.4208	1.7374	.49944	.15253	.23990
.16	.1899	.31349	.4382	.7517	.50378	.15783	.24346
.17	.2220	.31037	.4558	.7662	.50812	.16311	.24703
.18	.2544	.30728	.4735	.7808	.51247	.16836	.25062
.19	.2871	.30422	.4914	.7957	.51681	.17360	.25422
1.20	3.3201	.30119	1.5095	1.8107	.52115	.17882	.25784
.21	.3535	.29820	.5276	.8258	.52550	.18402	.26146
.22	.3872	.29523	.5460	.8412	.52984	.18920	.26510
.23	.4212	.29229	.5645	.8568	.53418	.19437	.26876
.24	.4556	.28938	.5831	.8725	.53853	.19951	.27242
1.25	3.4903	.28650	1.6019	1.8884	.54287	.20464	.27610
.26	.5254	.28365	.6209	.9045	.54721	.20975	.27979
.27	.5609	.28083	.6400	.9208	.55155	.21485	.28349
.28	.5966	.27804	.6593	.9373	.55590	.21993	.28721
.29	.6328	.27527	.6788	.9540	.56024	.22499	.29093
1.30	3.6693	.27253	1.6984	1.9709	.56458	.23004	.29467
.31	.7062	.26982	.7182	.9880	.56893	.23507	.29842
.32	.7434	.26714	.7381	2.0053	.57327	.24009	.30217
.33	.7810	.26448	.7583	.0228	.57761	.24509	.30594
.34	.8190	.26185	.7786	.0404	.58195	.25008	.30972
1.35	3.8574	.25924	1.7991	2.0583	.58630	.25505	.31352
.36	.8962	.25666	.8198	.0764	.59064	.26002	.31732
.37	.9354	.25411	.8406	.0947	.59498	.26496	.32113
.38	.9749	.25158	.8617	.1132	.59933	.26990	.32495
.39	4.0149	.24908	.8829	.1320	.60367	.27482	.32878
1.40	4.0552	.24660	1.9043	2.1509	.60801	.27974	.33262
.41	.0960	.24414	.9259	.1700	.61236	.28464	.33647
.42	.1371	.24171	.9477	.1894	.61670	.28952	.34033
.43	.1787	.23931	.9697	.2090	.62104	.29440	.34420
.44	.2207	.23693	.9919	.2288	.62538	.29926	.34807
1.45	4.2631	.23457	2.0143	2.2488	.62973	.30412	.35196
.46	.3060	.23224	.0369	.2691	.63407	.30896	.35585
.47	.3492	.22993	.0597	.2896	.63841	.31379	.35976
.48	.3929	.22764	.0827	.3103	.64276	.31862	.36367
.49	.4371	.22537	.1059	.3312	.64710	.32343	.36759
1.50	4.4817	.22313	2.1293	2.3524	.65144	.32823	.37151

x	e^x	e^{-x}	Sinh x	Cosh x	L e^x	L Sinh x	L Cosh x
1.50	4.4817	.22313	2.1293	2.3524	.65144	.32823	.37151
.51	.5267	.22091	.1529	.3738	.65578	.33303	.37545
.52	.5722	.21871	.1768	.3955	.66013	.33781	.37939
.53	.6182	.21654	.2008	.4174	.66447	.34258	.38334
.54	.6646	.21438	.2251	.4395	.66881	.34735	.38730
1.55	4.7115	.21225	2.2496	2.4619	.67316	.35211	.39126
.56	.7588	.21014	.2743	.4845	.67750	.35686	.39524
.57	.8066	.20805	.2993	.5073	.68184	.36160	.39921
.58	.8550	.20598	.3245	.5305	.68619	.36633	.40320
.59	.9037	.20393	.3499	.5538	.69053	.37105	.40719
1.60	4.9530	.20190	2.3756	2.5775	.69487	.37577	.41119
.61	5.0028	.19989	.4015	.6013	.69921	.38048	.41520
.62	.0531	.19790	.4276	.6255	.70356	.38518	.41921
.63	.1039	.19593	.4540	.6499	.70790	.38987	.42323
.64	.1552	.19398	.4806	.6746	.71224	.39456	.42725
1.65	5.2070	.19205	2.5075	2.6995	.71659	.39923	.43129
.66	.2593	.19014	.5346	.7247	.72093	.40391	.43532
.67	.3122	.18825	.5620	.7502	.72527	.40857	.43937
.68	.3656	.18637	.5896	.7760	.72961	.41323	.44341
.69	.4195	.18452	.6175	.8020	.73396	.41788	.44747
1.70	5.4739	.18268	2.6456	2.8283	.73830	.42253	.45153
.71	.5290	.18087	.6740	.8549	.74264	.42717	.45559
.72	.5845	.17907	.7027	.8818	.74699	.43180	.45966
.73	.6407	.17728	.7317	.9090	.75133	.43643	.46374
.74	.6973	.17552	.7609	.9364	.75567	.44105	.46782
1.75	5.7546	.17377	2.7904	2.9642	.76002	.44567	.47191
.76	.8124	.17204	.8202	.9922	.76436	.45028	.47600
.77	.8709	.17033	.8503	3.0206	.76870	.45488	.48009
.78	.9299	.16864	.8806	.0492	.77304	.45948	.48419
.79	.9895	.16696	.9112	.0782	.77739	.46408	.48830
1.80	6.0496	.16530	2.9422	3.1075	.78173	.46867	.49241
.81	.1104	.16365	.9734	.1371	.78607	.47325	.49652
.82	.1719	.16203	3.0049	.1669	.79042	.47783	.50064
.83	.2339	.16041	.0367	.1972	.79476	.48241	.50476
.84	.2965	.15882	.0689	.2277	.79910	.48698	.50889
1.85	6.3598	.15724	3.1013	3.2585	.80344	.49154	.51302
.86	.4237	.15567	.1340	.2897	.80779	.49610	.51716
.87	.4883	.15412	.1671	.3212	.81213	.50066	.52130
.88	.5535	.15259	.2005	.3530	.81647	.50521	.52544
.89	.6194	.15107	.2341	.3852	.82082	.50976	.52959
1.90	6.6859	.14957	3.2682	3.4177	.82516	.51430	.53374
.91	.7531	.14808	.3025	.4506	.82950	.51884	.53789
.92	.8210	.14661	.3372	.4838	.83385	.52338	.54205
.93	.8895	.14515	.3722	.5173	.83819	.52791	.54621
.94	.9588	.14370	.4075	.5512	.84253	.53244	.55038
1.95	7.0287	.14227	3.4432	3.5855	.84687	.53696	.55455
.96	.0993	.14086	.4792	.6201	.85122	.54148	.55872
.97	.1707	.13946	.5156	.6551	.85556	.54600	.56290
.98	.2427	.13807	.5523	.6904	.85990	.55051	.56707
.99	.3155	.13670	.5894	.7261	.86425	.55502	.57126
2.00	7.3891	.13534	3.6269	3.7622	.86859	.55953	.57544

x	e^x	e^{-x}	Sinh x	Cosh x	L e^x	L Sinh x	L Cosh x
2.00	7.3891	.13534	3.6269	3.7622	.86859	.55953	.57544
.01	.4633	.13399	.6647	.7987	.87293	.56403	.57963
.02	.5383	.13266	.7028	.8355	.87727	.56853	.58382
.03	.6141	.13134	.7414	.8727	.88162	.57303	.58802
.04	.6906	.13003	.7803	.9103	.88596	.57753	.59221
2.05	7.7679	.12873	3.8196	3.9483	.89030	.58202	.59641
.06	.8460	.12745	.8593	.9867	.89465	.58650	.60061
.07	.9248	.12619	.8993	4.0255	.89899	.59099	.60482
.08	8.0045	.12493	.9398	.0647	.90333	.59547	.60903
.09	.0849	.12369	.9806	.1043	.90768	.59995	.61324
2.10	8.1662	.12246	4.0219	4.1443	.91202	.60443	.61745
.11	.2482	.12124	.0635	.1847	.91636	.60890	.62167
.12	.3311	.12003	.1056	.2256	.92070	.61337	.62589
.13	.4149	.11884	.1480	.2669	.92505	.61784	.63011
.14	.4994	.11765	.1909	.3085	.92939	.62231	.63433
2.15	8.5849	.11648	4.2342	4.3507	.93373	.62677	.63856
.16	.6711	.11533	.2779	.3932	.93808	.63123	.64278
.17	.7583	.11418	.3221	.4362	.94242	.63569	.64701
.18	.8463	.11304	.3666	.4797	.94676	.64015	.65125
.19	.9352	.11192	.4116	.5236	.95110	.64460	.65548
2.20	9.0250	.11080	4.4571	4.5679	.95545	.64905	.65972
.21	.1157	.10970	.5030	.6127	.95979	.65350	.66396
.22	.2073	.10861	.5494	.6580	.96413	.65795	.66820
.23	.2999	.10753	.5962	.7037	.96848	.66240	.67244
.24	.3933	.10646	.6434	.7499	.97282	.66684	.67668
2.25	9.4877	.10540	4.6912	4.7966	.97716	.67128	.68093
.26	.5831	.10435	.7394	.8437	.98151	.67572	.68518
.27	.6794	.10331	.7880	.8914	.98585	.68016	.68943
.28	.7767	.10228	.8372	.9395	.99019	.68459	.69368
.29	.8749	.10127	.8868	.9881	.99453	.68903	.69794
2.30	9.9742	.10026	4.9370	5.0372	.99888	.69346	.70219
.31	10.074	.09926	.9876	.0868	.00322	.69789	.70645
.32	.176	.09827	5.0387	.1370	.00756	.70232	.71071
.33	.278	.09730	.0903	.1876	.01191	.70675	.71497
.34	.381	.09633	.1425	.2388	.01625	.71117	.71923
2.35	10.486	.09537	5.1951	5.2905	.02059	.71559	.72349
.36	.591	.09442	.2483	.3427	.02493	.72002	.72776
.37	.697	.09348	.3020	.3954	.02928	.72444	.73203
.38	.805	.09255	.3562	.4487	.03362	.72885	.73630
.39	.913	.09163	.4109	.5026	.03796	.73327	.74056
2.40	11.023	.09072	5.4662	5.5569	.04231	.73769	.74484
.41	.134	.08982	.5221	.6119	.04665	.74210	.74911
.42	.246	.08892	.5785	.6674	.05099	.74652	.75338
.43	.359	.08804	.6354	.7235	.05534	.75093	.75766
.44	.473	.08716	.6929	.7801	.05968	.75534	.76194
2.45	11.588	.08629	5.7510	5.8373	.06402	.75975	.76621
.46	.705	.08543	.8097	.8951	.06836	.76415	.77049
.47	.822	.08458	.8689	.9535	.07271	.76856	.77477
.48	.941	.08374	.9288	6.0125	.07705	.77296	.77906
.49	12.061	.08291	.9892	.0721	.08139	.77737	.78334
2.50	12.182	.08208	6.0502	6.1323	.08574	.78177	.78762

x	e^x	e^{-x}	Sinh x	Cosh x	L e^x	L Sinh x	L Cosh x
2.50	12.182	.08208	6.0502	6.1323	.08574	.78177	.78762
.51	.305	.08127	.1118	.1931	.09008	.78617	.79191
.52	.429	.08046	.1741	.2545	.09442	.79057	.79619
.53	.554	.07966	.2369	.3166	.09877	.79497	.80048
.54	.680	.07887	.3004	.3793	.10311	.79937	.80477
2.55	12.807	.07808	6.3645	6.4426	.10745	.80377	.80906
.56	.936	.07730	.4293	.5066	.11179	.80816	.81335
.57	13.066	.07654	.4946	.5712	.11614	.81256	.81764
.58	.197	.07577	.5607	.6365	.12048	.81695	.82194
.59	.330	.07502	.6274	.7024	.12482	.82134	.82623
2.60	13.464	.07427	6.6947	6.7690	.12917	.82573	.83052
.61	.599	.07353	.7628	.8363	.13351	.83012	.83482
.62	.736	.07280	.8315	.9043	.13785	.83451	.83912
.63	.874	.07208	.9008	.9729	.14219	.83890	.84341
.64	14.013	.07136	.9709	7.0423	.14654	.84329	.84771
2.65	14.154	.07065	7.0417	7.1123	.15088	.84768	.85201
.66	.296	.06995	.1132	.1831	.15522	.85206	.85631
.67	.440	.06925	.1854	.2546	.15957	.85645	.86061
.68	.585	.06856	.2583	.3268	.16391	.86083	.86492
.69	.732	.06788	.3319	.3998	.16825	.86522	.86922
2.70	14.880	.06721	7.4063	7.4735	.17260	.86960	.87352
.71	15.029	.06654	.4814	.5479	.17694	.87398	.87783
.72	.180	.06587	.5572	.6231	.18128	.87836	.88213
.73	.333	.06522	.6338	.6991	.18562	.88274	.88644
.74	.487	.06457	.7112	.7758	.18997	.88712	.89074
2.75	15.643	.06393	7.7894	7.8533	.19431	.89150	.89505
.76	.800	.06329	.8683	.9316	.19865	.89588	.89936
.77	.959	.06266	.9480	8.0106	.20300	.90026	.90367
.78	16.119	.06204	8.0285	.0905	.20734	.90463	.90798
.79	.281	.06142	.1098	.1712	.21168	.90901	.91229
2.80	16.445	.06081	8.1919	8.2527	.21602	.91339	.91660
.81	.610	.06020	.2749	.3351	.22037	.91776	.92091
.82	.777	.05961	.3586	.4182	.22471	.92213	.92522
.83	.945	.05901	.4432	.5022	.22905	.92651	.92953
.84	17.116	.05843	.5287	.5871	.23340	.93088	.93385
2.85	17.288	.05784	8.6150	8.6728	.23774	.93525	.93816
.86	.462	.05727	.7021	.7594	.24208	.93963	.94247
.87	.637	.05670	.7902	.8469	.24643	.94400	.94679
.88	.814	.05613	.8791	.9352	.25077	.94837	.95110
.89	.993	.05558	.9689	9.0244	.25511	.95274	.95542
2.90	18.174	.05502	9.0596	9.1146	.25945	.95711	.95974
.91	.357	.05448	.1512	.2056	.26380	.96148	.96405
.92	.541	.05393	.2437	.2976	.26814	.96584	.96837
.93	.728	.05340	.3371	.3905	.27248	.97021	.97269
.94	.916	.05287	.4315	.4844	.27683	.97458	.97701
2.95	19.106	.05234	9.5268	9.5791	.28117	.97895	.98133
.96	.298	.05182	.6231	.6749	.28551	.98331	.98565
.97	.492	.05130	.7203	.7716	.28985	.98768	.98997
.98	.688	.05079	.8185	.8693	.29420	.99205	.99429
.99	.886	.05029	.9177	.9680	.29854	.99641	.99861
3.00	20.086	.04979	10.018	10.068	.30288	.00078	.00293

x	e^x	e^{-x}	Sinh x	Cosh x	L e^x	L Sinh x	L Cosh x
3.00	20.086	.04979	10.018	10.068	.30288	.00078	.00293
.05	21.115	.04736	10.534	10.581	.32460	.02259	.02454
.10	22.198	.04505	11.076	11.122	.34631	.04440	.04616
.15	23.336	.04285	11.647	11.689	.36803	.06620	.06779
.20	24.533	.04076	12.246	12.287	.38974	.08799	.08943
3.25	25.790	.03877	12.876	12.915	.41146	.10977	.11108
.30	27.113	.03688	13.538	13.575	.43317	.13155	.13273
.35	28.503	.03508	14.234	14.269	.45489	.15332	.15439
.40	29.964	.03337	14.965	14.999	.47660	.17509	.17605
.45	31.500	.03175	15.734	15.766	.49832	.19685	.19772
3.50	33.115	.03020	16.543	16.573	.52003	.21860	.21940
.55	34.813	.02872	17.392	17.421	.54175	.24036	.24107
.60	36.598	.02732	18.285	18.313	.56346	.26211	.26275
.65	38.475	.02599	19.224	19.250	.58517	.28385	.28444
.70	40.447	.02472	20.211	20.236	.60689	.30559	.30612
3.75	42.521	.02352	21.249	21.272	.62860	.32733	.32781
.80	44.701	.02237	22.339	22.362	.65032	.34907	.34951
.85	46.993	.02128	23.486	23.507	.67203	.37081	.37120
.90	49.402	.02024	24.691	24.711	.69375	.39254	.39290
.95	51.935	.01925	25.958	25.977	.71546	.41427	.41459
4.00	54.598	.01832	27.290	27.308	.73718	.43600	.43629
.10	60.340	.01657	30.162	30.178	.78061	.47946	.47970
.20	66.686	.01500	33.336	33.351	.82404	.52291	.52310
.30	73.700	.01357	36.843	36.857	.86747	.56636	.56652
.40	81.451	.01228	40.719	40.732	.91090	.60980	.60993
4.50	90.017	.01111	45.003	45.014	.95433	.65324	.65335
.60	99.484	.01005	49.737	49.747	.99775	.69668	.69677
.70	109.95	.00910	54.969	54.978	.04118	.74012	.74019
.80	121.51	.00823	60.751	60.759	.08461	.78355	.78361
.90	134.29	.00745	67.141	67.149	.12804	.82699	.82704
5.00	148.41	.00674	74.203	74.210	.17147	.87042	.87046
.10	164.02	.00610	82.008	82.014	.21490	.91386	.91389
.20	181.27	.00552	90.633	90.639	.25833	.95729	.95731
.30	200.34	.00499	100.17	100.17	.30176	.00072	.00074
.40	221.41	.00452	110.70	110.71	.34519	.04415	.04417
5.50	244.69	.00409	122.34	122.35	.38862	.08758	.08760
.60	270.43	.00370	135.21	135.22	.43205	.13101	.13103
.70	298.87	.00335	149.43	149.44	.47548	.17444	.17445
.80	330.30	.00303	165.15	165.15	.51891	.21787	.21788
.90	365.04	.00274	182.52	182.52	.56234	.26130	.26131
6.00	403.43	.00248	201.71	201.72	.60577	.30473	.30474
6.25	518.01	.00193	259.01	259.01	.71434	.41331	.41331
6.50	665.14	.00150	332.57	332.57	.82291	.52188	.52189
6.75	854.06	.00117	427.03	427.03	.93149	.63046	.63046
7.00	1096.6	.00091	548.32	548.32	.04006	.73903	.73903
7.50	1808.0	.00055	904.02	904.02	.25721	.95618	.95618
8.00	2981.0	.00034	1490.5	1490.5	.47436	.17333	.17333
8.50	4914.8	.00020	2457.4	2457.4	.69150	.39047	.39047
9.00	8103.1	.00012	4051.5	4051.5	.90865	.60762	.60762
9.50	13360.	.00007	6679.9	6679.9	.12580	.82477	.82477
10.00	22026.	.00005	11013.	11013.	.34294	.04191	.04191

TABLE XVIII. MULTIPLIES OF M AND 1/M

Multiples of M				Multiples of 1/M			
0	0.00000 0	**50**	21.71472 4	**0**	0.00000 0	**50**	115.12925 5
1	0.43429 4	51	22.14901 9	1	2.30258 5	51	117.43184 0
2	0.86858 9	52	22.58331 3	2	4.60517 0	52	119.73442 5
3	1.30288 3	53	23.01760 8	3	6.90775 5	53	122.03701 0
4	1.73717 8	54	23.45190 2	4	9.21034 0	54	124.33959 5
5	2.17147 2	55	23.88619 7	5	11.51292 5	55	126.64218 0
6	2.60576 7	56	24.32049 1	6	13.81551 1	56	128.94476 5
7	3.04006 1	57	24.75478 5	7	16.11809 6	57	131.24735 0
8	3.47435 6	58	25.18908 0	8	18.42068 1	58	133.54993 5
9	3.90865 0	59	25.62337 4	9	20.72326 6	59	135.85252 0
10	4.34294 5	**60**	26.05766 9	**10**	23.02585 1	**60**	138.15510 6
11	4.77723 9	61	26.49196 3	11	25.32843 6	61	140.45769 1
12	5.21153 4	62	26.92625 8	12	27.63102 1	62	142.76027 6
13	5.64582 8	63	27.36055 2	13	29.93360 6	63	145.06286 1
14	6.08012 3	64	27.79484 7	14	32.23619 1	64	147.36544 6
15	6.51441 7	65	28.22914 1	15	34.53877 6	65	149.66803 1
16	6.94871 2	66	28.66343 6	16	36.84136 1	66	151.97061 6
17	7.38300 6	67	29.09773 0	17	39.14394 7	67	154.27320 1
18	7.81730 1	68	29.53202 5	18	41.44653 2	68	156.57578 6
19	8.25159 5	69	29.96631 9	19	43.74911 7	69	158.87837 1
20	8.68589 0	**70**	30.40061 4	**20**	46.05170 2	**70**	161.18095 7
21	9.12018 4	71	30.83490 8	21	48.35428 7	71	163.48354 2
22	9.55447 9	72	31.26920 3	22	50.65687 2	72	165.78612 7
23	9.98877 3	73	31.70349 7	23	52.95945 7	73	168.08871 2
24	10.42306 8	74	32.13779 2	24	55.26204 2	74	170.39129 7
25	10.85736 2	75	32.57208 6	25	57.56462 7	75	172.69388 2
26	11.29165 7	76	33.00638 1	26	59.86721 2	76	174.99646 7
27	11.72595 1	77	33.44067 5	27	62.16979 8	77	177.29905 2
28	12.16024 5	78	33.87497 0	28	64.47238 3	78	179.60163 7
29	12.59454 0	79	34.30926 4	29	66.77496 8	79	181.90422 2
30	13.02883 4	**80**	34.74355 9	**30**	69.07755 3	**80**	184.20680 7
31	13.46312 9	81	35.17785 3	31	71.38013 8	81	186.50939 3
32	13.89742 3	82	35.61214 8	32	73.68272 3	82	188.81197 8
33	14.33171 8	83	36.04644 2	33	75.98530 8	83	191.11456 3
34	14.76601 2	84	36.48073 6	34	78.28789 3	84	193.41714 8
35	15.20030 7	85	36.91503 1	35	80.59047 8	85	195.71973 3
36	15.63460 1	86	37.34932 5	36	82.89306 3	86	198.02231 8
37	16.06889 6	87	37.78362 0	37	85.19564 8	87	200.32490 3
38	16.50319 0	88	38.21791 4	38	87.49823 4	88	202.62748 8
39	16.93748 5	89	38.65220 9	39	89.80081 9	89	204.93007 3
40	17.37177 9	**90**	39.08650 3	**40**	92.10340 4	**90**	207.23265 8
41	17.80607 4	91	39.52079 8	41	94.40598 9	91	209.53524 3
42	18.24036 8	92	39.95509 2	42	96.70857 4	92	211.83782 9
43	18.67466 3	93	40.38938 7	43	99.01115 9	93	214.14041 4
44	19.10895 7	94	40.82368 1	44	101.31374 4	94	216.44299 9
45	19.54325 2	95	41.25797 6	45	103.61632 9	95	218.74558 4
46	19.97754 6	96	41.69227 0	46	105.91891 4	96	221.04816 9
47	20.41184 1	97	42.12656 5	47	108.22149 9	97	223.35075 4
48	20.84613 5	98	42.56085 9	48	110.52408 4	98	225.65333 9
49	21.28043 0	99	42.99515 4	49	112.82667 0	99	227.95592 4
50	21.71472 4	**100**	43.42944 8	**50**	115.12925 5	**100**	230.25850 9

$$\log_{10} N = M \log_e N; \quad \log_e N = (1/M) \log_{10} N$$

TABLE XIX. LOGARITHMS OF PRIME NUMBERS

N	Log N			N	Log N			N	Log N		
2	0.301	029	9957	269	2.429	752	2800	617	2.790	285	1640
3	.477	121	2547	271	.432	969	2909	619	.791	690	6490
5	.698	970	0043	277	.442	479	7691	631	.800	029	3592
7	.845	098	0400	281	.448	706	3199	641	.806	858	0295
11	1.041	392	6852	283	.451	786	4355	643	.808	210	9729
13	1.113	943	3523	293	2.466	867	6204	647	2.810	904	2807
17	.230	448	9214	307	.487	138	3755	653	.814	913	1813
19	.278	753	6010	311	.492	760	3890	659	.818	885	4146
23	.361	727	8360	313	.495	544	3375	661	.820	201	4595
29	.462	397	9979	317	.501	059	2622	673	.828	015	0642
31	1.491	361	6938	331	2.519	827	9938	677	2.830	588	6687
37	.568	201	7241	337	.527	629	9009	683	.834	420	7037
41	.612	783	8567	347	.540	329	4748	691	.839	478	0474
43	.633	468	4556	349	.542	825	4270	701	.845	718	0180
47	.672	097	8579	353	.547	774	7054	709	.850	646	2352
53	1.724	275	8696	359	2.555	094	4486	719	2.856	728	8904
59	.770	852	0116	367	.564	666	0643	727	.861	534	4109
61	.785	329	8350	373	.571	708	8318	733	.865	103	9746
67	.826	074	8027	379	.578	639	2100	739	.868	644	4384
71	.851	258	3487	383	.583	198	7740	743	.870	988	8138
73	1.863	322	8601	389	2.589	949	6013	751	2.875	639	9370
79	.897	627	0913	397	.598	790	5068	757	.879	095	8795
83	.919	078	0924	401	.603	144	3726	761	.881	384	6568
89	.949	390	0066	409	.611	723	3080	769	.885	926	3398
97	.986	771	7343	419	.622	214	0230	773	.888	179	4939
101	2.004	321	3738	421	2.624	282	0958	787	2.895	974	7324
103	.012	837	2247	431	.634	477	2702	797	.901	458	3214
107	.029	383	7777	433	.636	487	8964	809	.907	948	5216
109	.037	426	4979	439	.642	464	5202	811	.909	020	8542
113	.053	078	4435	443	.646	403	7262	821	.914	343	1571
127	2.103	803	7210	449	2.652	246	3410	823	2.915	399	8352
131	.117	271	2957	457	.659	916	2001	827	.917	505	5096
137	.136	720	5672	461	.663	700	9254	829	.918	554	5306
139	.143	014	8003	463	.665	580	9910	839	.923	761	9608
149	.173	186	2684	467	.669	316	8806	853	.930	949	0312
151	2.178	976	9473	479	2.680	335	5134	857	2.932	980	8219
157	.195	899	6524	487	.687	528	9612	859	.933	993	1638
163	.212	187	6044	491	.691	081	4921	863	.936	010	7957
167	.222	716	4711	499	.698	100	5456	877	.942	999	5934
173	.238	046	1031	503	.701	567	9851	881	.944	975	9084
179	2.252	853	0310	509	2.706	717	7823	883	2.945	960	7036
181	.257	678	5749	521	.716	837	7233	887	.947	923	6198
191	.281	033	3672	523	.718	501	6889	907	.957	607	2871
193	.285	557	3090	541	.733	197	2651	911	.959	518	3770
197	.294	466	2262	547	.737	987	3263	919	.963	315	5114
199	2.298	853	0764	557	2.745	855	1952	929	2.968	015	7140
211	.324	282	4553	563	.750	508	3949	937	.971	739	5909
223	.348	304	8630	569	.755	112	2664	941	.973	589	6234
227	.356	025	8572	571	.756	636	1082	947	.976	349	9790
229	.359	835	4823	577	.761	175	8132	953	.979	092	9006
233	2.367	355	9210	587	2.768	638	1012	967	2.985	426	4741
239	.378	397	9009	593	.773	054	6934	971	.987	219	2299
241	.382	017	0426	599	.777	426	8224	977	.989	894	5637
251	.399	673	7215	601	.778	874	4720	983	.992	553	5178
257	.409	933	1233	607	.783	188	6911	991	.996	073	6545
263	2.419	955	7485	613	2.787	460	4745	997	2.998	695	1583

PART III. FORMULAS AND GRAPHS FOR REFERENCE

I. FORMULAS AND THEOREMS FROM ALGEBRA

1. Laws of algebraic operations:

(a) Commutative law: $a + b = b + a,\quad ab = ba$;

(b) Associative law: $a + (b + c) = (a + b) + c,\quad a(bc) = (ab)c$;

(c) Distributive law: $c(a + b) = ca + cb$.

2. Operations with fractions:

$$\frac{a}{b} \pm \frac{c}{d} = \frac{ad \pm bc}{bd}, \quad \frac{a}{b} \times \frac{c}{d} = \frac{ac}{bd}, \quad \frac{a}{b} \div \frac{c}{d} = \frac{a}{b} \times \frac{d}{c} = \frac{ad}{bc}.$$

3. Laws of exponents:

$$a^m \times a^n = a^{m+n}, \quad (ab)^n = a^n b^n, \quad (a^m)^n = a^{mn}.$$

4. Zero and negative exponents:

$$a^0 = 1 \text{ if } a \neq 0, \quad a^{-n} = \frac{1}{a^n}, \quad a^m \div a^n = a^{m-n}.$$

5. Complex numbers:

$$i = \sqrt{-1}, \ i^2 = -1, \ i^3 = -i, \ i^4 = 1, \ i^5 = i, \ i^6 = -1, \cdots;$$

$$(a + bi) + (c + di) = (a + c) + (b + d)i;$$

$$(a + bi)(c + di) = (ac - bd) + (ad + bc)i;$$

$$\frac{a + bi}{c + di} = \frac{(a + bi)(c - di)}{(c + di)(c - di)} = \frac{ac + bd}{c^2 + d^2} + \frac{bc - ad}{c^2 + d^2} i.$$

6. Laws of logarithms (all logarithms with positive base k different from unity):

$$\log ab = \log a + \log b, \quad \log \frac{a}{b} = \log a - \log b,$$

$$\log a^n = n \log a, \quad \log \sqrt[n]{a} = \frac{1}{n} \log a,$$

$$\log \frac{1}{a} = - \log a, \quad \log k = 1, \quad \log 1 = 0.$$

Change of base of logarithms:

$$\log_k a = \log_m a \cdot \log_k m = \frac{\log_m a}{\log_m k}.$$

7. $n! = \lfloor n = 1 \cdot 2 \cdot 3 \cdot 4 \cdots (n-1) \cdot n$.

8. Binomial theorem ($n =$ positive integer):

$$(a+b)^n = a^n + na^{n-1}b + \frac{n(n-1)}{2!}\,a^{n-2}b^2 + \frac{n(n-1)(n-2)}{3!}\,a^{n-3}b^3 +$$
$$\cdots + nab^{n-1} + b^n.$$

9. The roots of the quadratic equation $ax^2 + bx + c = 0$ are

$$\frac{-b \pm \sqrt{b^2 - 4\,ac}}{2\,a}.$$

These roots are real and unequal when the discriminant $b^2 - 4\,ac$ is positive, real and equal when $b^2 - 4\,ac = 0$, and complex when $b^2 - 4\,ac$ is negative. The sum of these two roots is $-b/a$ and their product is c/a.

10. Factorization of $x^n \pm y^n$:

(a) $x^n - y^n = (x-y)(x^{n-1} + x^{n-2}y + x^{n-3}y^2 + \cdots + xy^{n-2} + y^{n-1})$,

(b) $x^{2n+1} + y^{2n+1} = (x+y)(x^{2n} - x^{2n-1}y + x^{2n-2}y^2 - \cdots - xy^{2n-1} + y^{2n})$.

11. Let s denote the sum of the arithmetical series a, $a+d$, $a+2\,d$, $\cdots$, l, of n terms. Then we have

$$l = a + (n-1)d, \quad s = \tfrac{1}{2}\,n(a+l) = \tfrac{1}{2}\,n[2\,a + (n-1)d].$$

12. Let s denote the sum of the geometrical series a, ar, ar^2, $\cdots$, l, of n terms. Then we have

$$l = ar^{n-1}, \quad s = \frac{rl - a}{r-1} = \frac{a(r^n - 1)}{r-1} = \frac{a}{1-r} - \frac{ar^n}{1-r}.$$

If $|r| < 1$ and the number of terms becomes infinite, we have

$$a + ar + ar^2 + ar^3 + \cdots = \frac{a}{1-r}.$$

13. Let $_nP_r$ and $_nC_r$ denote, respectively, the number of permutations and the number of combinations of n distinct things taken r at a time. Then we have

$$_nP_r = n(n-1)(n-2)\cdots(n-r+1) = \frac{n!}{(n-r)!},$$
$$_nP_n = n!,$$
$$_nC_r = \frac{n!}{r!\,(n-r)!} = \frac{_nP_r}{r!}.$$

14. Let $_nP_{n:p}$ denote the number of permutations of n things taken all at once, p of them being alike and the remaining ones being all different. Similarly, let $_nP_{n:p,q}$ denote the number of permutations when p are alike and q others are alike, the remaining (if any) being all different. Then we have

$$_nP_{n:p} = \frac{n!}{p!}, \quad _nP_{n:p,q} = \frac{n!}{p!\,q!}, \quad _nP_{n:r,\,n-r} = {}_nC_r.$$

15. Let D denote the determinant

$$D = \begin{vmatrix} a_{11} & a_{12} & \cdots & a_{1n} \\ a_{21} & a_{22} & \cdots & a_{2n} \\ \multicolumn{4}{c}{\cdots\cdots\cdots} \\ a_{n1} & a_{n2} & \cdots & a_{nn} \end{vmatrix}$$

of order n, that is, the sum

$$\Sigma \pm a_{1i}a_{2j}a_{3k}\cdots a_{nl}$$

of $n!$ terms, the plus or the minus sign (in a given term) being taken according as the number of inversions in the corresponding sequence i, j, k, $\cdots$, l (of the numbers 1, 2, 3, $\cdots$, n) is even or odd. Let A_{ij} denote the cofactor of the element a_{ij}, that is, the product of $(-1)^{i+j}$ by the determinant obtained from D by striking out the ith row and the jth column. Then we have the following propositions:

(a) The determinant D is unchanged by replacing the rows of D by the columns of D taken in order.

(b) An interchange of any two rows changes D to $-D$.

(c) If two rows are alike, then $D = 0$.

(d) If each element of a row of D is multiplied by m the new determinant has the value mD.

(e) The value of D is unchanged if to each element in a row of D is added m times the corresponding element in another row.

(f) $D = a_{1i}A_{1i} + a_{2i}A_{2i} + \cdots + a_{ni}A_{ni},$ \hfill $i = 1, 2, \cdots, n.$

(g) $a_{1i}A_{1j} + a_{2i}A_{2j} + \cdots + a_{ni}A_{nj} = 0$ if $i \neq j$.

(h) If $D \neq 0$ and if C_k is what D becomes when the elements in its kth column are replaced by $c_1, c_2, \cdots, c_n$ respectively, then the solution of the system

$$a_{i1}x_1 + a_{i2}x_2 + \cdots + a_{in}x_n = c_i, \qquad i = 1, 2, \cdots, n,$$

is unique and is given by the equations

$$Dx_1 = C_1, \quad Dx_2 = C_2, \cdots, \quad Dx_n = C_n.$$

16. If the polynomial $f(x)$ is divided by $x - a$ the remainder is $f(a)$. Hence, if a is a root of the equation $f(x) = 0$, then $f(x)$ is divisible by $x - a$. A polynomial of the nth degree may be factored into the form $c(x - c_1)$ $(x - c_2) \cdots (x - c_n)$; and then its (real or complex) zeros are $c_1, c_2, \cdots, c_n$. If $f(x)$ is a polynomial with real coefficients it may be separated into real factors whose degrees do not exceed two; and its complex roots enter in conjugate pairs.

17. The nth roots of unity are the following n numbers:

$$\cos\frac{2\,k\pi}{n} + i\sin\frac{2\,k\pi}{n}, \qquad k = 0, 1, 2, \cdots, n-1.$$

18. The n nth roots of any number are found by multiplying any given one of them by the n nth roots of unity.

19. The rational function

$$\frac{b_0x^m + b_1x^{m-1} + \cdots + b_m}{(x-a_1)^{k_1}(x-a_2)^{k_2}\cdots(x-a_t)^{k_t}}, \qquad m < k_1 + k_2 + \cdots + k_t,$$

where $a_1, a_2, \cdots, a_t$ are all distinct, may be written in the form

$$\sum_{i=1}^{t}\left(\frac{c_{i1}}{x-a_i} + \frac{c_{i2}}{(x-a_i)^2} + \cdots + \frac{c_{ik_i}}{(x-a_i)^{k_i}}\right)$$

where the symbols c_{ij} denote constants.

20. If $R(x)$ denotes the quotient of two polynomials with real coefficients, the divisor polynomial being of higher degree than the dividend, then $R(x)$ may be separated into partial fractions as follows:

(1) For a real root a of the divisor of multiplicity k there will be a part

$$\frac{c_1}{x-a} + \frac{c_2}{(x-a)^2} + \cdots + \frac{c_k}{(x-a)^k},$$

where $c_1, c_2, \cdots, c_k$ are constants.

(2) For a quadratic factor $x^2 + px + q$ of the divisor, corresponding to two conjugate imaginary roots of multiplicity l, there will be a part

$$\frac{c_1x + d_1}{x^2 + px + q} + \frac{c_2x + d_2}{(x^2 + px + q)^2} + \cdots + \frac{c_lx + d_l}{(x^2 + px + q)^l},$$

where the c_i and the d_i are constants.

Then $R(x)$ itself will be the sum of all such parts.

21. If S_n denotes the sum $\mu_1 + \mu_2 + \cdots + \mu_n$ of the first n terms of the infinite series $\mu_1 + \mu_2 + \mu_3 + \cdots$ and if $\lim_{n=\infty} S_n$ exists and is the finite quantity s, then the infinite series is said to converge and to have the sum s.

II. THEOREMS AND FORMULAS FROM ELEMENTARY GEOMETRY

1. If the sides and the hypotenuse of a right triangle are a, b, c, respectively, then $a^2 + b^2 = c^2$.

2. The area of a parallelogram (or a rectangle or a square) is the product of the base by the altitude.

3. The area of a triangle is half the product of the base by the altitude.

4. If a, b, c are the sides of a triangle and $2s = a + b + c$, then the area of the triangle is $\{s(s-a)(s-b)(s-c)\}^{\frac{1}{2}}$.

5. The area of a trapezoid is equal to half the product of the altitude by the sum of the parallel sides.

6. The area of a regular polygon is equal to half the product of its apothem by its perimeter.

7. The area of a circle of radius r is πr^2, and its circumference is $2\pi r$.

8. The area of a sector of a circle is equal to half the product of its radius by the arc length of the sector.

9. The volume of a rectangular parallepiped is equal to the product of its three dimensions.

10. The volume of a prism (or a cylinder) is equal to the product of the area of its base by its altitude.

11. The volume of a pyramid (or a cone) is equal to one third the product of the area of its base by its altitude.

12. The lateral area of a regular pyramid (or cone) is equal to one half the product of its slant height by the perimeter of its base.

13. The volume V of a frustum of a pyramid (or a cone) is

$$V = \tfrac{1}{3} h(B + b + \sqrt{Bb})$$

where h is the altitude and B and b are the areas of the two bases.

14. Using r for the radius and h for the altitude of a right circular cylinder, we have
    (a) Lateral area $= 2\,\pi rh$;
    (b) Total area $= 2\,\pi r(h + r)$;
    (c) Volume $= \pi r^2 h$.

15. Using r for the radius and h for the altitude and s for the slant height of a right circular cone, we have
    (a) Lateral area $= \pi rs$;
    (b) Total area $= \pi r(s + r)$;
    (c) Volume $= \tfrac{1}{3}\,\pi r^2 h$.

16. If r denotes the radius of a sphere, then
    (a) The area of its surface $= 4\,\pi r^2$;
    (b) Its volume $= \tfrac{4}{3}\,\pi r^3$.

III. FORMULAS FROM TRIGONOMETRY

FUNDAMENTAL FORMULAS

1. π radians $= 180° =$ two right angles.

2. 1 radian $= \dfrac{180°}{\pi} = 57.2957795$ degrees $= 57°\ 17'\ 44.806''$.

3. 1 degree $= \dfrac{\pi}{180}$ radian $= 0.01745329$ radian.

4. $\sin \alpha = \dfrac{y}{r}$, $\cos \alpha = \dfrac{x}{r}$, $\tan \alpha = \dfrac{y}{x}$, $\csc \alpha = \dfrac{r}{y}$, $\sec \alpha = \dfrac{r}{x}$, $\cot \alpha = \dfrac{x}{y}$.

5. $\sin^2 \alpha + \cos^2 \alpha = 1$, $1 + \tan^2 \alpha = \sec^2 \alpha$, $1 + \cot^2\alpha = \csc^2\alpha$.

6. $\sin \alpha \csc \alpha = 1$, $\cos \alpha \sec \alpha = 1$, $\tan \alpha \cot \alpha = 1$.

7. $\tan \alpha = \dfrac{\sin \alpha}{\cos \alpha} = \dfrac{\sec \alpha}{\csc \alpha}$, $\cot \alpha = \dfrac{\cos \alpha}{\sin \alpha} = \dfrac{\csc \alpha}{\sec \alpha}$.

8. $\operatorname{vers}\alpha = 1 - \cos\alpha$, $\operatorname{covers}\alpha = 1 - \sin\alpha$, $\operatorname{hav}\alpha = \tfrac{1}{2}(1 - \cos\alpha) = \tfrac{1}{2}\operatorname{vers}\alpha$.

REDUCTION FORMULAS

9. $\sin(-\alpha) = -\sin\alpha, \quad \cos(-\alpha) = \cos\alpha,$
$\tan(-\alpha) = -\tan\alpha, \quad \cot(-\alpha) = -\cot\alpha.$

10. $\sin(90° - \alpha) = \cos\alpha, \quad \cos(90° - \alpha) = \sin\alpha,$
$\tan(90° - \alpha) = \cot\alpha, \quad \cot(90° - \alpha) = \tan\alpha.$

11. $\sin(90° + \alpha) = \cos\alpha, \quad \cos(90° + \alpha) = -\sin\alpha,$
$\tan(90° + \alpha) = -\cot\alpha, \quad \cot(90° + \alpha) = -\tan\alpha.$

12. $\sin(180° - \alpha) = \sin\alpha, \quad \cos(180° - \alpha) = -\cos\alpha,$
$\tan(180° - \alpha) = -\tan\alpha, \quad \cot(180° - \alpha) = -\cot\alpha.$

13. A given trigonometric function of an angle of the form $n \cdot 90° \pm \alpha$, where n is an integer (positive or negative or zero), can be expressed as plus or minus a trigonometric function of α. For this purpose take the algebraic sign which belongs to the given function when α is a positive acute angle and take the given function or its cofunction according as n is even or odd.

FUNCTIONS OF SUMS AND DIFFERENCES OF ANGLES

14. $\sin(\alpha \pm \beta) = \sin\alpha\cos\beta \pm \cos\alpha\sin\beta.$

15. $\cos(\alpha \pm \beta) = \cos\alpha\cos\beta \mp \sin\alpha\sin\beta.$

16. $\tan(\alpha \pm \beta) = \dfrac{\tan\alpha \pm \tan\beta}{1 \mp \tan\alpha\tan\beta}.$

17. $\cot(\alpha \pm \beta) = \dfrac{\cot\alpha\cot\beta \mp 1}{\cot\beta \pm \cot\alpha}.$

18. $\sin\alpha + \sin\beta = 2\sin\frac{1}{2}(\alpha+\beta)\cos\frac{1}{2}(\alpha-\beta).$

19. $\sin\alpha - \sin\beta = 2\cos\frac{1}{2}(\alpha+\beta)\sin\frac{1}{2}(\alpha-\beta).$

20. $\cos\alpha + \cos\beta = 2\cos\frac{1}{2}(\alpha+\beta)\cos\frac{1}{2}(\alpha-\beta).$

21. $\cos\alpha - \cos\beta = -2\sin\frac{1}{2}(\alpha+\beta)\sin\frac{1}{2}(\alpha-\beta).$

22. $\dfrac{\sin\alpha + \sin\beta}{\sin\alpha - \sin\beta} = \dfrac{\tan\frac{1}{2}(\alpha+\beta)}{\tan\frac{1}{2}(\alpha-\beta)}.$

23. $\sin^2\alpha - \sin^2\beta = \sin(\alpha+\beta)\sin(\alpha-\beta).$

INVERSE FUNCTIONS

24. If $\sin\alpha = x$, then $\arcsin x = n\pi + (-1)^n\alpha.$

25. If $\cos\alpha = x$, then $\arccos x = 2n\pi \pm \alpha.$

26. If $\tan\alpha = x$, then $\arctan x = n\pi + \alpha.$

27. $\arcsin x \pm \arcsin y = \arcsin(x\sqrt{1-y^2} \pm y\sqrt{1-x^2}).$

28. $\arccos x \pm \arccos y = \arccos(xy \mp \sqrt{(1-x^2)(1-y^2)}).$

29. $\arctan x \pm \arctan y = \arctan\dfrac{x \pm y}{1 \mp xy}.$

FUNCTIONS OF HALF-ANGLES AND MULTIPLE ANGLES

30. $\sin 2\,\alpha = 2 \sin \alpha \cos \alpha.$

31. $\cos 2\,\alpha = \cos^2 \alpha - \sin^2 \alpha = 2 \cos^2 \alpha - 1 = 1 - 2 \sin^2 \alpha.$

32. $\tan 2\,\alpha = \dfrac{2 \tan \alpha}{1 - \tan^2 \alpha}, \quad \cot 2\,\alpha = \dfrac{\cot^2 \alpha - 1}{2 \cot \alpha}.$

33. $\sin \frac{1}{2}\,\alpha = \pm\sqrt{\frac{1}{2}(1 - \cos \alpha)}, \quad \cos \frac{1}{2}\,\alpha = \pm\sqrt{\frac{1}{2}(1 + \cos \alpha)}.$

34. $\tan \frac{1}{2}\,\alpha = \pm\sqrt{\dfrac{1 - \cos \alpha}{1 + \cos \alpha}} = \dfrac{\sin \alpha}{1 + \cos \alpha} = \dfrac{1 - \cos \alpha}{\sin \alpha}.$

35. $\sin 3\,\alpha = 3 \sin \alpha - 4 \sin^3 \alpha, \quad \cos 3\,\alpha = 4 \cos^3 \alpha - 3 \cos \alpha.$

36. $\tan 3\,\alpha = \dfrac{3 \tan \alpha - \tan^3 \alpha}{1 - 3 \tan^2 \alpha}.$

37. $\sin 4\,\alpha = \cos \alpha\,(4 \sin \alpha - 8 \sin^3 \alpha).$

38. $\cos 4\,\alpha = 8 \cos^4 \alpha - 8 \cos^2 \alpha + 1.$

DE MOIVRE'S AND RELATED IDENTITIES

39. $a + bi = r(\cos \theta + i \sin \theta),$ where

$$r = + (a^2 + b^2)^{\frac{1}{2}} \quad \text{and} \quad \theta = \text{arc } \cos \frac{a}{r} = \text{arc } \sin \frac{b}{r}.$$

40. $(\cos \theta + i \sin \theta)^n = \cos n\theta + i \sin n\theta.$

41. $\left(\cos \dfrac{2\,k\pi}{n} + i \sin \dfrac{2\,k\pi}{n}\right)^n = 1, \qquad\qquad k = 0, 1, \cdots, n - 1.$

42. $\sin \theta + \sin [\theta + \alpha] + \sin [\theta + 2\,\alpha] + \cdots + \sin [\theta + (n - 1)\alpha]$
$= \sin [\theta + \frac{1}{2}(n - 1)\alpha] \sin \frac{1}{2}\,n\alpha \csc \frac{1}{2}\,\alpha.$

43. $\cos \theta + \cos [\theta + \alpha] + \cos [\theta + 2\,\alpha] + \cdots + \cos [\theta + (n - 1)\alpha]$
$= \cos [\theta + \frac{1}{2}(n - 1)\alpha] \sin \frac{1}{2}\,n\alpha \csc \frac{1}{2}\,\alpha.$

44. $\sin^2 \theta + \sin^2 [\theta + \alpha] + \sin^2 [\theta + 2\,\alpha] + \cdots + \sin^2 [\theta + (n - 1)\alpha]$
$= \frac{1}{2}\,n - \frac{1}{2} \cos [2\,\theta + (n - 1)\alpha] \sin n\alpha \csc \alpha.$

45. $\cos^2 \theta + \cos^2 [\theta + \alpha] + \cos^2 [\theta + 2\,\alpha] + \cdots + \cos^2 [\theta + (n - 1)\,\alpha]$
$= \frac{1}{2}\,n + \frac{1}{2} \cos [2\,\theta + (n - 1)\,\alpha] \sin n\alpha \csc \alpha.$

PLANE TRIANGLES

Denote by α, β, γ the angles of a plane triangle ABC; and let a, b, c, respectively, denote the sides opposite these angles. Define s by the relation $a + b + c = 2\,s$. Let R denote the radius of the circumscribed circle, and let r denote the radius of the inscribed circle. Let K denote the area of the triangle. Then we have formulas 46–60.

46. (Law of Sines.) $\dfrac{\sin \alpha}{a} = \dfrac{\sin \beta}{b} = \dfrac{\sin \gamma}{c} = \dfrac{1}{2\,R}.$

47. (Law of Cosines.) $a^2 = b^2 + c^2 - 2\,bc \cos \alpha.$

48. (Law of Tangents.) $\dfrac{a - b}{a + b} = \dfrac{\tan \frac{1}{2}(\alpha - \beta)}{\tan \frac{1}{2}(\alpha + \beta)}.$

49. $\sin \frac{1}{2}\alpha = \sqrt{\dfrac{(s-b)(s-c)}{bc}}.$

50. $\cos \frac{1}{2}\alpha = \sqrt{\dfrac{s(s-a)}{bc}}.$

51. $\tan \frac{1}{2}\alpha = \sqrt{\dfrac{(s-b)(s-c)}{s(s-a)}} = \dfrac{r}{s-a}.$

52. $r = \sqrt{\dfrac{(s-a)(s-b)(s-c)}{s}}.$

53. $K = \frac{1}{2} bc \sin \alpha = \dfrac{c^2 \sin \alpha \sin \beta}{2 \sin \gamma}.$

54. $K = \sqrt{s(s-a)(s-b)(s-c)} = sr.$

55. $K = \dfrac{abc}{2R}.$

56. $\tan \alpha + \tan \beta + \tan \gamma = \tan \alpha \tan \beta \tan \gamma.$

57. $\sin \alpha + \sin \beta + \sin \gamma = 4 \cos \frac{1}{2}\alpha \cos \frac{1}{2}\beta \cos \frac{1}{2}\gamma.$

58. $\cos \alpha + \cos \beta + \cos \gamma = 4 \sin \frac{1}{2}\alpha \sin \frac{1}{2}\beta \sin \frac{1}{2}\gamma + 1.$

59. $\sin 2\alpha + \sin 2\beta + \sin 2\gamma = 4 \sin \alpha \sin \beta \sin \gamma.$

60. $\tan \frac{1}{2}\alpha \tan \frac{1}{2}\beta + \tan \frac{1}{2}\beta \tan \frac{1}{2}\gamma + \tan \frac{1}{2}\gamma \tan \frac{1}{2}\alpha = 1.$

RIGHT SPHERICAL TRIANGLES

Let γ be a right angle in a right spherical triangle ABC, and let α and β denote the angles at A and B respectively. Let the sides opposite the angles α, β, γ be respectively a, b, c, these sides being measured by the angle subtended at the center of the sphere.

61. $\sin a = \sin \alpha \sin c, \quad \sin b = \sin \beta \sin c.$

62. $\sin a = \tan b \cot \beta, \quad \sin b = \tan a \cot \alpha.$

63. $\cos \alpha = \cos a \sin \beta, \quad \cos \beta = \cos b \sin \alpha.$

64. $\cos \alpha = \tan b \cot c, \quad \cos \beta = \tan a \cot c.$

65. $\cos c = \cot \alpha \cot \beta, \quad \cos c = \cos a \cos b.$

66. $\sin a \cot \alpha = \sin c \cos \alpha = \cos a \sin b = \cos c \tan b.$

67. $\sin b \cot \beta = \sin c \cos \beta = \cos b \sin a = \cos c \tan a.$

68. *Napier's rules of circular parts.* Consider the five quantities a, b, co.α (complement of α), co.c, co.β, arranged in circular order as the corresponding parts appear in the triangle itself. If we call any one of these quantities a *middle* part then two of the other parts are *adjacent* to it and the other two are *opposite* to it. Then formulas 61 to 65 give rise to the following two easily remembered rules:

I. The *sine* of a *middle* part is equal to the product of the *tangents* of the *adjacent* parts.

II. The *sine* of a *middle* part is equal to the product of the *cosines* of the *opposite* parts.

SPHERICAL TRIANGLES IN GENERAL

Denote the angles of a spherical triangle by α, β, γ; and by a, b, c, respectively, denote the sides opposite these, these sides being measured by the angle subtended at the center of the sphere. Denote the corresponding parts of the polar triangle by α', β', γ', a', b', c'. Define s and σ by the relations $a + b + c = 2\,s$, $\alpha + \beta + \gamma = 2\,\sigma$. Let E denote the spherical excess of the triangle. Let R be the radius of the sphere on which the triangle is drawn.

69. $0° < a + b + c < 360°$, $180° < \alpha + \beta + \gamma < 540°$.

70. $\alpha = 180° - a'$, $\beta = 180° - b'$, $\gamma = 180° - c'$,

$\qquad a = 180° - \alpha'$, $b = 180° - \beta'$, $c = 180° - \gamma'$.

71. (Law of Sines.) $\dfrac{\sin \alpha}{\sin a} = \dfrac{\sin \beta}{\sin b} = \dfrac{\sin \gamma}{\sin c}$.

72. (Law of Cosines.) $\cos a = \cos b \cos c + \sin b \sin c \cos \alpha$.

73. (Law of Cosines.) $\cos \alpha = - \cos \beta \cos \gamma + \sin \beta \sin \gamma \cos a$.

74. $\sin \tfrac{1}{2} \alpha = \sqrt{\sin (s - b) \sin (s - c) \csc b \csc c}$.

75. $\cos \tfrac{1}{2} \alpha = \sqrt{\sin s \sin (s - a) \csc b \csc c}$.

76. $\tan \tfrac{1}{2} \alpha = \sqrt{\csc s \csc (s - a) \sin (s - b) \sin (s - c)}$.

77. $\sin \tfrac{1}{2} a = \sqrt{- \cos \sigma \cos (\sigma - \alpha) \csc \beta \csc \gamma}$.

78. $\cos \tfrac{1}{2} a = \sqrt{\cos (\sigma - \beta) \cos (\sigma - \gamma) \csc \beta \csc \gamma}$.

79. $\tan \tfrac{1}{2} a = \sqrt{- \cos \sigma \cos (\sigma - \alpha) \sec (\sigma - \beta) \sec (\sigma - \gamma)}$.

80. $\dfrac{\sin \tfrac{1}{2}(\alpha + \beta)}{\sin \tfrac{1}{2}(\alpha - \beta)} = \dfrac{\tan \tfrac{1}{2} c}{\tan \tfrac{1}{2}(a - b)}$.

81. $\dfrac{\sin \tfrac{1}{2}(a + b)}{\sin \tfrac{1}{2}(a - b)} = \dfrac{\cot \tfrac{1}{2} \gamma}{\tan \tfrac{1}{2}(\alpha - \beta)}$.

82. $\dfrac{\cos \tfrac{1}{2}(\alpha + \beta)}{\cos \tfrac{1}{2}(\alpha - \beta)} = \dfrac{\tan \tfrac{1}{2} c}{\tan \tfrac{1}{2}(a + b)}$.

83. $\dfrac{\cos \tfrac{1}{2}(a + b)}{\cos \tfrac{1}{2}(a - b)} = \dfrac{\cot \tfrac{1}{2} \gamma}{\tan \tfrac{1}{2}(\alpha + \beta)}$.

84. $\sin \tfrac{1}{2}(\alpha + \beta) \cos \tfrac{1}{2} c = \cos \tfrac{1}{2}(a - b) \cos \tfrac{1}{2} \gamma$.

85. $\cos \tfrac{1}{2}(\alpha + \beta) \cos \tfrac{1}{2} c = \cos \tfrac{1}{2}(a + b) \sin \tfrac{1}{2} \gamma$.

86. $\sin \tfrac{1}{2}(\alpha - \beta) \sin \tfrac{1}{2} c = \sin \tfrac{1}{2}(a - b) \cos \tfrac{1}{2} \gamma$.

87. $\cos \tfrac{1}{2}(\alpha - \beta) \sin \tfrac{1}{2} c = \sin \tfrac{1}{2}(a + b) \sin \tfrac{1}{2} \gamma$.

88. area of triangle $= \dfrac{\pi R^2 E}{180}$.

89. $\tan^2 \tfrac{1}{4} E = \tan \tfrac{1}{2} s \tan \tfrac{1}{2}(s - a) \tan \tfrac{1}{2}(s - b) \tan \tfrac{1}{2}(s - c)$.

HYPERBOLIC FUNCTIONS

Definitions of hyperbolic sines (sinh), hyperbolic cosines (cosh), etc., and of the corresponding inverse functions.

90. $\sinh u = \frac{1}{2}(e^u - e^{-u}) = \dfrac{1}{\operatorname{csch} u}$.

91. $\cosh u = \frac{1}{2}(e^u + e^{-u}) = \dfrac{1}{\operatorname{sech} u}$.

92. $\tanh u = \dfrac{e^u - e^{-u}}{e^u + e^{-u}} = \dfrac{1}{\coth u}$.

93. $\sinh^{-1} u = \log\left(u + \sqrt{u^2 + 1}\right)$.

94. $\cosh^{-1} u = \log\left(u + \sqrt{u^2 - 1}\right)$.

95. $\tanh^{-1} u = \frac{1}{2}\log\dfrac{1 + u}{1 - u}$.

96. $\coth^{-1} u = \frac{1}{2}\log\dfrac{u + 1}{u - 1}$.

97. $\operatorname{sech}^{-1} u = \log\dfrac{1 + \sqrt{1 - u^2}}{u}$.

98. $\operatorname{csch}^{-1} u = \log\dfrac{1 + \sqrt{1 + u^2}}{u}$.

99. $e^{iu} = \cos u + i \sin u, \quad e^{-iu} = \cos u - i \sin u$.

100. $\sin u = \dfrac{1}{2i}(e^{iu} - e^{-iu}), \quad \cos u = \dfrac{1}{2}(e^{iu} + e^{-iu})$.

101. $\sin iu = i \sinh u, \quad \cos iu = \cosh u$.

102. $\sin u = -i \sinh iu, \quad \cos u = \cosh iu$.

103. $\tanh u = \dfrac{\sinh u}{\cosh u}, \quad \cosh^2 u - \sinh^2 u = 1$.

104. $\log(x \pm yi) = \frac{1}{2}\log(x^2 + y^2) \pm i \arctan y/x$.

IV. FORMULAS AND THEOREMS FROM ANALYTIC GEOMETRY*

1. The distance from $P_1(x_1, y_1)$ to $P_2(x_2, y_2)$ is
$$\sqrt{(x_2 - x_1)^2 + (y_2 - y_1)^2}.$$

2. Let $P(x, y)$ be a point on the line containing the points $P_1(x_1, y_1)$ and $P_2(x_2, y_2)$; if P is such that $P_1P/PP_2 = r_1/r_2$, then
$$x = \frac{r_1 x_2 + r_2 x_1}{r_1 + r_2}, \quad y = \frac{r_1 y_2 + r_2 y_1}{r_1 + r_2}.$$
In particular, if P is the midpoint of P_1P_2 we have
$$x = \frac{1}{2}(x_1 + x_2), \quad y = \frac{1}{2}(y_1 + y_2).$$

---

* On pp. 231–240 will be found numerous standard graphs many of which are encountered by the student in his study of analytic geometry.

3. The slope m of the line segment from $P_1(x_1, y_1)$ to $P_2(x_2, y_2)$ is

$$m = \frac{y_2 - y_1}{x_2 - x_1}.$$

4. If two lines have the slopes m_1 and m_2 they are parallel if $m_1 = m_2$, and they are perpendicular if $m_1 m_2 = -1$.

5. Let θ be the angle measured from the line l_1 of slope m_1 to the line l_2 of slope m_2; then

$$\tan \theta = \frac{m_2 - m_1}{1 + m_1 m_2}.$$

6. Let the pole of a system of polar coördinates in the plane be at the origin of rectangular coördinates and let the polar axis coincide with the positive x-axis; then

$$x = \rho \cos \theta, \quad y = \rho \sin \theta, \quad \rho = (x^2 + y^2)^{\frac{1}{2}}, \quad \theta = \text{arc tan } (y/x).$$

7. As equations of straight lines we have the following:
    (a) General form: $Ax + By + C = 0$;
    (b) Slope-point form: $y - y_1 = m(x - x_1)$;
    (c) Two-point form: $y - y_1 = \dfrac{y_2 - y_1}{x_2 - x_1}(x - x_1)$;
    (d) Slope-intercept form: $y = mx + b$;
    (e) Intercept form: $\dfrac{x}{a} + \dfrac{y}{b} = 1$;
    (f) Normal form: $x \cos \omega + y \sin \omega = p$;
    (g) Polar equation: $\rho \cos (\theta - \omega) = p$.

8. On reducing the general equation $Ax + By + C = 0$ to the normal form $x \cos \omega + y \sin \omega - p = 0$ we have

$$\frac{Ax}{\pm \sqrt{A^2 + B^2}} + \frac{By}{\pm \sqrt{A^2 + B^2}} + \frac{C}{\pm \sqrt{A^2 + B^2}} = 0,$$

the sign of the radical being taken opposite to that of C when $C \neq 0$.

9. The distance d from the point (x_1, y_1) to the line

$$x \cos \omega + y \sin \omega - p = 0$$

is $d = x_1 \cos \omega + y_1 \sin \omega - p.$

10. Translation of axes: $x = x' + h, \ y = y' + k$.

11. Rotation of axes:

$$x = x' \cos \theta - y' \sin \theta, \quad y = x' \sin \theta + y' \cos \theta.$$

12. Standard equations of the circle:

(a) $x^2 + y^2 = r^2$;   (d) $\rho^2 + \rho_1^2 - 2\rho\rho_1 \cos (\theta - \theta_1) = r^2$;
(b) $(x - h)^2 + (y - k)^2 = r^2$;   (e) $\rho = 2r \cos \theta$;
(c) $x^2 + y^2 + ax + by + c = 0$;   (f) $\rho = 2r \sin \theta$.

13. Standard equations of the parabola:

(a) $y^2 = 2\,px$;

(b) $x^2 = 2\,py$;

(c) $(y - k)^2 = 2\,p(x - h)$;

(d) $(x - h)^2 = 2\,p(y - k)$;

(e) $\rho = p/(1 - \cos\theta)$.

14. Standard equations of the ellipse:

(a) $\dfrac{x^2}{a^2} + \dfrac{y^2}{b^2} = 1$;

(b) $\dfrac{(x - h)^2}{a^2} + \dfrac{(y - k)^2}{b^2} = 1$;

(c) $\rho = ep/(1 - e\cos\theta)$, $e < 1$.

15. Standard equations of the hyperbola:

(a) $\dfrac{x^2}{a^2} - \dfrac{y^2}{b^2} = 1$;

(c) $\dfrac{y^2}{a^2} - \dfrac{x^2}{b^2} = 1$;

(b) $\dfrac{(x - h)^2}{a^2} - \dfrac{(y - k)^2}{b^2} = 1$;

(d) $\dfrac{(y - k)^2}{a^2} - \dfrac{(x - h)^2}{b^2} = 1$;

(e) $\rho = ep/(1 - e\cos\theta)$, $e > 1$.

16. The asymptotes of the hyperbola 15 (a) are $y = \pm\, bx/a$.

17. With the equation of each of the following curves is given the equation of the tangent to that curve at the point (x_1, y_1) on the curve:

(a) $y^2 = 2\,px$, $yy_1 = p(x + x_1)$;

(b) $\dfrac{x^2}{a^2} + \dfrac{y^2}{b^2} = 1$, $\dfrac{xx_1}{a^2} + \dfrac{yy_1}{b^2} = 1$;

(c) $\dfrac{x^2}{a^2} - \dfrac{y^2}{b^2} = 1$, $\dfrac{xx_1}{a^2} - \dfrac{yy_1}{b^2} = 1$;

(d) $Ax^2 + 2\,Bxy + Cy^2 + 2\,Dx + 2\,Ey + F = 0$,

$Axx_1 + B(x_1y + xy_1) + Cyy_1 + D(x + x_1) + E(y + y_1) + F = 0$.

18. The line of slope m which is tangent to

(a) $y^2 = 2\,px$ is $y = mx + \dfrac{p}{2\,m}$;

(b) $\dfrac{x^2}{a^2} + \dfrac{y^2}{b^2} = 1$ is $y = mx \pm \sqrt{a^2m^2 + b^2}$;

(c) $\dfrac{x^2}{a^2} - \dfrac{y^2}{b^2} = 1$ is $y = mx \pm \sqrt{a^2m^2 - b^2}$.

19. For the same curves the loci of the middle points of the chords of slope m are respectively the straight lines

(a) $y = \dfrac{p}{m}$;   (b) $y = -\dfrac{p^2}{a^2m}\,x$;   (c) $y = \dfrac{b^2}{a^2m}\,x$.

20. For fitting empirical data one may often employ one of the following laws:

(a) The linear law: $y = mx + b$;

(b) The power law: $y = kx^n$;

(c) The compound interest law: $y = ab^x$;

(d) The hyperbolic law: $xy = ax + by$;

(e) The polynomial law: $y = a + bx + cx^2 + \cdots + kx^n$.

21. The distance d from $P_1(x_1, y_1, z_1)$ to $P_2(x_2, y_2, z_2)$ is

$$d = \sqrt{(x_2 - x_1)^2 + (y_2 - y_1)^2 + (z_2 - z_1)^2}.$$

22. The direction cosines of the same segment are

$$\cos \alpha = \frac{x_2 - x_1}{d}, \quad \cos \beta = \frac{y_2 - y_1}{d}, \quad \cos \gamma = \frac{z_2 - z_1}{d}.$$

23. The direction cosines of the line from the origin to the point (x, y, z) are

$$\cos \alpha = \frac{x}{(x^2 + y^2 + z^2)^{\frac{1}{2}}}, \quad \cos \beta = \frac{y}{(x^2 + y^2 + z^2)^{\frac{1}{2}}}, \quad \cos \gamma = \frac{z}{(x^2 + y^2 + z^2)^{\frac{1}{2}}}.$$

24. In both 22 and 23 we have

$$\cos^2\alpha + \cos^2\beta + \cos^2\gamma = 1.$$

25. If α, β, γ and α_1, β_1, γ_1 are the direction angles of two lines and if θ is the angle between them, we have

$$\cos \theta = \cos \alpha \cos \alpha_1 + \cos \beta \cos \beta_1 + \cos \gamma \cos \gamma_1.$$

When the lines are perpendicular each member of this equation is equal to zero.

26. If the line l from the origin perpendicular to a plane is of length p and has direction angles α, β, γ, then the equation of the plane is

$$x \cos \alpha + y \cos \beta + z \cos \gamma = p.$$

27. If A, B, C are not all zero, the equation

$$Ax + By + Cz + D = 0$$

represents a plane.

28. The line which has the direction angles α, β, γ and which passes through the point (x_1, y_1, z_1) has the equations

$$\frac{x - x_1}{\cos \alpha} = \frac{y - y_1}{\cos \beta} = \frac{z - z_1}{\cos \gamma}.$$

29. The line through the points (x_1, y_1, z_1) and (x_2, y_2, z_2) has the equations

$$\frac{x - x_1}{x_2 - x_1} = \frac{y - y_1}{y_2 - y_1} = \frac{z - z_1}{z_2 - z_1}.$$

30. If the distinct planes

$$Ax + By + Cz + D = 0 \quad \text{and} \quad A_1x + B_1y + C_1z + D_1 = 0$$

have a line in common, then the two given equations may serve as the equations of that line.

V. FORMULAS AND THEOREMS FROM CALCULUS*

DIFFERENTIAL CALCULUS

1. If $y = f(x)$ we write

$$y' = f'(x) = \frac{dy}{dx} = \lim_{\Delta x = 0} \frac{\Delta y}{\Delta x} = \lim_{\Delta x = 0} \frac{f(x + \Delta x) - f(x)}{\Delta x}.$$

2. General theorems on differentiation:

(a) $\dfrac{dc}{dx} = 0,$

$dc = 0;$

(b) $\dfrac{d(x)}{dx} = 1,$

$d(x) = dx;$

(c) $\dfrac{d}{dx}(u + v - w) = \dfrac{du}{dx} + \dfrac{dv}{dx} - \dfrac{dw}{dx},$

$d(u + v - w) = du + dv - dw;$

(d) $\dfrac{d}{dx}(cv) = c\dfrac{dv}{dx},$

$d(cv) = c\,dv;$

(e) $\dfrac{d}{dx}(uv) = u\dfrac{dv}{dx} + v\dfrac{du}{dx},$

$d(uv) = u\,dv + v\,du;$

(f) $\dfrac{d}{dx}(v^n) = nv^{n-1}\dfrac{dv}{dx},\quad \dfrac{d}{dx}(x^n) = nx^{n-1},$

$d(v^n) = nv^{n-1}\,dv;$

(g) $\dfrac{d}{dx}\left(\dfrac{u}{v}\right) = \dfrac{v\dfrac{du}{dx} - u\dfrac{dv}{dx}}{v^2},$

$d\left(\dfrac{u}{v}\right) = \dfrac{v\,du - u\,dv}{v^2};$

(h) $\dfrac{d}{dx}\left(\dfrac{c}{v}\right) = -\dfrac{c\dfrac{dv}{dx}}{v^2},$

$d\left(\dfrac{c}{v}\right) = -\dfrac{c\,dv}{v^2}.$

3. $\dfrac{d(v_1 v_2 \cdots v_n)}{v_1 v_2 \cdots v_n} = \dfrac{dv_1}{v_1} + \dfrac{dv_2}{v_2} + \cdots + \dfrac{dv_n}{v_n} = d\log(v_1 v_2 \cdots v_n).$

4. $\dfrac{dy}{dx} = \dfrac{dy}{dv}\dfrac{dv}{dx},\quad \dfrac{dy}{dx} = \dfrac{1}{\dfrac{dx}{dy}},\quad \dfrac{dy}{dx} = \dfrac{\dfrac{dy}{dt}}{\dfrac{dx}{dt}}.$

5. $\dfrac{d^n}{dx^n}(uv) = \dfrac{d^n u}{dx^n}v + n\dfrac{d^{n-1}u}{dx^{n-1}}\dfrac{dv}{dx} + \dfrac{n(n-1)}{2!}\dfrac{d^{n-2}u}{dx^{n-2}}\dfrac{d^2 v}{dx^2}$

$$+ \dfrac{n(n-1)(n-2)}{3!}\dfrac{d^{n-3}u}{dx^{n-3}}\dfrac{d^3 v}{dx^3} + \cdots + u\dfrac{d^n v}{dx^n}.$$

6. $\lim\limits_{x=0} \dfrac{\sin x}{x} = 1,\quad \lim\limits_{x=0}(1+x)^{\frac{1}{x}} = e = 2.71828\cdots = 1 + 1 + \dfrac{1}{2!} + \dfrac{1}{3!} + \cdots.$

7. $\dfrac{d}{dx}(\sin u) = \cos u\dfrac{du}{dx},\quad \dfrac{d}{dx}(\cos u) = -\sin u\dfrac{du}{dx}.$

8. $\dfrac{d}{dx}(\tan u) = \sec^2 u\dfrac{du}{dx},\quad \dfrac{d}{dx}(\cot u) = -\csc^2 u\dfrac{du}{dx}.$

9. $\dfrac{d}{dx}(\sec u) = \sec u\tan u\dfrac{du}{dx},\quad \dfrac{d}{dx}(\csc u) = -\csc u\cot u\dfrac{du}{dx}.$

---

*For the standard formulas of integration and for tables of integrals see pp. 241-265. For series see pp. 266-269.

10. $\dfrac{d}{dx}\left(\text{arc sin }\dfrac{u}{a}\right) = \dfrac{1}{\sqrt{a^2 - u^2}}\dfrac{du}{dx}, \quad \dfrac{d}{dx}\left(\text{arc cos }\dfrac{u}{a}\right) = -\dfrac{1}{\sqrt{a^2 - u^2}}\dfrac{du}{dx}.$

11. $\dfrac{d}{dx}\left(\text{arc tan }\dfrac{u}{a}\right) = \dfrac{a}{a^2 + u^2}\dfrac{du}{dx}, \quad \dfrac{d}{dx}\left(\text{arc cot }\dfrac{u}{a}\right) = -\dfrac{a}{a^2 + u^2}\dfrac{du}{dx}.$

12. $\dfrac{d}{dx}\left(\text{arc sec }\dfrac{u}{a}\right) = \dfrac{a}{u\sqrt{u^2 - a^2}}\dfrac{du}{dx}, \quad \dfrac{d}{dx}\left(\text{arc csc }\dfrac{u}{a}\right) = -\dfrac{a}{u\sqrt{u^2 - a^2}}\dfrac{du}{dx}.$

13. $\dfrac{d}{dx}\left(\text{arc vers }\dfrac{u}{a}\right) = \dfrac{1}{\sqrt{2\,au - u^2}}\dfrac{du}{dx}.$

14. $\dfrac{d}{dx}\log_a u = \dfrac{1}{\log a}\dfrac{1}{u}\dfrac{du}{dx}, \quad \dfrac{d}{dx}\log u = \dfrac{1}{u}\dfrac{du}{dx}.$

15. $\dfrac{d}{dx}(a^u) = \log a \cdot a^u\dfrac{du}{dx}, \quad \dfrac{d}{dx}(e^u) = e^u\dfrac{du}{dx}.$

16. $\dfrac{d}{dx}(u^v) = u^v\log u \cdot \dfrac{dv}{dx} + vu^{v-1}\dfrac{du}{dx}.$

17. The slope of the curve $y = f(x)$ at the point (x_1, y_1) on the curve $y = f(x)$ is $f'(x_1)$. Hence the equations of the tangent and the normal to this curve at the point (x_1, y_1) are respectively

$$y - y_1 = f'(x_1)(x - x_1) \quad \text{and} \quad y - y_1 = -\frac{1}{f'(x_1)}(x - x_1).$$

18. If a body moves the distance s along its path in time t and if v is its speed and a is its tangential acceleration, then we have

$$v = \frac{ds}{dt}, \quad a = \frac{dv}{dt} = \frac{d^2s}{dt^2}, \quad v\,dv = a\,ds.$$

19. If θ is the angle swept out by a revolving radius in time t and if ω is the angular speed and α is the angular acceleration, then we have

$$\omega = \frac{d\theta}{dt}, \quad \alpha = \frac{d\omega}{dt} = \frac{d^2\theta}{dt^2}, \quad \omega\,d\omega = \alpha\,d\theta.$$

20. If s denote the arc length of a plane curve, then

$$ds = \sqrt{dx^2 + dy^2} = \sqrt{1 + \left(\frac{dy}{dx}\right)^2}\,dx = \sqrt{1 + \left(\frac{dx}{dy}\right)^2}\,dy,$$

$$ds = \sqrt{d\rho^2 + \rho^2 d\theta^2} = \sqrt{\rho^2 + \left(\frac{d\rho}{d\theta}\right)^2}\,d\theta = \sqrt{1 + \rho^2\left(\frac{d\theta}{d\rho}\right)^2}\,d\rho.$$

21. If s denote the arc length of a space curve, then

$$ds = \sqrt{dx^2 + dy^2 + dz^2}.$$

22. If y and x are changing so that $y = f(x)$, then the time rate of change of y is equal to the time rate of change of x multiplied by the derivative of y with respect to x, that is,

$$\frac{dy}{dt} = f'(x)\frac{dx}{dt}.$$

23. In the same case we have

$$\frac{ds}{dt} = \sqrt{\left(\frac{dx}{dt}\right)^2 + \left(\frac{dy}{dt}\right)^2};$$

that is, the time rate of change along the arc is equal to the square root of the sum of the squares of the time rates of change of x and y.

24. For a curve in space the last relation becomes

$$\frac{ds}{dt} = \sqrt{\left(\frac{dx}{dt}\right)^2 + \left(\frac{dy}{dt}\right)^2 + \left(\frac{dz}{dt}\right)^2}.$$

25. Law of the mean (where $f(x)$ and $f'(x)$ are continuous) :

$$f(a + h) - f(a) = hf'(a + \theta h), \qquad 0 < \theta < 1.$$

26. If as x approaches a the functions $f(x)$ and $F(x)$ both approach zero while $F'(a) \neq 0$ and if $f'(x)$ and $F'(x)$ are continuous at $x = a$, then

$$\lim_{x=a} \frac{f(x)}{F(x)} = \frac{f'(a)}{F'(a)}.$$

27. If as x approaches a the functions $f(x)$ and $F(x)$, together with the first $n - 1$ derivatives of each of them, approach zero while $F^{(n)}(a) \neq 0$, and if $f^{(n)}(x)$ and $F^{(n)}(x)$ are continuous at $x = a$, then

$$\lim_{x=a} \frac{f(x)}{F(x)} = \frac{f^{(n)}(a)}{F^{(n)}(a)}.$$

28. Let K denote the curvature of a plane curve and R the radius of curvature. Then we have $K = 1/R$ and

$$R = \frac{\left[1 + \left(\frac{dy}{dx}\right)^2\right]^{\frac{3}{2}}}{\frac{d^2y}{dx^2}} = \frac{\left[\rho^2 + \left(\frac{d\rho}{d\theta}\right)^2\right]^{\frac{3}{2}}}{\rho^2 + 2\left(\frac{d\rho}{d\theta}\right)^2 - \rho\frac{d^2\rho}{d\theta^2}}.$$

29. If (α, β) is the center of curvature corresponding to the point (x_1, y_1) on the curve $y = f(x)$, then

$$\alpha = x_1 - \frac{f'(x_1)\{1 + [f'(x_1)]^2\}}{f''(x_1)},$$

$$\beta = y_1 + \frac{1 + [f'(x)]^2}{f''(x_1)}.$$

30. If $z = f(x, y, r, \cdots)$, then

(a) $\dfrac{dz}{dt} = \dfrac{\partial z}{\partial x}\dfrac{dx}{dt} + \dfrac{\partial z}{\partial y}\dfrac{dy}{dt} + \dfrac{\partial z}{\partial r}\dfrac{dr}{dt} + \cdots,$

(b) $\dfrac{dz}{dx} = \dfrac{\partial z}{\partial x} + \dfrac{\partial z}{\partial y}\dfrac{dy}{dx} + \dfrac{\partial z}{\partial r}\dfrac{dr}{dx} + \cdots,$

(c) $dz = \dfrac{\partial z}{\partial x}dx + \dfrac{\partial z}{\partial y}dy + \dfrac{\partial z}{\partial r}dr + \cdots.$

31. The condition that $M\,dx + N\,dy$ shall be exact is that

$$\frac{\partial M}{\partial y} = \frac{\partial N}{\partial x}.$$

32. The tangent plane to the surface $z = f(x, y)$ at the point $P(a, b, c)$ on the surface is

$$z - c = \frac{\partial z}{\partial x}\bigg]_P (x - a) + \frac{\partial z}{\partial y}\bigg]_P (y - b).$$

33. The tangent plane to the surface $F(x, y, z) = 0$ at the point $P(a, b, c)$ on the surface is

$$\frac{\partial F}{\partial x}\bigg]_P (x - a) + \frac{\partial F}{\partial y}\bigg]_P (y - b) + \frac{\partial F}{\partial z}\bigg]_P (z - c) = 0.$$

The normal at this point is

$$\frac{x - a}{\dfrac{\partial F}{\partial x}\bigg]_P} = \frac{y - b}{\dfrac{\partial F}{\partial y}\bigg]_P} = \frac{z - c}{\dfrac{\partial F}{\partial z}\bigg]_P}.$$

INTEGRAL CALCULUS

34. Definition of definite integral:

$$\int_a^b f(x)\,dx = \lim_{\text{Every } \Delta x_i = 0} \sum_a^b f(\xi_i)\Delta x_i,$$

where ξ_i is any value of x in the interval defined by Δx_i.

35. If $\phi(x)$ is a function whose derivative with respect to x is $f(x)$, then

$$\int_a^b f(x)\,dx = \phi(b) - \phi(a).$$

36.
$$\int_a^b cf(x)\,dx = c\int_a^b f(x)\,dx, \qquad\qquad c = \text{constant.}$$

37.
$$\int u\,dv = uv - \int v\,du.$$

38.
$$\int_b^a f(x)\,dx = -\int_a^b f(x)\,dx.$$

39.
$$\int_a^c f(x)\,dx + \int_c^b f(x)\,dx = \int_a^b f(x)\,dx.$$

40.
$$\int_a^\infty f(x)\,dx = \lim_{t=\infty} \int_a^t f(x)\,dx.$$

41.
$$\int_{-\infty}^b f(x)\,dx = \lim_{t=\infty} \int_{-t}^b f(x)\,dx.$$

42.
$$\int_{-\infty}^\infty f(x)\,dx = \int_{-\infty}^c f(x)\,dx + \int_c^\infty f(x)\,dx.$$

43. If $f(x)$ has a singular point at b and $b > a$,

$$\int_a^b f(x)dx = \lim_{\epsilon=0} \int_a^{b-\epsilon} f(x)dx.$$

44. If $f(x)$ has a singular point at a and $b > a$,

$$\int_a^b f(x)dx = \lim_{\epsilon=0} \int_{a+\epsilon}^b f(x)dx.$$

45. If $f(x)$ is not negative for any value of x, then the area bounded by $y = f(x)$, $y = 0$, $x = a$, $x = b$ is

$$\int_a^b f(x)dx.$$

46. The formula for plane areas in polar coördinates is

$$\tfrac{1}{2}\int_{\theta_1}^{\theta_2} \rho^2 d\theta.$$

47. For the length s of a plane curve from (a, c) to (b, d) we have

$$s = \int_a^b \sqrt{1 + \left(\frac{dy}{dx}\right)^2}\, dx = \int_c^d \sqrt{1 + \left(\frac{dx}{dy}\right)^2}\, dy.$$

In polar coördinates we have

$$s = \int_{\theta_1}^{\theta_2} \sqrt{\rho^2 + \left(\frac{d\rho}{d\theta}\right)^2}\, d\theta = \int_{\rho_1}^{\rho_2} \sqrt{\rho^2\left(\frac{d\theta}{d\rho}\right)^2 + 1}\, d\rho.$$

48. If for each x such that $a \leqq x \leqq b$ the plane perpendicular to the x-axis at $(x, 0, 0)$ cuts from a given solid a section whose area is $F(x)$, then the volume of that part of the solid which lies between the planes $x = a$ and $x = b$ is

$$\int_a^b F(x)dx.$$

49. The mean value of the function $f(x)$ on the interval (a, b) is

$$\frac{\int_a^b f(x)dx}{b - a}.$$

50. The volume of a solid of revolution when the curve is revolved

(a) about the x-axis is $\pi \int_a^b y^2 dx$;

(b) about the y-axis is $\pi \int_c^d x^2 dy.$

51. The surface of a solid of revolution when the curve is revolved

(a) about the x-axis is $2\pi \int_a^b y \sqrt{1 + \left(\frac{dy}{dx}\right)^2}\, dx$;

(b) about the y-axis is $2\pi \int_a^b \sqrt{1 + \left(\frac{dy}{dx}\right)^2}\, dx.$

These two volumes are respectively equal to

$$2\pi\int_c^d y\sqrt{1+\left(\frac{dx}{dy}\right)^2}\,dy \quad \text{and} \quad 2\pi\int_c^d x\sqrt{1+\left(\frac{dx}{dy}\right)^2}\,dy$$

where c and d are the values of y corresponding respectively to the values a and b of x.

52. If the curve $\rho = f(\theta)$ is revolved about the initial line the formula for the surface swept out is

$$2\pi\int_{\theta_1}^{\theta_2}\rho\sin\theta\sqrt{\rho^2+\left(\frac{d\rho}{d\theta}\right)^2}\,d\theta.$$

53. If a body is moved along the x-axis from $x = a$ to $x = b$ against a force $f(x)$ the work done is
$$W = \int_a^b f(x)\,dx.$$

54. Formulas for plane areas by means of double integration:
(a) rectangular coördinates,

$$\int_a^b\int_{\phi(x)}^{f(x)}dx\,dy \quad \text{or} \quad \int_c^d\int_{\xi(y)}^{\psi(y)}dy\,dx;$$

(b) polar coördinates,

$$\int_{\theta_1}^{\theta_2}\int_{f_1(\theta)}^{f_2(\theta)}\rho\,d\rho\,d\theta \quad \text{or} \quad \int_{\rho_1}^{\rho_2}\int_{\phi_1(\rho)}^{\phi_2(\rho)}\rho\,d\theta\,d\rho.$$

55. Formulas for volumes by means of double integration:

$$\int_a^b\int_{\phi(x)}^{\psi(x)}f(x,y)\,dy\,dx, \quad \int_c^d\int_{a(y)}^{\beta(y)}f(x,y)\,dx\,dy.$$

56. Formulas for volume by means of triple integration:
(a) rectangular coördinates,

$$\iiint dz\,dy\,dx;$$

(b) cylindrical coördinates,

$$\iiint \rho\,d\rho\,d\theta\,dz;$$

(c) spherical coördinates,

$$\iiint \rho^2\sin\phi\,d\theta\,d\phi\,d\rho.$$

In each of these cases there are additional formulas obtained by changing the order of integration. The limits of integration are to be supplied.

57. Formula for the area of a surface $z = f(x, y)$:

$$\iint\sqrt{\left(\frac{\partial z}{\partial x}\right)^2+\left(\frac{\partial z}{\partial y}\right)^2+1}\,dy\,dx.$$

58. If γ is the variable density and $\overline{\gamma}$ is the mean density of a solid of volume V, then

$$\overline{\gamma} = \frac{\int \gamma \, dV}{V}.$$

59. If a given solid of volume V is referred to rectangular coördinates, and if M_{xy}, M_{yz}, M_{zx} denote the first moments with respect to the coördinate planes, and if the centroid of the solid is $(\overline{x}, \overline{y}, \overline{z})$, then we have

$$\overline{x} = \frac{M_{yz}}{V} = \frac{\int x \, dV}{V}, \quad \overline{y} = \frac{M_{zx}}{V} = \frac{\int y \, dV}{V}, \quad \overline{z} = \frac{M_{xy}}{V} = \frac{\int z \, dV}{V}.$$

60. If γ is the density factor for a mass m of volume V and the centroid of mass (or center of gravity) is $(\overline{x}, \overline{y}, \overline{z})$, then

$$\overline{x} = \frac{\int \gamma x \, dV}{\int \gamma \, dV}, \quad \overline{y} = \frac{\int \gamma y \, dV}{\int \gamma \, dV}, \quad \overline{z} = \frac{\int \gamma z \, dV}{\int \gamma \, dV}.$$

61. If volumes $V_1, V_2, \cdots, V_n$ have $\overline{x}_1, \overline{x}_2, \cdots, \overline{x}_n$ respectively as the x-coördinates of their centroids, then the x-coördinate of the centroid of the several volumes is

$$\overline{x} = \frac{V_1 \overline{x}_1 + V_2 \overline{x}_2 + \cdots + V_n \overline{x}_n}{V_1 + V_2 + \cdots + V_n}.$$

There are like formulas for the other two coördinates.

62. Formulas for second moments:

(a) for a plane curve, $\quad I_x = \int y^2 \, ds$;

(b) for a plane area, $\quad I_x = \int y^2 \, dA$;

(c) for a volume, $\quad I_{xy} = \int z^2 \, dV$;

(d) then we have $\quad I_x = I_{xz} + I_{xy}$.

63. If I is the second moment of a volume V and k is the radius of gyration, we have $I = Vk^2$. There are similar formulas for areas and lengths.

64. If volumes (or areas or masses) $v_1, v_2, \cdots, v_n$ have respectively the radii of gyration $k_1, k_2, \cdots, k_n$ with respect to a line or a plane, then with respect to this line or plane the several volumes taken together have the radius of gyration k, where

$$k^2 = \frac{v_1 k_1{}^2 + v_2 k_2{}^2 + \cdots + v_n k_n{}^2}{v_1 + v_2 + \cdots + v_n}.$$

65. The surface of a solid of revolution is equal to the product of the length of the generating arc and the circumference of the circle described by the centroid of the generating arc.

66. The volume of a solid of revolution is equal to the product of the generating area and the circumference of the circle described by the centroid of this area.

67. Formulas for approximate values of integrals: Let $h = \dfrac{b-a}{n}$, $y_0 = f(a)$, $y_1 = f(a+h)$, $y_2 = f(a+2h)$, $\cdots$, $y_{n-1} = f(b-h)$, $y_n = f(b)$.

(a) Trapezoidal Rule,

$$\int_a^b f(x)dx = \frac{h}{2}[y_0 + 2y_1 + 2y_2 + \cdots + 2y_{n-1} + y_n];$$

(b) Simpson's Rule (n an even integer),

$$\int_a^b f(x)dx = \frac{h}{3}[y_0 + 4y_1 + 2y_2 + 4y_3 + 2y_4 + \cdots + 2y_{n-2} + 4y_{n-1} + y_n];$$

(c) Durand's Rule,

$$\int_a^b f(x)dx = h[.4y_0 + 1.1y_1 + y_2 + y_3 + \cdots + y_{n-2} + 1.1y_{n-1} + .4y_n];$$

(d) Newton's Rule ($n = 3$),

$$\int_a^b f(x)dx = \tfrac{3}{8}h[y_0 + 3y_1 + 3y_2 + y_3];$$

(e) Weddle's Rule ($n = 6$),

$$\int_a^b f(x)dx = \tfrac{3}{10}h[y_0 + 5y_1 + y_2 + 6y_3 + y_4 + 5y_5 + y_6].$$

VI. GRAPHS FOR REFERENCE

Circle

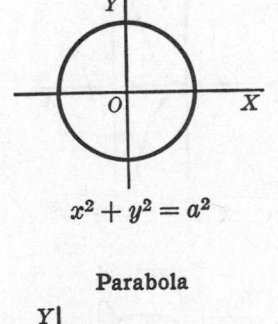

$$x^2 + y^2 = a^2$$

Ellipse

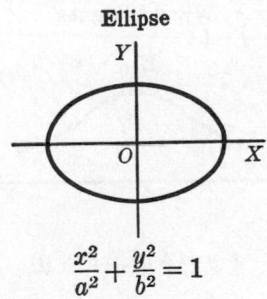

$$\frac{x^2}{a^2} + \frac{y^2}{b^2} = 1$$

Parabola

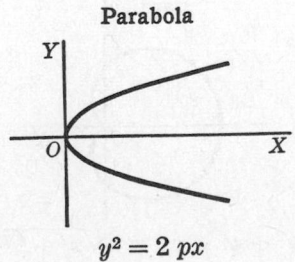

$$y^2 = 2px$$

Hyperbola

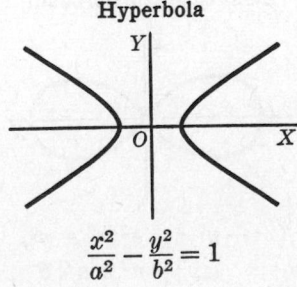

$$\frac{x^2}{a^2} - \frac{y^2}{b^2} = 1$$

Cubical Parabola

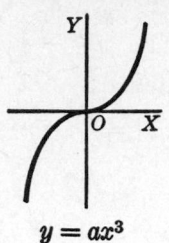

$$y = ax^3$$

Semicubical Parabola

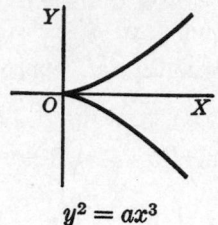

$$y^2 = ax^3$$

Equilateral Hyperbola

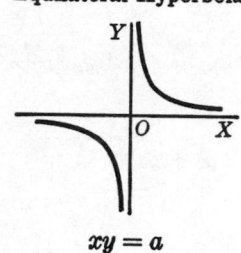

$$xy = a$$

Typical Rational Function

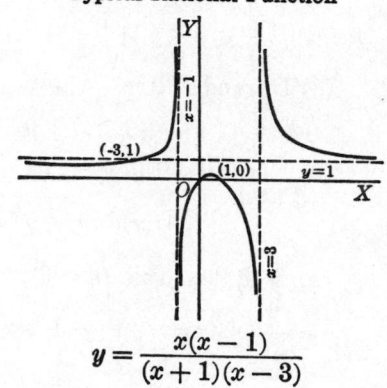

$$y = \frac{x(x-1)}{(x+1)(x-3)}$$

Witch of Agnesi

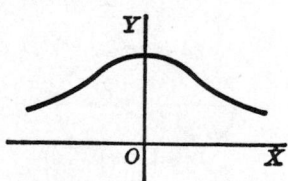

$$x^2 y = 4\,a^2(2\,a - y)$$

Cissoid of Diocles

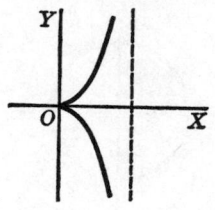

$$y^2(2\,a - x) = x^3$$

Lemniscate of Bernoulli

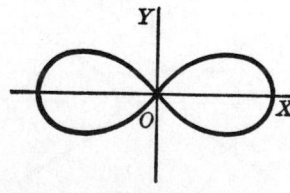

$$(x^2 + y^2)^2 = a^2(x^2 - y^2)$$
$$\rho^2 = a^2 \cos 2\,\theta$$

Cardioid

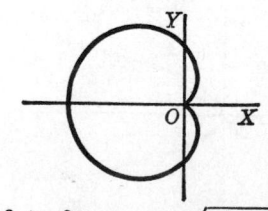

$$x^2 + y^2 + ax = a\sqrt{x^2 + y^2}$$
$$\rho = a(1 - \cos \theta)$$

Conchoid of Nicomedes

$a < b$ $a > b$

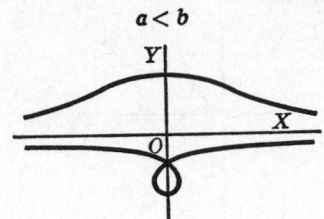 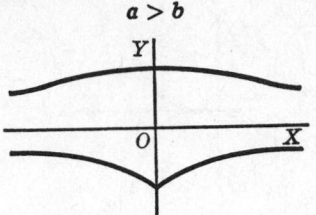

$$x^2y^2 = (y + a)^2(b^2 - y^2)$$
$$\rho = a \csc \theta \pm b$$

Folium of Descartes

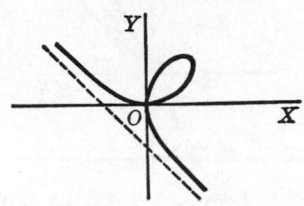

$$x^3 + y^3 - 3\,axy = 0$$
$$x = \frac{3\,at}{1 + t^3}, \quad y = \frac{3\,at^2}{1 + t^3}.$$

Parabola

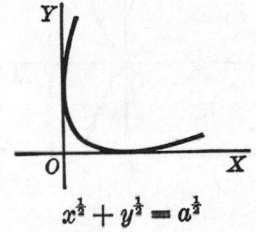

$$x^{\frac{1}{2}} + y^{\frac{1}{2}} = a^{\frac{1}{2}}$$

Hypocycloid of Four Cusps

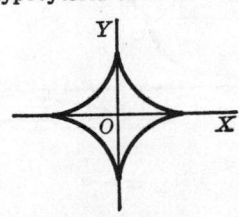

$$x^{\frac{2}{3}} + y^{\frac{2}{3}} = a^{\frac{2}{3}}$$

Evolute of Ellipse

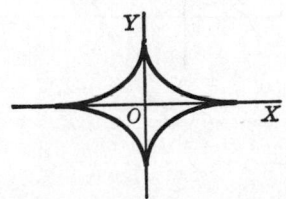

$$(ax)^{\frac{2}{3}} + (by^{\frac{2}{3}}) = (a^2 - b^2)^{\frac{2}{3}}$$

Curve with Isolated Point at the Origin

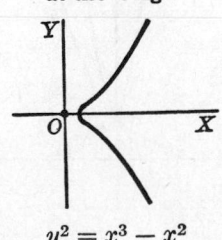

$$y^2 = x^3 - x^2$$

Curve with Tacnode at the Origin

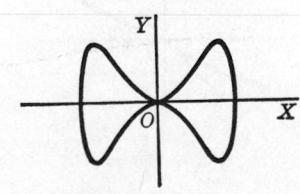

$$y^2 = x^4(1 - x^2)$$

Cusp of First Kind

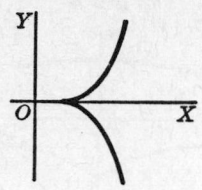

$$y^2 = (x - 1)^5$$

Cusp of Second Kind

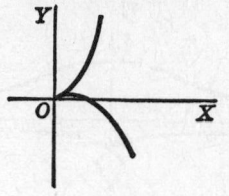

$$(y - x^2)^2 = x^5$$

Strophoid. Double Point at the Origin

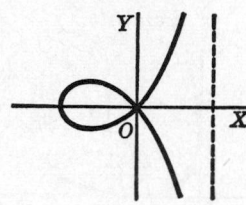

$$y^2 = x^2 \cdot \frac{a + x}{a - x}$$

Curve with Triple Point at the Origin

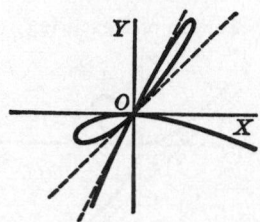

$$x^4 + 2\,x^2y - 3\,xy^2 + y^3 = 0$$
$$x = -\,t(t - 1)(t - 2), \quad y = tx$$

TRANSCENDENTAL CURVES

Curves with End Points

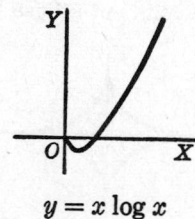

$$y = x \log x$$

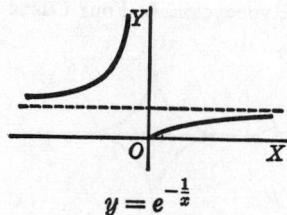

$$y = e^{-\frac{1}{x}}$$

Curve with Salient Point

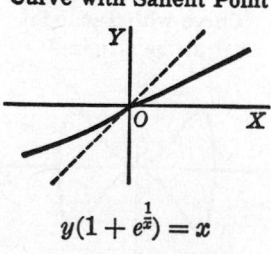

$$y(1 + e^{\frac{1}{x}}) = x$$

Exponential Curve

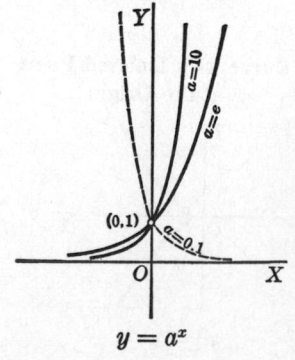

$$y = a^x$$

Logarithmic Curve

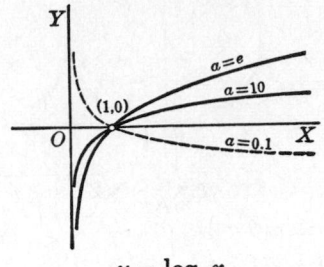

$$y = \log_a x$$

Probability Curve

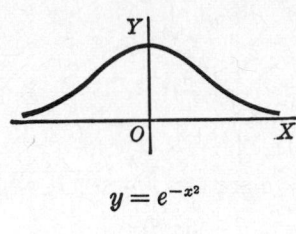

$$y = e^{-x^2}$$

Sine Curve

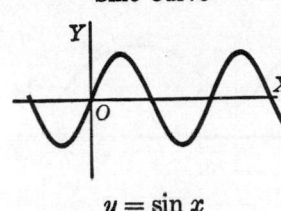

$$y = \sin x$$

Cosine Curve

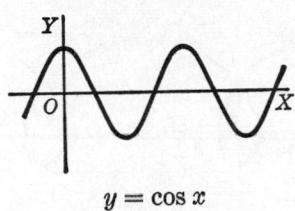

$$y = \cos x$$

Secant Curve

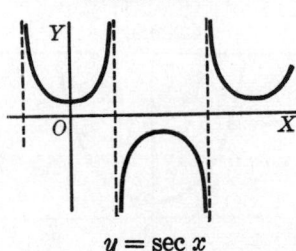

$$y = \sec x$$

Tangent Curve

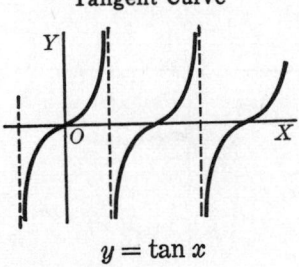

$$y = \tan x$$

Inverse Trigonometric Functions

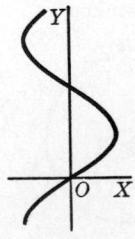

$$y = \text{arc } \sin x$$

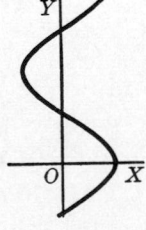

$$y = \text{arc } \cos x$$

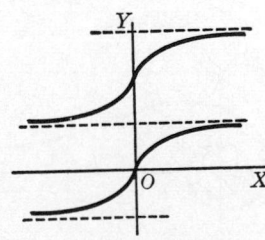

$$y = \text{arc } \tan x$$

Ordinary Cycloids

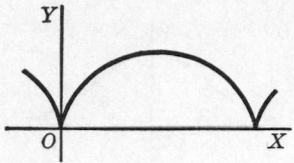

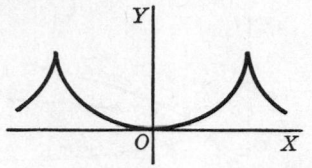

$x = a \text{ arc vers } \dfrac{y}{a} - \sqrt{2\,ay - y^2}$

$x = a(\theta - \sin\theta)$

$y = a(1 - \cos\theta)$

$x = a \text{ arc vers } \dfrac{y}{a} + \sqrt{2\,ay - y^2}$

$x = a(\theta + \sin\theta)$

$y = a(1 - \cos\theta)$

Prolate Cycloid

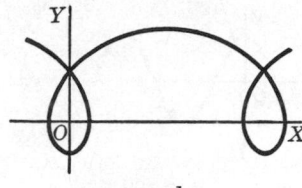

$a < b$

$x = a\theta - b\sin\theta$

Curtate Cycloid

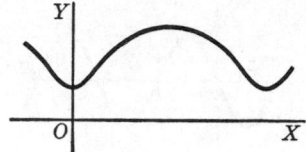

$a > b$

$y = a - b\cos\theta$

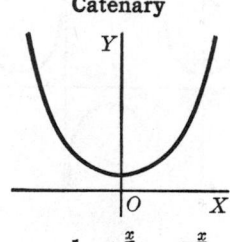

$y = e^{-x}\sin x$

Catenary

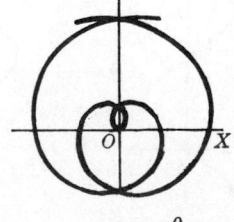

$y = \frac{1}{2}\,a(e^{\frac{x}{a}} + e^{-\frac{x}{a}})$

Lituus

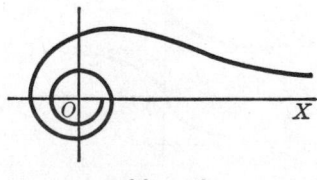

$\rho^2\theta = a^2$

Spiral of Archimedes

$\rho = a\theta$

Logarithmic Spiral

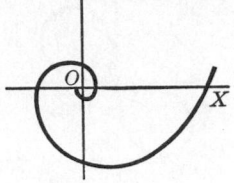

$$\log \rho = a\theta,$$
or, $$\rho = e^{a\theta}$$

Hyperbolic Spiral

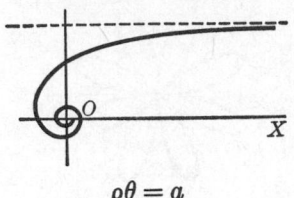

$$\rho\theta = a$$

Parabolic Spiral

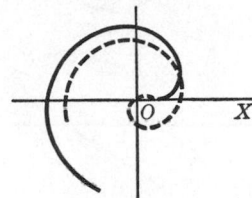

$$(\rho - a)^2 = 4\,ac\theta$$

Limaçon of Pascal

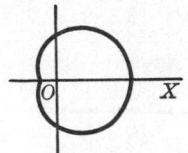

$$\rho = 2\,a\cos\theta + b$$
(Drawn for $b > 2\,a$)

Three-Leafed Rose

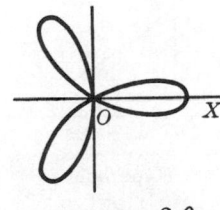

$$\rho = a\cos 3\,\theta$$

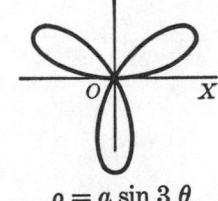

$$\rho = a\sin 3\,\theta$$

Four-Leafed Rose

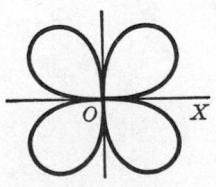

$$\rho = a\sin 2\,\theta$$

$$\rho = a\cos 2\,\theta$$

Eight-Leafed Rose

$$\rho = a \sin 4\,\theta$$

Lemniscate

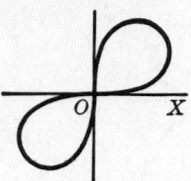

$$\rho^2 = a^2 \sin 2\,\theta$$

Circles

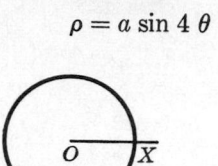

$$\rho = a$$

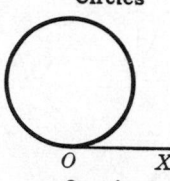

$$\rho = 2\,a \sin \theta$$

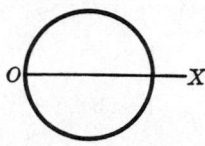

$$\rho = 2\,a \cos \theta$$

Parabolas

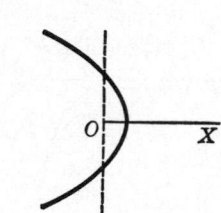

$$\rho = a \sec^2 \frac{\theta}{2} = \frac{2\,a}{1 + \cos \theta}$$

$$\rho = \frac{p}{1 - \cos \theta}$$

Ellipse

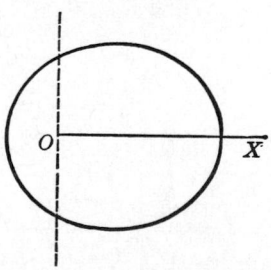

$$\rho = \frac{ep}{1 - e \cos \theta}, \ \ e < 1$$
(Drawn for $e = \frac{1}{2}$)

Hyperbola

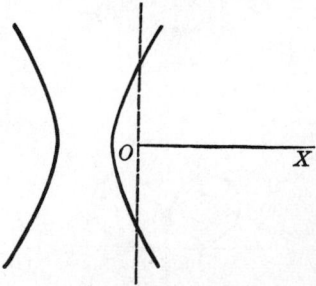

$$\rho = \frac{ep}{1 - e \cos \theta}, \ \ e > 1$$
(Drawn for $e = 2$)

FIGURES IN THREE DIMENSIONS

Plane

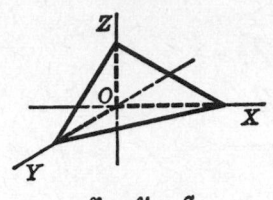

$$\frac{x}{a} + \frac{y}{b} + \frac{z}{c} = 1$$

Circular Cylinder*

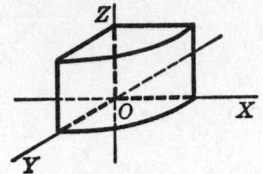

$$x^2 + y^2 = a^2$$

Sphere †

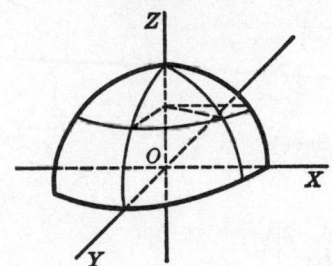

$$x^2 + y^2 + z^2 = a^2$$

Ellipsoid

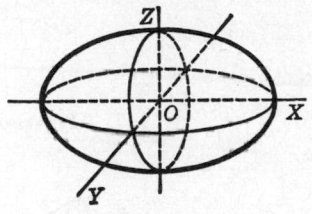

$$\frac{x^2}{a^2} + \frac{y^2}{b^2} + \frac{z^2}{c^2} = 1$$

Portion of Cone

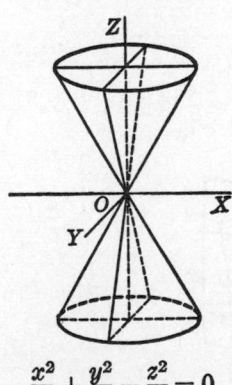

$$\frac{x^2}{a^2} + \frac{y^2}{b^2} - \frac{z^2}{c^2} = 0$$

Elliptic Paraboloid

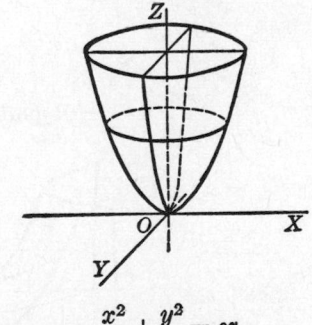

$$\frac{x^2}{a^2} + \frac{y^2}{b^2} = cz$$

*The drawing shows one fourth of the portion of the cylinder included between two planes parallel to the xy-plane.

† Only a part of the figure is shown.

Hyperboloid of One Sheet

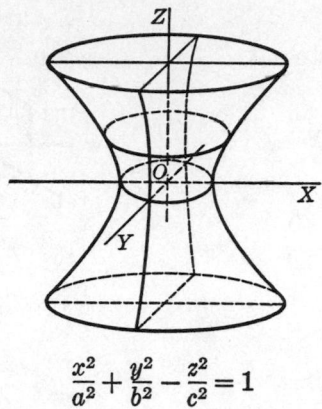

$$\frac{x^2}{a^2} + \frac{y^2}{b^2} - \frac{z^2}{c^2} = 1$$

Hyperboloid of Two Sheets

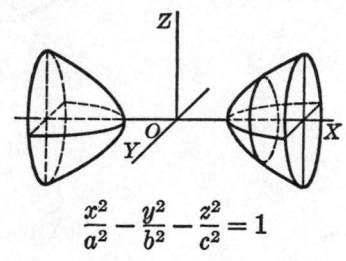

$$\frac{x^2}{a^2} - \frac{y^2}{b^2} - \frac{z^2}{c^2} = 1$$

Hyperbolic Paraboloid

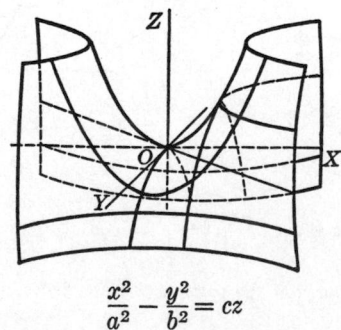

$$\frac{x^2}{a^2} - \frac{y^2}{b^2} = cz$$

VII. TABLE OF INTEGRALS *

STANDARD ELEMENTARY FORMS

(The symbols C and c denote constants.)

1. $\displaystyle\int 0\, dv = C.$ 2. $\displaystyle\int dv = v + C.$ 3. $\displaystyle\int c\, dv = c \int dv.$

4. $\displaystyle\int (du + dv - dw) = \int du + \int dv - \int dw.$

5. $\displaystyle\int u\, dv = uv - \int v\, du.$

6. $\displaystyle\int v^n\, dv = \frac{v^{n+1}}{n+1} + C$ if n is a constant and $\neq -1.$

7. $\displaystyle\int \frac{dv}{v} = \log v + C = \log cv$ if $\log c = C.$

8. $\displaystyle\int c^v\, dv = \frac{c^v}{\log c} + C.$ 9. $\displaystyle\int e^v\, dv = e^v + C.$

10. $\displaystyle\int \sin v\, dv = -\cos v + C.$ 11. $\displaystyle\int \cos v\, dv = \sin v + C.$

12. $\displaystyle\int \sec^2 v\, dv = \tan v + C.$ 13. $\displaystyle\int \csc^2 v\, dv = -\cot v + C.$

14. $\displaystyle\int \sec v \tan v\, dv = \sec v + C.$ 15. $\displaystyle\int \csc v \cot v\, dv = -\csc v + C.$

16. $\displaystyle\int \tan v\, dv = \log \sec v + C = -\log \cos v + C.$

17. $\displaystyle\int \cot v\, dv = \log \sin v + C = -\log \csc v + C.$

18. $\displaystyle\int \sec v\, dv = \log (\sec v + \tan v) + C = \log \tan \left(\frac{v}{2} + \frac{\pi}{4}\right) + C'.$

19. $\displaystyle\int \csc v\, dv = \log (\csc v - \cot v) + C = \log \tan \frac{v}{2} + C'.$

20. $\displaystyle\int \sin^2 v\, dv = \tfrac{1}{2} v - \tfrac{1}{2} \sin v \cos v + C.$

21. $\displaystyle\int \cos^2 v\, dv = \tfrac{1}{2} v + \tfrac{1}{2} \sin v \cos v + C.$

22. $\displaystyle\int \tan^2 v\, dv = \tan v - v + C.$

---

* Except in the case of the standard elementary forms, the constant of integration is uniformly omitted for the sake of simplicity in printing.

23. $\int \cot^2 v \, dv = - \cot v - v + C.$

24. $\int \dfrac{dv}{v^2 + a^2} = \dfrac{1}{a} \arctan \dfrac{v}{a} + C.$

25. $\int \dfrac{dv}{v^2 - a^2} = \dfrac{1}{2a} \log \dfrac{v - a}{v + a} + C$ if $v^2 > a^2,$

$\qquad = \dfrac{1}{2a} \log \dfrac{a - v}{a + v} + C$ if $v^2 < a^2.$

26. $\int \dfrac{dv}{\sqrt{a^2 - v^2}} = \arcsin \dfrac{v}{a} + C = - \arccos \dfrac{v}{a} + C'.$

27. $\int \dfrac{dv}{\sqrt{v^2 \pm a^2}} = \log(v + \sqrt{v^2 \pm a^2}) + C.$

28. $\int \dfrac{dv}{\sqrt{2 av - v^2}} = \text{arc vers} \dfrac{v}{a} + C.$

29. $\int \dfrac{dv}{v \sqrt{v^2 - a^2}} = \dfrac{1}{a} \text{arc sec} \dfrac{v}{a} + C = - \dfrac{1}{a} \text{arc csc} \dfrac{v}{a} + C'.$

30. $\int \sqrt{a^2 - v^2} \, dv = \tfrac{1}{2}\left(v \sqrt{a^2 - v^2} + a^2 \arcsin \dfrac{v}{a}\right) + C.$

31. $\int \sqrt{v^2 \pm a^2} \, dv = \tfrac{1}{2}[v \sqrt{v^2 \pm a^2} \pm a^2 \log (v + \sqrt{v^2 \pm a^2})] + C.$

HYPERBOLIC FUNCTIONS

32. $\int \sinh x \, dx = \cosh x, \quad \int \cosh x \, dx = \sinh x.$

33. $\int \text{sech}^2 x \, dx = \tanh x, \quad \int \text{csch}^2 x \, dx = - \coth x.$

34. $\int \text{sech} \, x \tanh x \, dx = - \text{sech} \, x.$

35. $\int \text{csch} \, x \coth x \, dx = - \text{csch} \, x.$

36. $\int \dfrac{dx}{\sqrt{x^2 + 1}} = \sinh^{-1} x = \log (x + \sqrt{x^2 + 1}).$

37. $\int \dfrac{dx}{\sqrt{x^2 - 1}} = \cosh^{-1} x = \log (x + \sqrt{x^2 - 1}).$

38. $\int \dfrac{dx}{1 - x^2} = \tanh^{-1} x = \tfrac{1}{2} \log \dfrac{1 + x}{1 - x}.$

39. $\int \dfrac{dx}{x \sqrt{1 - x^2}} = - \text{sech}^{-1} x = - \log \left(\dfrac{1}{x} + \sqrt{\dfrac{1}{x^2} - 1}\right).$

40. $\int \dfrac{dx}{x \sqrt{x^2 + 1}} = - \text{csch}^{-1} x = - \log \left(\dfrac{1}{x} + \sqrt{\dfrac{1}{x^2} + 1}\right).$

STANDARD ELEMENTARY PROCESSES

41. For the integration of $\int R(x)\,dx$, where $R(x)$ is a rational function of x, one proceeds as follows. Write $R(x)$ in the form of a fraction whose terms are polynomials. If the fraction is improper divide the numerator by the denominator and so write $R(x)$ as the sum of a quotient $Q(x)$ and a proper fraction $S(x)$. Then $Q(x)$ is readily integrated. Then integrate $S(x)$ by separating it into a sum of partial fractions (by the method indicated below) and integrating each summand separately.

If we write $S(x) = f(x)/\phi(x)$ where $f(x)$ and $\phi(x)$ are polynomials and

$$\phi(x) = (x - a)^p\ (x - b)^q\ (x - c)^r \cdots$$

and the constants $a, b, c, \cdots$ are all different, then constants $A_i, B_i, \cdots$ exist such that

$$\frac{f(x)}{\phi(x)} = \frac{A_1}{x - a} + \frac{A_2}{(x - a)^2} + \cdots + \frac{A_p}{(x - a)^p}$$

$$+ \frac{B_1}{x - b} + \frac{B_2}{(x - b)^2} + \cdots + \frac{B_q}{(x - b)^q} + \cdots.$$

In case $p = q = r = \cdots = 1$ we have

$$\frac{f(x)}{\phi(x)} = \frac{A}{x - a} + \frac{B}{x - b} + \frac{C}{x - c} + \cdots,$$

where $\qquad A = \dfrac{f(a)}{\phi'(a)}, \quad B = \dfrac{f(b)}{\phi'(b)}, \quad C = \dfrac{f(c)}{\phi'(c)}, \cdots.$

For integrating the separate terms we may employ the formulas

$$\int \frac{dx}{x - \alpha} = \log(x - \alpha), \quad \int \frac{dx}{(x - \alpha)^l} = \frac{-1}{(l - 1)(x - \alpha)^{l-1}} \text{ if } l > 1.$$

In case $f(x)$ and $\phi(x)$ are polynomials with real coefficients and the equation $\phi(x) = 0$ has imaginary roots, the foregoing method (while valid) leads to imaginary quantities, whereas in this case it is frequently desirable to avoid imaginary quantities in the processes and the result. This may be done by a separation into partial fractions in a different way. Corresponding to a p-fold real root a of $\phi(x)$ we use as summands the same expression as before, namely

$$\frac{A_1}{x - a} + \frac{A_2}{(x - a)^2} + \cdots + \frac{A_p}{(x - a)^p}.$$

If $x^2 + \alpha x + \beta$ is a real quadratic factor of $\phi(x)$ which does not separate into real linear factors, and if it occurs λ times as a factor of $\phi(x)$, then we take summands of the form

$$\frac{D_1 x + E_1}{x^2 + \alpha x + \beta} + \frac{D_2 x + E_2}{(x^2 + \alpha x + \beta)^2} + \cdots + \frac{D_\lambda x + E_\lambda}{(x^2 + \alpha x + \beta)^\lambda},$$

where the quantities D and E are constants. These are to replace the two sets of terms occurring in the first expansion of $f(x)/\phi(x)$ and depending on the conjugate complex roots of the equation $x^2 + \alpha x + \beta = 0$. These new summands are to be separately integrated by means of formulas 83 etc.

By the means here indicated every integral of the form $\int R(x)dx$ may be evaluated. But in special cases more convenient means are often available.

42. If R is a rational function of two arguments and p is an integer, then to integrate $\int R[x, (ax+b)^{\frac{1}{p}}]dx$ when no more convenient method is apparent, make the transformation

$$ax + b = t^p \quad \text{or} \quad x = \frac{t^p - b}{a}.$$

Then the integral reduces to $\int S(t)dt$, where $S(t)$ is a rational function of t. Then proceed as in 41.

43. If R is a rational function of two arguments, then to integrate $\int R[x, (x^2 + px + q)^{\frac{1}{2}}]dx$, when no more convenient method is apparent, make the transformation

$$(x^2 + px + q)^{\frac{1}{2}} = z - x,$$

whence

$$x = \frac{z^2 - q}{p + 2z}, \; dx = \frac{2(z^2 + pz + q)}{(p + 2z)^2} dz, \; (x^2 + px + q)^{\frac{1}{2}} = z - x = \frac{z^2 + pz + q}{p + 2z},$$

and thence proceed as in 41. [See also 212.]

44. If R is a rational function of two arguments, then to integrate $\int R[x, (-x^2 + px + q)^{\frac{1}{2}}]dx$, when no more convenient method is apparent, write $-x^2 + px + q$ in the factored form $(x - a)(b - x)$ and make one of the transformations

$$(-x^2 + px + q)^{\frac{1}{2}} = (x - a)z \quad \text{or} \quad (-x^2 + px + q)^{\frac{1}{2}} = (b - x)z.$$

If the former is made, then

$$x = \frac{az^2 + b}{z^2 + 1}, \; dx = \frac{2(a - b)zdz}{(z^2 + 1)^2}, \; (-x^2 + px + q)^{\frac{1}{2}} = \frac{(b - a)z}{z^2 + 1}.$$

After performing the transformation one proceeds as in 41. [See also 212.]

45. If R is a rational function of two arguments, then to integrate $\int R(\sin x, \cos x)dx$, when no more convenient method is apparent, make the transformation

$$\tan \frac{x}{2} = t.$$

Then
$$\sin x = \frac{2t}{1+t^2}, \quad \cos x = \frac{1-t^2}{1+t^2}, \quad \tan x = \frac{2t}{1-t^2}, \quad dx = \frac{2\,dt}{1+t^2}.$$

After performing this transformation one proceeds as in 41.

The same transformation may be applied to the case of the integral $\int T\,dx$ where T is a rational function of any one or more of the trigonometric functions $\sin x$, $\cos x$, $\tan x$, $\cot x$, $\sec x$, $\csc x$.

46. Expressions containing one of the radicals $(a^2 - x^2)^{\frac{1}{2}}$ and $(x^2 \pm a^2)^{\frac{1}{2}}$ can often be integrated by trigonometric substitution. In using this method,

when $(a^2 - x^2)^{\frac{1}{2}}$ occurs, put $x = a \sin t$;

when $(x^2 - a^2)^{\frac{1}{2}}$ occurs, put $x = a \sec t$;

when $(x^2 + a^2)^{\frac{1}{2}}$ occurs, put $x = a \tan t$.

47. The integral $\int \sin^m x \cos^n x\, dx$ can be evaluated by putting

(a) $\cos x = t$ when m is an odd integer,

(b) $\sin x = t$ when n is an odd integer.

48. When m and n are both even integers the integral $\int \sin^m x \cos^n x\, dx$ may be transformed by aid of the identities

$$2 \sin x \cos x = \sin 2x, \quad 2 \sin^2 x = 1 - \cos 2x, \quad 2 \cos^2 x = 1 + \cos 2x;$$

it may be necessary (in integrating) to employ also the method of 47 after making the transformation here suggested.

49. The integrals $\int \tan^n x\, dx$ and $\int \cot^n x\, dx$, where n is an integer, may be transformed by aid of the identities

$$\tan^2 x = \sec^2 x - 1, \quad \cot^2 x = \csc^2 x - 1.$$

One may treat similarly the integrals

$$\int \tan^n x \sec^m x\, dx \text{ and } \int \cot^n x \csc^m x\, dx$$

when m is an even positive integer or n is an odd positive integer.

50. The method of integration by parts, based on the formula

$$\int u\, dv = uv - \int v\, du,$$

is often effective in reducing a given integral to depend on one or more simpler integrals and also in obtaining numerous reduction formulas.

EXPRESSIONS CONTAINING $a + bx$

By writing $y = a + bx$ or $z = (a + bx)/x$ the following integrals may be reduced to standard forms and thus be easily evaluated, with the results given:

51. $\displaystyle\int \frac{dx}{a + bx} = \frac{1}{b} \log (a + bx).$

52. $\displaystyle\int \frac{dx}{(a + bx)^n} = \frac{1}{b(1 - n)(a + bx)^{n-1}}$ if $n \neq 1.$

53. $\displaystyle\int (a + bx)^n \, dx = \frac{(a + bx)^{n+1}}{b(n + 1)}$ if $n \neq -1.$

54. $\displaystyle\int \frac{x \, dx}{a + bx} = \frac{1}{b^2} [a + bx - a \log (a + bx)].$

55. $\displaystyle\int \frac{x^2 \, dx}{a + bx} = \frac{1}{b^3} [\tfrac{1}{2}(a + bx)^2 - 2\,a(a + bx) + a^2 \log (a + bx)].$

56. $\displaystyle\int \frac{x \, dx}{(a + bx)^2} = \frac{1}{b^2}\left[\log (a + bx) + \frac{a}{a + bx}\right].$

57. $\displaystyle\int \frac{x^2 \, dx}{(a + bx)^2} = \frac{1}{b^3}\left[a + bx - 2\,a \log (a + bx) - \frac{a^2}{a + bx}\right].$

58. $\displaystyle\int \frac{x \, dx}{(a + bx)^3} = \frac{1}{b^2}\left[\frac{a}{2(a + bx)^2} - \frac{1}{a + bx}\right].$

59. $\displaystyle\int \frac{dx}{x(a + bx)} = -\frac{1}{a} \log \frac{a + bx}{x}.$

60. $\displaystyle\int \frac{dx}{x(a + bx)^2} = \frac{1}{a(a + bx)} - \frac{1}{a^2} \log \frac{a + bx}{x}.$

61. $\displaystyle\int \frac{dx}{x^2(a + bx)} = -\frac{1}{ax} + \frac{b}{a^2} \log \frac{a + bx}{x}.$

62. $\displaystyle\int \frac{dx}{x^2(a + bx)^2} = \frac{-1}{ax(a + bx)} - \frac{2\,b}{a} \int \frac{dx}{x(a + bx)^2}.$

EXPRESSIONS CONTAINING $a + bx^n$ (INCLUDING $a^2 + x^2$, $a^2 - x^2$, AND $x^2 - a^2$)

63. $\displaystyle\int \frac{dx}{a^2 + x^2} = \frac{1}{a} \arctan \frac{x}{a} = \frac{1}{a} \arcsin \frac{x}{(a^2 + x^2)^{\frac{1}{2}}}.$

64. $\displaystyle\int \frac{dx}{x^2 - a^2} = \frac{1}{2\,a} \log \frac{x - a}{x + a}$ if $x^2 > a^2,$

$\qquad\qquad = \frac{1}{2\,a} \log \frac{a - x}{a + x}$ if $x^2 < a^2.$

65. $\int \dfrac{dx}{a + bx^2} = \dfrac{1}{\sqrt{ab}} \arctan\left(x\sqrt{\dfrac{b}{a}}\right)$ if $a > 0,\, b > 0,$

$\qquad = \dfrac{1}{2}\dfrac{1}{\sqrt{-ab}} \log \dfrac{\sqrt{a} + x\sqrt{-b}}{\sqrt{a} - x\sqrt{-b}}$ if $a > 0,\, b < 0.$

66. $\int \dfrac{dx}{a^2 - b^2 x^2} = \dfrac{1}{2\,ab} \log \dfrac{a + bx}{a - bx}.$

67. $\int \dfrac{dx}{(a + bx^2)^{m+1}} = \dfrac{x}{2\,ma(a + bx^2)^m} + \dfrac{2\,m - 1}{2\,ma} \int \dfrac{dx}{(a + bx^2)^m}.$

68. $\int \dfrac{x\,dx}{(a + bx^2)^{m+1}} = \dfrac{1}{2} \int \dfrac{dz}{(a + bz)^{m+1}}$ if $z = x^2.$

69. $\int \dfrac{x\,dx}{a + bx^2} = \dfrac{1}{2\,b} \log \left(x^2 + \dfrac{a}{b}\right).$

70. $\int \dfrac{x^2\,dx}{a + bx^2} = \dfrac{x}{b} - \dfrac{a}{b} \int \dfrac{dx}{a + bx^2}.$

71. $\int \dfrac{dx}{x(a + bx^2)} = \dfrac{1}{2\,a} \log \dfrac{x^2}{a + bx^2}.$

72. $\int \dfrac{dx}{x^2(a + bx^2)} = -\dfrac{1}{ax} - \dfrac{b}{a} \int \dfrac{dx}{a + bx^2}.$

73. $\int \dfrac{dx}{(a + bx^2)^2} = \dfrac{x}{2\,a(a + bx^2)} + \dfrac{1}{2\,a} \int \dfrac{dx}{a + bx^2}.$

74. $\int \dfrac{x^2\,dx}{(a + bx^2)^m} = \dfrac{1}{b} \int \dfrac{dx}{(a + bx^2)^{m-1}} - \dfrac{a}{b} \int \dfrac{dx}{(a + bx^2)^m}.$

75. $\int \dfrac{dx}{x^2(a + bx^2)^m} = \dfrac{1}{a} \int \dfrac{dx}{x^2(a + bx^2)^{m-1}} - \dfrac{b}{a} \int \dfrac{dx}{(a + bx^2)^m}.$

76. $\int \dfrac{dx}{a + bx^3} = \dfrac{k}{6\,a}\left[\log \dfrac{(k + x)^2}{k^2 - kx + x^2} + 2\sqrt{3} \arctan \dfrac{2\,x - k}{k\sqrt{3}}\right]$ if $bk^3 = a.$

77. $\int \dfrac{x\,dx}{a + bx^3} = \dfrac{1}{6\,bk}\left[\log \dfrac{k^2 - kx + x^2}{(k + x)^2} + 2\sqrt{3} \arctan \dfrac{2\,x - k}{k\sqrt{3}}\right]$ if $bk^3 = a.$

78. $\int \dfrac{dx}{x(a + bx^n)} = \dfrac{1}{an} \log \dfrac{x^n}{a + bx^n}.$

79. $\int \dfrac{dx}{(a + bx^n)^m} = \dfrac{1}{a} \int \dfrac{dx}{(a + bx^n)^{m-1}} - \dfrac{b}{a} \int \dfrac{x^n\,dx}{(a + bx^n)^m}.$

80. $\int x^{m-1}(a + bx^n)^p\,dx = \dfrac{1}{m + np}[x^m(a + bx^n)^p + npa \int x^{m-1}(a + bx^n)^{p-1}dx]$

$\qquad = \dfrac{1}{an\,(p+1)}[-x^m(a + bx^n)^{p+1} + (m + np + n) \int x^{m-1}(a + bx^n)^{p+1}dx]$

$\qquad = \dfrac{1}{b(m + np)}[x^{m-n}(a + bx^n)^{p+1} - (m - n)a \int x^{m-n-1}(a + bx^n)^p\,dx]$

$\qquad = \dfrac{1}{ma}[x^m(a + bx^n)^{p+1} - (m + np + n) \int x^{m+n-1}(a + bx^n)^p\,dx].$

81. $\int \dfrac{x^m \, dx}{(a + bx^n)^p} = \dfrac{1}{b} \int \dfrac{x^{m-n} \, dx}{(a + bx^n)^{p-1}} - \dfrac{a}{b} \int \dfrac{x^{m-n} \, dx}{(a + bx^n)^p}.$

82. $\int \dfrac{dx}{x^m (a + bx^n)^p} = \dfrac{1}{a} \int \dfrac{dx}{x^m (a + bx^n)^{p-1}} - \dfrac{b}{a} \int \dfrac{dx}{x^{m-n}(a + bx^n)^p}.$

EXPRESSIONS CONTAINING $ax^2 + bx + c$

For brevity we use the symbols X and q, where

$$X = ax^2 + bx + c, \quad q = b^2 - 4\,ac.$$

83. $\int \dfrac{dx}{X} = \dfrac{1}{\sqrt{q}} \log \dfrac{2\,ax + b - \sqrt{q}}{2\,ax + b + \sqrt{q}}$ if $q > 0,$

$\qquad = \dfrac{2}{\sqrt{-q}} \arctan \dfrac{2\,ax + b}{\sqrt{-q}}$ if $q < 0.$

84. $\int \dfrac{x \, dx}{X} = \dfrac{1}{2\,a} \log X - \dfrac{b}{2\,a} \int \dfrac{dx}{X}.$

85. $\int \dfrac{x^2 \, dx}{X} = \dfrac{x}{a} - \dfrac{b}{2\,a^2} \log X + \dfrac{b^2 - 2\,ac}{2\,a^2} \int \dfrac{dx}{X}.$

86. $\int \dfrac{dx}{xX} = \dfrac{1}{2\,c} \log \dfrac{x^2}{X} - \dfrac{b}{2\,c} \int \dfrac{dx}{X}.$

87. $\int \dfrac{dx}{x^2 \, X} = \dfrac{b}{2\,c^2} \log \dfrac{X}{x^2} - \dfrac{1}{cx} + \dfrac{b^2 - 2\,ac}{2\,c^2} \int \dfrac{dx}{X}.$

88. $\int \dfrac{dx}{X^2} = -\dfrac{2\,ax + b}{qX} - \dfrac{2\,c}{q} \int \dfrac{dx}{X}.$

89. $\int \dfrac{x \, dx}{X^2} = \dfrac{bx + 2\,c}{qX} + \dfrac{b}{q} \int \dfrac{dx}{X}.$

90. $\int \dfrac{x^2 \, dx}{X^2} = -\dfrac{(b^2 - 2\,ac)\dot{x} + bc}{aqX} - \dfrac{2\,a}{q} \int \dfrac{dx}{X}.$

91. $\int \dfrac{dx}{X^{n+1}} = -\dfrac{2\,ax + b}{nq \, X^n} - \dfrac{2(2\,n - 1)a}{qn} \int \dfrac{dx}{X^n}.$

92. $\int \dfrac{x \, dx}{X^{n+1}} = \dfrac{2\,c + bx}{nq \, X^n} + \dfrac{b(2\,n - 1)}{qn} \int \dfrac{dx}{X^n}.$

93. $\int \dfrac{x^m \, dx}{X^{n+1}} = -\dfrac{x^{m-1}}{(2\,n - m + 1)aX^n} - \dfrac{n - m + 1}{2\,n - m + 1} \dfrac{b}{a} \int \dfrac{x^{m-1} \, dx}{X^{n+1}}$

$\qquad\qquad\qquad + \dfrac{m - 1}{2\,n - m + 1} \dfrac{c}{a} \int \dfrac{x^{m-2} \, dx}{X^{n+1}}.$

94. $\int \dfrac{dx}{x^m \, X^{n+1}} = \dfrac{-1}{(m - 1)cx^{m-1}X^n} - \dfrac{n + m - 1}{m - 1} \dfrac{b}{c} \int \dfrac{dx}{x^{m-1} X^{n+1}}$

$\qquad\qquad\qquad - \dfrac{2\,n + m - 1}{m - 1} \dfrac{a}{c} \int \dfrac{dx}{x^{m-2} \, X^{n+1}}.$

EXPRESSIONS CONTAINING $(a + bx)^{\frac{1}{2}}$

See 42 for the general method of dealing with the integral

$$\int R[x, (a + bx)^{\frac{1}{2}}]\, dx.$$

95. $\displaystyle\int (a + bx)^{\frac{1}{2}n}\, dx = \frac{2(a + bx)^{\frac{1}{2}(n + 2)}}{b(n + 2)}.$

96. $\displaystyle\int x(a + bx)^{\frac{1}{2}n}\, dx = \frac{2}{b^2}\left[\frac{(a + bx)^{\frac{1}{2}(n + 4)}}{n + 4} - \frac{a(a + bx)^{\frac{1}{2}(n + 2)}}{n + 2}\right].$

97. $\displaystyle\int x^{-1}(a + bx)^{\frac{1}{2}n}\, dx = b\int (a + bx)^{\frac{1}{2}(n - 2)}\, dx + a\int x^{-1}(a + bx)^{\frac{1}{2}(n - 2)}\, dx.$

98. $\displaystyle\int x^{-1}(a + bx)^{-\frac{1}{2}n}\, dx = \frac{1}{a}\int x^{-1}(a + bx)^{-\frac{1}{2}(n - 2)}\, dx - \frac{b}{a}\int (a + bx)^{-\frac{1}{2}n}\, dx.$

99. $\displaystyle\int \frac{x^m\, dx}{(a + bx)^{\frac{1}{2}}} = \frac{2\, x^m(a + bx)^{\frac{1}{2}}}{(2\, m + 1)b} - \frac{2\, ma}{(2\, m + 1)b}\int \frac{x^{m-1}\, dx}{(a + bx)^{\frac{1}{2}}}\, dx.$

100. $\displaystyle\int \frac{dx}{x^n(a + bx)^{\frac{1}{2}}} = -\frac{(a + bx)^{\frac{1}{2}}}{(n - 1)ax^{n-1}} - \frac{(2\, n - 3)b}{(2\, n - 2)a}\int \frac{dx}{x^{n-1}(a + bx)^{\frac{1}{2}}}.$

EXPRESSIONS CONTAINING $(a^2 + x^2)^{\frac{1}{2}}$

101. $\displaystyle\int (a^2 + x^2)^{\frac{1}{2}}\, dx = \frac{1}{2}\, x(a^2 + x^2)^{\frac{1}{2}} + \frac{1}{2}\, a^2 \log\, [x + (a^2 + x^2)^{\frac{1}{2}}].$

102. $\displaystyle\int x(a^2 + x^2)^{\frac{1}{2}}\, dx = \frac{1}{3}(a^2 + x^2)^{\frac{3}{2}}.$

103. $\displaystyle\int x^2(a^2 + x^2)^{\frac{1}{2}}\, dx = \frac{x}{8}(2\, x^2 + a^2)(a^2 + x^2)^{\frac{1}{2}} - \frac{a^4}{8}\log\, [x + (a^2 + x^2)^{\frac{1}{2}}].$

104. $\displaystyle\int x^3(a^2 + x^2)^{\frac{1}{2}}\, dx = \frac{1}{5}(a^2 + x^2)^{\frac{5}{2}} - \frac{a^2}{3}(a^2 + x^2)^{\frac{3}{2}}.$

105. $\displaystyle\int (a^2 + x^2)^{\frac{3}{2}}\, dx = \frac{x}{8}(2\, x^2 + 5\, a^2)(a^2 + x^2)^{\frac{1}{2}} + \frac{3\, a^4}{8}\log\, [x + (a^2 + x^2)^{\frac{1}{2}}].$

106. $\displaystyle\int x(a^2 + x^2)^{\frac{3}{2}}\, dx = \frac{1}{5}(a^2 + x^2)^{\frac{5}{2}}.$

107. $\displaystyle\int (a^2 + x^2)^{\frac{1}{2}n}\, dx = \frac{x(a^2 + x^2)^{\frac{1}{2}n}}{n + 1} + \frac{na^2}{n + 1}\int (a^2 + x^2)^{\frac{1}{2}(n - 2)}\, dx.$

108. $\displaystyle\int x(a^2 + x^2)^{\frac{1}{2}n}\, dx = \frac{(a^2 + x^2)^{\frac{1}{2}(n + 2)}}{n + 2}.$

109. $\displaystyle\int x^2(a^2 + x^2)^{\frac{1}{2}n}\, dx = \int (a^2 + x^2)^{\frac{1}{2}(n + 2)}\, dx - a^2\int (a^2 + x^2)^{\frac{1}{2}n}\, dx.$

110. $\int x^3(a^2 + x^2)^{\frac{1}{2}n}\, dx = \int x(a^2 + x^2)^{\frac{1}{2}(n+2)}\, dx - a^2 \int x(a^2 + x^2)^{\frac{1}{2}n}\, dx.$

111. $\int \dfrac{dx}{(a^2 + x^2)^{\frac{1}{2}}} = \log\, [x + (a^2 + x^2)^{\frac{1}{2}}].$

112. $\int \dfrac{x\, dx}{(a^2 + x^2)^{\frac{1}{2}}} = (a^2 + x^2)^{\frac{1}{2}}.$

113. $\int \dfrac{x^2\, dx}{(a^2 + x^2)^{\frac{1}{2}}} = \int (a^2 + x^2)^{\frac{1}{2}}\, dx - a^2 \int \dfrac{dx}{(a^2 + x^2)^{\frac{1}{2}}}.$

114. $\int \dfrac{x^3\, dx}{(a^2 + x^2)^{\frac{1}{2}}} = \int x(a^2 + x^2)^{\frac{1}{2}}\, dx - a^2 \int \dfrac{x\, dx}{(a^2 + x^2)^{\frac{1}{2}}}.$

115. $\int \dfrac{dx}{(a^2 + x^2)^{\frac{3}{2}}} = \dfrac{x}{a^2(a^2 + x^2)^{\frac{1}{2}}}.$

116. $\int \dfrac{x\, dx}{(a^2 + x^2)^{\frac{3}{2}}} = \dfrac{-1}{(a^2 + x^2)^{\frac{1}{2}}}.$

117. $\int \dfrac{x^2\, dx}{(a^2 + x^2)^{\frac{3}{2}}} = \int \dfrac{dx}{(a^2 + x^2)^{\frac{1}{2}}} - a^2 \int \dfrac{dx}{(a^2 + x^2)^{\frac{3}{2}}}.$

118. $\int \dfrac{(a^2 + x^2)^{\frac{1}{2}}\, dx}{x} = (a^2 + x^2)^{\frac{1}{2}} - a \log \dfrac{a + (a^2 + x^2)^{\frac{1}{2}}}{x}.$

119. $\int \dfrac{(a^2 + x^2)^{\frac{1}{2}}\, dx}{x^2} = -\dfrac{(a^2 + x^2)^{\frac{1}{2}}}{x} + \log\, [x + (a^2 + x^2)^{\frac{1}{2}}].$

120. $\int \dfrac{dx}{x(a^2 + x^2)^{\frac{1}{2}}} = -\dfrac{1}{a} \log \dfrac{a + (a^2 + x^2)^{\frac{1}{2}}}{x}.$

121. $\int \dfrac{dx}{x^2(a^2 + x^2)^{\frac{1}{2}}} = -\dfrac{(a^2 + x^2)^{\frac{1}{2}}}{a^2 x}.$

122. $\int \dfrac{dx}{(a^2 + x^2)^{n + \frac{1}{2}}} = \dfrac{x}{a^2(2\,n - 1)(a^2 + x^2)^{n - \frac{1}{2}}} + \dfrac{2(n - 1)}{a^2(2\,n - 1)} \int \dfrac{dx}{(a^2 + x^2)^{n - \frac{1}{2}}}.$

123. $\int \dfrac{x\, dx}{(a^2 + x^2)^{n + \frac{1}{2}}} = \dfrac{-1}{(2\,n - 1)(a^2 + x^2)^{n - \frac{1}{2}}}.$

124. $\int \dfrac{x^2\, dx}{(a^2 + x^2)^{n + \frac{1}{2}}} = \int \dfrac{dx}{(a^2 + x^2)^{n - \frac{1}{2}}} - a^2 \int \dfrac{dx}{(a^2 + x^2)^{n + \frac{1}{2}}}.$

125. $\int \dfrac{dx}{x(a^2 + x^2)^{n + \frac{1}{2}}} = \dfrac{1}{(2\,n - 1)a^2(a^2 + x^2)^{n - \frac{1}{2}}} + \dfrac{1}{a^2} \int \dfrac{dx}{x(a^2 + x^2)^{n - \frac{1}{2}}}.$

126. $\int \dfrac{x^m\, dx}{(a^2 + x^2)^{n + \frac{1}{2}}} = \int \dfrac{x^{m-2}\, dx}{(a^2 + x^2)^{n - \frac{1}{2}}} - a^2 \int \dfrac{x^{m-2}\, dx}{(a^2 + x^2)^{n + \frac{1}{2}}}.$

127. $\int x^m(a^2+x^2)^{n-\frac{1}{2}}\,dx$

$$=\frac{x^{m-1}(a^2+x^2)^{n+\frac{1}{2}}}{2\,n+m}-\frac{(m-1)a^2}{2\,n+m}\int x^{m-2}(a^2+x^2)^{n-\frac{1}{2}}\,dx.$$

128. $\int \dfrac{dx}{x^m(a^2+x^2)^{n+\frac{1}{2}}}$

$$=\frac{-1}{(m-1)a^2x^{m-1}(a^2+x^2)^{n-\frac{1}{2}}}-\frac{2\,n+m-2}{(m-1)a^2}\int\frac{dx}{x^{m-2}(a^2+x^2)^{n+\frac{1}{2}}}.$$

129. $\int \dfrac{(a^2+x^2)^{n+\frac{1}{2}}}{x}\,dx=\dfrac{(a^2+x^2)^{n+\frac{1}{2}}}{2\,n+1}+a^2\int\dfrac{(a^2+x^2)^{n-\frac{1}{2}}}{x}\,dx.$

130. $\int \dfrac{(a^2+x^2)^{n+\frac{1}{2}}\,dx}{x^m}=-\dfrac{(a^2+x^2)^{n+\frac{1}{2}}}{(m-1)x^{m-1}}+\dfrac{2\,n+1}{m-1}\int\dfrac{(a^2+x^2)^{n-\frac{1}{2}}\,dx}{x^{m-2}}.$

EXPRESSIONS CONTAINING $(a^2-x^2)^{\frac{1}{2}}$

131. $\int (a^2-x^2)^{\frac{1}{2}}\,dx=\frac{1}{2}\,x(a^2-x^2)^{\frac{1}{2}}+\frac{1}{2}\,a^2\arcsin\dfrac{x}{a}.$

132. $\int x(a^2-x^2)^{\frac{1}{2}}\,dx=-\frac{1}{3}(a^2-x^2)^{\frac{3}{2}}.$

133. $\int x^2(a^2-x^2)^{\frac{1}{2}}\,dx=\dfrac{x}{8}\,(2\,x^2-a^2)(a^2-x^2)^{\frac{1}{2}}+\dfrac{a^4}{8}\arcsin\dfrac{x}{a}.$

134. $\int x^3(a^2-x^2)^{\frac{1}{2}}\,dx=\frac{1}{5}(a^2-x^2)^{\frac{5}{2}}-\dfrac{a^2}{3}\,(a^2-x^2)^{\frac{3}{2}}.$

135. $\int (a^2-x^2)^{\frac{3}{2}}\,dx=\dfrac{x}{8}\,(5\,a^2-2\,x^2)(a^2-x^2)^{\frac{1}{2}}+\dfrac{3\,a^4}{8}\arcsin\dfrac{x}{a}.$

136. $\int x(a^2-x^2)^{\frac{3}{2}}\,dx=-\frac{1}{5}(a^2-x^2)^{\frac{5}{2}}.$

137. $\int (a^2-x^2)^{\frac{1}{2}n}\,dx=\dfrac{x(a^2-x^2)^{\frac{1}{2}n}}{n+1}+\dfrac{a^2\,n}{n+1}\int(a^2-x^2)^{\frac{1}{2}(n-2)}\,dx.$

138. $\int x(a^2-x^2)^{\frac{1}{2}n}\,dx=-\dfrac{(a^2-x^2)^{\frac{1}{2}(n+2)}}{n+2}.$

139. $\int x^2(a^2-x^2)^{\frac{1}{2}n}\,dx=-\int(a^2-x^2)^{\frac{1}{2}(n+2)}\,dx+a^2\int(a^2-x^2)^{\frac{1}{2}n}\,dx.$

140. $\int x^3(a^2-x^2)^{\frac{1}{2}n}\,dx=-\int x(a^2-x^2)^{\frac{1}{2}(n+2)}\,dx+a^2\int x(a^2-x^2)^{\frac{1}{2}n}\,dx.$

141. $\int \dfrac{dx}{(a^2-x^2)^{\frac{1}{2}}}=\arcsin\dfrac{x}{a}.$

142. $\int \dfrac{x\,dx}{(a^2-x^2)^{\frac{1}{2}}}=-(a^2-x^2)^{\frac{1}{2}}.$

143. $\int \dfrac{x^2\,dx}{(a^2-x^2)^{\frac{1}{2}}}=-\frac{1}{2}\,x(a^2-x^2)^{\frac{1}{2}}+\frac{1}{2}\,a^2\arcsin\dfrac{x}{a}.$

144. $\int \dfrac{x^3\,dx}{(a^2 - x^2)^{\frac{1}{2}}} = -\int x(a^2 - x^2)^{\frac{1}{2}}\,dx + a^2 \int \dfrac{x\,dx}{(a^2 - x^2)^{\frac{1}{2}}}.$

145. $\int \dfrac{dx}{(a^2 - x^2)^{\frac{3}{2}}} = \dfrac{x}{a^2(a^2 - x^2)^{\frac{1}{2}}}.$

146. $\int \dfrac{x\,dx}{(a^2 - x^2)^{\frac{3}{2}}} = \dfrac{1}{(a^2 - x^2)^{\frac{1}{2}}}.$

147. $\int \dfrac{x^2\,dx}{(a^2 - x^2)^{\frac{3}{2}}} = -\int \dfrac{dx}{(a^2 - x^2)^{\frac{1}{2}}} + a^2 \int \dfrac{dx}{(a^2 - x^2)^{\frac{3}{2}}}.$

148. $\int \dfrac{(a^2 - x^2)^{\frac{1}{2}}\,dx}{x} = (a^2 - x^2)^{\frac{1}{2}} - a \log \dfrac{a + (a^2 - x^2)^{\frac{1}{2}}}{x}.$

149. $\int \dfrac{(a^2 - x^2)^{\frac{1}{2}}\,dx}{x^2} = -\dfrac{(a^2 - x^2)^{\frac{1}{2}}}{x} - \arcsin \dfrac{x}{a}.$

150. $\int \dfrac{dx}{x(a^2 - x^2)^{\frac{1}{2}}} = -\dfrac{1}{a} \log \dfrac{a + (a^2 - x^2)^{\frac{1}{2}}}{x}.$

151. $\int \dfrac{dx}{x^2(a^2 - x^2)^{\frac{1}{2}}} = -\dfrac{(a^2 - x^2)^{\frac{1}{2}}}{a^2 x}.$

152. $\int \dfrac{dx}{(a^2 - x^2)^{n+\frac{1}{2}}} = \dfrac{x}{(2n-1)a^2(a^2 - x^2)^{n-\frac{1}{2}}} + \dfrac{2(n-1)}{a^2(2n-1)} \int \dfrac{dx}{(a^2 - x^2)^{n-\frac{1}{2}}}.$

153. $\int \dfrac{x\,dx}{(a^2 - x^2)^{n+\frac{1}{2}}} = \dfrac{1}{(2n-1)(a^2 - x^2)^{n-\frac{1}{2}}}.$

154. $\int \dfrac{x^2\,dx}{(a^2 - x^2)^{n+\frac{1}{2}}} = -\int \dfrac{dx}{(a^2 - x^2)^{n-\frac{1}{2}}} + a^2 \int \dfrac{dx}{(a^2 - x^2)^{n+\frac{1}{2}}}.$

155. $\int \dfrac{dx}{x(a^2 - x^2)^{n+\frac{1}{2}}} = \dfrac{1}{(2n-1)a^2(a^2 - x^2)^{n-\frac{1}{2}}} + \dfrac{1}{a^2} \int \dfrac{dx}{x(a^2 - x^2)^{n-\frac{1}{2}}}.$

156. $\int \dfrac{x^m\,dx}{(a^2 - x^2)^{n+\frac{1}{2}}} = -\int \dfrac{x^{m-2}\,dx}{(a^2 - x^2)^{n-\frac{1}{2}}} + a^2 \int \dfrac{x^{m-2}\,dx}{(a^2 - x^2)^{n+\frac{1}{2}}}.$

157. $\int x^m(a^2 - x^2)^{n-\frac{1}{2}}\,dx$

$\qquad = -\dfrac{x^{m-1}(a^2 - x^2)^{n+\frac{1}{2}}}{2n + m} + \dfrac{(m-1)a^2}{2n + m} \int x^{m-2}(a^2 - x^2)^{n-\frac{1}{2}}\,dx.$

158. $\int \dfrac{dx}{x^m(a^2 - x^2)^{n+\frac{1}{2}}}$

$\qquad = \dfrac{-1}{(m-1)a^2 x^{m-1}(a^2 - x^2)^{n-\frac{1}{2}}} + \dfrac{2n + m - 2}{(m-1)a^2} \int \dfrac{dx}{x^{m-2}(a^2 - x^2)^{n+\frac{1}{2}}}.$

159. $\int \dfrac{(a^2 - x^2)^{n+\frac{1}{2}}\,dx}{x} = \dfrac{(a^2 - x^2)^{n+\frac{1}{2}}}{2n + 1} + a^2 \int \dfrac{(a^2 - x^2)^{n-\frac{1}{2}}\,dx}{x}.$

160. $\int \dfrac{(a^2 - x^2)^{n+\frac{1}{2}}\,dx}{x^m} = -\dfrac{(a^2 - x^2)^{n+\frac{1}{2}}}{(m-1)x^{m-1}} - \dfrac{2n + 1}{m - 1} \int \dfrac{(a^2 - x^2)^{n-\frac{1}{2}}\,dx}{x^{m-2}}.$

EXPRESSIONS CONTAINING $(x^2 - a^2)^{\frac{1}{2}}$

161. $\int (x^2 - a^2)^{\frac{1}{2}} \, dx = \frac{1}{2} \, x(x^2 - a^2)^{\frac{1}{2}} - \frac{1}{2} \, a^2 \log [x + (x^2 - a^2)^{\frac{1}{2}}].$

162. $\int x(x^2 - a^2)^{\frac{1}{2}} \, dx = \frac{1}{3}(x^2 - a^2)^{\frac{3}{2}}.$

163. $\int x^2(x^2 - a^2)^{\frac{1}{2}} \, dx = \frac{x}{8}(2 \, x^2 - a^2)(x^2 - a^2)^{\frac{1}{2}} - \frac{a^4}{8} \log [x + (x^2 - a^2)^{\frac{1}{2}}].$

164. $\int x^3(x^2 - a^2)^{\frac{1}{2}} \, dx = \frac{1}{5}(x^2 - a^2)^{\frac{5}{2}} + \frac{a^2}{3}(x^2 - a^2)^{\frac{3}{2}}.$

165. $\int (x^2 - a^2)^{\frac{3}{2}} \, dx = \frac{x}{8}(2 \, x^2 - 5 \, a^2)(x^2 - a^2)^{\frac{1}{2}} + \frac{3 \, a^4}{8} \log [x + (x^2 - a^2)^{\frac{1}{2}}].$

166. $\int x(x^2 - a^2)^{\frac{3}{2}} \, dx = \frac{1}{5}(x^2 - a^2)^{\frac{5}{2}}.$

167. $\int (x^2 - a^2)^{\frac{1}{2}n} \, dx = \frac{x(x^2 - a^2)^{\frac{1}{2}n}}{n + 1} - \frac{na^2}{n + 1} \int (x^2 - a^2)^{\frac{1}{2}(n-2)} \, dx.$

168. $\int x(x^2 - a^2)^{\frac{1}{2}n} \, dx = \frac{(x^2 - a^2)^{\frac{1}{2}(n+2)}}{n + 2}.$

169. $\int x^2(x^2 - a^2)^{\frac{1}{2}n} \, dx = \int (x^2 - a^2)^{\frac{1}{2}(n+2)} \, dx + a^2 \int (x^2 - a^2)^{\frac{1}{2}n} \, dx.$

170. $\int x^3(x^2 - a^2)^{\frac{1}{2}n} \, dx = \int x(x^2 - a^2)^{\frac{1}{2}(n+2)} \, dx + a^2 \int x(x^2 - a^2)^{\frac{1}{2}n} \, dx.$

171. $\int \dfrac{dx}{(x^2 - a^2)^{\frac{1}{2}}} = \log [x + (x^2 - a^2)^{\frac{1}{2}}].$

172. $\int \dfrac{x \, dx}{(x^2 - a^2)^{\frac{1}{2}}} = (x^2 - a^2)^{\frac{1}{2}}.$

173. $\int \dfrac{x^2 dx}{(x^2 - a^2)^{\frac{1}{2}}} = \frac{1}{2} \, x(x^2 - a^2)^{\frac{1}{2}} + \frac{1}{2} \, a^2 \log [x + (x^2 - a^2)^{\frac{1}{2}}].$

174. $\int \dfrac{x^3 dx}{(x^2 - a^2)^{\frac{1}{2}}} = \int x(x^2 - a^2)^{\frac{1}{2}} \, dx + a^2 \int \dfrac{x \, dx}{(x^2 - a^2)^{\frac{1}{2}}}.$

175. $\int \dfrac{dx}{(x^2 - a^2)^{\frac{3}{2}}} = -\dfrac{x}{a^2(x^2 - a^2)^{\frac{1}{2}}}.$

176. $\int \dfrac{x \, dx}{(x^2 - a^2)^{\frac{3}{2}}} = \dfrac{-1}{(x^2 - a^2)^{\frac{1}{2}}}.$

177. $\int \dfrac{x^2 dx}{(x^2 - a^2)^{\frac{3}{2}}} = \int \dfrac{dx}{(x^2 - a^2)^{\frac{1}{2}}} + a^2 \int \dfrac{dx}{(x^2 - a^2)^{\frac{3}{2}}}.$

178. $\int \dfrac{(x^2 - a^2)^{\frac{1}{2}} \, dx}{x} = (x^2 - a^2)^{\frac{1}{2}} - a \arccos \dfrac{a}{x}.$

179. $\displaystyle\int \frac{(x^2 - a^2)^{\frac{1}{2}}\,dx}{x^2} = -\frac{(x^2 - a^2)^{\frac{1}{2}}}{x} + \log\,[x + (x^2 - a^2)^{\frac{1}{2}}].$

180. $\displaystyle\int \frac{dx}{x(x^2 - a^2)^{\frac{1}{2}}} = \frac{1}{a}\,\text{arc sec}\,\frac{x}{a}.$

181. $\displaystyle\int \frac{dx}{x^2(x^2 - a^2)^{\frac{1}{2}}} = \frac{(x^2 - a^2)^{\frac{1}{2}}}{a^2 x}.$

182. $\displaystyle\int \frac{dx}{(x^2 - a^2)^{n+\frac{1}{2}}} = \frac{-x}{a^2(2\,n-1)(x^2 - a^2)^{n-\frac{1}{2}}} - \frac{n-1}{a^2(2\,n-1)}\int \frac{dx}{(x^2 - a^2)^{n-\frac{1}{2}}}.$

183. $\displaystyle\int \frac{x\,dx}{(x^2 - a^2)^{n+\frac{1}{2}}} = \frac{-1}{(2\,n-1)(x^2 - a^2)^{n-\frac{1}{2}}}.$

184. $\displaystyle\int \frac{x^2\,dx}{(x^2 - a^2)^{n+\frac{1}{2}}} = \frac{-x}{(2\,n-1)(x^2 - a^2)^{n-\frac{1}{2}}} + \frac{1}{2\,n-1}\int \frac{dx}{(x^2 - a^2)^{n-\frac{1}{2}}}.$

185. $\displaystyle\int \frac{dx}{x(x^2 - a^2)^{n+\frac{1}{2}}} = \frac{-1}{(2\,n-1)a^2(x^2 - a^2)^{n-\frac{1}{2}}} - \frac{1}{a^2}\int \frac{dx}{x(x^2 - a^2)^{n-\frac{1}{2}}}.$

186. $\displaystyle\int \frac{x^m\,dx}{(x^2 - a^2)^{n+\frac{1}{2}}} = \int \frac{x^{m-2}\,dx}{(x^2 - a^2)^{n-\frac{1}{2}}} + a^2\int \frac{x^{m-2}\,dx}{(x^2 - a^2)^{n+\frac{1}{2}}}.$

187. $\displaystyle\int x^m(x^2 - a^2)^{n-\frac{1}{2}}\,dx$
$$= \frac{x^{m-1}(x^2 - a^2)^{n+\frac{1}{2}}}{2\,n+m} + \frac{(m-1)a^2}{2\,n+m}\int x^{m-2}(x^2 - a^2)^{n-\frac{1}{2}}\,dx.$$

188. $\displaystyle\int \frac{dx}{x^m(x^2 - a^2)^{n+\frac{1}{2}}}$
$$= \frac{1}{(m-1)a^2 x^{m-1}(x^2 - a^2)^{n-\frac{1}{2}}} + \frac{2\,n+m-2}{(m-1)a^2}\int \frac{dx}{x^{m-2}(x^2 - a^2)^{n+\frac{1}{2}}}.$$

189. $\displaystyle\int \frac{(x^2 - a^2)^{n+\frac{1}{2}}\,dx}{x} = \frac{(x^2 - a^2)^{n+\frac{1}{2}}}{2\,n+1} - a^2\int \frac{(x^2 - a^2)^{n-\frac{1}{2}}\,dx}{x}.$

190. $\displaystyle\int \frac{(x^2 - a^2)^{n+\frac{1}{2}}\,dx}{x^m} = -\frac{(x^2 - a^2)^{n+\frac{1}{2}}}{(m-1)x^{m-1}} + \frac{2\,n+1}{m-1}\int \frac{(x^2 - a^2)^{n-\frac{1}{2}}\,dx}{x^{m-2}}.$

EXPRESSIONS CONTAINING $(2\,ax \pm x^2)^{\frac{1}{2}}$

191. $\displaystyle\int (2\,ax - x^2)^{\frac{1}{2}}\,dx = \frac{1}{2}(x - a)(2\,ax - x^2)^{\frac{1}{2}} + \frac{1}{2}\,a^2\,\text{arc vers}\,\frac{x}{a}$
$$\text{or } \frac{1}{2}(x - a)(2\,ax - x^2)^{\frac{1}{2}} + \frac{1}{2}\,a^2\,\text{arc sin}\,\frac{x-a}{a}.$$

192. $\displaystyle\int \frac{dx}{(2\,ax - x^2)^{\frac{1}{2}}} = \text{arc vers}\,\frac{x}{a}\ \text{or arc sin}\,\frac{x-a}{a}.$

193. $\displaystyle\int (2\,ax + x^2)^{\frac{1}{2}}\,dx = \frac{1}{2}(x + a)(2\,ax + x^2)^{\frac{1}{2}} - \frac{1}{2}\,a^2\,\log\,[x + a + (2\,ax + x^2)^{\frac{1}{2}}].$

194. $\int \dfrac{dx}{(2\,ax+x^2)^{\frac{1}{2}}} = \log\,[x+a+(2\,ax+x^2)^{\frac{1}{2}}].$

195. $\int F[x,\,(2\,ax-x^2)^{\frac{1}{2}}]dx = \int F[z+a,\,(a^2-z^2)^{\frac{1}{2}}]dz$ if $z=x-a.$

196. $\int F[x,\,(2\,ax+x^2)^{\frac{1}{2}}]dx = \int F[z-a,\,(z^2-a^2)^{\frac{1}{2}}]dz$ if $z=x+a.$

197. $\int x^n(2\,ax-x^2)^{\frac{1}{2}}\,dx$
$$=-\frac{x^{n-1}(2\,ax-x^2)^{\frac{3}{2}}}{n+2}+\frac{(2\,n+1)a}{n+2}\int x^{n-1}(2\,ax-x^2)^{\frac{1}{2}}\,dx.$$

198. $\int \dfrac{(2\,ax-x^2)^{\frac{1}{2}}\,dx}{x^n} = -\dfrac{(2\,ax-x^2)^{\frac{3}{2}}}{(2\,n-3)ax^n}+\dfrac{n-3}{(2\,n-3)a}\int\dfrac{(2\,ax-x^2)^{\frac{1}{2}}\,dx}{x^{n-1}}.$

199. $\int \dfrac{x^n dx}{(2\,ax-x^2)^{\frac{1}{2}}} = -\dfrac{x^{n-1}(2\,ax-x^2)^{\frac{1}{2}}}{n}+\dfrac{(2\,n-1)a}{n}\int\dfrac{x^{n-1}\,dx}{(2\,ax-x^2)^{\frac{1}{2}}}.$

200. $\int \dfrac{dx}{x^n(2\,ax-x^2)^{\frac{1}{2}}} = -\dfrac{(2\,ax-x^2)^{\frac{1}{2}}}{(2\,n-1)ax^n}+\dfrac{n-1}{(2\,n-1)a}\int\dfrac{dx}{x^{n-1}(2\,ax-x^2)^{\frac{1}{2}}}.$

201. $\int \dfrac{dx}{(2\,ax-x^2)^{\frac{3}{2}}} = \dfrac{x-a}{a^2(2\,ax-x^2)^{\frac{1}{2}}}.$

202. $\int \dfrac{x\,dx}{(2\,ax-x^2)^{\frac{3}{2}}} = \dfrac{x}{a(2\,ax-x^2)^{\frac{1}{2}}}.$

MISCELLANEOUS ALGEBRAIC EXPRESSIONS

203. $\int \dfrac{dx}{(a+bx)^{\frac{1}{2}}(\alpha+\beta x)^{\frac{1}{2}}} = \dfrac{2}{(-b\beta)^{\frac{1}{2}}}\,\text{arc tan}\Big(\dfrac{-\beta(a+bx)}{b(\alpha+\beta x)}\Big)^{\frac{1}{2}}$

or $\qquad\qquad \dfrac{2}{(b\beta)^{\frac{1}{2}}}\tanh^{-1}\Big(\dfrac{\beta(a+bx)}{b(\alpha+\beta x)}\Big).$

204. $\int (a+bx)^{\frac{1}{2}}(\alpha+\beta x)^{\frac{1}{2}}\,dx$
$$=\frac{(2\,b\beta x+a\beta+b\alpha)(a+bx)^{\frac{1}{2}}(\alpha+\beta x)^{\frac{1}{2}}}{4\,b\beta}$$
$$-\frac{(a\beta-\alpha b)^2}{8\,b\beta}\int\frac{dx}{(a+bx)^{\frac{1}{2}}(\alpha+\beta x)^{\frac{1}{2}}}.$$

205. $\int \Big(\dfrac{\alpha+\beta x}{a+bx}\Big)^{\frac{1}{2}}dx = \dfrac{(a+bx)^{\frac{1}{2}}(\alpha+\beta x)^{\frac{1}{2}}}{b}-\dfrac{a\beta-\alpha b}{2\,b}\int\dfrac{dx}{(a+bx)^{\frac{1}{2}}(\alpha+\beta x)^{\frac{1}{2}}}.$

206. $\int \Big(\dfrac{\alpha+x}{a+x}\Big)^{\frac{1}{2}}dx = (a+x)^{\frac{1}{2}}(\alpha+x)^{\frac{1}{2}}+(\alpha-a)\log\,[(a+x)^{\frac{1}{2}}+(\alpha+x)^{\frac{1}{2}}].$

207. $\int \Big(\dfrac{1+x}{1-x}\Big)^{\frac{1}{2}}dx = \text{arc sin }x-(1-x^2)^{\frac{1}{2}}.$

208. $\int \Big(\dfrac{a-x}{b+x}\Big)^{\frac{1}{2}}dx = (a-x)^{\frac{1}{2}}(b+x)^{\frac{1}{2}}+(a+b)\,\text{arc sin}\,\Big(\dfrac{x+b}{a+b}\Big)^{\frac{1}{2}}.$

209. $\int \left(\dfrac{a+x}{b-x}\right)^{\frac{1}{2}} dx = -(a+x)^{\frac{1}{2}}(b-x)^{\frac{1}{2}} + (a+b) \arcsin \left(\dfrac{b-x}{a+b}\right)^{\frac{1}{2}}.$

210. $\int \dfrac{dx}{(x-a)^{\frac{1}{2}}(b-x)^{\frac{1}{2}}} = 2 \arcsin \left(\dfrac{x-a}{b-a}\right)^{\frac{1}{2}}.$

211. $\int F\left[x, \left(\dfrac{a+bx}{\alpha+\beta x}\right)^{\frac{1}{2}}\right] dx = n(\alpha b - a\beta) \int F\left[\dfrac{a-\alpha z^n}{\beta z^n - b}, z\right] \dfrac{z^{n-1}\,dz}{(\beta z^n - b)^2}$

if $\qquad\qquad\qquad z^n(\alpha + \beta x) = a + bx.$

212. If $xz + a^{\frac{1}{2}} = (a + bx + cx^2)^{\frac{1}{2}}$, then

$$\int F[x, (a+bx+cx^2)^{\frac{1}{2}}]\,dx = 2\int F\left[\dfrac{2a^{\frac{1}{2}}z - b}{1-z^2}, \dfrac{a^{\frac{1}{2}}z^2 - bz + a^{\frac{1}{2}}}{1-z^2}\right] \dfrac{a^{\frac{1}{2}}z^2 - bz + a^{\frac{1}{2}}}{(1-z^2)^2}\,dz.$$

[See also 43 and 44.]

EXPRESSIONS CONTAINING $(ax^2 + bx + c)^{\frac{1}{2}}$

For brevity we use the symbols X and q, where

$$X = ax^2 + bx + c, \quad q = b^2 - 4ac.$$

For methods of rationalizing $\int R[x, (ax^2 + bx + c)^{\frac{1}{2}}]\,dx$, where R is a rational function of its two arguments, see numbers 43, 44, and 212. By rationalization or by means of integration by parts the following formulas may be established.

213. $\int \dfrac{dx}{X^{\frac{1}{2}}} = \dfrac{1}{a^{\frac{1}{2}}} \log\left(X^{\frac{1}{2}} + a^{\frac{1}{2}}x + \dfrac{b}{2a^{\frac{1}{2}}}\right)$ if $a > 0$

$\qquad = \dfrac{1}{(-a)^{\frac{1}{2}}} \arcsin \dfrac{2ax - b}{q^{\frac{1}{2}}}$ if $a < 0.$

214. $\int X^{\frac{1}{2}}\,dx = \dfrac{(2ax+b)X^{\frac{1}{2}}}{4a} - \dfrac{q}{8a}\int \dfrac{dx}{X^{\frac{1}{2}}}.$

215. $\int xX^{\frac{1}{2}}\,dx = \dfrac{X^{\frac{3}{2}}}{3a} - \dfrac{b}{2a}\int X^{\frac{1}{2}}\,dx.$

216. $\int \dfrac{x\,dx}{X^{\frac{1}{2}}} = \dfrac{X^{\frac{1}{2}}}{a} - \dfrac{b}{2a}\int \dfrac{dx}{X^{\frac{1}{2}}}.$

217. $\int x^2 X^{\frac{1}{2}}\,dx = \left(x - \dfrac{5b}{6a}\right)\dfrac{X^{\frac{3}{2}}}{4a} + \dfrac{5b^2 - 4ac}{16a^2}\int X^{\frac{1}{2}}\,dx.$

218. $\int \dfrac{x^2\,dx}{X^{\frac{1}{2}}} = \left(\dfrac{x}{2a} - \dfrac{3b}{4a^2}\right)X^{\frac{1}{2}} + \dfrac{3b^2 - 4ac}{8a^2}\int \dfrac{dx}{X^{\frac{1}{2}}}.$

219. $\int x^3 X^{\frac{1}{2}}\,dx = \left(x^2 - \dfrac{7bx}{8a} + \dfrac{35b^2}{48a^2} - \dfrac{2c}{3a}\right)\dfrac{X^{\frac{3}{2}}}{5a} + \left(\dfrac{3bc}{8a^2} - \dfrac{7b^3}{32a^3}\right)\int X^{\frac{1}{2}}\,dx.$

220. $\int \dfrac{x^3\,dx}{X^{\frac{1}{2}}} = \left(\dfrac{x^2}{3a} - \dfrac{5bx}{12a^2} + \dfrac{5b^2}{8a^3} - \dfrac{2c}{3a^2}\right)X^{\frac{1}{2}} + \left(\dfrac{3bc}{4a^2} - \dfrac{5b^3}{16a^3}\right)\int \dfrac{dx}{X^{\frac{1}{2}}}.$

221. $\int \dfrac{dx}{x\,X^{\frac{1}{2}}} = -\dfrac{1}{c^{\frac{1}{2}}} \log\left(\dfrac{X^{\frac{1}{2}}+c^{\frac{1}{2}}}{x}+\dfrac{b}{2\,c^{\frac{1}{2}}}\right)$ if $c > 0$,

$\qquad = \dfrac{1}{(-c)^{\frac{1}{2}}} \text{ arc sin}\left(\dfrac{bx+2\,c}{x\,q^{\frac{1}{2}}}\right)$ if $c < 0$,

$\qquad = -\dfrac{2\,X^{\frac{1}{2}}}{bx}$ if $c = 0$.

222. $\int \dfrac{X^{\frac{1}{2}}\,dx}{x} = X^{\frac{1}{2}} + \dfrac{b}{2}\int \dfrac{dx}{X^{\frac{1}{2}}} + c\int \dfrac{dx}{x\,X^{\frac{1}{2}}}\cdot$

223. $\int \dfrac{dx}{x^2\,X^{\frac{1}{2}}} = -\dfrac{X^{\frac{1}{2}}}{cx} - \dfrac{b}{2\,c}\int \dfrac{dx}{x\,X^{\frac{1}{2}}}\cdot$

224. $\int \dfrac{X^{\frac{1}{2}}\,dx}{x^2} = -\dfrac{X^{\frac{1}{2}}}{x} + \dfrac{b}{2}\int \dfrac{dx}{x\,X^{\frac{1}{2}}} + a\int \dfrac{dx}{X^{\frac{1}{2}}}\cdot$

225. $\int \dfrac{dx}{X^{\frac{3}{2}}} = -\dfrac{2(2\,ax+b)}{q\,X^{\frac{1}{2}}}\cdot$

226. $\int X^{\frac{3}{2}}\,dx = \dfrac{(2\,ax-b)\,X^{\frac{1}{2}}}{8\,a}\left(X - \dfrac{3\,q}{8\,a}\right) + \dfrac{3\,q^2}{128\,a^2}\int \dfrac{dx}{X^{\frac{1}{2}}}\cdot$

227. $\int x\,X^{\frac{3}{2}}\,dx = \dfrac{X^{\frac{5}{2}}}{5\,a} - \dfrac{b}{2\,a}\int X^{\frac{3}{2}}\,dx.$

228. $\int \dfrac{x\,dx}{X^{\frac{3}{2}}} = \dfrac{2(bx+2\,c)}{q\,X^{\frac{1}{2}}}\cdot$

229. $\int x^2\,X^{\frac{3}{2}}\,dx = \dfrac{x\,X^{\frac{5}{2}}}{6\,a} - \dfrac{7\,b}{12\,a}\int x\,X^{\frac{3}{2}}\,dx - \dfrac{c}{6\,a}\int X^{\frac{3}{2}}\,dx.$

230. $\int \dfrac{x^2\,dx}{X^{\frac{3}{2}}} = -\dfrac{(2\,b^2-4\,ac)x+2\,bc}{aq\,X^{\frac{1}{2}}} + \dfrac{1}{a}\int \dfrac{dx}{X^{\frac{1}{2}}}\cdot$

231. $\int \dfrac{dx}{x\,X^{\frac{3}{2}}} = \dfrac{1}{c\,X^{\frac{1}{2}}} + \dfrac{1}{c}\int \dfrac{dx}{x\,X^{\frac{1}{2}}} - \dfrac{b}{2\,c}\int \dfrac{dx}{X^{\frac{3}{2}}}\cdot$

232. $\int \dfrac{X^{\frac{3}{2}}\,dx}{x} = \dfrac{X^{\frac{3}{2}}}{3} + c\int \dfrac{X^{\frac{1}{2}}\,dx}{x} + \dfrac{b}{2}\int X^{\frac{1}{2}}\,dx.$

233. $\int \dfrac{dx}{x^2\,X^{\frac{3}{2}}} = \dfrac{-1}{cx\,X^{\frac{1}{2}}} - \dfrac{3\,b}{2\,c}\int \dfrac{dx}{x\,X^{\frac{3}{2}}} - \dfrac{2\,a}{c}\int \dfrac{dx}{X^{\frac{3}{2}}}\cdot$

234. $\int \dfrac{X^{\frac{3}{2}}\,dx}{x^2} = -\dfrac{X^{\frac{3}{2}}}{x} + \dfrac{3\,b}{2}\int \dfrac{X^{\frac{1}{2}}\,dx}{x} + 3\,a\int X^{\frac{1}{2}}\,dx.$

235. (a) $\int \dfrac{dx}{X^{n+\frac{1}{2}}} = -\dfrac{2(2\,ax+b)}{(2\,n-1)q\,X^{n-\frac{1}{2}}} - \dfrac{8\,a(n-1)}{(2\,n-1)q}\int \dfrac{dx}{X^{n-\frac{1}{2}}}\cdot$

(b) $\int X^{n+\frac{1}{2}}\,dx = \dfrac{(2\,ax+b)\,X^{n+\frac{1}{2}}}{4\,a(n+1)} - \dfrac{(2\,n+1)q}{8\,a(n+1)}\int X^{n-\frac{1}{2}}\,dx.$

236. $\int x\,X^{n+\frac{1}{2}}\,dx = \dfrac{X^{n+\frac{3}{2}}}{(2n+3)a} - \dfrac{b}{2\,a}\int X^{n+\frac{1}{2}}\,dx.$

237. $\int \dfrac{x\,dx}{X^{n+\frac{1}{2}}} = \dfrac{-1}{(2\,n-1)a\,X^{n-\frac{1}{2}}} - \dfrac{b}{2\,a}\int \dfrac{dx}{X^{n+\frac{1}{2}}}.$

238. $\int x^2\,X^{n+\frac{1}{2}}\,dx$

$$= \dfrac{x\,X^{n+\frac{3}{2}}}{2\,a(n+2)} - \dfrac{(2\,n+5)b}{4\,a(n+2)}\int x\,X^{n+\frac{1}{2}}\,dx - \dfrac{c}{2\,a(n+2)}\int X^{n+\frac{1}{2}}\,dx.$$

239. $\int \dfrac{x^2\,dx}{X^{n+\frac{1}{2}}} = -\dfrac{(2\,b^2-4\,ac)x + 2\,ab}{(2\,n-1)aq\,X^{n-\frac{1}{2}}} - \dfrac{4\,ac+(2\,n-3)b^2}{(2\,n-1)aq}\int \dfrac{dx}{X^{n-\frac{1}{2}}}.$

240. $\int \dfrac{dx}{x\,X^{n+\frac{1}{2}}} = \dfrac{1}{(2\,n-1)c\,X^{n-\frac{1}{2}}} + \dfrac{1}{c}\int \dfrac{dx}{x\,X^{n-\frac{1}{2}}} - \dfrac{b}{2\,c}\int \dfrac{dx}{X^{n+\frac{1}{2}}}.$

241. $\int \dfrac{X^{n+\frac{1}{2}}\,dx}{x} = \dfrac{X^{n+\frac{1}{2}}}{2\,n+1} + c\int \dfrac{X^{n-\frac{1}{2}}\,dx}{x} + \dfrac{b}{2}\int X^{n-\frac{1}{2}}\,dx.$

242. $\int \dfrac{dx}{x^2\,X^{n+\frac{1}{2}}} = \dfrac{-1}{cx\,X^{n-\frac{1}{2}}} - \dfrac{(2\,n-1)b}{2\,c}\int \dfrac{dx}{x\,X^{n+\frac{1}{2}}} - \dfrac{2\,an}{c}\int \dfrac{dx}{X^{n+\frac{1}{2}}}.$

243. $\int \dfrac{X^{n+\frac{1}{2}}\,dx}{x^2} = -\dfrac{X^{n+\frac{1}{2}}}{x} + \dfrac{(2\,n+1)b}{2}\int \dfrac{X^{n-\frac{1}{2}}\,dx}{x} + (2\,n+1)a\int X^{n-\frac{1}{2}}\,dx.$

244. $\int \dfrac{x^m\,dx}{X^{n+\frac{1}{2}}} = \dfrac{1}{a}\int \dfrac{x^{m-2}\,dx}{X^{n-\frac{1}{2}}} - \dfrac{b}{a}\int \dfrac{x^{m-1}\,dx}{X^{n+\frac{1}{2}}} - \dfrac{c}{a}\int \dfrac{x^{m-2}\,dx}{X^{n+\frac{1}{2}}}.$

245. $\int x^m\,X^{n+\frac{1}{2}}\,dx = \dfrac{x^{m-1}\,X^{n+\frac{3}{2}}}{a(2\,n+m+2)} - \dfrac{(2\,n+2\,m+1)b}{2\,a(2\,n+m+2)}\int x^{m-1}\,X^{n+\frac{1}{2}}\,dx.$

246. $\int \dfrac{dx}{x^m\,X^{n+\frac{1}{2}}} = \dfrac{-1}{(m-1)cx^{m-1}\,X^{n-\frac{1}{2}}} - \dfrac{(2\,n+2\,m-3)b}{2\,c(m-1)}\int \dfrac{dx}{x^{m-1}\,X^{n+\frac{1}{2}}}$

$$- \dfrac{(2\,n+m-2)a}{(m-1)c}\int \dfrac{dx}{x^{m-2}\,X^{n+\frac{1}{2}}}.$$

247. $\int \dfrac{X^{n+\frac{1}{2}}\,dx}{x^m}$

$$= -\dfrac{X^{n+\frac{1}{2}}}{(m-1)x^{m-1}} + \dfrac{(2\,n+1)b}{2(m-1)}\int \dfrac{X^{n-\frac{1}{2}}\,dx}{x^{m-1}} + \dfrac{(2\,n+1)a}{m-1}\int \dfrac{X^{n-\frac{1}{2}}\,dx}{x^{m-2}}.$$

248. $\int \dfrac{dx}{(\alpha+\beta x)X^{\frac{1}{2}}} = \dfrac{1}{(-h)^{\frac{1}{2}}}\text{ arc tan }\dfrac{2\,h+m(\alpha+\beta x)}{2\,\beta(-hX)^{\frac{1}{2}}}$

$$\text{or }\dfrac{1}{h^{\frac{1}{2}}}\log \dfrac{2\,h+m(\alpha+\beta x) - 2\,\beta(hX)^{\frac{1}{2}}}{\alpha+\beta x},$$

where

$$h = c\beta^2 - \alpha\beta b + a\alpha^2 \quad\text{and}\quad m = b\beta - 2\,a\alpha.$$

If $h=0$ the value of the integral is

$$\dfrac{-2\,\beta X^{\frac{1}{2}}}{m(\alpha+\beta x)}.$$

TRIGONOMETRIC EXPRESSIONS

(Not involving exponential functions)

249. $\int \sin x \, dx = - \cos x, \qquad \int \cos x \, dx = \sin x.$

250. $\int \sin^2 x \, dx = \frac{1}{2} x - \frac{1}{2} \sin x \cos x = \frac{1}{2} x - \frac{1}{4} \sin 2x.$

251. $\int \cos^2 x \, dx = \frac{1}{2} x + \frac{1}{2} \sin x \cos x = \frac{1}{2} x + \frac{1}{4} \sin 2x.$

252. $\int \sin^3 x \, dx = - \frac{1}{3} (\sin^2 x + 2) \cos x.$

253. $\int \cos^3 x \, dx = \frac{1}{3} (\cos^2 x + 2) \sin x.$

254. $\int \sin^n x \, dx = - \frac{\sin^{n-1} x \cos x}{n} + \frac{n-1}{n} \int \sin^{n-2} x \, dx.$

255. $\int \cos^n x \, dx = \frac{\cos^{n-1} x \sin x}{n} + \frac{n-1}{n} \int \cos^{n-2} x \, dx.$

256. $\int \sin^m x \cos x \, dx = \frac{\sin^{m+1} x}{m+1}.$

257. $\int \sin x \cos^m x \, dx = - \frac{\cos^{m+1} x}{m+1}.$

258. $\int \sin^2 x \cos^2 x \, dx = - \frac{1}{8} (\frac{1}{4} \sin 4x - x).$

259. $\int \cos^m x \sin^n x \, dx = \frac{\cos^{m-1} x \sin^{n+1} x}{m+n} + \frac{m-1}{m+n} \int \cos^{m-2} x \sin^n x \, dx.$

260. $\int \cos^m x \sin^n x \, dx = - \frac{\sin^{n-1} x \cos^{m+1} x}{m+n} + \frac{n-1}{m+n} \int \cos^m x \sin^{n-2} x \, dx.$

261. $\int \frac{dx}{\sin^m x} = - \frac{1}{m-1} \frac{\cos x}{\sin^{m-1} x} + \frac{m-2}{m-1} \int \frac{dx}{\sin^{m-2} x}.$

262. $\int \frac{dx}{\cos^m x} = \frac{1}{m-1} \frac{\sin x}{\cos^{m-1} x} + \frac{m-2}{m-1} \int \frac{dx}{\cos^{m-2} x}.$

263. $\int \frac{dx}{\sin x \cos x} = \log \tan x.$

264. $\int \frac{dx}{\sin^m x \cos^n x} = \frac{1}{n-1} \frac{1}{\sin^{m-1} x \cos^{n-1} x} + \frac{m+n-2}{n-1} \int \frac{dx}{\sin^m x \cos^{n-2} x}.$

265. $\int \frac{dx}{\sin^m x \cos^n x} = - \frac{1}{m-1} \frac{1}{\sin^{m-1} x \cos^{n-1} x} + \frac{m+n-2}{m-1} \int \frac{dx}{\sin^{m-2} x \cos^n x}.$

266. $\displaystyle\int \frac{\cos^m x \, dx}{\sin^n x} = -\frac{\cos^{m+1} x}{(n-1)\sin^{n-1} x} - \frac{m-n+2}{n-1} \int \frac{\cos^m x \, dx}{\sin^{n-2} x}.$

267. $\displaystyle\int \frac{\cos^m x \, dx}{\sin^n x} = \frac{\cos^{m-1} x}{(m-n)\sin^{n-1} x} + \frac{m-1}{m-n} \int \frac{\cos^{m-2} x \, dx}{\sin^n x}.$

268. $\displaystyle\int \frac{\sin^n x \, dx}{\cos^m x} = -\int \frac{\cos^n(\frac{1}{2}\pi - x) d(\frac{1}{2}\pi - x)}{\sin^m(\frac{1}{2}\pi - x)}.$

269. $\displaystyle\int \tan x \, dx = -\log\cos x, \quad \int \cot x \, dx = \log\sin x.$

270. $\displaystyle\int \tan^2 x \, dx = \tan x - x, \quad \int \cot^2 x \, dx = -\cot x - x.$

271. $\displaystyle\int \tan^n x \, dx = \frac{\tan^{n-1} x}{n-1} - \int \tan^{n-2} x \, dx.$

272. $\displaystyle\int \cot^n x \, dx = -\frac{\cot^{n-1} x}{n-1} - \int \cot^{n-2} x \, dx.$

273. $\displaystyle\int \sec x \, dx = \log(\sec x + \tan x) \text{ or } \log\tan(\tfrac{1}{4}\pi + \tfrac{1}{2}x).$

274. $\displaystyle\int \csc x \, dx = \log(\csc x - \cot x) \text{ or } \log\tan\tfrac{1}{2}x.$

275. $\displaystyle\int \sec^2 x \, dx = \tan x, \quad \int \csc^2 x \, dx = -\cot x.$

276. $\displaystyle\int \sec^n x \, dx = \int \frac{dx}{\cos^n x}, \quad \int \csc^n x \, dx = \int \frac{dx}{\sin^n x}.$

277. $\displaystyle\int \frac{dx}{a + b\cos x} = \frac{-1}{(a^2 - b^2)^{\frac{1}{2}}} \arcsin\frac{b + a\cos x}{a + b\cos x} \text{ if } a > b > 0,$

$\displaystyle\qquad\qquad = \frac{1}{(b^2 - a^2)^{\frac{1}{2}}} \log\frac{b + a\cos x + (b^2 - a^2)^{\frac{1}{2}}\sin x}{a + b\cos x} \text{ if } a > 0, b^2 > a^2,$

$\displaystyle\qquad\qquad = \frac{1}{(a^2 - b^2)^{\frac{1}{2}}} \arctan\frac{(a^2 - b^2)^{\frac{1}{2}}\sin x}{b + a\cos x} \text{ if } a > b > 0.$

278. $\displaystyle\int \frac{dx}{a + b\sin x} = -\int \frac{d(\frac{1}{2}\pi - x)}{a + b\cos(\frac{1}{2}\pi - x)}.$

279. $\displaystyle\int \frac{dx}{a + b\cos x + c\sin x}$

$\displaystyle\qquad = \frac{-1}{(a^2 - b^2 - c^2)^{\frac{1}{2}}} \arcsin\frac{b^2 + c^2 + a(b\cos x + c\sin x)}{(b^2 + c^2)^{\frac{1}{2}}(a + b\cos x + c\sin x)}.$

$\displaystyle\qquad \text{or } \frac{1}{(b^2 + c^2 - a^2)^{\frac{1}{2}}} \times$

$\displaystyle\qquad \log\frac{b^2 + c^2 + a(b\cos x + c\sin x) + (b^2 + c^2 - a^2)^{\frac{1}{2}}(b\sin x - c\cos x)}{(b^2 + c^2)^{\frac{1}{2}}(a + b\cos x + c\sin x)}.$

280. $\int \dfrac{dx}{a^2 \cos^2 x + b^2 \sin^2 x} = \dfrac{1}{ab} \text{ arc tan } \dfrac{b \tan x}{a}.$

281. $\int x \sin x \, dx = \sin x - x \cos x.$

282. $\int x \cos x \, dx = \cos x + x \sin x.$

283. $\int x^2 \sin x \, dx = 2 \, x \sin x - (x^2 - 2) \cos x.$

284. $\int x^2 \cos x \, dx = 2 \, x \cos x + (x^2 - 2) \sin x.$

285. $\int x^m \sin x \, dx = - \, x^m \cos x + m \int x^{m-1} \cos x \, dx.$

286. $\int x^m \cos x \, dx = x^m \sin x - m \int x^{m-1} \sin x \, dx.$

287. $\int \dfrac{\sin x \, dx}{x} = x - \dfrac{x^3}{3 \cdot 3!} + \dfrac{x^5}{5 \cdot 5!} - \dfrac{x^7}{7 \cdot 7!} + \cdots.$

288. $\int \dfrac{\cos x \, dx}{x} = \log x - \dfrac{x^2}{2 \cdot 2!} + \dfrac{x^4}{4 \cdot 4!} - \dfrac{x^6}{6 \cdot 6!} + \cdots.$

289. $\int \dfrac{\sin x \, dx}{x^m} = - \dfrac{1}{m-1} \dfrac{\sin x}{x^{m-1}} + \dfrac{1}{m-1} \int \dfrac{\cos x \, dx}{x^{m-1}}.$

290. $\int \dfrac{\cos x \, dx}{x^m} = - \dfrac{1}{m-1} \dfrac{\cos x}{x^{m-1}} - \dfrac{1}{m-1} \int \dfrac{\sin x \, dx}{x^{m-1}}.$

291. $\int \text{arc } \sin x \, dx = x \text{ arc } \sin x + (1 - x^2)^{\frac{1}{2}}.$

292. $\int \text{arc } \cos x \, dx = x \text{ arc } \cos x - (1 - x^2)^{\frac{1}{2}}.$

293. $\int (\text{arc } \sin x)^2 \, dx = x(\text{arc } \sin x)^2 - 2 \, x + 2(1 - x^2)^{\frac{1}{2}} \text{ arc } \sin x.$

294. $\int (\text{arc } \cos x)^2 \, dx = x(\text{arc } \cos x)^2 - 2 \, x - 2(1 - x^2)^{\frac{1}{2}} \text{ arc } \cos x.$

295. $\int \text{arc } \tan x \, dx = x \text{ arc } \tan x - \frac{1}{2} \log (x^2 + 1).$

296. $\int \text{arc } \cot x \, dx = x \text{ arc } \cot x + \frac{1}{2} \log (x^2 + 1).$

297. $\int x \text{ arc } \sin x \, dx = \frac{1}{4}[(2 \, x^2 - 1) \text{ arc } \sin x + x(1 - x^2)^{\frac{1}{2}}].$

298. $\int x \arccos x \, dx = \frac{1}{4}[(2\,x^2 - 1) \arccos x - x(1 - x^2)^{\frac{1}{2}}].$

299. $\int x \arctan x \, dx = \frac{1}{2}(x^2 + 1) \arctan x - \frac{1}{2}\,x.$

300. $\int x^n \arcsin x \, dx = \dfrac{x^{n+1} \arcsin x}{n + 1} - \dfrac{1}{n + 1} \int \dfrac{x^{n+1}\, dx}{(1 - x^2)^{\frac{1}{2}}}.$

301. $\int x^n \arccos x \, dx = \dfrac{x^{n+1} \arccos x}{n + 1} + \dfrac{1}{n + 1} \int \dfrac{x^{n+1}\, dx}{(1 - x^2)^{\frac{1}{2}}}.$

302. $\int x^n \arctan x \, dx = \dfrac{x^{n+1} \arctan x}{n + 1} - \dfrac{1}{n + 1} \int \dfrac{x^{n+1}\, dx}{1 + x^2}.$

303. $\int x^m \sin ax \, dx = \dfrac{x^{m-1}}{a^2}(-ax\cos ax + m\sin ax) - \dfrac{m(m-1)}{a^2} \int x^{m-2} \sin ax \, dx.$

304. $\int x^m \cos ax \, dx = \dfrac{x^{m-1}}{a^2}(ax\sin ax + m\cos ax) - \dfrac{m(m-1)}{a^2} \int x^{m-2} \cos ax \, dx.$

EXPRESSIONS CONTAINING EXPONENTIAL FUNCTIONS

305. $\int e^{ax} \, dx = \dfrac{e^{ax}}{a}, \quad \int a^x \, dx = \dfrac{a^x}{\log a}.$

306. $\int x e^{ax} \, dx = \dfrac{e^{ax}}{a^2}(ax - 1).$

307. $\int x^m e^{ax} \, dx = \dfrac{x^m e^{ax}}{a} - \dfrac{m}{a} \int x^{m-1}\, e^{ax} \, dx.$

308. $\int \dfrac{e^{ax}\, dx}{x^m} = -\dfrac{1}{m - 1} \dfrac{e^{ax}}{x^{m-1}} + \dfrac{a}{m - 1} \int \dfrac{e^{ax}\, dx}{x^{m-1}}.$

309. $\int e^x \sin x \, dx = \frac{1}{2}\, e^x(\sin x - \cos x).$

310. $\int e^x \cos x \, dx = \frac{1}{2}\, e^x(\sin x + \cos x).$

311. $\int e^{ax} \sin bx \, dx = \dfrac{e^{ax}(a \sin bx - b \cos bx)}{a^2 + b^2}.$

312. $\int e^{ax} \cos bx \, dx = \dfrac{e^{ax}(b \sin bx + a \cos bx)}{a^2 + b^2}.$

313. $\int e^{ax} \cos^n x \, dx = \dfrac{e^{ax} \cos^{n-1} x(a\cos x + n\sin x)}{a^2 + n^2} + \dfrac{n(n-1)}{a^2 + n^2} \int e^{ax} \cos^{n-2} x \, dx.$

314. $\int e^{ax} \sin^n x \, dx = \dfrac{e^{ax} \sin^{n-1} x(a\sin x - n\cos x)}{a^2 + n^2} + \dfrac{n(n-1)}{a^2 + n^2} \int e^{ax} \sin^{n-2} x \, dx.$

315. $\int \dfrac{e^x}{x} \, dx = \log x + x + \dfrac{x^2}{2 \cdot 2!} + \dfrac{x^3}{3 \cdot 3!} + \dfrac{x^4}{4 \cdot 4!} + \cdots.$

EXPRESSIONS CONTAINING LOGARITHMS

316. $\int \log x \, dx = x \log x - x.$

317. $\int x^n \log x \, dx = x^{n+1} \left[\dfrac{\log x}{n+1} - \dfrac{1}{(n+1)^2} \right].$

318. $\int x^n (\log x)^m \, dx = \dfrac{x^{n+1}}{n+1} (\log x)^m - \dfrac{m}{n+1} \int x^n (\log x)^{m-1} \, dx.$

319. $\int \dfrac{(\log x)^n}{x} \, dx = \dfrac{1}{n+1} (\log x)^{n+1}.$

320. $\int \dfrac{dx}{x \log x} = \log (\log x).$

321. $\int \dfrac{dx}{x(\log x)^n} = - \dfrac{1}{(n-1)(\log x)^{n-1}}.$

322. $\int \dfrac{x^n dx}{(\log x)^m} = - \dfrac{x^{n+1}}{(m-1)(\log x)^{m-1}} + \dfrac{n+1}{m-1} \int \dfrac{x^n dx}{(\log x)^{m-1}}.$

323. $\int e^{ax} \log x \, dx = \dfrac{e^{ax} \log x}{a} - \dfrac{1}{a} \int \dfrac{e^{ax}}{x} \, dx.$

DEFINITE INTEGRALS

324. $\displaystyle\int_0^\infty \dfrac{dx}{a^2 + x^2} = \dfrac{\pi}{a}.$

325. $\displaystyle\int_0^\infty x^{n-1} e^{-x} \, dx = \Gamma(n)$, where

$\Gamma(n+1) = n \, \Gamma(n)$ if $n > 0,$
$\Gamma(n+1) = n!$ if n is a positive integer,
$\Gamma(2) = \Gamma(1) = 1,$
$\Gamma(\tfrac{1}{2}) = \pi^{\frac{1}{2}}.$

326. $\displaystyle\int_0^1 \left(\log \dfrac{1}{x} \right)^n dx = \Gamma(n+1).$

327. $\displaystyle\int_0^\infty e^{-zx} z^n x^{n-1} \, dx = \Gamma(n).$

328. $\displaystyle\int_0^1 x^{m-1}(1-x)^{n-1} \, dx = \int_0^\infty \dfrac{x^{m-1} \, dx}{(1+x)^{m+n}} = \dfrac{\Gamma(m)\Gamma(n)}{\Gamma(m+n)}.$

329. $\displaystyle\int_0^\infty \dfrac{x^{n-1}}{1+x} \, dx = \dfrac{\pi}{\sin n\pi}.$

330. $\displaystyle\int_0^\infty e^{-a^2x^2} \, dx = \dfrac{\pi^{\frac{1}{2}}}{2\,a}.$

331. $\displaystyle\int_0^\infty x^n e^{-ax} = \frac{\Gamma(n+1)}{a^{n+1}},$

$$= \frac{n!}{a^{n+1}} \text{ if } n \text{ is a positive integer.}$$

332. $\displaystyle\int_0^\infty x^{2n} e^{-ax^2} dx = \frac{1 \cdot 3 \cdot 5 \cdots (2n-1)}{2^{n+1} a^n} \left(\frac{\pi}{a}\right)^{\frac{1}{2}}.$

333. $\displaystyle\int_0^\infty e^{-x^2 - \frac{a^2}{x^2}} dx = \frac{1}{2} e^{-2a} \pi^{\frac{1}{2}}.$

334. $\displaystyle\int_0^{\frac{\pi}{2}} \sin^n x \, dx = \int_0^{\frac{\pi}{2}} \cos^n x \, dx,$

$$= \frac{1}{2} \pi^{\frac{1}{2}} \frac{\Gamma(\frac{1}{2} n + \frac{1}{2})}{\Gamma(\frac{1}{2} n + 1)} \text{ if } n > -1,$$

$$= \frac{2 \cdot 4 \cdot 6 \cdots (n-1)}{1 \cdot 3 \cdot 5 \cdots n} \text{ if } n \text{ is an odd integer,}$$

$$= \frac{1 \cdot 3 \cdot 5 \cdots (n-1)}{2 \cdot 4 \cdot 6 \cdots n} \frac{\pi}{2} \text{ if } n \text{ is an even integer.}$$

335. $\displaystyle\int_0^\infty \frac{\sin^2 x \, dx}{x^2} = \frac{\pi}{2}.$

336. $\displaystyle\int_0^\infty \cos(x^2) dx = \int_0^\infty \sin(x^2) dx = \frac{1}{2} (\frac{1}{2} \pi)^{\frac{1}{2}}.$

337. $\displaystyle\int_0^\infty \frac{\sin mx \, dx}{x} = \frac{\pi}{2} \text{ if } m > 0.$

338. $\displaystyle\int_0^\infty e^{-ax} \cos bx \, dx = \frac{a}{a^2 + b^2}.$

339. $\displaystyle\int_0^\infty e^{-ax} \sin bx \, dx = \frac{b}{a^2 + b^2}.$

340. $\displaystyle\int_0^\infty e^{-a^2 x^2} \cos bx \, dx = \frac{\pi^{\frac{1}{2}}}{2a} e^{-\frac{b^2}{4a^2}}.$

341. $\displaystyle\int_{-\infty}^\infty \frac{dx}{a + 2bx + cx^2} = \frac{\pi}{(ac - b^2)^{\frac{1}{2}}} \text{ if } ac > b^2.$

342. $\displaystyle\int_0^\infty \frac{e^{-ax} - e^{-bx}}{x} dx = \log \frac{b}{a}.$

343. $\displaystyle\int_0^\infty \frac{\arctan ax - \arctan bx}{x} dx = \frac{\pi}{2} \log \frac{a}{b}.$

344. $\displaystyle\int_0^\infty \frac{\cos ax - \cos bx}{x} dx = \log \frac{b}{a}.$

345. $\displaystyle\int_0^{\frac{\pi}{2}} \frac{dx}{a^2 \cos^2 x + b^2 \sin^2 x} = \frac{\pi}{2ab}.$

346. $\int_0^{\frac{\pi}{2}} \dfrac{dx}{(a^2 \cos^2 x + b^2 \sin^2 x)^2} = \dfrac{\pi(a^2 + b^2)}{4\,a^3 b^3}.$

347. $\int_0^{\pi} \dfrac{dx}{a + b \cos x} = \dfrac{\pi}{(a^2 - b^2)^{\frac{1}{2}}}$ if $a > b > 0$.

348. $\int_0^{\infty} \dfrac{\cos mx \, dx}{1 + x^2} = \dfrac{1}{2}\,\pi e^{-m}.$

349. $\int_0^{\infty} \dfrac{\cos x \, dx}{x^{\frac{1}{2}}} = \int_0^{\infty} \dfrac{\sin x \, dx}{x^{\frac{1}{2}}} = (\tfrac{1}{2}\,\pi)^{\frac{1}{2}}.$

350. $\int_0^{\infty} \dfrac{\sin ax \sin bx}{x^2}\, dx = \dfrac{1}{2}\,\pi a$ if $a < b$.

351. $\int_0^{\pi} \dfrac{(a - b \cos x)dx}{a^2 - 2\,ab \cos x + b^2} = 0$ if $a^2 < b^2$,

$$= \dfrac{\pi}{a} \text{ if } a^2 > b^2,$$

$$= \dfrac{\pi}{2\,a} \text{ if } a = b.$$

352. $\int_0^{\pi} \dfrac{\log\,(1 + \sin a \cos x)}{\cos x}\, dx = \pi a.$

353. $\int_0^1 \dfrac{x^b - x^a}{\log x}\, dx = \log \dfrac{1 + b}{1 + a}.$

354. $\int_0^1 \dfrac{\log x}{1 - x}\, dx = -\dfrac{\pi^2}{6}.$

355. $\int_0^1 \dfrac{\log x}{1 + x}\, dx = -\dfrac{\pi^2}{12}.$

356. $\int_0^1 \dfrac{\log x}{1 - x^2}\, dx = -\dfrac{\pi^2}{8}.$

357. $\int_0^1 \log\left(\dfrac{1 + x}{1 - x}\right)\dfrac{dx}{x} = \dfrac{\pi^2}{4}.$

358. $\int_0^{\infty} \log\left(\dfrac{e^x + 1}{e^x - 1}\right)dx = \dfrac{\pi^2}{4}.$

359. $\int_0^{\frac{\pi}{2}} \log \sin x \, dx = \int_0^{\frac{\pi}{2}} \log \cos x \, dx = -\dfrac{1}{2}\,\pi \log 2.$

360. $\int_0^{\pi} x \log \sin x \, dx = -\dfrac{1}{2}\,\pi^2 \log 2.$

(For integrals expressed as infinite series see page 269.)

VIII. INFINITE SERIES

1. $e^x = 1 + x + \dfrac{x^2}{2!} + \dfrac{x^3}{3!} + \cdots,$ $-\infty < x < \infty.$

2. $e^{kx} = 1 + kx + \dfrac{k^2x^2}{2!} + \dfrac{k^3x^3}{3!} + \cdots,$ $-\infty < x < \infty.$

3. $a^x = 1 + x \log a + \dfrac{(x \log a)^2}{2!} + \dfrac{(x \log a)^3}{3!} + \cdots,$ $-\infty < x < \infty.$

4. $\log (1 + x) = x - \dfrac{x^2}{2} + \dfrac{x^3}{3} - \dfrac{x^4}{4} + \cdots,$ $-1 < x \leqq 1.$

5. $\log (1 - x) = - x - \dfrac{x^2}{2} - \dfrac{x^3}{3} - \dfrac{x^4}{4} - \cdots,$ $-1 \leqq x < 1.$

6. $\log (y + 1) = \log y + 2 \left[\dfrac{1}{2y+1} + \dfrac{1}{3} \left(\dfrac{1}{2y+1} \right)^3 + \dfrac{1}{5} \left(\dfrac{1}{2y+1} \right)^5 + \cdots \right],$ $y > 0.$

7. $\sin x = x - \dfrac{x^3}{3!} + \dfrac{x^5}{5!} - \dfrac{x^7}{7!} + \cdots,$ $-\infty < x < \infty.$

8. $\cos x = 1 - \dfrac{x^2}{2!} + \dfrac{x^4}{4!} - \dfrac{x^6}{6!} + \cdots,$ $-\infty < x < \infty.$

9. $\arctan x = x - \dfrac{x^3}{3} + \dfrac{x^5}{5} - \dfrac{x^7}{7} + \dfrac{x^9}{9} - \cdots,$ $-1 \leqq x \leqq 1.$

10. $(1 - x)^{-\frac{1}{2}} = 1 + \dfrac{1}{2} x + \dfrac{1 \cdot 3}{2 \cdot 4} x^2 + \dfrac{1 \cdot 3 \cdot 5}{2 \cdot 4 \cdot 6} x^3 + \dfrac{1 \cdot 3 \cdot 5 \cdot 7}{2 \cdot 4 \cdot 6 \cdot 8} x^4 + \cdots,$
                                          $-1 < x < 1.$

11. $(1 - x^2)^{-\frac{1}{2}} = 1 + \dfrac{1}{2} x^2 + \dfrac{1 \cdot 3}{2 \cdot 4} x^4 + \dfrac{1 \cdot 3 \cdot 5}{2 \cdot 4 \cdot 6} x^6 + \dfrac{1 \cdot 3 \cdot 5 \cdot 7}{2 \cdot 4 \cdot 6 \cdot 8} x^8 + \cdots,$
                                          $-1 < x < 1.$

12. $\arcsin x = x + \dfrac{1}{2} \dfrac{x^3}{3} + \dfrac{1 \cdot 3}{2 \cdot 4} \dfrac{x^5}{5} + \dfrac{1 \cdot 3 \cdot 5}{2 \cdot 4 \cdot 6} \dfrac{x^7}{7} + \cdots,$ $-1 \leqq x \leqq 1.$

13. $(1 - x)^{-m} = 1 + mx + \dfrac{m(m + 1)}{2!} x^2 + \dfrac{m(m + 1)(m + 2)}{3!} x^3 + \cdots,$
                                          $-1 < x < 1.$

14. $(1 + x)^m = 1 + mx + \dfrac{m(m - 1)}{2!} x^2 + \dfrac{m(m - 1)(m - 2)}{3!} x^3 + \cdots,$
                                          $-1 < x < 1.$

15. $e^{\sin x} = 1 + x + \dfrac{x^2}{2} - \dfrac{x^4}{8} + \cdots,$ $-\infty < x < \infty.$

16. $\tan x = x + \dfrac{x^3}{3} + \dfrac{2 x^5}{15} + \cdots,$ $-\dfrac{\pi}{2} < x < \dfrac{\pi}{2}.$

17. $\sec x = 1 + \dfrac{x^2}{2} + \dfrac{5 x^4}{24} + \cdots,$ $-\dfrac{\pi}{2} < x < \dfrac{\pi}{2}.$

18. $\log \cos x = - \dfrac{x^2}{2} - \dfrac{x^4}{12} - \dfrac{x^6}{45} - \cdots,$ $-\dfrac{\pi}{2} < x < \dfrac{\pi}{2}.$

19. $\sin^2 x = x^2 - \dfrac{2^3 x^4}{4!} + \dfrac{2^5 x^6}{6!} - \dfrac{2^7 x^8}{8!} + \cdots,$ $-\infty < x < \infty.$

20. $(\arc\sin x)^2 = x^2 + \dfrac{2}{3}\dfrac{x^4}{2} + \dfrac{2\cdot 4}{3\cdot 5}\dfrac{x^6}{3} + \dfrac{2\cdot 4\cdot 6}{3\cdot 5\cdot 7}\dfrac{x^8}{4} + \cdots,\quad -1 < x < 1.$

21. $\dfrac{\arc\sin x}{(1-x^2)^{\frac{1}{2}}} = x + \dfrac{2}{3}x^3 + \dfrac{2\cdot 4}{3\cdot 5}x^5 + \dfrac{2\cdot 4\cdot 6}{3\cdot 5\cdot 7}x^7 + \cdots,\qquad -1 < x < 1.$

22. $\arc\tan x = \dfrac{x}{1+x^2}\left[1 + \dfrac{2}{3}\dfrac{x^2}{1+x^2} + \dfrac{2\cdot 4}{3\cdot 5}\left(\dfrac{x^2}{1+x^2}\right)^2 + \dfrac{2\cdot 4\cdot 6}{3\cdot 5\cdot 7}\left(\dfrac{x^2}{1+x^2}\right)^3 + \cdots\right],$
$$-\infty < x < \infty.$$

23. If for $m > 1$ we use U_m, V_m, W_m to denote the sums of the series

$$U_m = 1 + \frac{1}{2^m} + \frac{1}{3^m} + \frac{1}{4^m} + \cdots,$$

$$V_m = 1 + \frac{1}{3^m} + \frac{1}{5^m} + \frac{1}{7^m} + \cdots,$$

$$W_m = 1 - \frac{1}{3^m} + \frac{1}{5^m} - \frac{1}{7^m} + \cdots,$$

then we have

$$\log(x\csc x) = U_2\left(\frac{x}{\pi}\right)^2 + \frac{1}{2}U_4\left(\frac{x}{\pi}\right)^4 + \frac{1}{3}U_6\left(\frac{x}{\pi}\right)^6 + \cdots,\qquad -\pi < x < \pi,$$

$$\log(\sec x) = V_2\left(\frac{2\,x}{\pi}\right)^2 + \frac{1}{2}V_4\left(\frac{2\,x}{\pi}\right)^4 + \frac{1}{3}V_6\left(\frac{2\,x}{\pi}\right)^6 + \cdots,\quad -\frac{\pi}{2} < x < \frac{\pi}{2},$$

$$\frac{\pi}{2}\left(\cot x - \frac{1}{x}\right) = -U_2\left(\frac{x}{\pi}\right) + U_4\left(\frac{x}{\pi}\right)^3 - U_6\left(\frac{x}{\pi}\right)^5 + \cdots,\qquad -\pi < x < \pi,$$

$$\frac{\pi}{2}\tan x = 2^2 V_2\left(\frac{x}{\pi}\right) + 2^4 V_4\left(\frac{x}{\pi}\right)^3 + 2^6 V_6\left(\frac{x}{\pi}\right)^5 + \cdots,\quad -\frac{\pi}{2} < x < \frac{\pi}{2},$$

$$\frac{\pi}{2}\sec x = 2\,W_1 + 2^3 W_3\left(\frac{x}{\pi}\right)^2 + 2^5 W_5\left(\frac{x}{\pi}\right)^4 + \cdots,\qquad -\frac{\pi}{2} < x < \frac{\pi}{2}.$$

(See formulas 38.)

24. $\csc^2 x = \dfrac{1}{x^2} + \displaystyle\sum_{n=1}^{\infty}\left(\dfrac{1}{(x+n\pi)^2} + \dfrac{1}{(x-n\pi)^2}\right).$

25. $\cot x = \dfrac{1}{x} + \displaystyle\sum_{n=1}^{\infty}\left(\dfrac{1}{x+n\pi} + \dfrac{1}{x-n\pi}\right)$

$$= \frac{1}{x} + \sum_{n=1}^{\infty}\frac{2\,x}{x^2 - n^2\pi^2}.$$

26. $\dfrac{\pi}{2}\tan\dfrac{\pi x}{2} = \dfrac{2\,x}{1-x^2} + \dfrac{2\,x}{3^2-x^2} + \dfrac{2\,x}{5^2-x^2} + \dfrac{2\,x}{7^2-x^2} + \cdots.$

27. $\dfrac{\pi}{\sin\pi x} = \dfrac{1}{x} + \dfrac{2\,x}{1-x^2} - \dfrac{2\,x}{2^2-x^2} + \dfrac{2\,x}{3^2-x^2} - \dfrac{2\,x}{4^2-x^2} + \cdots.$

28. $\dfrac{\pi}{\cos\frac{1}{2}\pi x} = 4\left\{\dfrac{1}{1-x^2} - \dfrac{3}{3^2-x^2} + \dfrac{5}{5^2-x^2} - \dfrac{7}{7^2-x^2} + \cdots\right\}.$

29. $\dfrac{\pi}{2\,x}\coth\pi x - \dfrac{1}{2\,x^2} = \dfrac{1}{1+x^2} + \dfrac{1}{2^2+x^2} + \dfrac{1}{3^2+x^2} + \cdots.$

30. $\dfrac{\pi}{4\,x}\tanh\dfrac{\pi x}{2} = \dfrac{1}{1+x^2} + \dfrac{1}{3^2+x^2} + \dfrac{1}{5^2+x^2} + \cdots.$

31. $\dfrac{\sin \pi x}{\pi} = x \displaystyle\prod_{n=1}^{\infty}\left(1 - \dfrac{x^2}{n^2}\right).$

32. $\cos \pi x = \displaystyle\prod_{n=0}^{\infty}\left(1 - \dfrac{x^2}{(n + \frac{1}{2})^2}\right).$

33. $e = 1 + 1 + \dfrac{1}{2!} + \dfrac{1}{3!} + \dfrac{1}{4!} + \dfrac{1}{5!} + \cdots.$

34. $\dfrac{\pi}{6} = \dfrac{1}{2} + \dfrac{1}{2}\cdot\dfrac{1}{2^3 \cdot 3} + \dfrac{1 \cdot 3}{2 \cdot 4}\cdot\dfrac{1}{2^5 \cdot 5} + \dfrac{1 \cdot 3 \cdot 5}{2 \cdot 4 \cdot 6}\dfrac{1}{2^7 \cdot 7} + \cdots.$

35. $\dfrac{\pi}{4} = 2 \arctan \dfrac{1}{3} + \arctan \dfrac{1}{7}$

$$= \dfrac{3}{5}\left[1 + \dfrac{2}{3}\cdot\dfrac{1}{10} + \dfrac{2 \cdot 4}{3 \cdot 5}\cdot\dfrac{1}{10^2} + \dfrac{2 \cdot 4 \cdot 6}{3 \cdot 5 \cdot 7}\cdot\dfrac{1}{10^3} + \cdots\right]$$

$$+ \dfrac{7}{50}\left[1 + \dfrac{2}{3}\left(\dfrac{2}{100}\right) + \dfrac{2 \cdot 4}{3 \cdot 5}\left(\dfrac{2}{100}\right)^2 + \dfrac{2 \cdot 4 \cdot 6}{3 \cdot 5 \cdot 7}\left(\dfrac{2}{100}\right)^3 + \cdots\right].$$

36. Other series for calculating π may be obtained from formula 22 and the following relations:

$$\dfrac{\pi}{4} = 4 \arctan \dfrac{1}{5} - \arctan \dfrac{1}{239},$$

$$\dfrac{\pi}{4} = 4 \arctan \dfrac{1}{5} - \arctan \dfrac{1}{70} + \arctan \dfrac{1}{99},$$

$$\dfrac{\pi}{4} = \arctan \dfrac{1}{2} + \arctan \dfrac{1}{5} + \arctan \dfrac{1}{8}.$$

37. $\dfrac{\pi}{2} = \dfrac{2}{1}\cdot\dfrac{2}{3}\cdot\dfrac{4}{3}\cdot\dfrac{4}{5}\cdot\dfrac{6}{5}\cdot\dfrac{6}{7}\cdot\dfrac{8}{7}\cdot\dfrac{8}{9}\cdots.$

38. If we write

$$\dfrac{x}{2}\cot \dfrac{x}{2} = 1 - B_1 \dfrac{x^2}{2!} - B_2 \dfrac{x^4}{4!} - B_3 \dfrac{x^6}{6!} - B_4 \dfrac{x^8}{8!} - \cdots,$$

then we have

$$B_1 = \tfrac{1}{6},\ B_2 = \tfrac{1}{30},\ B_3 = \tfrac{1}{42},\ B_4 = \tfrac{1}{30},\ B_5 = \tfrac{5}{66},\ \cdots$$

while the numbers B_i satisfy the recurrence relation

$$\dfrac{1}{2^{2n+1}\cdot(2n)!} - \dfrac{1}{2^{2n+1}\cdot(2n+1)!} - \dfrac{B_1}{2^{2n-1}\cdot 2!(2n-1)!} + \dfrac{B_2}{2^{2n-3}\cdot 4!(2n-3)!}$$

$$- \dfrac{B_3}{2^{2n-5}\cdot 6!\cdot(2n-5)!} + \cdots = 0,$$

the series closing with the term containing B_n. We also have

$$\dfrac{t}{e^t - 1} = 1 - \dfrac{t}{2} + \dfrac{B_1 t^2}{2!} - \dfrac{B_2 t^4}{4!} + \dfrac{B_3 t^6}{6!} - \cdots.$$

If for integral values of n we denote by S_{2n} the sum of the series

$$S_{2n} = 1 + \dfrac{1}{2^{2n}} + \dfrac{1}{3^{2n}} + \dfrac{1}{4^{2n}} + \cdots,$$

then we have $\qquad S_{2n} = \dfrac{2^{2n-1}\,\pi^{2n}}{(2n)!}\,B_n.$

In particular,
$$\frac{\pi^2}{6} = 1 + \frac{1}{2^2} + \frac{1}{3^2} + \frac{1}{4^2} + \cdots,$$

$$\frac{\pi^4}{90} = 1 + \frac{1}{2^4} + \frac{1}{3^4} + \frac{1}{4^4} + \cdots.$$

(See formulas 23.)

39. $1 + \frac{1}{3} - \frac{1}{5} - \frac{1}{7} + \frac{1}{9} + \frac{1}{11} - \frac{1}{13} - \frac{1}{15} + \frac{1}{17} + \cdots$

$$= \int_0^1 (1 + x^2 - x^4 - x^6 + x^8 + x^{10} - x^{12} - x^{14} + x^{16} + \cdots)\, dx$$

$$= \int_0^1 \frac{1 + x^2}{1 + x^4}\, dx = \frac{\pi}{4}\sqrt{2}.$$

40. $1 - \frac{1}{7} + \frac{1}{9} - \frac{1}{15} + \frac{1}{17} - \frac{1}{23} + \frac{1}{25} - \frac{1}{31} + \cdots = \int_0^1 \frac{1 - x^6}{1 - x^8}\, dx = \frac{\pi}{4}\frac{1 + \sqrt{2}}{2}.$

41. $\displaystyle\int \frac{\sin x\, dx}{x} = c + x - \frac{x^3}{3 \cdot 3!} + \frac{x^5}{5 \cdot 5!} - \frac{x^7}{7 \cdot 7!} + \cdots.$

42. $\displaystyle\int \frac{\cos x\, dx}{x} = c + \log x - \frac{x^2}{2 \cdot 2!} + \frac{x^4}{4 \cdot 4!} - \frac{x^6}{6 \cdot 6!} + \cdots.$

43. $\displaystyle\int \frac{e^x\, dx}{x} = c + \log x + x + \frac{x^2}{2 \cdot 2!} + \frac{x^3}{3 \cdot 3!} + \frac{x^4}{4 \cdot 4!} + \cdots.$

44. $\displaystyle\int_0^{\frac{\pi}{2}} \frac{dx}{(1 - k^2 \sin^2 x)^{\frac{1}{2}}} = \frac{\pi}{2}\left[1 + \left(\frac{1}{2}\right)^2 k^2 + \left(\frac{1 \cdot 3}{2 \cdot 4}\right)^2 \frac{k^4}{3} + \left(\frac{1 \cdot 3 \cdot 5}{2 \cdot 4 \cdot 6}\right)^2 \frac{k^6}{5} + \cdots\right],$
if $k^2 < 1$.

45. $\displaystyle\int_0^{\frac{\pi}{2}} (1 - k^2 \sin^2 x)^{\frac{1}{2}}\, dx = \frac{\pi}{2}\left[1 - \left(\frac{1}{2}\right)^2 k^2 - \left(\frac{1 \cdot 3}{2 \cdot 4}\right)^2 k^4 - \left(\frac{1 \cdot 3 \cdot 5}{2 \cdot 4 \cdot 6}\right)^2 k^6 - \cdots\right],$
if $k^2 < 1$.

46. $\displaystyle\int_0^1 \frac{\log (1 + x)}{x}\, dx = \frac{1}{1^2} - \frac{1}{2^2} + \frac{1}{3^2} - \frac{1}{4^2} + \cdots = \frac{\pi^2}{12}.$

47. $\displaystyle\int_\infty^1 \frac{e^{-xu}}{u}\, du = \gamma + \log x - x + \frac{x^2}{2 \cdot 2!} - \frac{x^3}{3 \cdot 3!} + \frac{x^4}{4 \cdot 4!} - \frac{x^5}{5 \cdot 5!} + \cdots,$

where $\gamma = \lim_{t = \infty}\left(1 + \frac{1}{2} + \frac{1}{3} + \cdots + \frac{1}{t} - \log t\right) = 0.5772157 \cdots.$

48. $\displaystyle\int_\infty^1 \frac{\cos xu}{u}\, du = \gamma + \log x - \frac{x^2}{2 \cdot 2!} + \frac{x^4}{4 \cdot 4!} - \frac{x^6}{6 \cdot 6!} + \cdots,$
where γ has the value given in 47.

49. $\displaystyle\int_0^1 \frac{1 - e^{-xu}}{u}\, du = x - \frac{x^2}{2 \cdot 2!} + \frac{x^3}{3 \cdot 3!} - \frac{x^4}{4 \cdot 4!} + \cdots.$

50. $\displaystyle\int_0^1 \frac{e^{xu} - e^{-xu}}{u}\, du = 2\left(x + \frac{x^3}{3 \cdot 3!} + \frac{x^5}{5 \cdot 5!} + \cdots\right).$

A CATALOG OF SELECTED
DOVER BOOKS
IN ALL FIELDS OF INTEREST

A CATALOG OF SELECTED DOVER
BOOKS IN ALL FIELDS OF INTEREST

DRAWINGS OF REMBRANDT, edited by Seymour Slive. Updated Lippmann, Hofstede de Groot edition, with definitive scholarly apparatus. All portraits, biblical sketches, landscapes, nudes. Oriental figures, classical studies, together with selection of work by followers. 550 illustrations. Total of 630pp. 9⅛ × 12¼.
21485-0, 21486-9 Pa., Two-vol. set $25.00

GHOST AND HORROR STORIES OF AMBROSE BIERCE, Ambrose Bierce. 24 tales vividly imagined, strangely prophetic, and decades ahead of their time in technical skill: "The Damned Thing," "An Inhabitant of Carcosa," "The Eyes of the Panther," "Moxon's Master," and 20 more. 199pp. 5⅜ × 8½. 20767-6 Pa. $3.95

ETHICAL WRITINGS OF MAIMONIDES, Maimonides. Most significant ethical works of great medieval sage, newly translated for utmost precision, readability. Laws Concerning Character Traits, Eight Chapters, more. 192pp. 5⅜ × 8½.
24522-5 Pa. $4.50

THE EXPLORATION OF THE COLORADO RIVER AND ITS CANYONS, J. W. Powell. Full text of Powell's 1,000-mile expedition down the fabled Colorado in 1869. Superb account of terrain, geology, vegetation, Indians, famine, mutiny, treacherous rapids, mighty canyons, during exploration of last unknown part of continental U.S. 400pp. 5⅜ × 8½. 20094-9 Pa. $6.95

HISTORY OF PHILOSOPHY, Julián Marías. Clearest one-volume history on the market. Every major philosopher and dozens of others, to Existentialism and later. 505pp. 5⅜ × 8½. 21739-6 Pa. $8.50

ALL ABOUT LIGHTNING, Martin A. Uman. Highly readable non-technical survey of nature and causes of lightning, thunderstorms, ball lightning, St. Elmo's Fire, much more. Illustrated. 192pp. 5⅜ × 8½. 25237-X Pa. $5.95

SAILING ALONE AROUND THE WORLD, Captain Joshua Slocum. First man to sail around the world, alone, in small boat. One of great feats of seamanship told in delightful manner. 67 illustrations. 294pp. 5⅜ × 8½. 20326-3 Pa. $4.95

LETTERS AND NOTES ON THE MANNERS, CUSTOMS AND CONDITIONS OF THE NORTH AMERICAN INDIANS, George Catlin. Classic account of life among Plains Indians: ceremonies, hunt, warfare, etc. 312 plates. 572pp. of text. 6⅛ × 9¼. 22118-0, 22119-9 Pa. Two-vol. set $15.90

ALASKA: The Harriman Expedition, 1899, John Burroughs, John Muir, et al. Informative, engrossing accounts of two-month, 9,000-mile expedition. Native peoples, wildlife, forests, geography, salmon industry, glaciers, more. Profusely illustrated. 240 black-and-white line drawings. 124 black-and-white photographs. 3 maps. Index. 576pp. 5⅜ × 8½. 25109-8 Pa. $11.95

THE BOOK OF BEASTS: Being a Translation from a Latin Bestiary of the Twelfth Century, T. H. White. Wonderful catalog real and fanciful beasts: manticore, griffin, phoenix, amphivius, jaculus, many more. White's witty erudite commentary on scientific, historical aspects. Fascinating glimpse of medieval mind. Illustrated. 296pp. 5⅜ × 8¼. (Available in U.S. only) 24609-4 Pa. $5.95

FRANK LLOYD WRIGHT: ARCHITECTURE AND NATURE With 160 Illustrations, Donald Hoffmann. Profusely illustrated study of influence of nature—especially prairie—on Wright's designs for Fallingwater, Robie House, Guggenheim Museum, other masterpieces. 96pp. 9¼ × 10¾. 25098-9 Pa. $7.95

FRANK LLOYD WRIGHT'S FALLINGWATER, Donald Hoffmann. Wright's famous waterfall house: planning and construction of organic idea. History of site, owners, Wright's personal involvement. Photographs of various stages of building. Preface by Edgar Kaufmann, Jr. 100 illustrations. 112pp. 9¼ × 10.
23671-4 Pa. $7.95

YEARS WITH FRANK LLOYD WRIGHT: Apprentice to Genius, Edgar Tafel. Insightful memoir by a former apprentice presents a revealing portrait of Wright the man, the inspired teacher, the greatest American architect. 372 black-and-white illustrations. Preface. Index. vi + 228pp. 8¼ × 11. 24801-1 Pa. $9.95

THE STORY OF KING ARTHUR AND HIS KNIGHTS, Howard Pyle. Enchanting version of King Arthur fable has delighted generations with imaginative narratives of exciting adventures and unforgettable illustrations by the author. 41 illustrations. xviii + 313pp. 6⅛ × 9¼. 21445-1 Pa. $5.95

THE GODS OF THE EGYPTIANS, E. A. Wallis Budge. Thorough coverage of numerous gods of ancient Egypt by foremost Egyptologist. Information on evolution of cults, rites and gods; the cult of Osiris; the Book of the Dead and its rites; the sacred animals and birds; Heaven and Hell; and more. 956pp. 6⅛ × 9¼.
22055-9, 22056-7 Pa., Two-vol. set $21.90

A THEOLOGICO-POLITICAL TREATISE, Benedict Spinoza. Also contains unfinished *Political Treatise*. Great classic on religious liberty, theory of government on common consent. R. Elwes translation. Total of 421pp. 5⅜ × 8½.
20249-6 Pa. $6.95

INCIDENTS OF TRAVEL IN CENTRAL AMERICA, CHIAPAS, AND YUCATAN, John L. Stephens. Almost single-handed discovery of Maya culture; exploration of ruined cities, monuments, temples; customs of Indians. 115 drawings. 892pp. 5⅜ × 8½. 22404-X, 22405-8 Pa., Two-vol. set $15.90

LOS CAPRICHOS, Francisco Goya. 80 plates of wild, grotesque monsters and caricatures. Prado manuscript included. 183pp. 6⅛ × 9⅜. 22384-1 Pa. $4.95

AUTOBIOGRAPHY: The Story of My Experiments with Truth, Mohandas K. Gandhi. Not hagiography, but Gandhi in his own words. Boyhood, legal studies, purification, the growth of the Satyagraha (nonviolent protest) movement. Critical, inspiring work of the man who freed India. 480pp. 5⅜ × 8½. (Available in U.S. only)
24593-4 Pa. $6.95

ILLUSTRATED DICTIONARY OF HISTORIC ARCHITECTURE, edited by Cyril M. Harris. Extraordinary compendium of clear, concise definitions for over 5,000 important architectural terms complemented by over 2,000 line drawings. Covers full spectrum of architecture from ancient ruins to 20th-century Modernism. Preface. 592pp. 7½ × 9⅝. 24444-X Pa. $14.95

THE NIGHT BEFORE CHRISTMAS, Clement Moore. Full text, and woodcuts from original 1848 book. Also critical, historical material. 19 illustrations. 40pp. 4⅝ × 6. 22797-9 Pa. $2.50

THE LESSON OF JAPANESE ARCHITECTURE: 165 Photographs, Jiro Harada. Memorable gallery of 165 photographs taken in the 1930's of exquisite Japanese homes of the well-to-do and historic buildings. 13 line diagrams. 192pp. 8⅜ × 11¼. 24778-3 Pa. $8.95

THE AUTOBIOGRAPHY OF CHARLES DARWIN AND SELECTED LETTERS, edited by Francis Darwin. The fascinating life of eccentric genius composed of an intimate memoir by Darwin (intended for his children); commentary by his son, Francis; hundreds of fragments from notebooks, journals, papers; and letters to and from Lyell, Hooker, Huxley, Wallace and Henslow. xi + 365pp. 5⅜ × 8. 20479-0 Pa. $5.95

WONDERS OF THE SKY: Observing Rainbows, Comets, Eclipses, the Stars and Other Phenomena, Fred Schaaf. Charming, easy-to-read poetic guide to all manner of celestial events visible to the naked eye. Mock suns, glories, Belt of Venus, more. Illustrated. 299pp. 5¼ × 8¼. 24402-4 Pa. $7.95

BURNHAM'S CELESTIAL HANDBOOK, Robert Burnham, Jr. Thorough guide to the stars beyond our solar system. Exhaustive treatment. Alphabetical by constellation: Andromeda to Cetus in Vol. 1; Chamaeleon to Orion in Vol. 2; and Pavo to Vulpecula in Vol. 3. Hundreds of illustrations. Index in Vol. 3. 2,000pp. 6⅛ × 9¼. 23567-X, 23568-8, 23673-0 Pa., Three-vol. set $37.85

STAR NAMES: Their Lore and Meaning, Richard Hinckley Allen. Fascinating history of names various cultures have given to constellations and literary and folkloristic uses that have been made of stars. Indexes to subjects. Arabic and Greek names. Biblical references. Bibliography. 563pp. 5⅜ × 8½. 21079-0 Pa. $7.95

THIRTY YEARS THAT SHOOK PHYSICS: The Story of Quantum Theory, George Gamow. Lucid, accessible introduction to influential theory of energy and matter. Careful explanations of Dirac's anti-particles, Bohr's model of the atom, much more. 12 plates. Numerous drawings. 240pp. 5⅜ × 8½. 24895-X Pa. $4.95

CHINESE DOMESTIC FURNITURE IN PHOTOGRAPHS AND MEASURED DRAWINGS, Gustav Ecke. A rare volume, now affordably priced for antique collectors, furniture buffs and art historians. Detailed review of styles ranging from early Shang to late Ming. Unabridged republication. 161 black-and-white drawings, photos. Total of 224pp. 8⅜ × 11¼. (Available in U.S. only) 25171-3 Pa. $12.95

VINCENT VAN GOGH: A Biography, Julius Meier-Graefe. Dynamic, penetrating study of artist's life, relationship with brother, Theo, painting techniques, travels, more. Readable, engrossing. 160pp. 5⅜ × 8½. (Available in U.S. only) 25253-1 Pa. $3.95

CATALOG OF DOVER BOOKS

HOW TO WRITE, Gertrude Stein. Gertrude Stein claimed anyone could understand her unconventional writing—here are clues to help. Fascinating improvisations, language experiments, explanations illuminate Stein's craft and the art of writing. Total of 414pp. 4⅜ × 6⅜. 23144-5 Pa. $5.95

ADVENTURES AT SEA IN THE GREAT AGE OF SAIL: Five Firsthand Narratives, edited by Elliot Snow. Rare true accounts of exploration, whaling, shipwreck, fierce natives, trade, shipboard life, more. 33 illustrations. Introduction. 353pp. 5⅜ × 8½. 25177-2 Pa. $7.95

THE HERBAL OR GENERAL HISTORY OF PLANTS, John Gerard. Classic descriptions of about 2,850 plants—with over 2,700 illustrations—includes Latin and English names, physical descriptions, varieties, time and place of growth, more. 2,706 illustrations. xlv + 1,678pp. 8½ × 12¼. 23147-X Cloth. $75.00

DOROTHY AND THE WIZARD IN OZ, L. Frank Baum. Dorothy and the Wizard visit the center of the Earth, where people are vegetables, glass houses grow and Oz characters reappear. Classic sequel to *Wizard of Oz*. 256pp. 5⅜ × 8. 24714-7 Pa. $4.95

SONGS OF EXPERIENCE: Facsimile Reproduction with 26 Plates in Full Color, William Blake. This facsimile of Blake's original "Illuminated Book" reproduces 26 full-color plates from a rare 1826 edition. Includes "The Tyger," "London," "Holy Thursday," and other immortal poems. 26 color plates. Printed text of poems. 48pp. 5¼ × 7. 24636-1 Pa. $3.50

SONGS OF INNOCENCE, William Blake. The first and most popular of Blake's famous "Illuminated Books," in a facsimile edition reproducing all 31 brightly colored plates. Additional printed text of each poem. 64pp. 5¼ × 7. 22764-2 Pa. $3.50

PRECIOUS STONES, Max Bauer. Classic, thorough study of diamonds, rubies, emeralds, garnets, etc.: physical character, occurrence, properties, use, similar topics. 20 plates, 8 in color. 94 figures. 659pp. 6⅛ × 9¼. 21910-0, 21911-9 Pa., Two-vol. set $15.90

ENCYCLOPEDIA OF VICTORIAN NEEDLEWORK, S. F. A. Caulfeild and Blanche Saward. Full, precise descriptions of stitches, techniques for dozens of needlecrafts—most exhaustive reference of its kind. Over 800 figures. Total of 679pp. 8⅛ × 11. Two volumes. Vol. 1 22800-2 Pa. $11.95
Vol. 2 22801-0 Pa. $11.95

THE MARVELOUS LAND OF OZ, L. Frank Baum. Second Oz book, the Scarecrow and Tin Woodman are back with hero named Tip, Oz magic. 136 illustrations. 287pp. 5⅜ × 8½. 20692-0 Pa. $5.95

WILD FOWL DECOYS, Joel Barber. Basic book on the subject, by foremost authority and collector. Reveals history of decoy making and rigging, place in American culture, different kinds of decoys, how to make them, and how to use them. 140 plates. 156pp. 7⅞ × 10¾. 20011-6 Pa. $8.95

HISTORY OF LACE, Mrs. Bury Palliser. Definitive, profusely illustrated chronicle of lace from earliest times to late 19th century. Laces of Italy, Greece, England, France, Belgium, etc. Landmark of needlework scholarship. 266 illustrations. 672pp. 6⅛ × 9¼. 24742-2 Pa. $14.95

ILLUSTRATED GUIDE TO SHAKER FURNITURE, Robert Meader. All furniture and appurtenances, with much on unknown local styles. 235 photos. 146pp. 9 × 12. 22819-3 Pa. $7.95

WHALE SHIPS AND WHALING: A Pictorial Survey, George Francis Dow. Over 200 vintage engravings, drawings, photographs of barks, brigs, cutters, other vessels. Also harpoons, lances, whaling guns, many other artifacts. Comprehensive text by foremost authority. 207 black-and-white illustrations. 288pp. 6 × 9. 24808-9 Pa. $8.95

THE BERTRAMS, Anthony Trollope. Powerful portrayal of blind self-will and thwarted ambition includes one of Trollope's most heartrending love stories. 497pp. 5⅜ × 8½. 25119-5 Pa. $8.95

ADVENTURES WITH A HAND LENS, Richard Headstrom. Clearly written guide to observing and studying flowers and grasses, fish scales, moth and insect wings, egg cases, buds, feathers, seeds, leaf scars, moss, molds, ferns, common crystals, etc.—all with an ordinary, inexpensive magnifying glass. 209 exact line drawings aid in your discoveries. 220pp. 5⅜ × 8½. 23330-8 Pa. $4.50

RODIN ON ART AND ARTISTS, Auguste Rodin. Great sculptor's candid, wide-ranging comments on meaning of art; great artists; relation of sculpture to poetry, painting, music; philosophy of life, more. 76 superb black-and-white illustrations of Rodin's sculpture, drawings and prints. 119pp. 8⅜ × 11¼. 24487-3 Pa. $6.95

FIFTY CLASSIC FRENCH FILMS, 1912–1982: A Pictorial Record, Anthony Slide. Memorable stills from Grand Illusion, Beauty and the Beast, Hiroshima, Mon Amour, many more. Credits, plot synopses, reviews, etc. 160pp. 8¼ × 11. 25256-6 Pa. $11.95

THE PRINCIPLES OF PSYCHOLOGY, William James. Famous long course complete, unabridged. Stream of thought, time perception, memory, experimental methods; great work decades ahead of its time. 94 figures. 1,391pp. 5⅜ × 8½. 20381-6, 20382-4 Pa., Two-vol. set $19.90

BODIES IN A BOOKSHOP, R. T. Campbell. Challenging mystery of blackmail and murder with ingenious plot and superbly drawn characters. In the best tradition of British suspense fiction. 192pp. 5⅜ × 8½. 24720-1 Pa. $3.95

CALLAS: PORTRAIT OF A PRIMA DONNA, George Jellinek. Renowned commentator on the musical scene chronicles incredible career and life of the most controversial, fascinating, influential operatic personality of our time. 64 black-and-white photographs. 416pp. 5⅜ × 8¼. 25047-4 Pa. $7.95

GEOMETRY, RELATIVITY AND THE FOURTH DIMENSION, Rudolph Rucker. Exposition of fourth dimension, concepts of relativity as Flatland characters continue adventures. Popular, easily followed yet accurate, profound. 141 illustrations. 133pp. 5⅜ × 8½. 23400-2 Pa. $3.50

HOUSEHOLD STORIES BY THE BROTHERS GRIMM, with pictures by Walter Crane. 53 classic stories—Rumpelstiltskin, Rapunzel, Hansel and Gretel, the Fisherman and his Wife, Snow White, Tom Thumb, Sleeping Beauty, Cinderella, and so much more—lavishly illustrated with original 19th century drawings. 114 illustrations. x + 269pp. 5⅜ × 8½. 21080-4 Pa. $4.50

CHRISTMAS CUSTOMS AND TRADITIONS, Clement A. Miles. Origin, evolution, significance of religious, secular practices. Caroling, gifts, yule logs, much more. Full, scholarly yet fascinating; non-sectarian. 400pp. 5⅜ × 8½.
23354-5 Pa. $6.50

THE HUMAN FIGURE IN MOTION, Eadweard Muybridge. More than 4,500 stopped-action photos, in action series, showing undraped men, women, children jumping, lying down, throwing, sitting, wrestling, carrying, etc. 390pp. 7⅞ × 10⅝.
20204-6 Cloth. $19.95

THE MAN WHO WAS THURSDAY, Gilbert Keith Chesterton. Witty, fast-paced novel about a club of anarchists in turn-of-the-century London. Brilliant social, religious, philosophical speculations. 128pp. 5⅜ × 8½.
25121-7 Pa. $3.95

A CEZANNE SKETCHBOOK: Figures, Portraits, Landscapes and Still Lifes, Paul Cezanne. Great artist experiments with tonal effects, light, mass, other qualities in over 100 drawings. A revealing view of developing master painter, precursor of Cubism. 102 black-and-white illustrations. 144pp. 8¾ × 6⅝.
24790-2 Pa. $5.95

AN ENCYCLOPEDIA OF BATTLES: Accounts of Over 1,560 Battles from 1479 B.C. to the Present, David Eggenberger. Presents essential details of every major battle in recorded history, from the first battle of Megiddo in 1479 B.C. to Grenada in 1984. List of Battle Maps. New Appendix covering the years 1967–1984. Index. 99 illustrations. 544pp. 6½ × 9¼.
24913-1 Pa. $14.95

AN ETYMOLOGICAL DICTIONARY OF MODERN ENGLISH, Ernest Weekley. Richest, fullest work, by foremost British lexicographer. Detailed word histories. Inexhaustible. Total of 856pp. 6½ × 9¼.
21873-2, 21874-0 Pa., Two-vol. set $17.00

WEBSTER'S AMERICAN MILITARY BIOGRAPHIES, edited by Robert McHenry. Over 1,000 figures who shaped 3 centuries of American military history. Detailed biographies of Nathan Hale, Douglas MacArthur, Mary Hallaren, others. Chronologies of engagements, more. Introduction. Addenda. 1,033 entries in alphabetical order. xi + 548pp. 6½ × 9¼. (Available in U.S. only)
24758-9 Pa. $11.95

LIFE IN ANCIENT EGYPT, Adolf Erman. Detailed older account, with much not in more recent books: domestic life, religion, magic, medicine, commerce, and whatever else needed for complete picture. Many illustrations. 597pp. 5⅜ × 8½.
22632-8 Pa. $8.95

HISTORIC COSTUME IN PICTURES, Braun & Schneider. Over 1,450 costumed figures shown, covering a wide variety of peoples: kings, emperors, nobles, priests, servants, soldiers, scholars, townsfolk, peasants, merchants, courtiers, cavaliers, and more. 256pp. 8⅜ × 11¼.
23150-X Pa. $7.95

THE NOTEBOOKS OF LEONARDO DA VINCI, edited by J. P. Richter. Extracts from manuscripts reveal great genius; on painting, sculpture, anatomy, sciences, geography, etc. Both Italian and English. 186 ms. pages reproduced, plus 500 additional drawings, including studies for *Last Supper, Sforza* monument, etc. 860pp. 7⅞ × 10¾. (Available in U.S. only) 22572-0, 22573-9 Pa., Two-vol. set $25.90

CATALOG OF DOVER BOOKS

SIR HARRY HOTSPUR OF HUMBLETHWAITE, Anthony Trollope. Incisive, unconventional psychological study of a conflict between a wealthy baronet, his idealistic daughter, and their scapegrace cousin. The 1870 novel in its first inexpensive edition in years. 250pp. 5⅜ × 8½. 24953-0 Pa. $4.95

LASERS AND HOLOGRAPHY, Winston E. Kock. Sound introduction to burgeoning field, expanded (1981) for second edition. Wave patterns, coherence, lasers, diffraction, zone plates, properties of holograms, recent advances. 84 illustrations. 160pp. 5⅜ × 8¼. (Except in United Kingdom) 24041-X Pa. $3.50

INTRODUCTION TO ARTIFICIAL INTELLIGENCE: SECOND, ENLARGED EDITION, Philip C. Jackson, Jr. Comprehensive survey of artificial intelligence—the study of how machines (computers) can be made to act intelligently. Includes introductory and advanced material. Extensive notes updating the main text. 132 black-and-white illustrations. 512pp. 5⅜ × 8½. 24864-X Pa. $8.95

HISTORY OF INDIAN AND INDONESIAN ART, Ananda K. Coomaraswamy. Over 400 illustrations illuminate classic study of Indian art from earliest Harappa finds to early 20th century. Provides philosophical, religious and social insights. 304pp. 6⅜ × 9⅜. 25005-9 Pa. $8.95

THE GOLEM, Gustav Meyrink. Most famous supernatural novel in modern European literature, set in Ghetto of Old Prague around 1890. Compelling story of mystical experiences, strange transformations, profound terror. 13 black-and-white illustrations. 224pp. 5⅜ × 8½. (Available in U.S. only) 25025-3 Pa. $5.95

ARMADALE, Wilkie Collins. Third great mystery novel by the author of *The Woman in White* and *The Moonstone*. Original magazine version with 40 illustrations. 597pp. 5⅜ × 8½. 23429-0 Pa. $7.95

PICTORIAL ENCYCLOPEDIA OF HISTORIC ARCHITECTURAL PLANS, DETAILS AND ELEMENTS: With 1,880 Line Drawings of Arches, Domes, Doorways, Facades, Gables, Windows, etc., John Theodore Haneman. Sourcebook of inspiration for architects, designers, others. Bibliography. Captions. 141pp. 9 × 12. 24605-1 Pa. $6.95

BENCHLEY LOST AND FOUND, Robert Benchley. Finest humor from early 30's, about pet peeves, child psychologists, post office and others. Mostly unavailable elsewhere. 73 illustrations by Peter Arno and others. 183pp. 5⅜ × 8½. 22410-4 Pa. $3.95

ERTÉ GRAPHICS, Erté. Collection of striking color graphics: *Seasons, Alphabet, Numerals, Aces* and *Precious Stones*. 50 plates, including 4 on covers. 48pp. 9⅜ × 12¼. 23580-7 Pa. $6.95

THE JOURNAL OF HENRY D. THOREAU, edited by Bradford Torrey, F. H. Allen. Complete reprinting of 14 volumes, 1837–61, over two million words; the sourcebooks for *Walden*, etc. Definitive. All original sketches, plus 75 photographs. 1,804pp. 8½ × 12¼. 20312-3, 20313-1 Cloth., Two-vol. set $80.00

CASTLES: THEIR CONSTRUCTION AND HISTORY, Sidney Toy. Traces castle development from ancient roots. Nearly 200 photographs and drawings illustrate moats, keeps, baileys, many other features. Caernarvon, Dover Castles, Hadrian's Wall, Tower of London, dozens more. 256pp. 5⅜ × 8¼. 24898-4 Pa. $5.95

AMERICAN CLIPPER SHIPS: 1833–1858, Octavius T. Howe & Frederick C. Matthews. Fully-illustrated, encyclopedic review of 352 clipper ships from the period of America's greatest maritime supremacy. Introduction. 109 halftones. 5 black-and-white line illustrations. Index. Total of 928pp. 5⅜ × 8½.
25115-2, 25116-0 Pa., Two-vol. set $17.90

TOWARDS A NEW ARCHITECTURE, Le Corbusier. Pioneering manifesto by great architect, near legendary founder of "International School." Technical and aesthetic theories, views on industry, economics, relation of form to function, "mass-production spirit," much more. Profusely illustrated. Unabridged translation of 13th French edition. Introduction by Frederick Etchells. 320pp. 6⅛ × 9¼.
(Available in U.S. only) 25023-7 Pa. $8.95

THE BOOK OF KELLS, edited by Blanche Cirker. Inexpensive collection of 32 full-color, full-page plates from the greatest illuminated manuscript of the Middle Ages, painstakingly reproduced from rare facsimile edition. Publisher's Note. Captions. 32pp. 9⅜ × 12¼. 24345-1 Pa. $4.95

BEST SCIENCE FICTION STORIES OF H. G. WELLS, H. G. Wells. Full novel *The Invisible Man,* plus 17 short stories: "The Crystal Egg," "Aepyornis Island," "The Strange Orchid," etc. 303pp. 5⅜ × 8½. (Available in U.S. only)
21531-8 Pa. $4.95

AMERICAN SAILING SHIPS: Their Plans and History, Charles G. Davis. Photos, construction details of schooners, frigates, clippers, other sailcraft of 18th to early 20th centuries—plus entertaining discourse on design, rigging, nautical lore, much more. 137 black-and-white illustrations. 240pp. 6⅛ × 9¼.
24658-2 Pa. $5.95

ENTERTAINING MATHEMATICAL PUZZLES, Martin Gardner. Selection of author's favorite conundrums involving arithmetic, money, speed, etc., with lively commentary. Complete solutions. 112pp. 5⅜ × 8½. 25211-6 Pa. $2.95

THE WILL TO BELIEVE, HUMAN IMMORTALITY, William James. Two books bound together. Effect of irrational on logical, and arguments for human immortality. 402pp. 5⅜ × 8½. 20291-7 Pa. $7.50

THE HAUNTED MONASTERY and THE CHINESE MAZE MURDERS, Robert Van Gulik. 2 full novels by Van Gulik continue adventures of Judge Dee and his companions. An evil Taoist monastery, seemingly supernatural events; overgrown topiary maze that hides strange crimes. Set in 7th-century China. 27 illustrations. 328pp. 5⅜ × 8½. 23502-5 Pa. $5.95

CELEBRATED CASES OF JUDGE DEE (DEE GOONG AN), translated by Robert Van Gulik. Authentic 18th-century Chinese detective novel; Dee and associates solve three interlocked cases. Led to Van Gulik's own stories with same characters. Extensive introduction. 9 illustrations. 237pp. 5⅜ × 8½.
23337-5 Pa. $4.95

Prices subject to change without notice.
Available at your book dealer or write for free catalog to Dept. GI, Dover Publications, Inc., 31 East 2nd St., Mineola, N.Y. 11501. Dover publishes more than 175 books each year on science, elementary and advanced mathematics, biology, music, art, literary history, social sciences and other areas.